ALCOHOLS AS MOTOR FUELS

(Selected papers through 1980)

Prepared under the auspices of the
Fuels and Lubricants Activity

Published by:
Society of Automotive Engineers, Inc.
400 Commonwealth Dr.
Warrendale, PA 15096

ISBN 0-89883-107-5
SAE/PT-80/19

Library of Congress Catalog Card Number: 80-52454
Printed in U.S.A.

PREFACE

THE KNOWLEDGE THAT ALCOHOLS can be used as internal combustion engine fuels is as old as the engines themselves. Periodically, interest has developed in various parts of the world regarding use of both methyl and ethyl alcohol in blends with petroleum-derived fuels, or as pure fuels. This interest has increased over the past ten years in response to a diversity of causes, including: real or perceived threats of shortages of traditionally less costly petroleum fuels, opportunities to reduce emissions from gasoline fueled vehicles, and response to political pressures from diverse constituencies who feel they stand to benefit from use of alcohols as fuels.

SAE literature records the considerable technology which has developed in step with this interest, a literature which is especially rich in the technology of the application of alcohols as fuels for prime movers. In collecting a definitive sample of this literature from the last 20 years, we attempt to serve the needs of those active in the field today.

An effort has been made to include only those papers which present significant new information, or significant new observations and/or conclusions based on previously known information. The temptation to include earlier references, some of which afford fascinating views of the mutual adaptation of engines and fuels as they have developed over the years, has been resisted in deference to the more immediate purpose of adequately representing a technology which has experienced its most important development in recent years. Even so, some difficult choices have been made to meet the demands of available space. An extensive bibliography, coupled with literature references contained in the papers themselves, will serve to identify references not included here, as well as the earlier historical material.

In addition, many papers on alcohols as fuels have been presented at other technical society meetings. Prime among these have been the three International Alcohol Fuels Technology Symposia which have been held in recent years. (A fourth is scheduled to be held in Brazil during October 1980). It is suggested that the reader refer to the publications from these Symposia which contain a wealth of materials on alcohol production, economics, use, etc., especially in areas outside of the United States.

In selecting the papers which appear in this volume, we have encountered inevitable problems arising from author's positions of bias and advocacy, varying bases for technical comparison, and varying degrees of diligence in defining terms and test conditions. Although we have had to wrestle with these problems in making what amounts to judgmental decisions, their final resolution remains the responsibility of the reader, who in the end must deal with them according to his own values and requirements.

J. H. Freeman
Joseph M. Colucci
R. W. Hurn
Editorial Advisory Committee

EDITORIAL ADVISORY COMMITTEE PREPARING PT-19

Joseph M. Colucci
General Motors Research Laboratories

J. H. Freeman
Sun Company

R. W. Hurn
R-H Technical Services, Inc.

TABLE OF CONTENTS

Comparative Performance of Alcohol and Hydrocarbon Fuels*

E. S. Starkman, H. K. Newhall
and R. D. Sutton
University of California
Berkeley, CA

PERIODICALLY THE ALCOHOLS, particularly methyl and ethyl, attract interest as engine fuels. With certain exceptions, the level of activity seems to vary with potential supply and demand for the alcohols and with social pressures concerned with the particular part of the economy which such utilization might influence. The one notable exception to this variable interest is the continuing application of methyl alcohol as a racing fuel component.

A companion paper to this by Prof. J. A. Bolt (1)* is entirely devoted to a history and review of the literature on alcohols as motor fuels. This paper, therefore, will be devoted to what is an extension of some of that literature. Due apology is also made to prior authors in this field, who might otherwise believe that their efforts are not being properly recognized by specific reference.

Regardless of the reasons, the alcohols are interesting because they are potentially capable of being used as engine fuels. This is particularly so for spark ignition applications. A review of this potential from time to time is wise. It has been a full generation since the last large flurry of interest. In the interim period a bit more has been added to the general fund of knowledge in the combustion engine field. It is in the light of this knowledge, conceived in part with the application of computers to combustion calculations, and in part to an enlarged understanding of combustion and of thermodynamics, that this paper is written. It is the purpose here to present calculated and experimental comparative engine performance characteristics of the two lowest alcohols and of two representative hydrocarbons, octane and benzene.

*Numbers in parentheses designate References at end of paper.

The information also follows as an extension to, and in natural sequence to, a number of other papers on the same general subject and which have been presented and published in the literature in the past few years (2-4). As pointed out in Ref. 4, which was devoted to the presentation of thermodynamic charts for engine calculation using isooctane as fuel, a similar series of calculations and charts were being made for other hydrocarbon and nonhydrocarbon fuels. Some of these other charts form a part of this paper.

Edson (2) discussed the influence of compression ratio -- up to 300:1 -- on the efficiency of the ideal Otto cycle. He showed, among other things, that increasing pressure offsets increasing temperature in the effect which dissociation has on efficiency. In this study he examined three of the fuels also included here. One of the discussers of Edson's paper suggested a more expanded examination of the compression ratio range representative of present practice.

In the process of presenting calculations and experimental results, an attempt will be made to demonstrate that some generally accepted theories, with respect to the reasons for enhanced performance with alcohols as compared to hydrocarbons, are subject to revision.

COMBUSTION PRODUCT CHARACTERISTICS AND RATIO OF MOLES OF PRODUCT TO MOLES OF REACTANT

When the fuel and air react in an engine cylinder to produce products of combustion, the number of moles of product is different than the number of moles of reactant. This can be illustrated easily. Consider the reaction between oc-

*This paper contained in SP-254, "Alcohols and Hydrocarbons as Motor Fuels". Presented at the Summer Meeting, Chicago, Illinois, June 1964.

ABSTRACT

Three factors are of consequence when considering the comparative performance of alcohols and hydrocarbons as spark ignition engine fuels. These are: relative amounts of products of combustion produced per unit of inducted charge, energy inducted per unit of charge, and latent heat differences among the fuels. Simple analysis showed significant increases in output can be expected from the use of methyl alcohol as compared to hydrocarbon and somewhat lesser improvement can be expected from ethyl alcohol. Attendant increases in fuel consumption, disproportionate to the power increase, can also be predicted.

More sophisticated analysis, based upon thermodynamic charts of combustion products, do not necessarily improve correspondence between prediction and engine results.

SAE/PT-80/19/$02.50

tane and air at chemically correct mixture ratio. For the moment, the effects of dissociation will be ignored. The reaction is thus:

$$C_8H_{18} + 12.5\ O_2 + 47.0\ N_2 \longrightarrow 8\ CO_2 + 9\ H_2O + 47.0\ N_2 \quad (1)$$

Which shows that 64.0 moles of product result from 60.5 moles of reactant, a ratio of 1.058. This is of course a trivial, academic exercise and normally would be considered unworthy of transcription. However, the principles involved and the influence of this ratio of moles of product to moles of reactant, are important to a comparison between fuels of such differing characteristics as alcohols and hydrocarbons. To continue, therefore, consider benzene as example on the same basis:

$$C_6H_6 + 7.5\ O_2 + 28.2\ N_2 \longrightarrow 6\ CO_2 + 3\ H_2O + 28.2\ N_2 \quad (2)$$

Here 37.2 moles of product are produced per 36.7 moles of reactant, a ratio of 1.014.

Further, for methyl alcohol, the reaction:

$$CH_3OH + 1.5\ O_2 + 5.65\ N_2 \longrightarrow CO_2 + 2\ H_2O + 5.65\ N_2 \quad (3)$$

indicates that 8.65 moles of product are produced for every 8.15 moles reactant. This is a ratio of 1.061, and for ethyl alcohol, the reaction:

$$C_2H_5OH + 3.0\ O_2 + 11.3\ N_2 \longrightarrow 2\ CO_2 + 3\ H_2O + 11.3\ N_2 \quad (4)$$

gives 16.3 moles product per 15.3 moles reactant, a ratio of 1.065.

Considering the reciprocating engine as a machine which regardless of fuel would induct one volume of reactant per unit of time, and assuming that all the fuels noted above would enter the engine completely evaporated, it follows that there are some differences in the fuels considered. The fuel giving the largest number of moles of product per mole of reactant should produce the greatest pressure in the cylinder after combustion, as compared to the others, all other factors being equal (which incidentally they are not, but this will be discussed in other sections). The greater pressure taken alone would result in an increase in engine power; but of course the output would not be modified in direct proportion to the ratio of the numbers quoted above.

On the other hand, a modern engine may not, and usually does not ingest its mixture with the fuel already evaporated. The fuel may even be entirely unevaporated, as extreme example, in a direct injection engine and only a little, if any, evaporated in the port injection engine. Under such conditions, the number of moles of product should be examined on the basis of the number of moles of air inducted since the fuel occupies very little of the volume. Table 1, therefore, illustrates the "dry" as well as the "wet" ratio of moles of product per mole of reactant, and per mole of air, respectively.

Considering the fuel to enter the cylinder in the liquid state points to a somewhat enhanced power output from methyl alcohol and a slightly increased output from ethyl alcohol on this rather simple basis. (Note, however, in Eqs. 1-4 that the amount of fuel per unit volume of air is changing radically, and thus also the fuel-air ratio. Further, the heating values of the fuels must be taken into consideration.)

Nevertheless, to complete this aspect of the analysis, it is necessary also to consider the effects of dissociation on the comparative production of products of combustion. This equilibrium calculation has been made for all the fuels considered here and over a range of pressures and temperatures representative of those occurring at completion of the combustion process in modern engines. An example of the results is shown in Fig. 1. This is for octane and illustrates the concurrent influences of pressure, temperature, and mixture strength (here expressed as equivalence ratio) on the ratio of moles of product to moles of reactant.

Table 1 - Comparative Moles of Product per Mole of Reactant or per Mole of Air at Chemically Correct Mixture Ratio and Neglecting Dissociation

	Dry Basis		Wet Basis	
Fuel	Ratio	Compared to Octane	Ratio	Compared to Octane
Octane	1.058	1.000	1.075	1.000
Benzene	1.014	0.959	1.042	0.969
Methyl Alcohol	1.061	1.004	1.210	1.126
Ethyl Alcohol	1.065	1.008	1.140	1.061

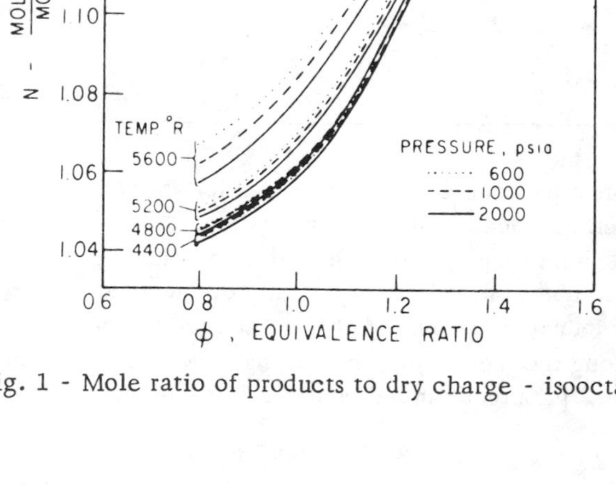

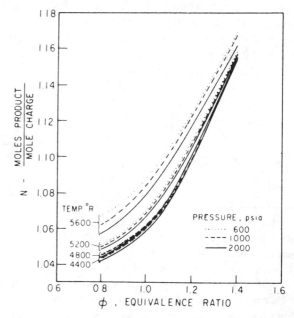

Fig. 1 - Mole ratio of products to dry charge - isooctane

Fig. 2 is a similar plot of the ratio of moles of product to moles of air (the fuel being considered liquid). Figs. 3-8 are similar to Figs. 1 and 2 and are for benzene, methyl alcohol (methanol), and ethyl alcohol (ethanol), respectively.

To compare the above four fuels to each other over the range of pressures and temperatures shown in Figs 1-8 would result in an extremely complex representation. As a consequence, only one set of representative pressure and temperature, 800 psia and 5000 R, has been chosen for this purpose. Additionally, the comparison has been based on the ratio of products to reactants for octane at an equivalence ratio of 1.0. The results are shown in Figs. 9 and 10 for "dry" and "wet" mixtures respectively. These results may be compared to the nondissociated results of Table 1.

Fig. 9, for "dry" mixture, shows in the same manner as Table 1 that there is very little difference in the ratio of moles of products produced by the fuels considered. The largest differential is between benzene and ethyl alcohol and amounts to approximately 5%. The rank is: ethyl alcohol, about 2% more product than octane, methyl alcohol 1% more, and benzene 3% less.

On a liquid fuel or "wet" basis, as in Fig. 10, the differences are larger, and the two alcohols change rank. Methyl

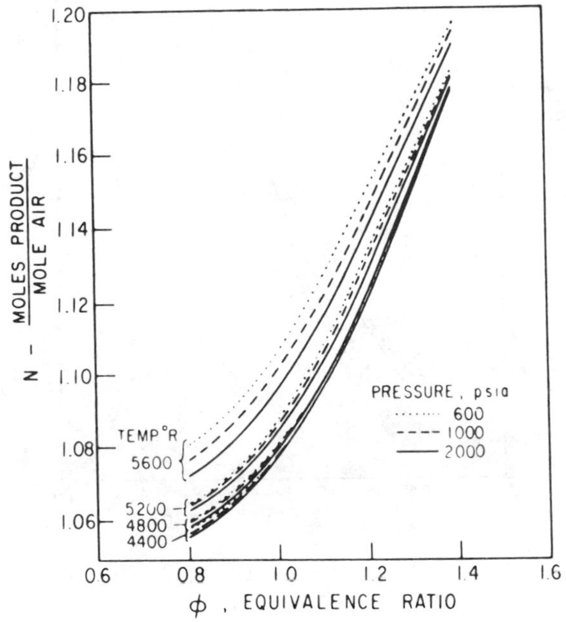

Fig. 2 - Mole ratio of products to air (wet charge) - isooctane

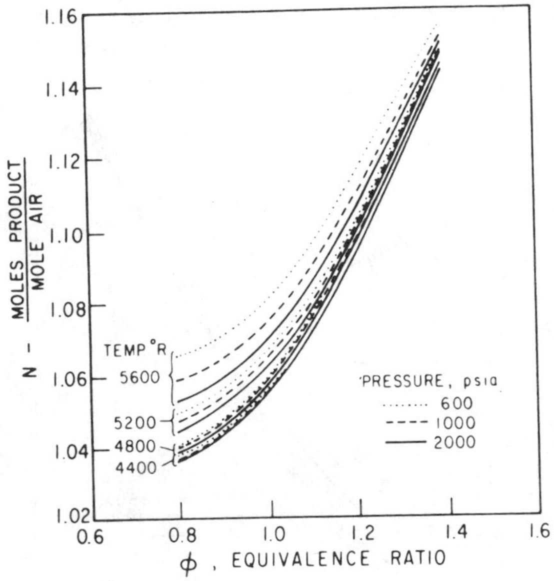

Fig. 4 - Mole ratio of products to air (wet change) - benzene

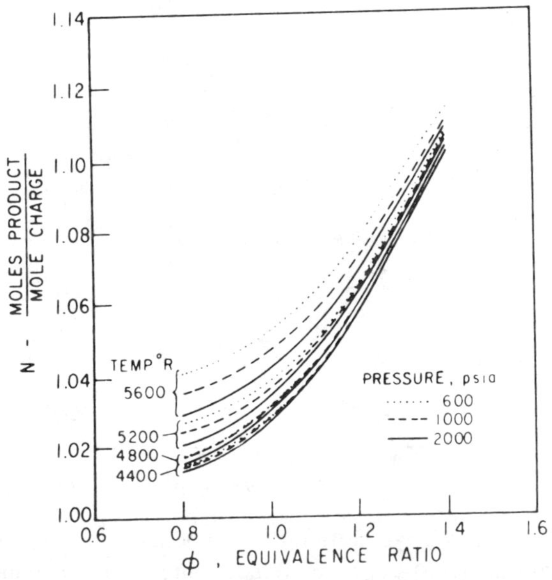

Fig. 3 - Mole ratio of products to dry charge - benzene

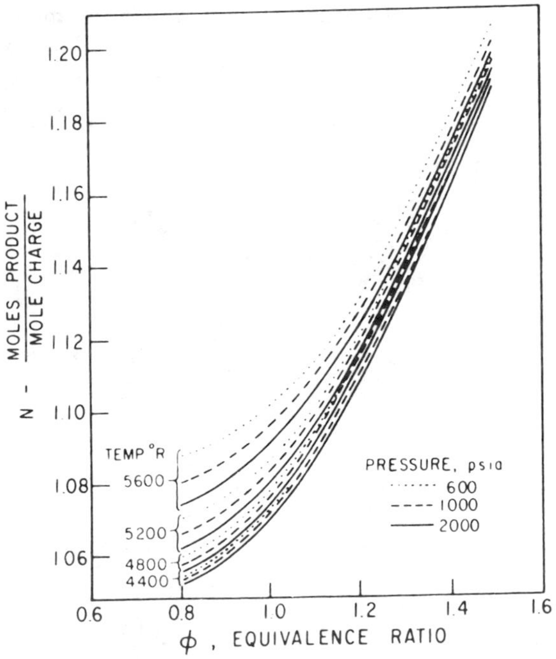

Fig. 5 - Mole ratio of products to dry charge - methanol

alcohol produces 13% more product than octane; ethyl alcohol about 7% more, and benzene 2-3% less.

Energy in the Charge - The four fuels being considered have very largely differing chemically correct fuel-air ratios. They also have very widely different energies of combustion. These characteristics are illustrated in Table 2. Because the isomers of octane differ slightly from one another thermochemically, even though they are the same stoichiometrically, it will be noted that isooctane (2,2,4-trimethylpentane) was specified.

It is perhaps worth the effort to look at these four fuels from the standpoint of the energies they potentially carry into the reaction per unit charge of air. This is both an oversimplification of the situation as well as erroneous in a thermodynamic sense. Fuel performance cannot be compared in such a simple manner because the energy carried into the charge of air as heat of combustion of the fuel distributes itself differently in the resulting products of one fuel compared to another. The engine output depends upon the pressure being exerted on the piston. That pressure, in turn,

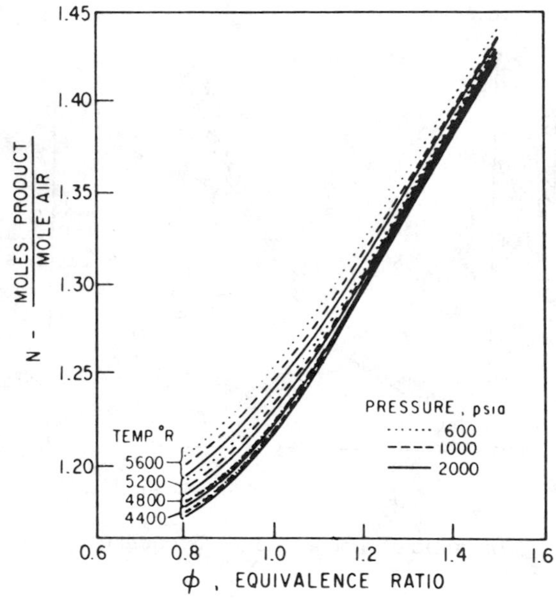

Fig. 6 - Mole ratio of products to air (wet charge) - methanol

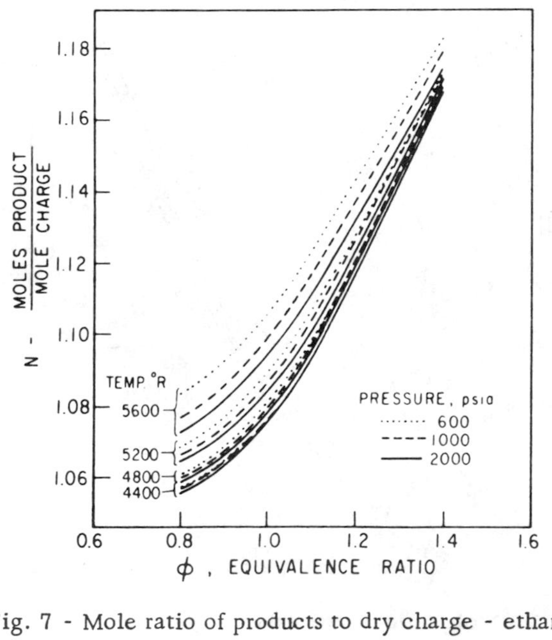

Fig. 7 - Mole ratio of products to dry charge - ethanol

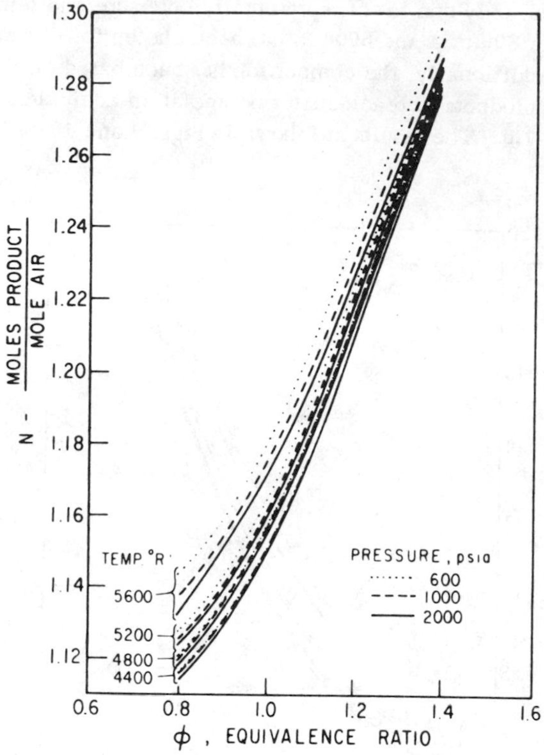

Fig. 8 - Mole ratio of products to air (wet charge) - ethanol

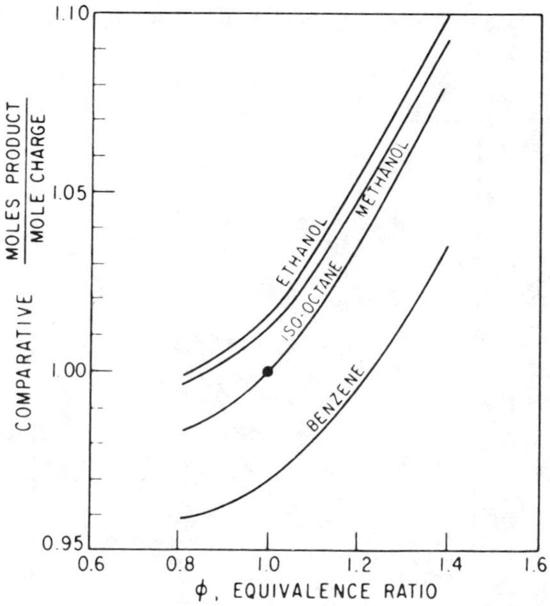

Fig. 9 - Comparative mole ratio of products to dry charge at 5000 R and 800 psia, normalized about ratio of isooctane at $\phi = 1.0$

is a function of the composition of the resulting gases as well as their temperature. The effects of dissociation and equilibrium are thus also of importance in addition to the carbon to hydrogen ratio and oxygen content of the fuel. Notwithstanding all of the above, a simple calculation of the potential energy per unit charge of air, compared to isooctane at an equivalence ratio of 1 is shown in Fig. 11.

To place the comparative contributions of energy per unit charge on a more defensible basis necessitates a choice of the particular conditions under which such comparison will be made. The important consequences of the energy density of the charge, the equilibrium temperature and pressure after combustion, will depend upon even other factors than those outlined in the previous paragraph. As example, compression ratio and intake state of the fuel-air mixture will have an influence. A number of calculations are carried out in a later section of this paper which illustrate the comparative pressures and temperatures obtained as a function of fuel type and compression ratio.

In any event, the rank of the fuels shown in Fig. 11 does not change as a consequence of consideration of the particulars of the combustion process, and as will be shown, the rough approximations of Fig. 11 are a good guide to the comparative performance when taken into consideration with Figs. 9 and 10.

Latent Heat of Evaporation - As pointed out previously, the alcohols exhibit a number of differing characteristics from hydrocarbon fuels. The most prominent is the latent heat of evaporation. Table 3 illustrates this comparison.

Observations of increased output from engines using alcohols have generally been explained by the disproportionate cooling effect which the alcohols exhibit and the supposition that the resulting chilled air-fuel charge, being more dense, was supplying the engine with an increased mass flow resulting in higher volumetric efficiency. In an engine having a relatively high intake air temperature and good atom-

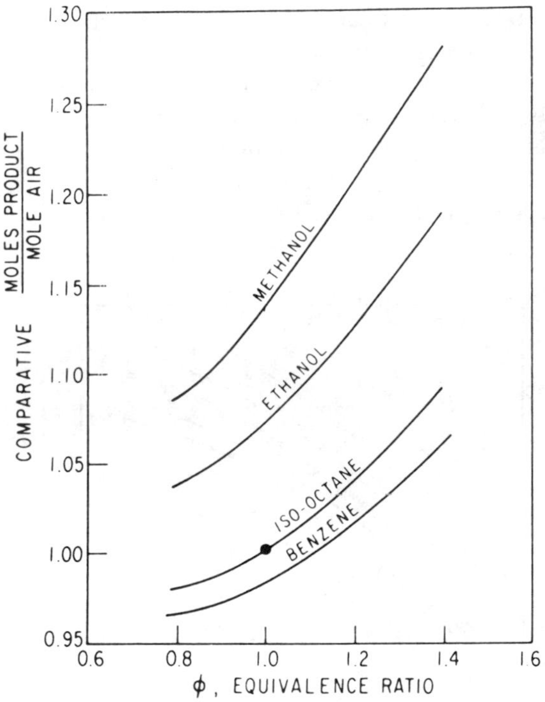

Fig. 10 - Comparative mole ratio of products to air (wet charge) at 5000 R and 800 psia, normalized about ratio of isooctane at $\phi = 1.0$

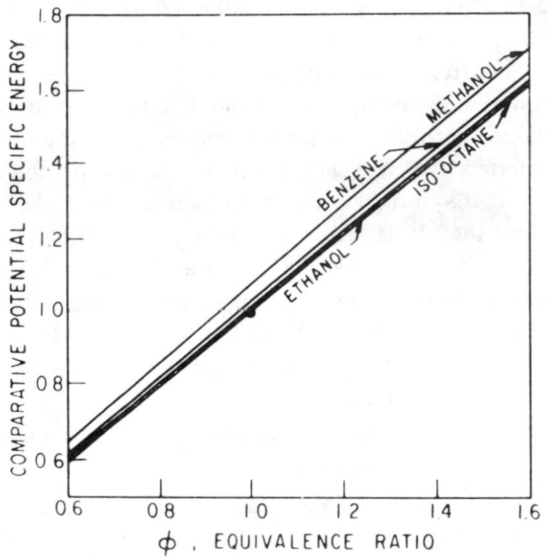

Fig. 11 - Comparative specific energy per unit quantity of air inducted (wet charge)

Table 2 - Lower Heating Value at Constant Volume of the Alcohols and Hydrocarbons (Fuel in Liquid State)

Fuel	Btu/lb Fuel(5) 537 R	Btu/lb Air for Stoichiometric Mixture 537 R
Isooctane	19,159	1268.33
Benzene	17,190	1303.00
Ethyl Alcohol	11,604	1289.20
Methyl Alcohol	8,644	1354.18

Table 3 - Latent Heat of Vaporization of the Alcohols and Hydrocarbons

Fuel	Btu/lb Fuel (5) 537 R	Btu/lb Air for Stoichiometric Mixture 537 R
Isooctane	132.28	8.76
Benzene	185.57	14.06
Ethyl Alcohol	395.40	43.93
Methyl Alcohol	502.25	78.50

ization and evaporation of the fuel, such might have appeared to be the case. However, this so-called charge chilling hypothesis applied to high latent heat fuels has usually not been substantiated by airflow rate measurements. As a matter of fact, the results usually reported, with surprise, show that the volumetric efficiency increased very little, if at all, with the alcohol or alcohol-gasoline blends.

These observations, by other authors as well as these, have led to the question of what effect the latent heat would have on engine performance if it was assumed that no evaporation of fuel took place prior to the sealing off of the charge in the cylinder, but rather during the compression stroke. This is the so-called wet basis. Calculations to predict comparative performance, as well as engine tests, were made on t this wet mixture basis. The result is to cause a comparison in which the airflow rate remains constant between fuels, or conversely the volumetric efficiency is constant.

THERMODYNAMIC PROPERTIES OF METHYL ALCOHOL AND AIR, ETHYL ALCOHOL AND AIR, AND BENZENE AND AIR

The preceding sections have considered separately the influences of the ratio of moles of products to reactants, the energy density of the charge, and the latent heat of evaporation. In a first approximation manner it has been demonstrated that these factors can influence engine output. Furthermore, it was shown that the factors may be additive or that they might counter each other when taken together. An example is benzene, which gives a slightly higher energy density in an engine than isooctane, and a slightly lower ratio of products to reactants, 4% on a gaseous fuel basis and 2% on a liquid fuel basis. The net result is an expectation that the power output from benzene would approximate that from isooctane. The engine results show this to be true for stoichiometric and rich mixtures but not for lean mixtures.

The more sophisticated approach recognizes, in addition, that fuels of widely differing chemical composition, heat of combustion, heat of vaporization, and chemical stoichiometry may also have product compositions which are quite different from each other. It is necessary, therefore, to go beyond the superficial analysis outlined above. This necessitates equilibrium calculation of the products and is today most easily done on a computer. The resulting data can then either be further incorporated into a computer programmed cycle, or charts can be prepared for those not fortunate enough to have a computer or computer technique available. One of the purposes of this presentation is to make available the charts for a limited number of fuels and fuel-air ratios.

The basis and procedures for calculating the properties of a fuel and air mixture and of the products resulting from the reaction of such mixture were described in a previous paper (4). This previous paper which was devoted to octane and air included the information that extension of the study was underway and would appear in further publications. This section of this paper is thus a continuation of Ref. 4.

The fuels and fuel-air equivalence ratios included here are tabulated in Table 4.

The thermodynamic properties charts resulting from calculations made on these fuels will be found in the Appendix.

Calculation of Comparative Performance - Using the charts for isooctane of Ref. 4 and those included here, a series of calculations was carried out to compare expected performance. The methods are relatively straightforward. Both wet and dry mixtures were assumed. Some of the more important results are shown in Figs. 12-19. Also of importance are peak temperatures and pressure associated with the cycles used in calculating the results. These are shown in Table 5.

The cycles analyzed were all ideal. Compression and expansion were assumed isentropic and combustion was assumed to take place at constant volume. Thus the effects of finite combustion time and of heat transfer have not been included. Intake pressure was taken as 1 atm and temperature as 100 F. Compression ratios ranged from 7 to 15.

Because the number of combustion charts constructed for any particular fuel is limited, the amount of information that can be calculated is also limited. The influence of

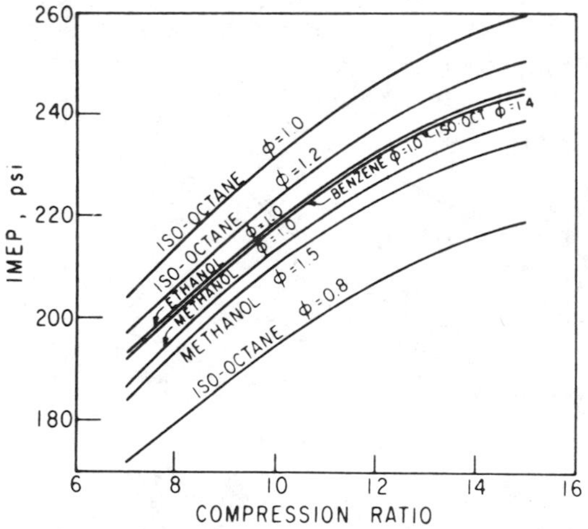

Fig. 12 - Calculated performance comparison (imep) for isooctane-benzene, ethanol and methanol (dry charge)

Table 4 - Fuel-Air Equivalence Ratios

Fuel	φ Fuel-Air Equivalence Ratio
Benzene	1.0
Methyl Alcohol	1.0
Methyl Alcohol	1.5
Ethyl Alcohol	1.0

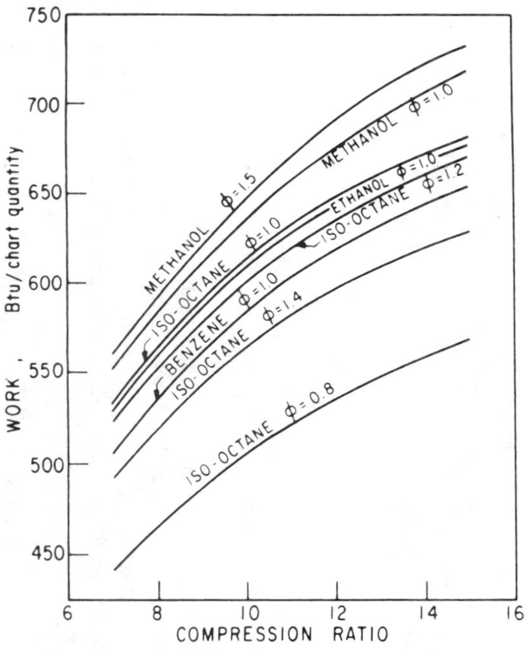

Fig. 13 - Calculated work comparison for isooctane-benzene, ethanol, and methanol (dry charge)

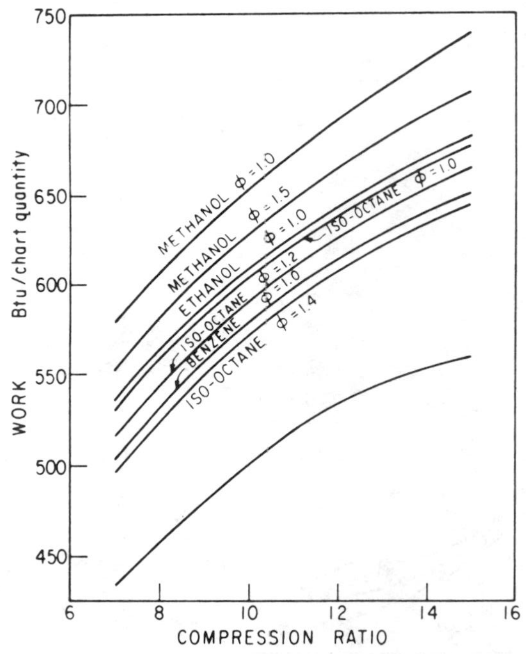

Fig. 15 - Calculated work comparison for isooctane-benzene, ethanol, and methanol (wet charge)

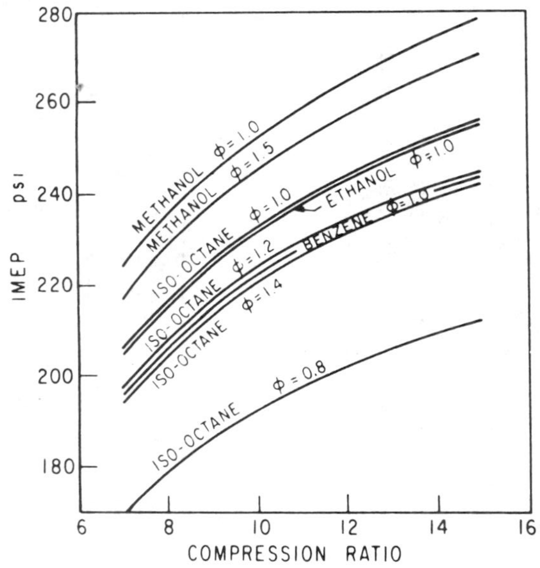

Fig. 14 - Calculated performance comparison (imep) for isooctane-benzene, ethanol, and methanol (wet charge)

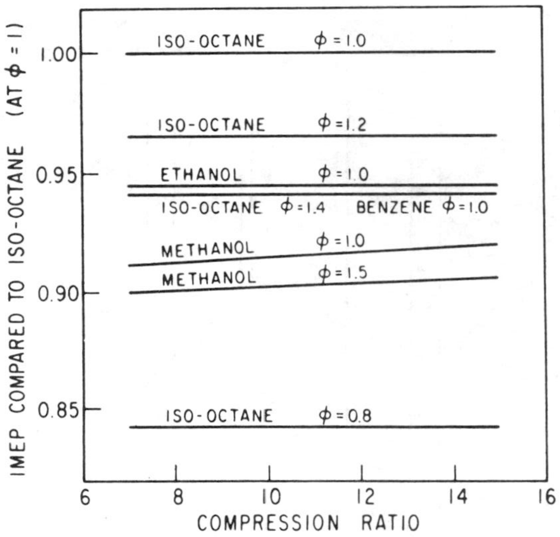

Fig. 16 - Comparative calculated performance (imep) for isooctane-benzene, ethanol, and methanol normalized about imep of isooctane at $\phi = 1.0$ (dry charge)

Table 5 - Chart Calculations of Cycle Characteristics
C.R. = 9:1, Intake Air = 100 F, 14.7 psia

A. Dry Charge

Fuel	Peak Temperature, R	Peak Pressure, psia	Cu Ft of Mix/ lb Air	Compression Work (Btu/lb)
Isooctane $\phi = 0.8$	4840	1030	13.85	131.5
Isooctane $\phi = 1.0$	5220	1260	14.07	134
Isooctane $\phi = 1.2$	5230	1240	14.29	134
Isooctane $\phi = 1.4$	4975	1200	14.06	133.5
Methanol $\phi = 1.0$	5100	1240	16.04	160
Methanol $\phi = 1.5$	4750	1230	16.50	170.5
Ethanol $\phi = 1.0$	5135	1200	14.87	144
Benzene $\phi = 1.0$	5355	1240	14.28	148

B. Wet Charge

Fuel	Peak Temperature, R	Peak Pressure, psia	Cu Ft of Mix/ lb Air	Compression Work (Btu/lb)
Isooctane $\phi = 0.8$	4820	1000	13.89	130
Isooctane $\phi = 1.0$	5200	1240	13.92	130
Isooctane $\phi = 1.2$	5170	1200	14.24	126
Isooctane $\phi = 1.4$	4910	1260	13.90	125
Methanol $\phi = 1.0$	4870	1300	13.92	101
Methanol $\phi = 1.5$	4360	1420	13.82	109
Ethanol $\phi = 1.0$	5000	1255	14.09	111
Benzene $\phi = 1.0$	5315	1265	13.98	128

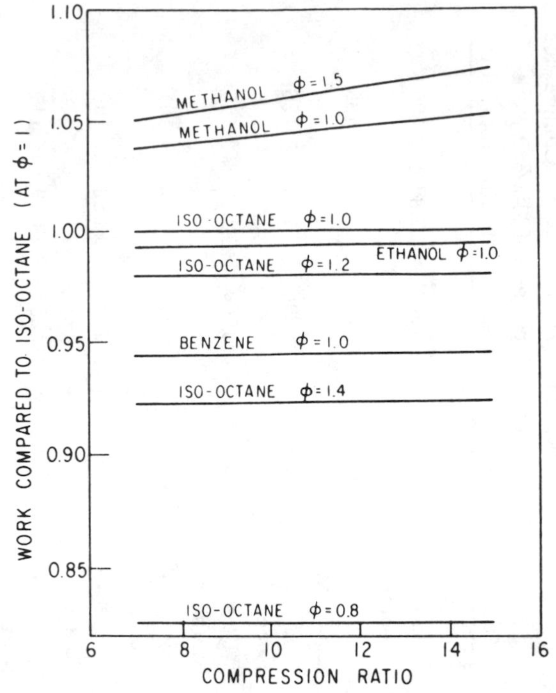

Fig. 17 - Comparative calculated work for isooctane-benzene, ethanol, and methanol normalized about work of isooctane at $\phi = 1.0$ (dry charge)

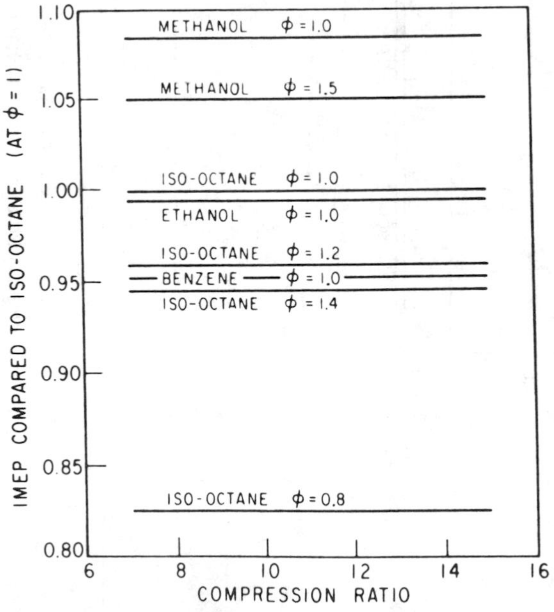

Fig. 18 - Comparative calculated performance (imep) for isooctane-benzene, ethanol, and methanol normalized about imep of isooctane at $\phi = 1.0$ (wet charge)

fuel-air equivalence ratio is, as a consequence, not capable of being determined in this study except for methanol. Even in the case of methanol, for which there are presented charts at equivalence ratios of 1.0 and 1.5, the trend of performance is not satisfactorily determined as a function of equivalence ratio. This is an admitted limitation of the use of

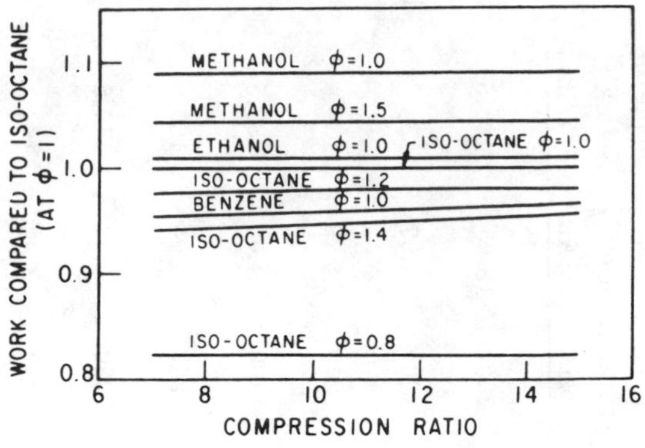

Fig. 19 - Comparative calculated work for isooctane-benzene, ethanol, and methanol normalized about work of isooctane at $\phi = 1.0$ (wet charge)

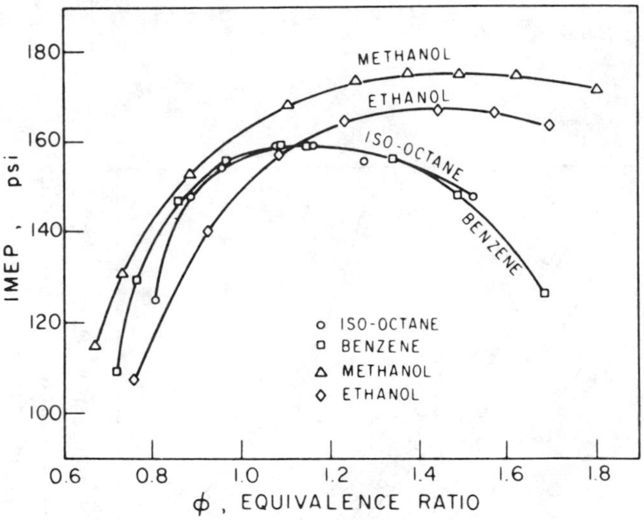

Fig. 20 - Performance results (imep) from engine

charts for such application.

An exhaustive description and explanation of these calculated results will not be attempted here or at this time. Further computer calculations are needed to verify anomalous behavior between certain fuels, particularly between benzene and isooctane and between methyl alcohol at $\phi = 1.0$ and $\phi = 1.5$, when compared to engine results.

ENGINE RESULTS

As indicated previously, the engine used to obtain comparative performance on the four fuels was operated at constant airflow rate. It was possible to maintain this condition by injecting the fuel into the inlet manifold just ahead of the intake valve, thereby minimizing the effect of fuel evaporation on airflow rate. The engine, a CFR equipped for supercharged operation, was run at a nominal 1800 rpm,

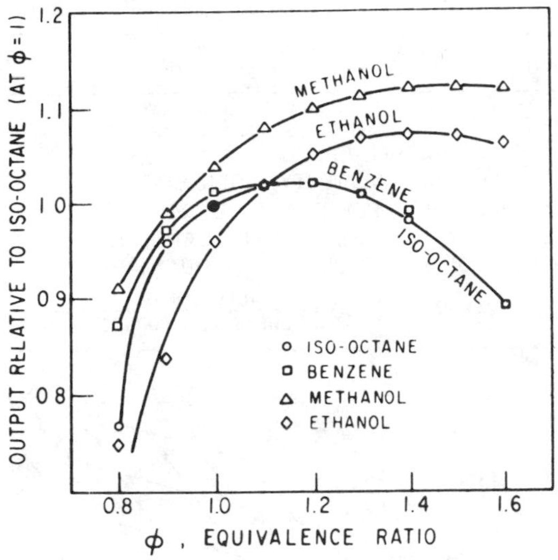

Fig. 21 - Comparative performance results (imep) from engine normalized about imep of isooctane at $\phi = 1.0$

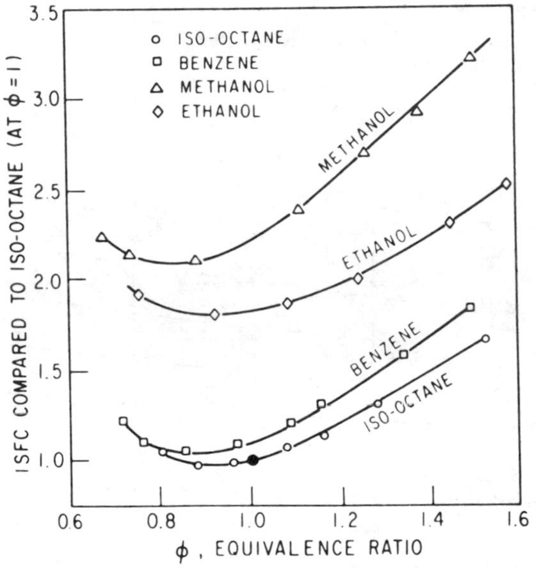

Fig. 22 - Comparative isfc from engine results normalized about isfc of isooctane at $\phi = 1.0$

100 F, and 14.7 psia inlet conditions and over a limited compression ratio range. The results included here are restricted to those obtained at a compression ratio of 9:1.

Fig. 20 shows the engine output as a function of fuel-air equivalence ratio. Equivalence ratio is used to show the effect of varying fuel-air ratio when comparing fuels of widely differing chemically correct ratios. This has the effect of better orienting the performance curves.

With certain exceptions, the results illustrate what has already been predicted. The exception is that isooctane and benzene are not as largely different from each other with respect to engine power as predicted. However, ethyl alcohol is slightly superior, if used at quite rich mixtures, and methanol is significantly superior, particularly at very rich mixtures, such as 50% more fuel than chemically correct.

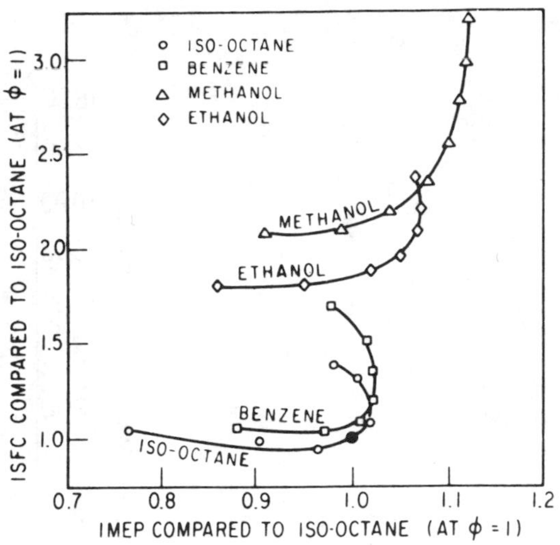

Fig. 23 - Fuel consumption loops from engine results normalized about consumption loop of isooctane at $\phi = 1.0$

A further adjustment of the results to illustrate better the comparative power is shown in Fig. 21. This places power output on the basis of the per cent of that produced by isooctane at chemically correct mixture ratio. Note that ethanol produces about 6% more output and methanol about 12% more than the two hydrocarbon fuels.

Fig. 22 indicates the specific fuel consumption rates experienced for the data of Figs. 20 and 21. Finally, the fuel consumption and power data have been combined on this normalized comparison basis in Fig. 23. Illustrated are the concurrent advantages and penalties associated with the use of alcohols as engine fuels.

Whereas Fig. 23 shows that gross output can be increased by 6% through the use of ethyl alcohol, approximately twice the fuel flow rate is necessary. Similarly, the output may be increased by 12% if methanol is used, but this will require about three times the fuel flow rate.

CONCLUSIONS

1. Differences in the stoichiometry and thermochemistry of hydrocarbons and alcohols led to a confirmation of the observed increase in output experienced with methyl and ethyl alcohols when used as spark ignition engine fuels.

2. Accompanying any increase in output is a disproportionate rise in fuel consumption.

3. A principal factor influencing the output is the relative volume of product produced per volume of inducted charge. An equally important factor is the influence of fuel latent heat, but not necessarily as the latent heat may affect volumetric efficiency, but rather as it may influence compression work.

4. Engine results are apparently better predicted by a very simple analysis based on first order approximations than they are by second order approximations which neglect the effects of heat transfer and irreversibility.

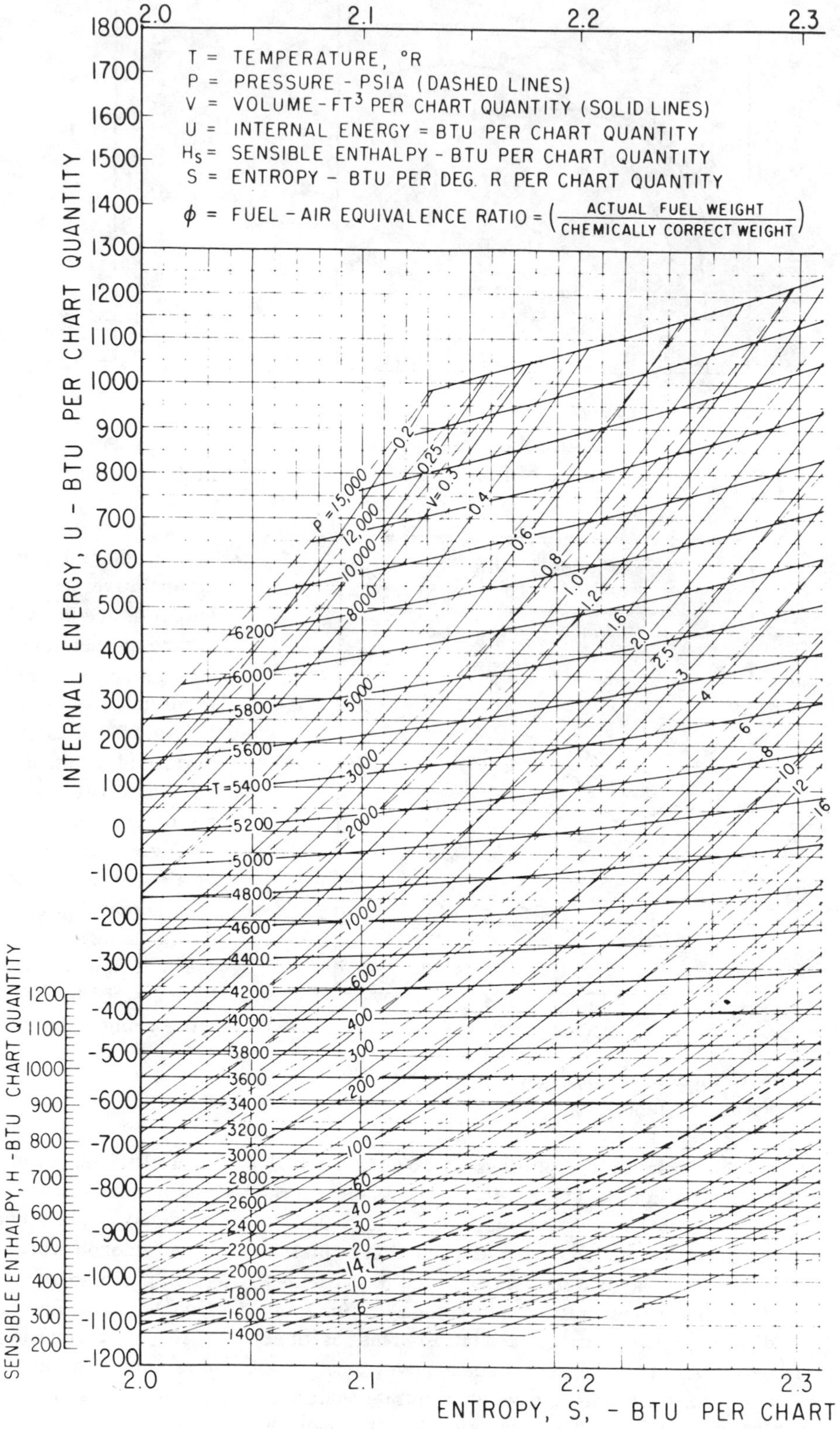

Chart 1 - Burned mixture chart for benzene - air,

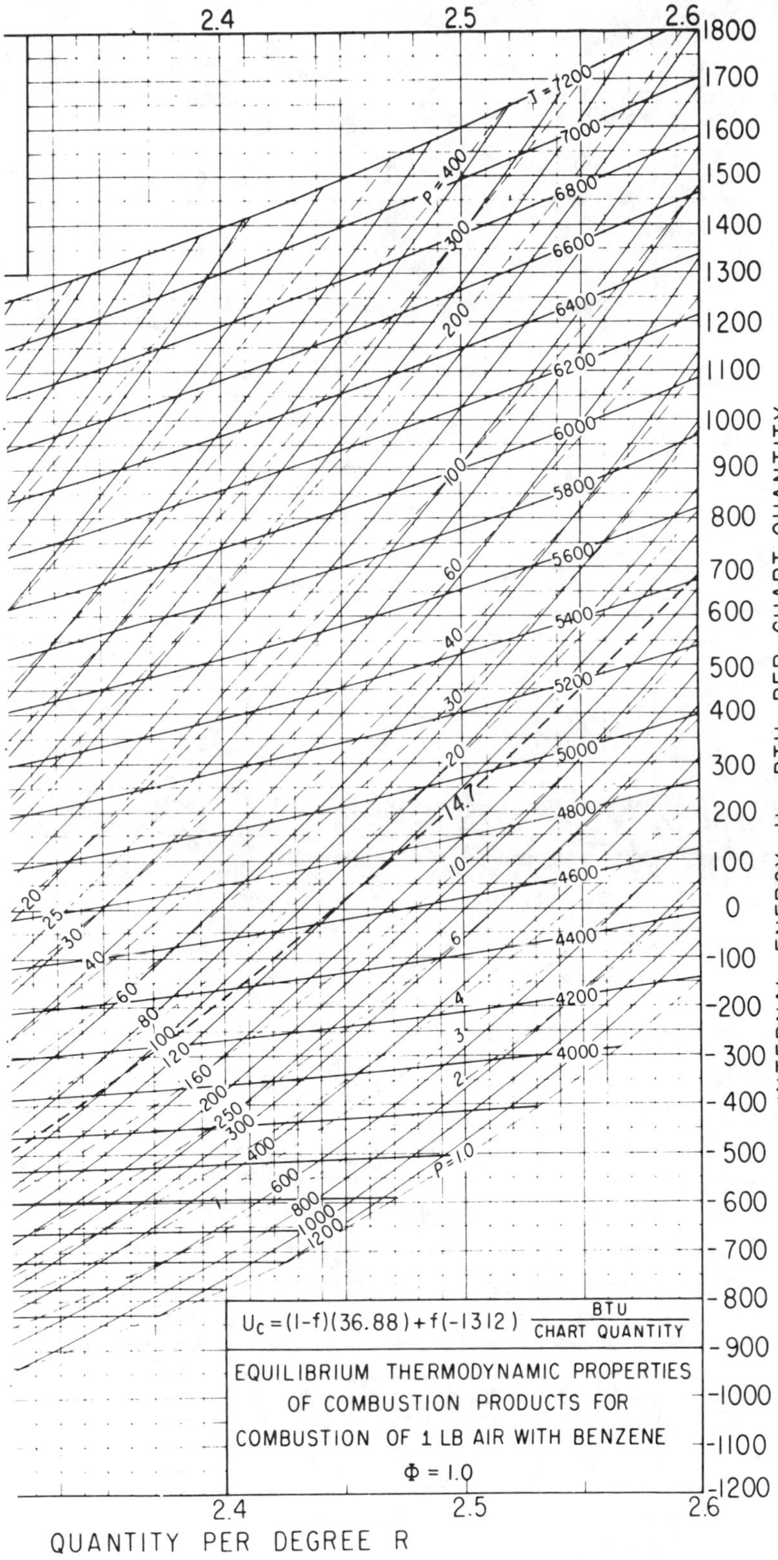

$\phi = 1.0$ (chemically correct mixture)

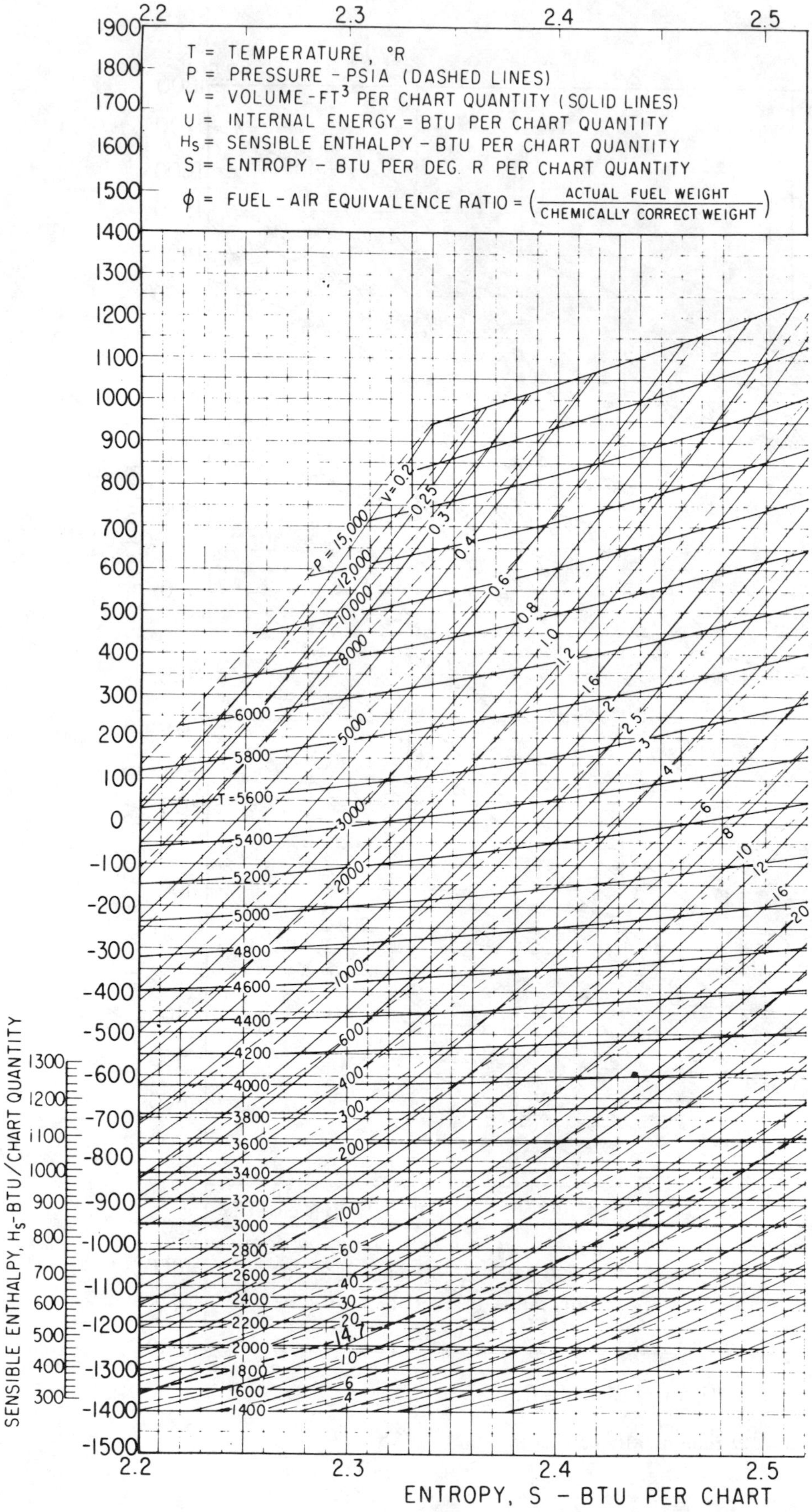

Chart 2 - Burned mixture chart for ethanol-air,

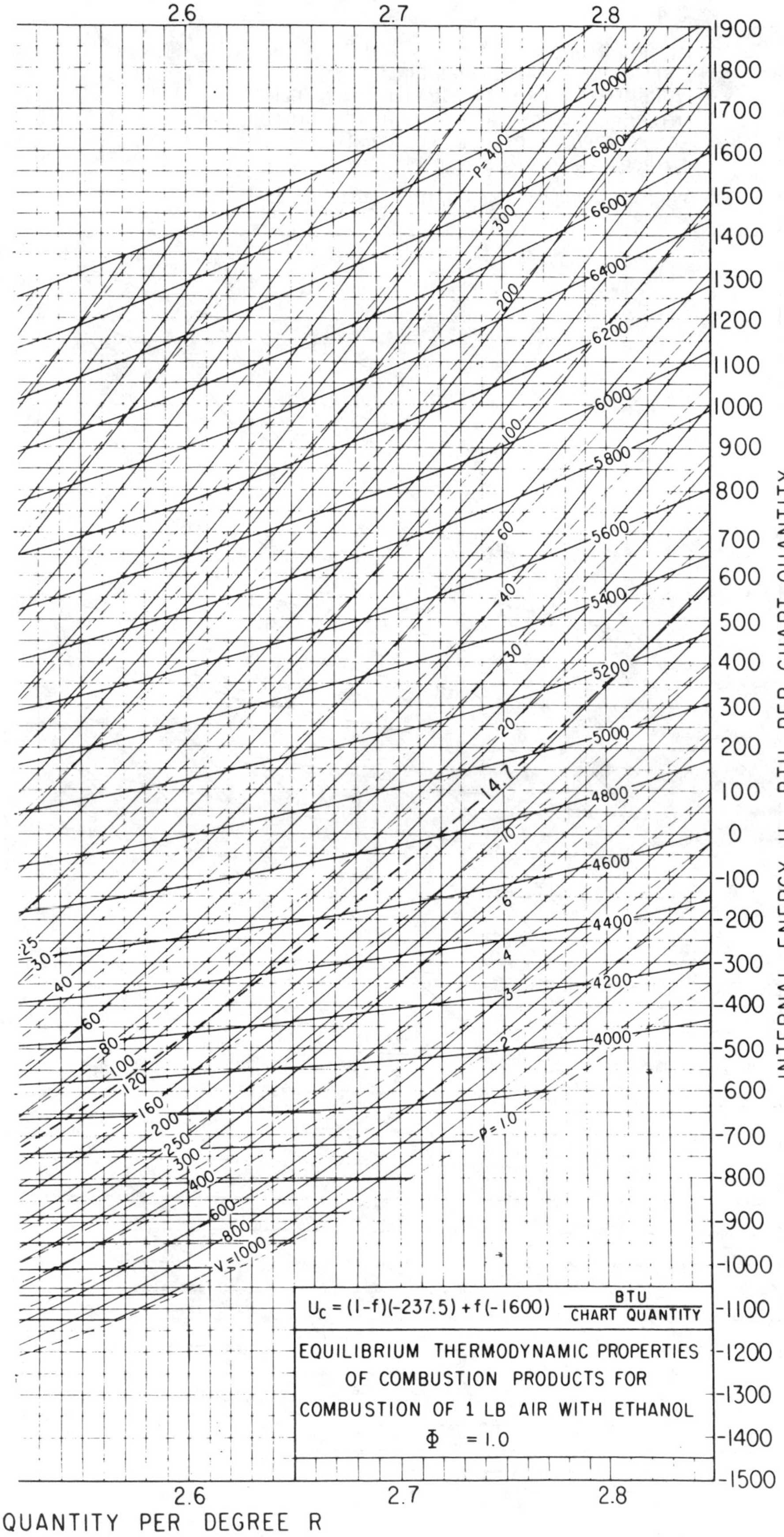

ϕ = 1.0 (chemically correct mixture)

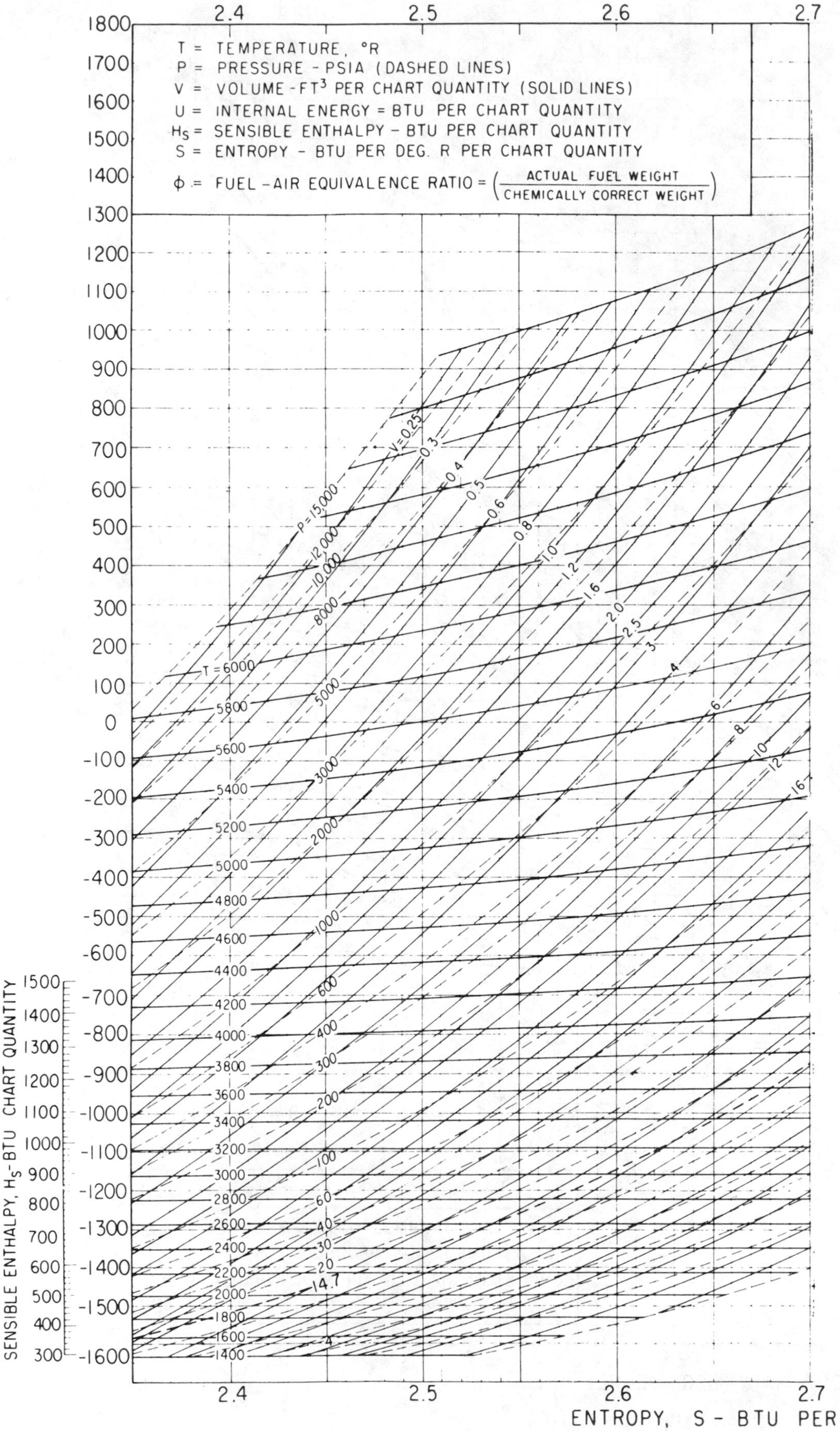

Chart 3 - Burned mixture chart for methanol-air,

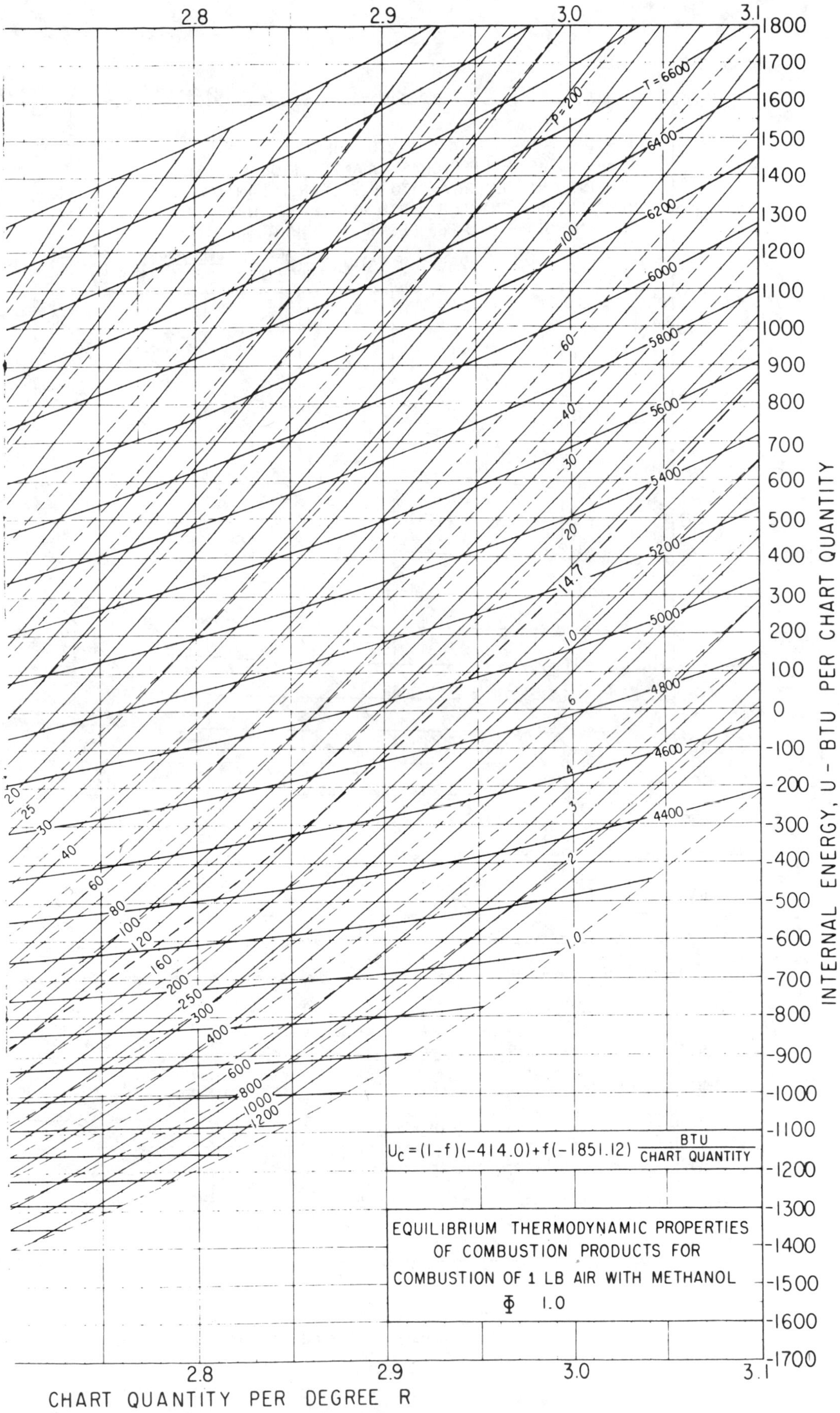

ϕ = 1.0 (chemically correct mixture)

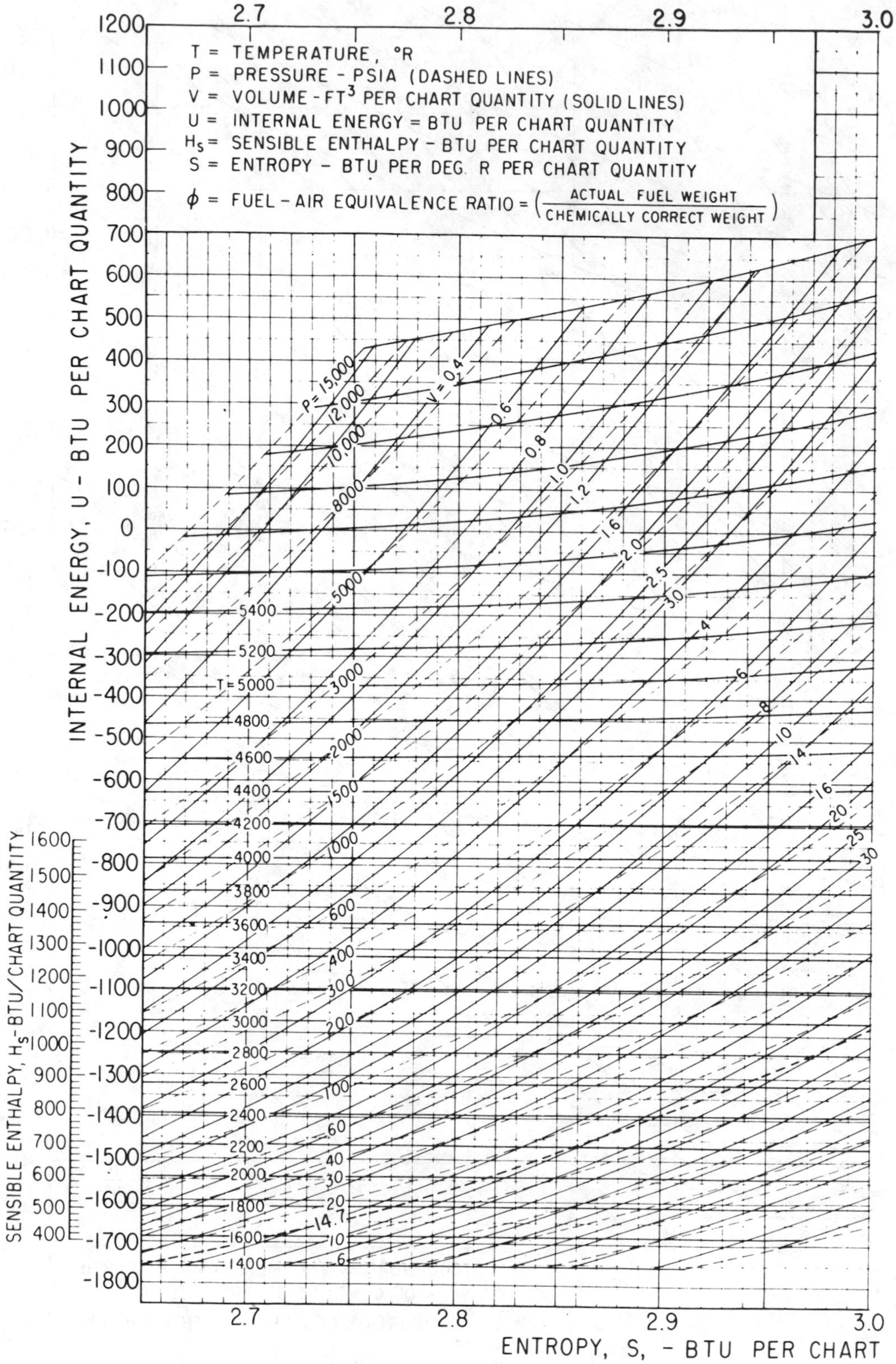

Chart 4 - Burned mixture chart for methanol-air,

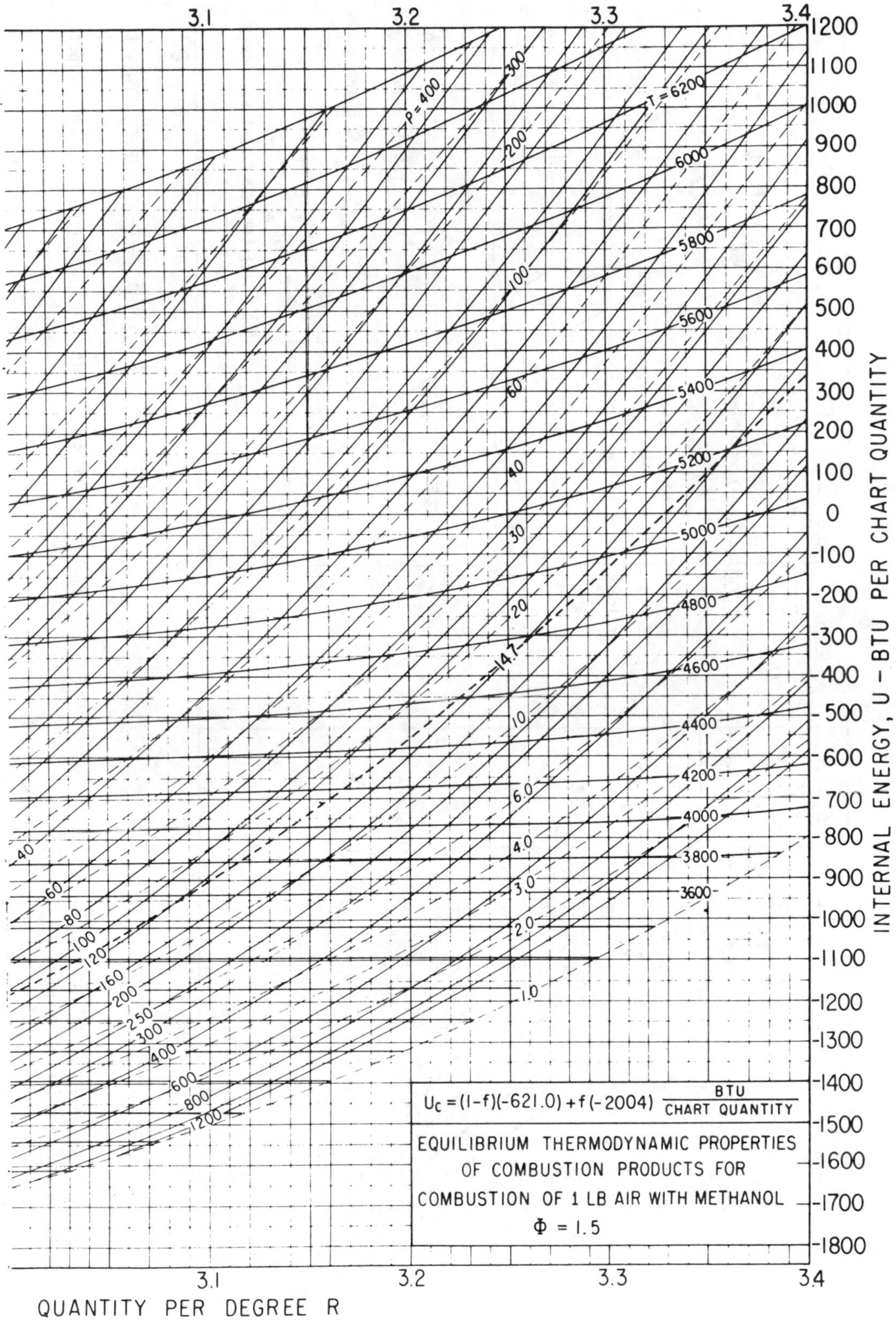

$\phi = 1.5$ (50% fuel rich mixture)

1.) ENTER CHART AT TEMPERATURE OF INDUCED CHARGE, T_1, READING $M\int_{520}^{T_1} C \frac{dT}{T}$ FROM ABSCISSA
2.) COMPUTE VALUE OF $\int_{T_1}^{T_2} C \frac{dT}{T}$ DUE TO COMPRESSION USING VOLUME OR PRESSURE RATIO TERMS
3.) COMBINE ABOVE TERMS TO OBTAIN $M\int_{520}^{T_2} C \frac{dT}{T}$ FOR THE COMPRESSED CHARGE
4.) ENTER ABSCISSA OF CHART WITH TOTAL INTEGRAL AND READ COMPRESSION TEMPERATURE FROM ORDINATE
5.) COMPRESSION PRESSURE OR VOLUME MAY BE DETERMINED FROM FORMULAE GIVEN BELOW AND INTERNAL ENERGY FROM AUXILIARY TEMPERATURE-ENERGY CHART

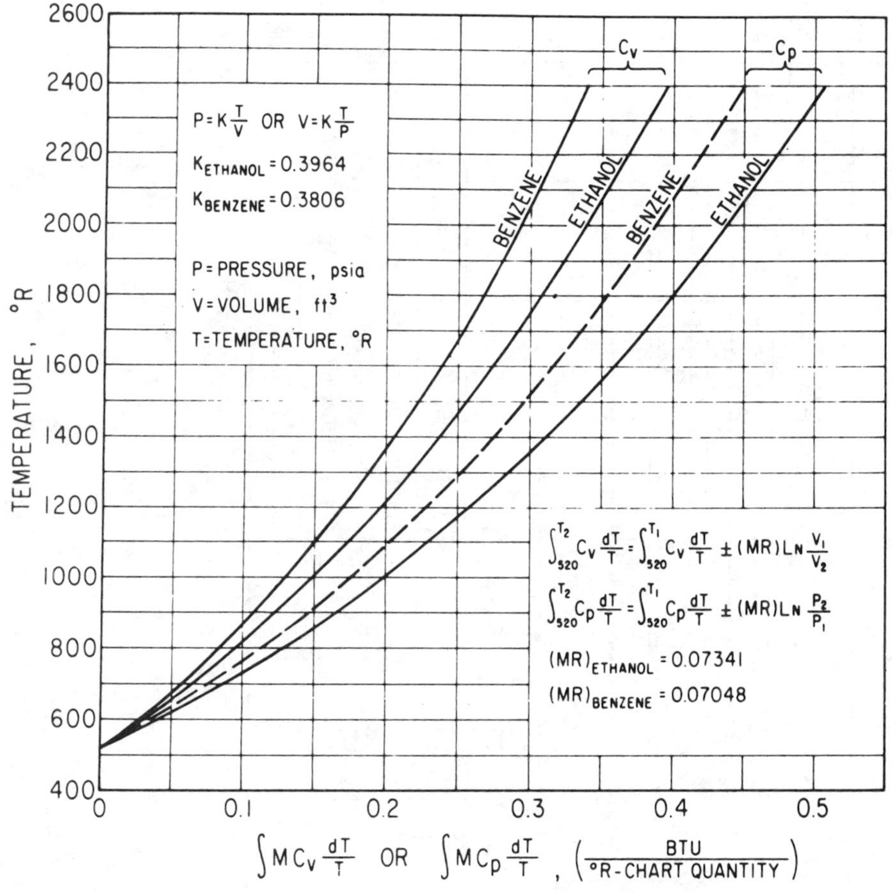

Chart 5 - Unburned mixture chart for determination of temperature of compressed ethanol-air and benzene-air mixtures

1.) ENTER CHART AT TEMPERATURE OF INDUCED CHARGE, T_1, READING $M\int_{520}^{T_1} C \frac{dT}{T}$ FROM ABSCISSA
2.) COMPUTE VALUE OF $\int_{T_1}^{T_2} C \frac{dT}{T}$ DUE TO COMPRESSION USING VOLUME OR PRESSURE RATIO TERMS
3.) COMBINE ABOVE TERMS TO OBTAIN $M\int_{520}^{T_2} C \frac{dT}{T}$ FOR THE COMPRESSED CHARGE
4.) ENTER ABSCISSA OF CHART WITH TOTAL INTEGRAL AND READ COMPRESSION TEMPERATURE FROM ORDINATE
5.) COMPRESSION PRESSURE OR VOLUME MAY BE DETERMINED FROM FORMULAE GIVEN BELOW AND INTERNAL ENERGY FROM AUXILIARY TEMPERATURE-ENERGY CHART

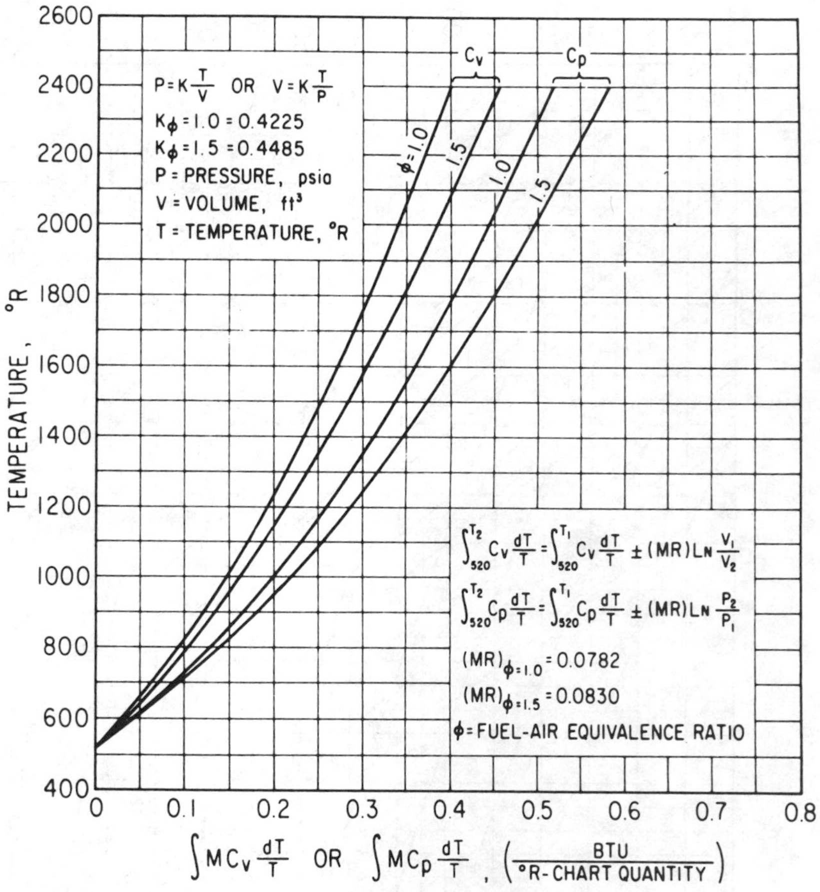

Chart 6 - Unburned mixture chart for determination of temperature of compressed methanol-air mixtures

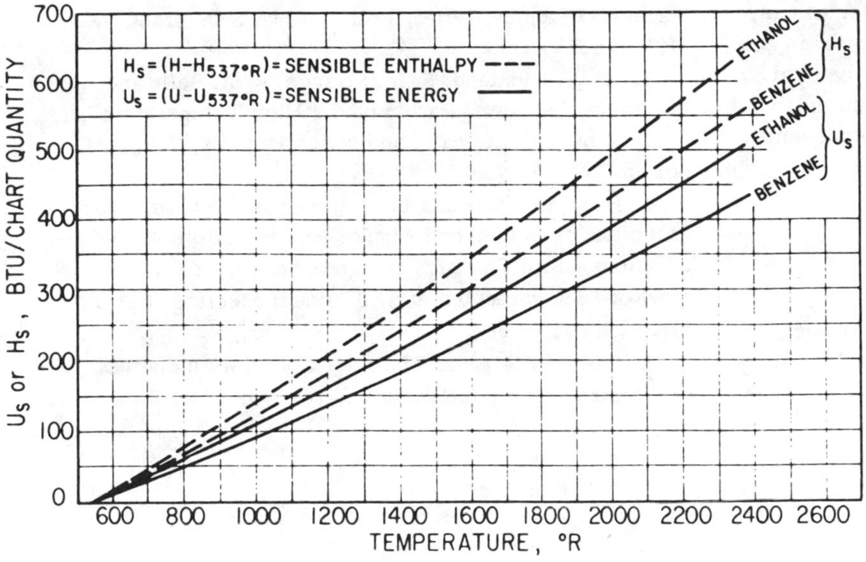

Chart 7 - Unburned mixture chart for determination of internal energy and enthalpy of compressed ethanol-air and benzene-air mixtures

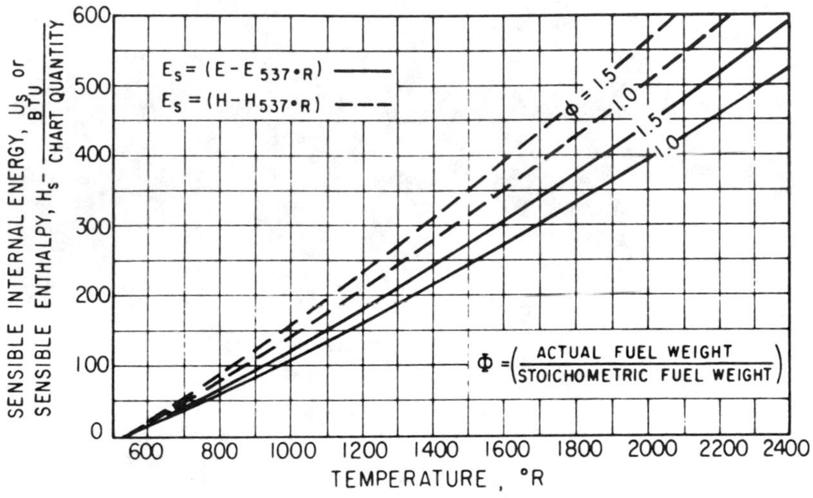

Chart 8 - Unburned mixture chart for determination of internal energy and enthalpy of compressed methanol-air mixtures

ACKNOWLEDGMENT

The authors are indebted to Commercial Solvents Corp. for aid in carrying out this research. Assistance was in the form of a grant to provide support for one of the junior authors during pursuit of a higher degree and in the generous donation of the fuels and some of the engine components used in the study.

REFERENCES

1. J. A. Bolt, "A Survey of Alcohol as a Motor Fuel," SP 254 presented at SAE Summer Meeting, June 1964, Chicago, Ill.

2. M. H. Edson, "The Influence of Compression Ratio and Dissociation on Ideal Otto Cycle Thermal Efficiency," SAE Technical Progress Series, Vol. 7, 1964, SAE Trans., Vol. 70, (1962), 665.

3. C. W. Vickland, F. M. Strange, R. A. Bell, and E. S. Starkman, "A Consideration of the High Temperature Thermodynamics of Internal Combustion Engines," SAE Transactions, Vol. 70 (1962), 785.

4. H. K. Newhall and E. S. Starkman, "Thermodynamic Properties of Octane and Air For Engine Performance Calculations," SAE Technical Progress Series, Vol. 7, 1964, Paper No. 633G presented at SAE Annual Meeting, Detroit, Mich., January 1963.

5. Edward F. Obert, "Internal Combustion Engines," 2nd Ed. Scranton, Pa.: International Textbook Co., 1953.

A Survey of Alcohol as a Motor Fuel *

Jay A. Bolt
The University of Michigan

IT HAS BEEN KNOWN since the invention of the internal combustion engine that alcohol could be used as a motor fuel. Ethyl alcohol has been produced in large quantities by fermentation of potatoes, sugar cane, and corn and can be produced from almost any of the produce from orchard and field. However, most industrial alcohol is now produced from petroleum. Many countries and people have tried for a variety of reasons to promote the use of ethyl alcohol as a motor fuel.

In the United States ethyl alcohol as a fuel for engines (power alcohol) has been promoted primarily by a desire to develop new outlets for farm produce. A further reason was a fear that our petroleum supply would be exhausted and that alternate energy sources must be developed.

In Europe, where large quantities of ethyl alcohol were used as a motor fuel before World War II, there were additional incentives:

1. A desire to build and maintain a large alcohol production capacity because of its importance to the manufacture of explosives and other materials of warfare.

2. Many European countries have no oil, and they desired to substitute domestic alcohol partially to improve their balance of trade.

Since World War II the use of alcohol as a motor fuel has been reduced to less than one-tenth of the pre-World War II levels, and continues to decline. Only in countries where there is a large supply of sugar cane, such as Cuba and the Phillipines, has there been an increase in the use of alcohol as a motor fuel in the post World War II period.

In 1906 the U.S. Congress passed the Industrial Denatured Alcohol Act, which freed alcohol from tax when used for industrial purposes. Many predicted this would lead to the wide use of power alcohol, and much work was done to accomplish this objective. The U.S. Dept. of Agriculture, the U.S. Bureau of Mines, the automotive industry, and the universities have all contributed to this work. However, two main factors in the past have prevented the general use of alcohol as a motor fuel, except under special circumstances, or when required by legislation: excessive cost of alcohol as compared with gasoline, and the absence of technical advantages to justify a higher cost.

During World War II water-alcohol (usually water and methanol) injection was used to obtain a dramatic increase in power output from supercharged piston aircraft engines. Following World War II there was promotional effort to use water-alcohol injection in gasoline automobile and truck engines. However, this application for unsupercharged engines has not provided enough advantages to justify the added complexity and dual fuel supply. Methyl alcohol has been popular as a fuel constituent for racing.

The present interest in alcohol as a fuel results from a hope that alcohol may reduce undesirable exhaust emissions. In the past the question of exhaust composition was not considered important and, therefore, relatively little data concerning exhaust composition is available. Several references contain data on the influence of alcohol on carbon monoxide composition. Data concerning unburned hydrocarbons and oxides of nitrogen, which is of particular interest, is sparse. Some additional information concerning these constituents is available from unpublished sources. Included in the bibliography are all the significant references which could be found that have information concerning alcohol and its exhaust composition.

The literature concerning alcohol is enormous, and spans a half century. Much of it is authoritative and the conclusions are well documented. Some of it is less well documented, and some represents a prejudiced viewpoint.

I have summarized most of the significant and authoritative literature under a series of topics which I hope will be of interest to the automotive industry. References are cited as much as practical so that those interested can pursue the subject in greater detail in the original sources, and to give credit to the original authors.

I have tried to choose and emphasize the more recent

*This paper contained in SP-254, "Alcohols and Hydrocarbons as Motor Fuels". Presented at the Summer Meeting, Chicago, Illinois, June 1964.

ABSTRACT

Alcohol has been promoted and used as a motor fuel for more than 50 years. However, United States ethyl alcohol production is small compared with gasoline production.

High latent heat of vaporization of alcohol makes possible some increase of power over gasoline. The heating value of alcohol is low and energy content of alcohol blends is less than that of gasoline; fuel consumption of blends is therefore increased.

The ability of ethanol to improve the octane number of gasoline has diminished as the octane number of gasoline has improved.

There is no published evidence that alcohols can appreciably reduce air pollution problems.

SAE/PT-80/19/$02.50

information. Literature on this subject is also available in other languages than English, and a few references in French and German have been included. The references are listed alphabetically according to first author. Any attempt to classify them would be difficult because many of the references discuss many aspects of the subject. A chronological key to the references has also been included for convenience.

CHEMISTRY OF COMBUSTION

The alcohol molecule is characterized by the presence of an OH radical in addition to the usual carbon and hydrogen of petroleum. The properties of two alcohols are shown in Table 1, together with isooctane, which may be taken as typical of gasoline in many respects.

Methyl alcohol, or methanol, is commonly called wood alcohol, because it was produced from the distillation of wood. It is now produced synthetically. It is very poisonous. Its heating value is the lowest of the alcohols because it contains the most oxygen (50% by weight). It is therefore a poor fuel for air-breathing engines because they can take in oxygen from the atmosphere.

The production and use of ethyl alcohol, or ethanol, dates back to some of man's earliest discoveries. Its common name, grain alcohol, is derived from its production by fermentation from grain. Ethyl alcohol is presently manufactured by fermentation, and by synthetic processes. It is readily miscible with water and is commonly sold commercially mixed with water. The "proof" of an alcohol is equal to twice the per cent by volume of alcohol in the water mixture.

Table 1 - Properties of Octane and Alcohol

	Octane	Alcohol Methyl	(Anhydrous) Ethyl
Chemical Formula	C_8H_{18}	CH_3OH	C_2H_5OH
Molecular Weight	114	32	46
Carbon %, by weight	84.0	37.5	52.0
Hydrogen %, by weight	16.0	12.5	13.0
Oxygen %, by weight	Nil	50.0	35.0
Heating Value			
Higher, Btu/lb	20,570	9,770	12,780
Lower, Btu/lb	19,080	8,640	11,550
Latent Heat of Vaporization	141	474	361
Specific Gravity, (60 F)	0.702	0.796	0.794
Stoichiometric Mass Ratio	15.1	6.45	9.0
Boiling Temperature, F	258	149	172
Octane Number, Research Method	100	106	106
Octane Number, Motor Method	100	92	89
Energy - Btu/ft^3 of Standard Stoichiometric Mixture	95.4	94.5	94.7

The complete combustion of ethyl alcohol with the stoichiometric (chemically correct) amount of air is as follows:

$$C_2H_5OH + 3\,O_2 + 3\left(\frac{79}{21}\right)N_2 \rightarrow 2\,CO_2 + 3\,H_2O + 3 \times 3.76\,N_2$$

$$\text{Mols} \quad \underbrace{1 \quad 3 \quad 11.3}_{15.3} \rightarrow \underbrace{2 + 3 + 11.3}_{16.3}$$

$$\text{Weights} \quad \underbrace{46 \quad 96 \quad 316}_{458} \rightarrow \underbrace{88 + 54 + 316}_{458}$$

$$\text{The } \frac{\text{lb air}}{\text{lb fuel}} = \frac{412}{46} = 9.0$$

The combustion reaction shown is exothermic and results in the liberation of (11,550 + 361) = 11,911 Btu/lb of alcohol burned from the vapor state. For the quantities represented by the combustion equation, the energy release by the alcohol in vapor state is 11,911 Btu/lb × 46 lb/mol. = 548,000 Btu/mol. The energy release per standard cubic foot of stoichiometric volume, will be:

$$\frac{548{,}000 \text{ Btu/mol. alcohol}}{(1 + 3 + 11.3) \text{ mols reactants} \times 378 \dfrac{\text{std. ft}^3}{\text{mol.}}}$$

$$= 94.7 \text{ Btu/ft}^3$$

Note from Table 1 that the corresponding energy per ft^3 of methyl alcohol mixture is nearly the same. The value for octane vapor is 95.4 Btu/ft^3. This quantity for gasoline will vary some for different gasolines, but is almost the same as that for ethyl alcohol. It is interesting that gasoline and alcohol, which in many respects are quite different, have nearly equal energy of combustion per unit volume of stoichiometric mixture. From this we may conclude that these fuels used in an engine under the same conditions with the same fraction of the stoichiometric mixture for both, and with fully vaporized fuel, will produce nearly the same power. This is also generally true of all hydrocarbon fuels. Therefore, the power of an engine cannot be greatly changed by changing these fuels, for similar charge conditions.

To generate equal amounts of energy and power will require: $\frac{19{,}080 + 141}{11{,}550 + 361} = \frac{19{,}221}{11{,}911} = 1.61$ or the use of approximately 60% greater weight of ethyl alcohol than gasoline. Engine data will be shown in a later section which bears out this conclusion.

VAPORIZATION AND VOLATILITY

In comparison with gasoline, the alcohols have a very high latent heat of vaporization (see Table 1); for methyl alcohol it is three and one-third times that of gasoline.

In contrast to the computations made earlier, fuels are usually supplied to the engine manifold and air stream in liquid form. The vaporization of the gasoline in a stoichiometric mixture of liquid gasoline and air (without external heating) results in an air temperature reduction of approximately 40 F. For the alcohols, this temperature drop will be greater. However, with the same heat addition per unit mass of charge, the per cent of the alcohol evaporated in the manifold will also be less. A discussion of these relations, with numerical illustrations, is given in Refs. 24 and 43.

The use of alcohol, with its greater evaporative cooling effect, reduces the charge temperature and thus usually improves the engine volumetric efficiency. However, the alcohols have a much smaller molecular weight than gasoline, and therefore their vapor occupies proportionately more volume than equal mass of gasoline vapor. It is the high latent heat of vaporization which is primarily responsible for the increased power outputs with alcohol, and which leads to the popularity of methyl alcohol as a blending constituent for racing fuels.

Volatility is also an important factor in determining the amount of fuel which will vaporize. Gasoline is composed of a myriad of molecular forms having boiling points ranging from approximately 100 to 400 F. Alcohol, in contrast, is composed of like molecules, with a single boiling point, as shown in Table 1. The alcohols lack the light ends with boiling points near 100 F which are essential for severe cold starting of spark-ignited engines. These differences are illustrated by Fig. 1 taken from Ref. 24.

The distillation curves of several blends of ethyl alcohol and gasoline are shown in Fig. 2, taken from Ref. 30. Most of the alcohol vaporizes at its boiling temperature of 172 F, giving the so-called "alcohol flat" in the distillation curves.

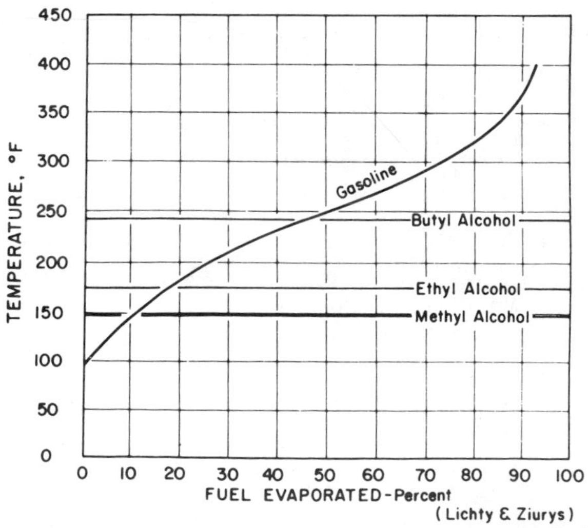

Fig. 1 - ASTM distillation curves for gasoline and alcohol

ENGINE PERFORMANCE - ALCOHOL

Having examined a few of the basic factors which influence the performance of fuels in an engine, let us now turn to actual engine tests.

From the many engine tests with alcohol reported, those of Sir Harry Ricardo are particularly complete and authoritative (42)*. He tested a great variety of fuels in a very comprehensive program, using a variable compression ratio, single-cylinder engine. Fig. 3 is a replot of Ricardo's engine test results for ethyl alcohol of 198 proof, and gasoline, taken from Ref. 43.

The increased mean effective pressure of alcohol at all mixture ratios, shown in Fig. 3, is the most noticeable dif-

---

*Numbers in parentheses designate References at end of paper.

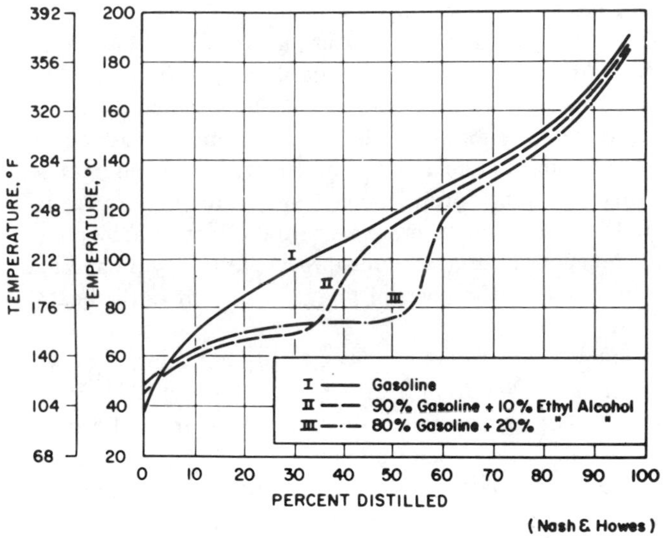

Fig. 2 - ASTM distillation curves for gasoline and ethyl alcohol blends

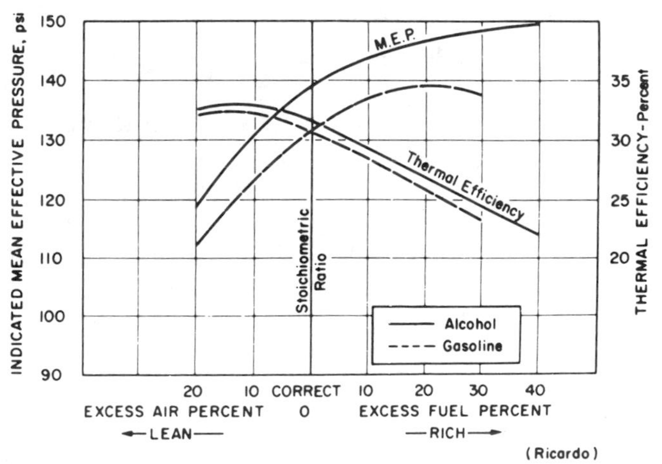

Fig. 3 - Engine performance with ethyl alcohol and gasoline

ference between the two fuels. This increase in mean pressure is due principally to the greater volumetric efficiency. This results from the high latent heat of vaporization of alcohol and the greater mass of fuel per unit mass of air. The intake manifold temperature is reduced, with resulting increase in the air density and engine volumetric efficiency. Much of the alcohol is probably evaporated during the compression stroke, with resulting reduction of the work on compression (43). It is interesting that for alcohol the mep increases with mixtures having up to at least 40% excess fuel, whereas for gasoline the maximum pressure is reached with 20% excess fuel. To obtain maximum power, therefore, there would be a temptation to use a greater per cent of excess fuel with alcohol. It will be noted that the very rich mixtures required to obtain the maximum mep with alcohol are accompanied by additional incomplete burning and the resulting reduced thermal efficiency.

The hydrocarbon fuels all have about the same mixture ratio at the lean limit of ignitability and combustion. The lean limit is the more important value and Fig. 3 shows that alcohol has the same lean mixture limitation as gasoline. Both fuels develop their maximum thermal efficiency with about 15% excess air. With mixtures leaner than this, the burning velocity for both is reduced and offsets other advantages of lean mixtures. The small increment of increased thermal efficiency of the alcohol in relation to gasoline is mainly due to the evaporative cooling effect of alcohol. This reduces the temperatures throughout the cycle and thus reduces the specific heats of the gases as well as the heat losses.

Fig. 4, taken from Ref. 24 of Lichty and Ziurys, shows similar data taken from tests with a 1935 6-cyl Chevrolet engine. The fuels in this case were 190 proof ethyl alcohol, and gasoline with a specific gravity of 0.745. Air was supplied at 100 F and the carburetor was fitted with a fuel needle valve to obtain the desired mixture ratios. The 2/3 and 1/3 loads were established by adjusting the throttle to give the same manifold pressure for both fuels.

The smaller values of air-fuel ratio for ethanol in comparison with gasoline are evident. For these tests at full throttle and the two part-throttle conditions the increase in power for the multicylinder engine with alcohol is much less than in the case of the single cylinder engine. The amount of increase in power with alcohol in comparison with gasoline is a function of the amount of heat added to the intake manifold, as suggested earlier. Lichty and Ziurys conclude that with complete evaporation and the same mixture temperature, which requires 184% more heat for the correct alcohol-air mixture, the chemically correct gasoline-air mixture should produce 2% more power. With the same amount of heat input, the chemically correct alcohol-air mixture should give 8.6% more power.

Table 2, taken from the summary information of Ref. 24, compares the fuel consumption with gasoline and with 190 proof ethyl alcohol. The authors conclude that the increased fuel consumption of the multicylinder engine in comparison with the single-cylinder engine is probably due both to fuel distribution problems and to differences in com-

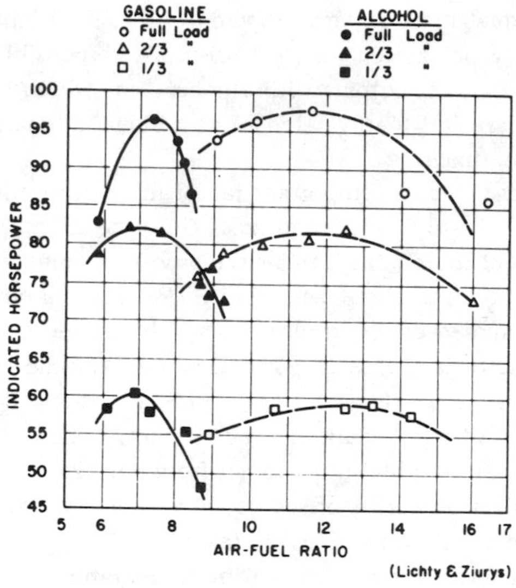

Fig. 4 - Comparison of engine performance with ethyl alcohol and gasoline

Table 2 - Comparison of Fuel Economy at Maximum Power with Gasoline and with Alcohol

		Single-Cyl Engine		Multi-Cyl Engine		
Fuel	Air-Fuel Ratio	Indicated Sp. Fuel Consumption Btu/Bhp-hr	Increase Over Gasoline	Indicated Sp. Fuel Consumption	Increase Over Gasoline	Increase of Multi-Cyl Engine Over Single-Cyl
Ethyl Alcohol (190 Proof)	7	0.75	59%	0.86	62%	14.7%
Gasoline	12	0.47	-	0.53	-	11.3%

(Lichty & Ziurys)

bustion chamber design. They also conclude the greater increase in fuel consumption from single cylinder to multicylinder in the case of alcohol may indicate slightly more distribution difficulty with alcohol than with gasoline.

Brooks (8) stated that for multicylinder engines the mixture distribution with 190 proof ethyl alcohol is poor and would probably require a special intake manifold.

I have recently run tests with students at the University of Michigan using 200 proof anhydrous ethyl alcohol in a 6-cyl Nash engine with glass windows set in the manifold runners and riser. Observation of the pulsing mixture in the manifolds under a high intensity light enables one to see the greater amount of liquid present in the mixture, in comparison with gasoline. It is reasonable to expect that the presence of more liquid in the manifold would make good mixture distribution more difficult.

OCTANE QUALITY

The octane numbers of methyl and ethyl alcohol (2) are listed in Table 1. The difference between the research and motor octane numbers is commonly taken as an indication of fuel sensitivity. By this measure (106 - 92 = 14 octane numbers for methyl alcohol) these fuels are sensitive to changes in engine conditions. Methanol displays a tendency to backfire under the knock rating conditions, the backfiring being an indication of preignition. Several authors, including Banks (1) and Ricardo (43) have noted this tendency of alcohol to induce preignition.

The principal interest in the alcohols as fuels lies in their use as blends with gasoline. Therefore, the octane number of such blends is of great importance.

The influence of alcohol additions on four base stocks -- straight run, catalytically cracked, thermally cracked, and polymer gasoline -- have been reported by Porter and Wiebe (40). Some of their results are shown in Fig. 5. In addition to the octane data shown, the curves reveal that the greatest improvement in octane number from alcohol addition is obtained for gasoline stocks of the lowest octane number, as would be expected.

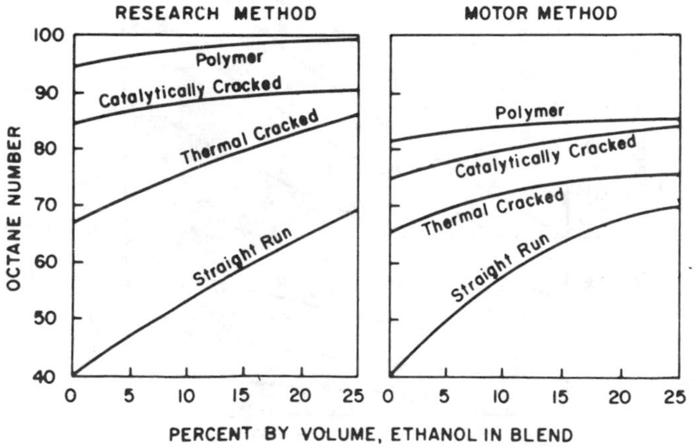

Fig. 5 - Increase of octane ratings of several gasoline stocks with alcohol addition

Southwest Research Institute (65) has reported the Research and Motor octane ratings of three regular and three premium commercial leaded gasolines, blended with 5, 10, and 25% by volume of anhydrous ethyl alcohol. These results are shown for a regular fuel in Fig. 6 and for a premium fuel in Fig. 7. As in Fig. 5, the octane numbers of the unblended gasolines can be read at the left edge of the charts. Road ratings of the base fuels and of the 10% alcohol blends are also reported, run in accordance with the modified CRC-F-8B borderline knock procedure. Fig. 6 reveals that the addition of up to 25% ethanol to regular gasoline improved the Research and Motor octane numbers in a near linear manner. The road octane numbers are also shown as a shaded band. The upper limit line of the road rating corresponds to low (1600) rpm and the lower borderline to high (3000) rpm. As often stated, here also the Research method rating corresponds quite well with the low speed road rating, and the Motor method nearly parallels

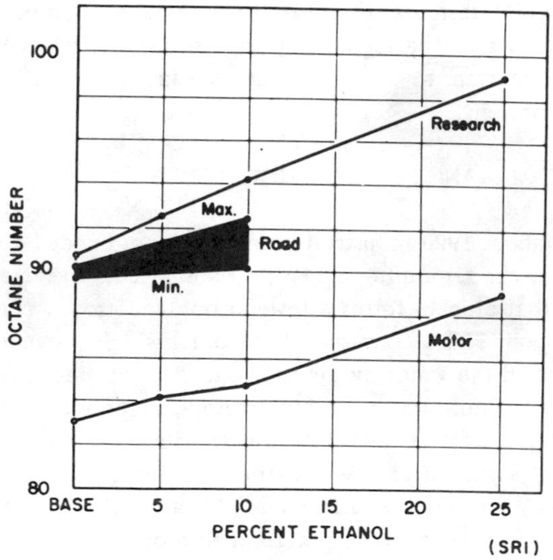

Fig. 6 - Octane ratings of regular gasoline-alcohol blends

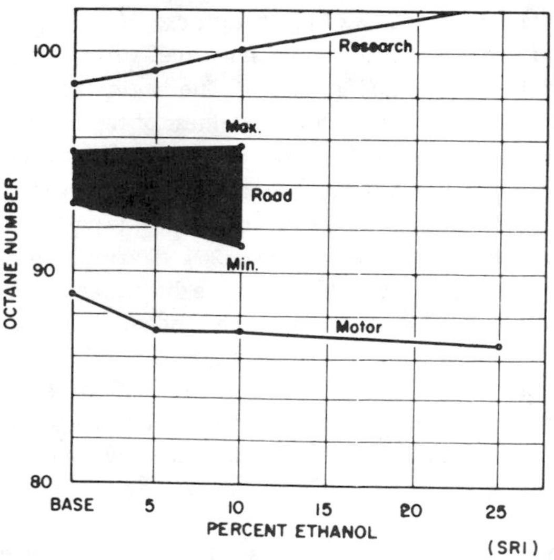

Fig. 7 - Octane ratings of premium gasoline-alcohol blends

the high speed road rating. Ethanol improves the high speed road rating only a small amount for this typical regular gasoline. Fig. 7 reveals that for the premium fuel the Research octane number shows less increase, as would be expected. The Motor method octane number for premium fuels is reduced by addition of ethanol. More important, the high engine speed road rating for this typical premium fuel is also depreciated by the addition of ethanol.

The older literature refers frequently to the high antiknock quality of alcohol. However, it must be remembered that this is in comparison with the gasoline of at least 20 years ago, and alcohol used as a blending agent improved the octane number of these earlier fuels. The Ethyl Corp. surveys (61) of motor fuels in the United States reveal that the average Research and Motor method octane numbers of premium fuels in the United States in 1963 were 99.8 and 90.8, respectively. The corresponding average values for regular fuels were 93.0 and 85.0. If the octane number of motor gasoline continues to rise, as it has in the past, ethyl alcohol will lose more of its former advantage as an octane improver. As emphasized in a Congressional report (64) it is more important to try to evaluate the future fuel situation than the present.

WATER TOLERANCE OF ALCOHOL BLENDS

Until about 1920 industrial alcohol did not exceed about 190 proof, the remaining 5% being water. This was due to the fact that alcohol forms a low-boiling azeotrope, containing about 95% alcohol and 5% water, which prevents separation of the water by distillation. The earlier difficulty of producing water-free (anhydrous) alcohol explains why the older engine tests were usually run with alcohol containing some water. For example, much of Ricardo's testing (42) was done with 90% and 95% alcohol. In more recent years it has become practical to produce alcohol with less than 0.1% water at little greater cost than the 95% product.

Gasoline and water-free alcohol are miscible in all proportions over a wide range of temperatures. However, even small additions of water to this blended fuel will cause separation of the alcohol and gasoline. The beginning of this separation is characterized by a cloudiness of the mixture. Fig. 8, taken from Ref. 15, shows that the ability of the blend to carry moisture without separation increases when more alcohol is present, and with increase in temperature. The water that can be tolerated by a 25% alcohol blend at room temperature is about 1%. If twice this amount is added to a sample of 25% blend, most of the alcohol will separate from the gasoline in a few seconds and settle to the bottom of the container. The interface between the alcohol and gasoline will be sharply defined. Other blending agents can be added to the mixture which will increase the water tolerance. Among these are benzol, benzene, acetone, and butyl alcohol (6, 15).

Separation of the alcohol and gasoline in the presence of water can be one of the most difficult problems attending the use of these blends as motor fuels. The condensation of moisture from the air in vented and partially filled tanks, and the accidental addition of water to gasoline storage tanks are well known. However, the fact that large quantities of alcohol-gasoline fuels have been supplied and used in Europe is evidence that the problems are not insuperable. The difficulties due to water separation have commonly led to the use of either 20-25% blends of alcohol alone or 10-15% alcohol and 10-15% benzol to reduce separation troubles (62).

The fact that water may be used as an agent to separate grain alcohol and gasoline also suggests that this could be done to obtain tax free alcohol from an automobile fuel tank for drinking. The taste and odor of gasoline in the separated grain alcohol could be removed by shaking with activated carbon (15). It is obvious that denaturing of the alcohol would be essential. The effects of denaturing fluids upon the characteristics of the alcohol-gasoline blends as motor fuels are unknown.

Alcohol is a good solvent for gum and many other materials which deposit from gasoline. A dirty gasoline fuel system is a source of trouble when filled with an alcohol-gasoline blend, since dirt is commonly loosened and plugs the system at critical points.

METERING CHARACTERISTICS OF ALCOHOL BLENDS

Alcohol blends are commonly recommended and used without changes in carburetors which have settings intended for gasoline. It is, therefore, of interest to note any changes in metering which will occur. Since the heating value of ethyl alcohol is about 60% of that of gasoline, blends of these fluids will also have less energy than gasoline. Table 3, taken from Ref. 15 of Egloff, shows the relative reduc-

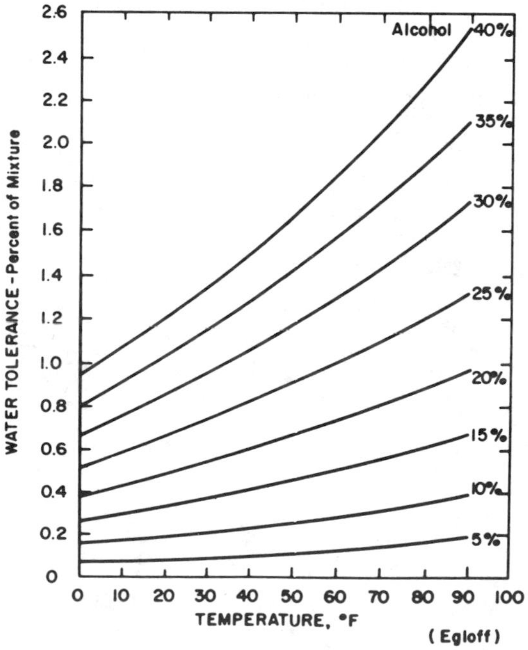

Fig. 8 - Water tolerance of alcohol - gasoline blends

tion of energy of the blend on a volume basis. As stated earlier, reduced energy per unit of volume will require proportionately increased fuel flow rates for proper engine operation.

For a given carburetor setting, the specific gravity and viscosity of the fluid will also affect the volume and mass rate of flow. The specific gravity of ethyl alcohol is 0.794, about 10% greater than that of present day gasolines. The alcohol-gasoline blend has a slightly lower specific gravity than that calculated from its components because the mixture displays a slight expansion, and cooling effects, when mixed. Table 4, also from Ref. 15, gives specific gravity and viscosity values of a 10% alcohol blend. It will be noted that the viscosity of ethyl alcohol is more than three times that of gasoline. The true viscosity values for blends should be determined experimentally for greatest accuracy since, for most proportions, the viscosity is less than that computed from arithmetic averages. The magnitude of the influence of viscosity upon flow in carburetors depends considerably on metering jet configuration, being least for sharp-edged orifices, and cammot be readily predicted. In general, for a given carburetor condition, the increased specific gravity of alcohol will increase its mass rate of flow, and the increased viscosity of the alcohol will reduce its flow rate.

Brown and Christenson (22) conducted tests on a truck engine with gasoline in comparison with a 10% ethanol-gasoline blend. They report that these two fuels gave practically identical air-fuel mass ratios at the same carburetor conditions.

The results of other unpublished tests indicate that for a 25% alcohol-gasoline blend, in comparison with gasoline, there will be only small changes in the air-fuel mass ratio. The changes that result will be partly dependent upon the particular carburetor involved, and upon the flow rate.

Therefore, the principal effect upon the engine of the change to an alcohol-gasoline blend with no carburetor setting changes will result from the reduced energy content of unit mass of the alcohol. When fuels of different heating value are being considered, the energy of the fuel per pound is equally important as the air-fuel mass ratio. An example, using a blend of 25% ethyl alcohol and 75% gasoline will illustrate the points and quantities involved. If gasoline having a specific gravity of 0.73 is used in the blend, computation indicates that the blend will have an energy content (Btu/lb of air) about 10% less than that of gasoline. This is equivalent to changing the fuel/air ratio by 10%, the alcohol blend being the leaner. This result is also indicated by Table 3.

If an alcohol blend is substituted for gasoline in an automobile, larger metering jets are required to maintain the same per cent of the stoichiometric air for combustion, or the same equivalence ratio.

ENGINE PERFORMANCE - ALCOHOL BLENDS

Most of the interest in alcohol centers on the use of mixtures of ethyl alcohol and gasoline, and this section of the paper is restricted to this blend. Comprehensive tests of anhydrous ethyl alcohol blends were run by Lichty and Phelps (26) using 5, 10, and 20% ethyl alcohol in commercial gasolines of 0.738 specific gravity. Tests were run in a single cylinder CFR engine at various compression ratios, and in a 1935 6-cyl Chevrolet engine. These comprehensive tests, reported in 1938, can best be summarized by quoting from the conclusions drawn by the authors, which follow:

"Volumetric efficiency data do not show a consistent increase with the addition of ethyl alcohol to gasoline either in the single or multicylinder tests.* Considering only the single-cylinder engine data at 6.2:1 and 6.8:1 compression ratio, the addition of 10 and 20% alcohol increases the volumetric efficiency about 1 and 2%, respectively.

"The power output, thermal efficiency, and heat loss to the cooling water, with comparable mixture conditions, does not change appreciably with the addition of ethyl alcohol to the gasoline, except where this addition reduced detonation and permits the use of optimum spark advance.

"An increase in indicated specific fuel consumption of 7 and 13% for the single-cylinder engine and an increase in

*The change in volumetric efficiency with alcohol blends is dependent upon the amount of alcohol in the blend and the heat added to the induction system.

Table 3 - Calorific Values of Alcohol-Gasoline Blends

Per Cent by Volume: Alcohol (99.5%)	Gasoline	Heating Value Btu/gal	Relative Heating Value (Gasoline = 1)
0	100	135,000	1.000
10	90	129,000	0.962
20	80	124,800	0.924
30	70	119,700	0.887

(Egloff)

Table 4 - Specific Gravity and Viscosity of Alcohol and Gasoline

Per Cent by Volume: Gasoline	Alcohol	Specific Gravity	ABS Viscosity Poise
100	0	0.742	0.00525
90	10	0.746	0.00552
0	100	0.794	0.01730

(Egloff)

brake specific fuel consumption of 5 and 9% for the multicylinder engine result from the use of 10 and 20% blends of ethyl alcohol compared to gasoline, respectively, at the same compression ratio and with the carburetor adjusted to give air-fuel ratios comparable for each fuel in regard to maximum power or maximum economy.

"Applying the multicylinder power and fuel consumption data to motor vehicles on the highway (that is, based on the same power output and with adjustment to maximum power or maximum economy air-fuel ratios for each fuel), the substitution of alcohol blends for gasoline should result in an increase in volumetric fuel consumption of about 5 and 7% with comparable mixture ratios for the 10 and 20% blends, respectively. Using air-fuel ratios equal to or richer than maximum power for gasoline and without adjustment of air-fuel ratio on substitution of the 10 and 20% blends, a decrease in volumetric fuel consumption of about 2 and 3%, respectively, should be obtained. However, the lowest fuel consumption can be obtained with gasoline, rather than alcohol blends, by adjusting the carburetor for maximum economy mixture."

Duck and Bruce (13) made tests of various nonpetroleum fuels in Plymouth, Chevrolet, and Ford automobiles. They found mixture distribution was somewhat less uniform with fuels containing ethanol, in comparison with gasoline. Fuel consumption was inversely proportional to the heating value of the fuel blend used.

A blend of 25% anhydrous ethanol and 75% regular gasoline was tried at The University of Michigan in 1963 and 1964 cars in March, when the temperature was near freezing. Starting and performance of the engines were quite normal. A hesitation could be felt following quick throttle opening during the warmup period. With the carburetor set closer to the lean limit of satisfactory performance with gasoline, the 25% alcohol blend gave unsatisfactory acceleration, and lean surging in the cruise condition was evident. This is to be expected, since the alcohol blend had, in effect, a leaner mixture, as discussed in the section on Metering Characteristics.

ANTIKNOCK FLUIDS FOR INCREASED OUTPUT

Kerosene burning farm tractors have used water to prevent combustion knock and pre-ignition since World War I. This made possible the use of increased compression ratio with the low octane kerosene, and reduced fuel consumption. These tractors commonly were equipped with a carburetor which permitted starting on gasoline, and which then could be manually changed over to kerosene and water after the engine was heated.

During World War II antidetonant fluids were supplied to supercharged piston aircraft engines. In this case the antidetonant fluid was commonly 50% methanol and 50% water. Because of the very high latent heat of water (1000 Btu/lb) and of methanol, this mixture provides an optimum internal cooling effect, and is a powerful knock suppressor. The methanol also prevents freezing of the fluid mixture. The water, however, caused many corrosion problems in the metering equipment. The methanol-water mixture was used only for high power operation, commonly called war emergency rating. Provision was made in the gasoline carburetor equipment to derich the gasoline and air mixture to the ratio for maximum power when the antidetonant fluid was used, and to again enrich the gasoline mixture if the methanol-water supply was all used. The latter helped to avoid damage to the engine. Obert (31) and Rowe, et al (45) ascribe the great antiknock value of water-alcohol primarily to the high latent heat of vaporization of these fluids, which provides cooling of the combustion chamber walls and valves, and reduces the temperatures of the cycle.

EXHAUST COMPOSITION

Throughout the long history of alcohol-blend fuels, little attention was given to the exhaust gas composition since it was not a point of great concern. The techniques for accurately measuring the small quantities of unburned hydrocarbons have also been developed only in recent years. Information on exhaust emissions with alcohol fuels is therefore scarce.

Nash and Howes (30) comment on the toxicity of engine exhaust, as follows: "Referring to the toxicity of alcohol fuels and their combustion products, all investigators on this subject agree that the type of fuel used, whether gasoline, benzol, or alcohol has no influence on the toxicity of the normal exhaust products."

Professors Lichty and Phelps (25) made tests on a CFR engine and a 6-cyl Chevrolet engine to determine the carbon monoxide in the exhaust with gasoline and with 10 and 20% ethyl alcohol blends with gasoline. Table 5, taken from their report, shows the per cent of carbon monoxide in the exhaust gases of the CFR test engine for the gasoline and alcohol blends. The carbon monoxide was determined with Orsat equipment.

The stoichiometric mass ratio for the gasoline was 15:1;

Table 5 - Carbon Monoxide in Exhaust As Function of Air-Fuel Ratio

Air-Fuel Ratio	Gasoline Per Cent	10% Alcohol Blend, Per Cent	20% Alcohol Blend, Per Cent
15.1	0	0	0
14.5	1.1	0	0
14.0	2.1	0.8	0
13.5	3.2	1.7	0.9
13.0	4.3	2.8	1.9
12.5	5.4	3.9	3.0
12.0	6.8	5.1	4.2

(Lichty & Phelps)

for the 10% alcohol blend, it was 14.4; and for the 20% blend, it was 13.8. These are the air-fuel ratios containing the required air for complete combustion and are indicated in Table 6 as 100% required air. Thus Table 6, like Table 5, shows the carbon monoxide from the several fuels. In Table 6, however, the mixture ratios are expressed in terms of per cent of stoichiometric air. This tabulation reveals that at the same fraction of stoichiometric mixture the carbon monoxide in the exhaust is almost independent of the fuel used, except that the 20% alcohol blend shows slightly more CO. Referring again to Table 5, it is apparent that the greater CO in the exhaust with gasoline is due to the fact that at any of the air fuel ratios the gasoline is burning with less available air, relative to alcohol blends. Brown and Christensen (9) also drew this same conclusion. Neglecting the small effects of specific gravity and viscosity differences upon flow for these blends, these are the relative conditions that prevail when these fuels are metered in a carburetor without any setting changes.

The Air Polution Foundation has sponsored research to determine the exhaust composition of alcohol blends in comparison with gasoline. The following is quoted from its report, (60), taken from Ref. 28: "The data show that no great changes which would be of help in solving the air pollution problem may be expected from using ethanol-gasoline mixtures... "

The clear indication from these references is that the amount of air available for combustion of the several fuels is the very significant factor affecting the quantity of carbon monoxide in the exhaust, and dwarfs any differences in the fuels. Jackson, et al (22) and Hagen, et al (17) conclude that air-fuel ratio has a most significant effect on unburned hydrocarbons. The minimum unburned hydrocarbon emission was found to occur at an air-fuel ratio between 16 and 18 (22), which was also the ratio for best economy. This brings out the fact that for reduction of the carbon monoxide and unburned hydrocarbons, engines should be operated with mixtures as lean as possible with all cylinders firing.

Many references have mentioned that alcohol is a "clean burning" fuel. For example, Van Hartesveldt (56) states that the lower molecular weight alcohols burn with less soot or deposit than hydrocarbons. No data is given to support the claim. Hobbs (20) also mentioned that his test engine showed reduced deposits when using water injection. This action of water is usually attributed to thermal shock, and is of doubtful significance in relation to alcohol.

Our student tests with 200 proof ethyl alcohol in comparison with gasoline in a Nash engine at the University, mentioned earlier, convinced us that the exhaust of the engine when running on the alcohol smells less obnoxious than with gasoline.

It is now well established that deposits in the combustion chambers cause quite marked increase in unburned hydrocarbons (22 and 17). This is perhaps due to the fact that charge is squeezed into the rather porous deposits during the compression stroke, and thus avoids being burned. The deposits may also add to the effective surface quenching area of the chamber. In any case, the nature of the deposits and reduction of deposit formation take on new urgency in view of the problems of exhaust emissions and air pollution. More work is needed to understand the phenomena and quantitative relationships between chamber deposits and undesirable exhaust emissions.

SOURCES AND PRODUCTION OF ALCOHOL

Alcohol has often been described as the most versatile of all chemical compounds. It is a solvent, a germicide, and an antifreeze; it is a combustible liquid and a most versatile building block for other organic chemicals. Alcohol can be readily made from grains and a host of other farm products, but most foodstuffs are relatively expensive materials. To be used as a large volume constituent of motor fuels in the United States there must be an unfailing and large source of inexpensive raw materials for its production. From a cost standpoint, it is almost essential that the raw material be transportable by pipe line.

Table 6 - Carbon Monoxide in Exhaust As Function of Per Cent Stoichiometric Ratio

	Air-Fuel Ratio			CO in Exhaust		
Required Air, %	Gasoline	10% Alcohol	20% Alcohol	Gasoline	10% Alcohol	20% Alcohol
100	15.1	14.4	13.8	0.1	0.1	0.3
95	14.3	13.7	13.1	1.6	1.3	1.8
90	13.6	13.0	12.4	2.9	2.7	3.2
85	12.8	12.2	11.7	4.0	4.5	4.8
80	12.1	11.5	11.0	6.5	6.5	6.8

(Lichty & Phelps)

Alcohol has an advantage over gasoline in that it can be produced (although at high cost) without using up irreplaceable petroleum. To the best of our knowledge, the conditions that created petroleum no longer exist. Although our sources of petroleum now seem large, they must inevitably be exhausted. We do not know if atomic energy sources will be available to power personal vehicles before this day comes. As long as the sun shines, and plants grow, alcohol will be unique because it possesses the attribute of perennial renewal.

For some countries, with little or no petroleum, and an abundant supply of farm products, such as sugar cane, alcohol may continue to be an important fuel. Dr. Donald Katz of The University of Michigan has also suggested that for many countries alcohol could be an important source of fuel for highway and farm use in a period of great national emergency. Given the necessary technical information, alcohol could be produced from available farm produce in a very decentralized manner, and essential machinery kept running.

It is not the intent of this survey to examine the economic aspects of alcohol blends as motor fuel, but a few facts and figures will help give perspective.

Most domestic industrial alcohol production before World War II came from blackstrap molasses, which was mostly imported (64). During World War II ethyl alcohol was produced by fermentation of grain, from molasses, and from petroleum. The cost of this production varied from $0.58 to $1.25 per gallon. The production in this country reached a maximum in 1945, and was 600,000,000 gal.

Much of the synthetic rubber produced during World War II came from grain alcohol. The butadiene plants using petroleum as a raw material came into operation late in the war. Quoting from Ref. 64, "Following the war the synthetic production of alcohol from petroleum gradually invaded the market and with the recent high prices for blackstrap molasses, petroleum alcohol has all but taken over the market."

In the year ending June 30, 1956, 255,700,000 gal of industrial alcohol was produced in the United States, 71% from petroleum. The available United States plant capacity (1957) for industrial alcohol from grain is 200,000,000 gal/year, and much of this capacity was idle in 1957. It is estimated (62) that anhydrous ethyl alcohol made from corn costing $1.40 a bushel would be about $0.65/gal. From corn at $0.50 a bushel the cost of alcohol would still be about $0.30/gal. In Ref. 64, p 95, it is estimated that synthetic industrial alcohol made from petroleum or natural gas would cost between $0.136 and $0.331/gal. There is agreement that as a raw material for industrial alcohol, grain cannot compete on an economic basis with petroleum.

In 1957, the year in which the above alcohol estimates were made, the production of gasoline was approximately 60,000,000,000 gal/year (62).. Taking the figure of 200,000,000 gal/year as the alcohol capacity from fermentation of grain, it is evident that only about 0.3 of 1% of the United States motor fuel demand could be met by available grain distillery capacity.

SUMMARY

1. The use of alcohol as a fuel for spark-ignited engines is old.

2. Ethyl alcohol as a fuel makes possible up to about 8% more power output, the increase varying down to zero, depending on the amount of heat added to the engine induction system.

3. The value of ethyl alcohol as an octane improver has been greatly reduced, and may be expected to diminish further if the octane number of gasoline continues to increase.

4. A 25% ethyl alcohol blend substituted for gasoline without carburetor changes, results in a mixture about 10% leaner on the basis of Btu/lb of air; in other words, there will be 10% more air available for the combustion of the alcohol. Since exhaust emissions are very dependent on the quantity of air available for combustion, it is most important that comparative tests of emissions be conducted on the basis of an equivalent amount of air for each. This requires different carburetor settings for alcohol and gasoline.

5. More ethyl alcohol has been used as a motor fuel in Europe than anywhere else, where it reached a maximum in 1936. This resulted mainly from government regulation. In the post World War II period its use has been reduced to less than 10% of the amount used in 1936.

6. No published evidence has been found to indicate that the use of ethyl alcohol as a constituent of motor fuel will help appreciably in solving the air pollution problem.

7. Because of the low water tolerance of alcohol-gasoline blends, anhydrous ethanol must be used and great care must be exercised to avoid water contamination. For the 25% alcohol blend, less than 2% of water will cause separation.

8. The use of alcohol blends for motor fuel will cost more than gasoline.

9. The production facility for making ethyl alcohol from grain is about 1/3 of 1% of United States production of gasoline.

REFERENCES

1. F. R. Banks, "Some Problems of Modern High-Duty Aero-Engines and Their Fuels," Inst. Petrol. Technol. Jour., Vol. 23, (1937), 63-177.

2. B. Brewster and R. V. Kerley, "Automotive Fuels and Combustion Problems." SAE paper, August 1963.

3. O. C. Bridgeman, "Alcohol-Gasoline Blends as Motor Fuels," Indus. and Engin. Chem, Vol. 11, (1933), 139-140.

4. O. C. Bridgeman and D. W. Querfeld, "Critical Solution Temperatures of Mixtures of Gasoline, Ethyl Alcohol and Water," Nat. Bur. Stds. J. Res., Vol. 10, (1933), 693-704.

5. O. C. Bridgeman and D. W. Querfeld, "Solubility

of Ethyl Alcohol in Gasoline," Indus. and Engin. Chem., Indus. Ed. Vol. 25, (1933), 523-525.

6. O. C. Bridgeman, "Utilization of Ethanol-Gasoline Blends as Motor Fuels," Indus. and Engin. Chem., Vol. 28, (1936), 1102-1112.

7. D. B. Brooks, "Single-Cylinder Engine Tests of Substitute Motor Fuels," U.S. Nat'l Bur. Stds. Jour. Res., Vol. 35, (1945), 1-37.

8. D. B. Brooks, "An Analysis of the Effect of Fuel Distribution on Engine Performance," U.S. Nat'l Bur. Stds. Jour. Res., Vol. 36, (1946), 425-439.

9. L. T. Brown and L. M. Christensen, "Gasoline and Alcohol-Gasoline Blends," Ind. and Eng. Chem., Vol. 28, (1936), 650-652.

10. L. M. Christensen, "Alcohol-Gasoline Blends," Ind. Engr. Chem. Vol. 28, 1089-1094, (1936), Vol. 28, 650-652, (1936).

11. W. T. David and A. S. Leah, "Fuel Economy in Gasoline Engines," Inst. Mech. Engin. (London), Jour and Proc, Vol. 143, (1940), 289-312P.

12. D. Downs, S. T. Griffiths, and R. W. Wheller, "Pre-Flame Reactions in the Spark-Ignition Engine: The Influence of Tetraethyl Lead and Other Antiknocks," Jour. of Inst. of Petroleum, Vol. 49, (January 1963), 8-25.

13. J. T. Duck and C. S. Bruce, "Utilization of Nonpetroleum Fuels in Automotive Engines," U.S. Nat'l Bur. Stds. Jour. Res., Vol. 35, (1945), 439-465.

14. A. E. Dunstan, A. W. Nash, B. T. Brooks, and H. Rizard, "The Science of Petroleum," 4 vol., 3192 pp, London: Oxford Univ. Press, 1938.

15. G. Egloff and J. C. Morrell, "Alcohol-Gasoline as Motor Fuel," Ind. and Eng. Chem., Vol. 28, (1936), 1080.

16. Emil Fischer, "Alcohol as Automotive Fuel," Motorwagen, Vol. 29, (1926), 487-493.

17. D. S. Hagen, G. W. Holiday, "The Effects of Engine Operating and Design Variables on Exhaust Emissions," SAE preprint 486C, March 1962. This paper is included in SAE Tech. Progress Book TP-6, 1964.

18. H. A. Havemann, M. R. K. Rao, et al, "Alcohol with Normal Diesel Fuels," Gas and Oil Power, Vol. 50, (January 1955), 15-19, ibid., (February 1955), 45-48, 50.

19. S. D. Heron, and H. A. Beatty, "Aircraft Fuels," Jour. Aeronaut. Sci., Vol. 5, (1938), 463-479.

20. G. W. Hobbs and M. L. Fast, "The Use of Water as an Anti-Detonant," Eng. Exp. Station Bul. No. 31, 1930, Mich State Col.

21. C.P. Hopkins and M. S. Kuhring, "The Use of Grain Alcohol in Motor Fuel in Canada, Engine Tests of Alcohol-Gasoline Blends," Canada Nat'l Res. Council, Rpt. 77 pp., Apr. 21, 1933, revised June 19, 1933.

22. M. W. Jackson, W. M. Wiese, J. T. Wentworth, "The Influence of Air-Fuel Ratio, Spark Timing, and Combustion Chamber Deposits on Exhaust Hydrocarbon Emission," SAE preprint 486A, March 1962. This paper is included in SAE Tech. Progress Book TP-6, 1964.

23. R. T. Jackson, "Five Hundred Mile Race Results with an Analysis of the New Developments Introduced this Year for Stamina and Speed," Automotive Ind., Vol. 82, (1940), 549-551, 590-592.

24. L. C. Lichty and E. J. Ziurys, "Engine Performance with Gasoline and Alcohol," Ind. and Eng. Chem. Indus. Ed. Vol. 28: (1936), 1094-1101. Also Yale Univ. School of Eng. Pub. No. 16, New Haven, Conn.

25. Lichty and Phelps, "Carbon Monoxide in Engine Exhaust Using Alcohol Blends," Yale Univ. Publ. Ser. No. 22, also Ind. and Eng. Chem., (May 1937), 495.

26. L. C. Lichty and C. W. Phelps, "Gasoline-Alcohol Blends in Internal Combustion Engines," Yale Univ. School of Engr., Publ. No. 31, New Haven, Conn., 1938, Ind. Eng. Chem., Vol. 30, (1938), 222-230.

27. A. J. Meyer and R. E. Davis, "Development of A Practical Method for Burning Alcohol in a Gasoline Tractor - Calculation and Charting of Thermodynamic Properties of Ethyl Alcohol-Air Mixture and its Combustion Products," Kentucky Univ. Eng. Exp. Station, Bul. No. 8, June 1948.

28. F. V. Morriss, R. Modrell, G. Atkinson, and C. Bolze, "The Exhaust Content of Automobiles Burning Ethanol-Gasoline Mixtures," presented at ACS Meeting, Minneapolis, Minn., September 1955, (preprint 77).

29. F. V. Morriss, et al, "Smog Chamber Studies of Unleaded and Leaded Fuels," Ind. and Eng. Chem., Vol. 50, No. 4, (April 1958), 673-676.

30. A. W. Nash and D. A. Howes, "Principles of Motor Fuel Preparation and Application," 538 pp Alcohol Fuels, Vol. 1, John Wiley & Sons, 1935.

31. E. F. Obert, "Detonation and Internal Coolants," SAE Quarterly Trans., Vol. 2, No. 1, (January 1948), 52-58.

32. C. O. Ostwald, "Alcohol Motor Fuel," Automobiltech. Ztschr. Vol. 36, (1933), 129-132, 157-158.

33. T. C. Owtram, "Economic Aspects of Alcohol in Motor Fuel," Inst. of Petrol. Jour., Vol. 38, (October 1952), 820-834; disc., 834-844.

34. J. P. Pfeiffer, "The Technical and Economical Advantages and Disadvantages of the Addition of Alcohol to Gasoline as a Fuel for Engines," Congr. Int'l Tech. et chim. des. Indus. Agr., Compt. Rend. V. Cong. Vol. 2, (1937), 615-632; C. A. (1938), 32, 3576.

35. S. J. W. Pleeth, "Reid Vapor Pressure of Alcohol Blends," Inst. Petrol. Jour., Vol. 28, (1942), 113-114.

36. S. J. W. Pleeth, "Alcohol -- A Fuel for Internal Combustion Engines," London: Chapman and Hall, Ltd., 1950, 260 pages.

37. S. J. W. Pleeth, "Alcohol Motor Fuels," Automobile Engineer, Vol. 42, (April 1952), 137-140.

38. S. J. W. Pleeth, "Alcohol Motor Fuels, Production and Use," Inst. of Petrol. Jour., Vol. 38, (October 1952), 805-819; disc., p 834-844.

39. J. C. Porter, W. B. Roth, and R. Wiebe, "Boosting Engine Performance with Alcohol-Water Injection," Automotive Inds., Vol. 98, (May 1, 1948), 34-37, 60.

40. J. C. Porter and R. Wiebe, "Alcohol as an Antiknock Agent in Automotive Engines," Ind. and Eng. Chem., Vol. 44, (May 1952), 1098-1104.

41. H. R. Ricardo, "Recent Research Work on the Internal-Combustion Engines," SAE Journal, Vol. 10, (1922), 305-336.

42. H. R. Ricardo, "Report of the Empire Motor Fuels Committee," The Inst. of Auto Engineers, XVIII, 1923-1924.

43. H. R. Ricardo, "The High-Speed Internal Combustion Engine," New York: Interscience Publishers, 434 pp., 1941.

44. A. R. Rogowski and C. F. Taylor, "Comparative Performance of Alcohol-Gasoline Blends in a Gasoline Engine," J. Aeronaut. Sci., Vol. 8, 384-392, (1941); C.A. 35, 8268 (1941).

45. M. R. Rowe and G. T. Ladd, "Water Injection for Aircraft Engines," SAE Jour., Vol. 54, No. 1, (January 1946), 26.

46. W. Shulman, "Physical Properties of Ethanol in Rapid Round-Up," Chem. Engr., Vol. 68, (May 15, 1961), 186.

47. James Small, "The Thermal Aspects of Carburetion with Special Reference to the Vaporization of Ethyl Alcohol," Phil. Mag. and Jour. Sci., Vol. 16, (1933), 641-656.

48. E. W. Steinitz, "Alcohol in Motor-Car Operation," Petroleum, London, Vol. 7, (1944), 202-203.

49. A. H. Stuart, "Conditions of Miscibility," Petroleum, London, (1950), 13, 85-86.

50. C. F. Taylor and E. S. Taylor, The Internal Combustion Engine, Int'l Textbook Co., 1961.

51. E. Terres and F. Wehrmann, "The Combustion of Liquid Fuels in Motors with Special Consideration of the Study of the Exhaust Gas," Ztschr. f. Electrochem., Vol. 27, (1921), 423-441.

52. T. R. Thoren, "The Physical and Anti-Knock Properties of Gasoline Alcohol Blends," Iowa Univ. Studies in Engin. Bul. No. 4, 32 pp., 1934.

53. H. T. Tizard and D. R. Pye, "Character of Various Fuels for Internal-Combustion Engines," p 1-47, Empire Motor Fuels Comm. Rpt., The Inst. of Auto. Engr., London, Sess. 1923-1924.

54. W. Traupel, "The Influence of the Type of Fuel on the Efficiency of Combustion," Allgem, Warmetechnik, Vol. 3, (1952), 1-9.

55. A. R. Ubbelohde, J. W. Drinkwater, and A. Egerton, "Pro-Knocks and Hydrocarbon Combustion," Royal Society, London, Proc., Ser. A, Vol. 193, (1935), 103-115.

56. C. H. Van Hartesveldt, "Antidetonant Injection," SAE Trans., Vol. 3, No. 2, (1949), 277-287.

57. Fritz Wehrmann, "The Combustion of Liquid Fuels in Motors with Special Consideration of the Study of the Exhaust Gas," Ztschr. f. Electrochem., Vol. 27, (1921), 379-393.

58. R. Wiebe and J. Nowakowska, "The Technical Literature of Agricultural Motor Fuels," U.S. Dept. Agr. Bibliography Bul. No. 10, 1949.

59. M. K. Wolfson, "Experience with Water-Alcohol Injection of J47 Engine," Automotive Ind. Vol. 116, (Jan. 1, 1957), 104.

60. Air Pollution Foundation (Los Angeles), Report No. 12, November 1955.

61. Annual Review of Gasoline Quality, Ethyl Corp., 1963.

62. A.P.I.C. Alcohol Motor Fuel Technical Advisory Committee Interim Report, Mar. 20, 1957, (unpublished).

63. Knocking Characteristics of Pure Hydrocarbons, ASTM Special Tech. Publ. 225 (Amer. Petrol. Inst. Res. Proj. 45), Phil. (1958), 66-67.

64. Report to the Congress from the Commission on Increased Industrial Use of Agricultural Products, Document No. 45, 1957, U.S. Government Printing Office.

65. Summary Report of Performance of Commercial Gasolines Blended with Ethanol, prepared for API committee by Southwest Research Inst., San Antonio, Texas, December 1956.

CHRONOLOGICAL KEY TO BIBLIOGRAPHY

Year	Author	Reference
1921	E. Terres and F. Wehrmann	51
1921	H. T. Tizard and D. R. Pye	53
1921	Fritz Wehrmann	57
1922	H. R. Ricardo	41
1923-1924	H. R. Ricardo	42
1926	Emil Fischer	16
1930	Hobbs and M. L. Fast	20
1933	O. C. Bridgeman	3
1933	O. C. Bridgeman and D. W. Querfeld	4
1933	O. C. Bridgeman and D. W. Querfeld	5
1933	C. P. Hopkins and Kuhring	21
1933	C. O. Ostwald, et al	32
1933	James Small	47
1934	T. R. Thoren	52
1935	Ubbelohde, Drinkwater, and Egerton	55
1935	Nash and D. A. Howes	30
1936	O. C. Bridgeman	6
1936	L. T. Brown and Christensen	9
1936	L. M. Christensen	10
1936	Egloff and J. C. Morrell	15
1936	L. C. Lichty and E. J. Ziurys	24
1937	F. R. Banks	1
1937	L. C. Lichty and C. W. Phelps	25
1937	J. P. Pfeiffer	34
1938	Dunstan, Nash, Brooks, Rizard	14
1938	S. D. Heron and H. A. Beatty	19
1938	L. C. Lichty and C. W. Phelps	26
1940	David and A. S. Leah	11
1940	R. T. Jackson	23

1941	H. R. Ricardo	43
1941	A. R. Rogowski and C. F. Taylor	44
1942	S. J. W. Pleeth	35
1944	E. W. Steinitz	48
1945	J. T. Duck and C. S. Bruce	13
1945	D. B. Brooks	7
1946	D. B. Brooks	8
1946	M. R. Rowe and G. T. Ladd	45
1948	A. J. Meyer and R. E. Davis	27
1948	E. F. Obert	31
1948	Porter, Roth, Wiebe	39
1949	C. H. Van Hartesveldt	56
1949	R. Wiebe and J. Nowakowska	58
1950	S. J. W. Pleeth	36
1950	A. H. Stuart	49
1952	T. C. Owtram	33
1952	S. J. W. Pleeth	37
1952	S. J. W. Pleeth	38
1952	J. C. Porter and R. Wiebe	40
1952	W. Traupel	54
1955	H. A. Havemann, Rao, et al	18
1955	F. V. Morriss, et al	28
1955	Air Pollution Report	60
1956	Report of Perf. of Commercial Gasoline	65
1957	M. K. Wolfson	59
1957	APIC Advisory Committee Report	62
1957	Report to Congress	64
1958	F. V. Morriss, et al	29
1958	ASTM Special Tech. Publ.	63
1961	W. Shulman	46
1961	Taylor and Taylor	50
1963	B. Brewster and R. V. Kerley	2
1963	Downs, S. T. Griffiths, and Wheeler	12
1963	Annual Review of Gasoline, Ethyl	61
1964	M. W. Jackson, et al	22
1964	Hagen, et al	17

Engine Performance and Exhaust Emissions: Methanol versus Isooctane *

G. D. Ebersole
Research and Development Dept.
Phillips Petroleum Co.

F. S. Manning
Dept. of Chemical Engineering
The University of Tulsa

PERFORMANCE AND EXHAUST emission characteristics of internal combustion ingines operating on alcohol-gasoline blends have been studied and discussed (1, 2).* These studies generally conclude that, given identical stoichiometric mixtures, engines operating on alcohol-gasoline blends do not perform significantly better than those fueled by the base gasoline. Nevertheless, at least one patent (3) covering alcohol-gasoline blends implies that significant reductions in exhaust carbon monoxide and hydrocarbons are realized when small percentages of specific alcohols are added to a base gasoline.

The literature contains very little information on performance and exhaust emission characteristics of an engine operating solely on an alcohol. However, engine performance with an alcohol fuel compared to a hydrocarbon fuel has been well documented in three studies: Ricardo (4) and Lichty and Ziurys (5) used ethanol and gasoline, and Starkman et al. (6) experimented with ethanol, methanol, and isooctane. These studies showed that slightly more engine output can be attained with an alcohol but that such gains were offset by a considerable increase in specific fuel consumption. Exhaust emission data only have been reported in two other studies. Starkman et al. (7) estimate carbon monoxide and nitric concentrations in engine exhausts, based on theoretical calculations for numerous single component fuels, each at a wide variety of stoichiometric mixtures. The data suggest that lower concentrations of carbon monoxide and nitric oxide may be realized with an alcohol fuel than with isooctane. Fitch and Kilgore (8) studied exhaust emissions from a methanol-fueled engine, but because of major difficulties encountered during the experimental program, could not properly assess the potential of methanol compared to gasoline.

*Numbers in parentheses designate References at end of paper.

*Paper 720692 presented at the National West Coast Meeting, San Francisco, California, August 1972.

ABSTRACT

Operating characteristics of a single-cylinder, spark-ignition engine fueled by both methanol and isooctane were determined. Engine output, indicated specific fuel consumption, and specific emissions of hydrocarbon, carbon monoxide, nitric oxide, and aldehydes were measured for both fuels and compared using performance maps.

The engine output comparisons showed that lean misfire limits occurred at leaner mixtures with methanol than with isooctane and that maximum engine output levels were nearly equal for both fuels.

Comparison of the specific parameters of each fuel at equivalent power levels obtained with maximum power spark timing permits the following conclusions: Use of methanol results in higher indicated specific fuel consumption, greater emission of aldehydes, but lower emissions of hydrocarbon and nitric oxide; the two fuels showed similar trends of carbon monoxide emission.

SAE/PT-80/19/$02.50

Since the potential impact of alcohol fuels on exhaust emissions is not adequately covered in the literature, this study determines performance and exhaust emission characteristics of an internal combustion engine fueled with alcohol and with gasoline. The fuels selected for this study were 2,2,4-trimethyl-pentane, representing gasolines, and methanol, representing the alcohol family.

The pure hydrocarbon, 2,2,4-trimethyl pentane, hereafter referred to as isooctane, was selected as the gasoline reference fuel to minimize any change due to gasoline fuel composition during the study and any effects due to deposit buildup in the combustion chamber of the engine. Moreover, because its molecular weight is close to the average molecular weight of a typical gasoline (9), isooctane has been used as a reference fuel in a large number of previous laboratory and theoretical studies involving fuel comparison.

Methanol was selected as the alcohol fuel because its characteristics, more than those of any other commercially available alcohol fuels, are different from those of isooctane.

Characteristics of the two fuels are shown in Table 1.

Table 1 - Properties of Isooctane and Methanol

Item	Isooctane	Methanol	Ref.
Formula	C_8H_{18}	CH_4O	
Molecular weight	114.224	32.042	
Carbon to hydrogen weight ratio	5.25	3.0	
Carbon, % by weight	84.0	37.5	
Hydrogen, % by weight	16.0	12.5	
Oxygen, % by weight	0.0	50.0	
Boiling point, F at 1 atm	210.63	148.1	10
Freezing point, F at 1 atm	-161.28	-144.0	10
Vapor pressure, psia at 100 F	1.708	4.6	10
Specific gravity, 60 F/60 F	0.6963	0.796	10
Liquid density, lb/gal at 60 F and 1 atm	5.795	6.637	10
Coefficient of expansion, 1/F at 60 F and 1 atm	0.00065	0.00065	10
Surface tension, dynes/cm at 68 F and 1 atm	18.77	22.61	10
Viscosity, centipoises at 68 F and 1 atm	0.503	0.596	10
Specific heat of liquid, Btu/lb-F at 77 F and 1 atm	0.5	0.6	11
Heat of vaporization, Btu/lb at boiling point and 1 atm	116.69	473.0	10
Heat of vaporization, Btu/lb at 77 F and 1 atm	132	503.3	9
Heat of combustion, Btu/lb at 77 F			
Gaseous fuel-liquid H_2O	20688	10279	9
Liquid fuel-liquid H_2O	20556	9776	*
Gaseous fuel-gaseous H_2O	19197	9099	9
Liquid fuel-gaseous H_2O	19065	8593	*
Stoichiometric mixture, lb air/lb	15.13	6.463	10

*Calculated from data in table.

EXPERIMENTAL CONSIDERATIONS

Engine output for a specific configuration depends on quantity of air and fuel inducted into the combustion chamber on the intake stroke, combustion timing relative to piston location, coolant temperature, compression ratio, and shaft speed. All these parameters must be surveyed to determine the optimum operating conditions for each load.

Experience shows that coolant temperature and inlet air conditions influence engine output to a lesser extent and may be regarded as secondary control parameters; compression ratio is usually maximized on the basis of the fuel available to the engine; and combustion timing for maximum power with minimum fuel consumption is single-valued. The response parameters—engine output and fuel consumption—are therefore studied as functions of the remaining control parameters (airflow, fuel flow, and engine speed) by first fixing the secondary control parameters, then establishing the compression ratio, and finally imposing the constraint of maximum power combustion timing at each combination of the primary control parameters. However, the primary control parameters also influence the composition of the exhaust (9). Hence, the number of response parameters is increased by the number of exhaust gas components considered.

EXPERIMENTAL DESIGN - Characterization of engine response parameters in terms of the control parameters is complicated because these relationships are usually not linear. Satisfactory results can be obtained by mapping the response parameters as functions of all combinations of the control parameters. To this end, sequential observations are made while traversing over the ranges of each primary control parameter, one at a time, until the response matrices have been obtained.

An experimental design of this type may become unwieldy because of the large number of combinations of the primary control parameters. However, the limitations imposed on the combustion process by the engine configuration tend to reduce the number of operational combinations of speed, airflow, and fuel flow.

Two limitations are the lean and rich misfire limits imposed by the specific engine and fuel. As a rule, engines are not operated at or near the rich limit because of excessive specific fuel consumption. On the other hand, engine operation on gasoline near the lean limit appears desirable because of potential reductions in fuel consumption and certain exhaust emissions (12). The lean limit forms a boundary of minimum fuel flow relative to airflow from which to increment toward, but never to reach, the rich limit.

Another limitation is engine speed. Unusually low speeds do not provide adequate power; on the other hand, unusually high speeds are detrimental to engine structures.

The following secondary control parameters were selected: coolant temperature, 149 F; fuel-air mixture temperature, 125 F; and water vapor content of inlet air, saturated at 32 F.

Two engine speeds were selected: 1000 rpm, representing low speeds; and 1800 rpm, representing a medium speed.

The maximum compression ratio for knock-free operation is taken from a knock-limited compression ratio versus fuel-air ratio curve (9). Data points for these curves are obtained by operating the engine at wide open throttle and maximum power combustion timing and by increasing the compression ratio until incipient detonation is observed. These data points are determined at several air-fuel ratios at and around the stoichiometric mixture. The compression ratio versus equivalence ratio* relationships are developed for each fuel-speed combination.

The lean equipment flammability limits are taken from a limiting equivalence ratio versus airflow curve. These data are generated by decreasing the fuel flow until incipient misfire is observed while operating the engine at a given throttle setting and maximum power combustion timing. Various throttle settings covering the entire airflow range are studied and the lean limit curves are determined for each fuel-speed combination.

With all these values established, the experimental design can be used for observation of the response parameters.

TEST ENGINE - The engine selected for a complex study should allow each control parameter to be varied independently over the range of interest. A multicylinder engine, such as that installed in current passenger cars, is not convenient because it is difficult to vary fuel flow (at constant throttle) over a wide range. This and other problems are overcome when a single-cylinder CFR engine (13) is used.

The basic test unit for this study was valve-in-head, single-cylinder CFR engine with appropriate temperature and spark-timing controls. The mechanical specifications of the engine are given in Table 2. Speed control and power absorption were provided by a d-c dynamometer. Intake air was controlled with a ball-valve throttle and measured with an Alcock viscous-flow meter. Liquid fuel flow was measured volumetrically from a nitrogen-pressurized fuel-feed system.

The intake system was equipped with a baffled fuel-air mixing and vaporizing chamber to produce as uniform a fuel-air mixture as possible. Liquid fuel was added to 180 F air just ahead of the mixing chamber by impinging a high-velocity stream on a flat target to produce a highly atomized fuel spray.

During some runs, the presence of liquid fuel in the intake mixture was observed through windows installed in the intake manifold just upstream of the intake port. In all cases, liquid fuel was either absent or present only in minor quantities as a fine fog. These measures, together with a 180-deg shroud on the intake valve, ensured that the mixture of fuel, air, and residual gas eventually burned was substantially without concentration gradients.

The test unit is shown in Fig. 1. The engine was given a complete upper-cylinder overhaul before any test runs were made.

POWER MEASUREMENT AND MISFIRE DETECTION - Indicated power is the output of an engine before the mechanical losses of the engine are subtracted. Indicated mean effective pressure (imep) is defined as the indicated power divided by the product of engine speed, piston displacement, and a conversion constant; the equation for a 4-stroke engine can be written as

$$\text{imep} = \frac{(\text{ihp}) \quad (792{,}000)}{(V)\,(N)}$$

where:

imep = indicated mean effective pressure, psi
ihp = indicated horsepower
V = swept volume of combustion chamber in^3
N = engine speed, rpm

Thus, the imep is the theoretical constant pressure exerted during each power stroke of the engine to produce power equal to the indicated power. The parameter shows how well

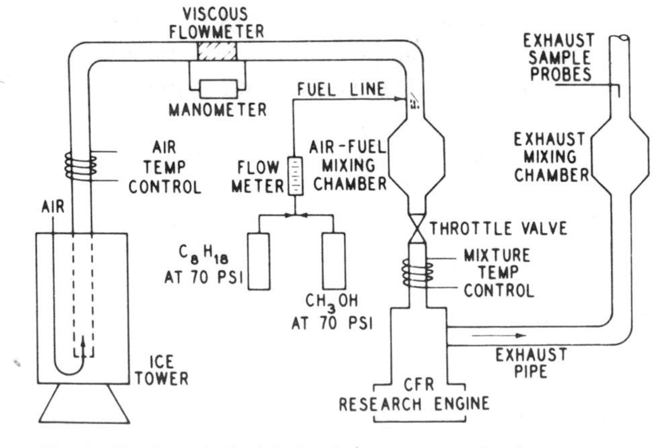

Fig. 1 - Engine, air-fuel induction system, and exhaust system

*A meaningful parameter used in comparing fuels is the fuel-air equivalence ratio, ϕ. It has been widely adopted in chemical and in engineering investigations and is defined as

$$\phi = \frac{(\text{fuel-air ratio})_{\text{actual}}}{(\text{fuel-air ratio})_{\text{stoichiometric}}}$$

or

$$\phi = \frac{(\text{air-fuel ratio})_{\text{stoichiometric}}}{(\text{air-fuel ratio})_{\text{actual}}}$$

Then, $\phi = 1.0$ is a chemically correct or stoichiometric mixture, $\phi = 0.8$ is a 20% (fuel) lean mixture, and $\phi = 1.2$ is a 20% (fuel) rich mixture.

Table 2 - Mechanical Specifications of Research Engine

Compression ratio	4-10
Bore, in	3.25
Stroke, in	4.50
Displacement, in^3	37.33
Valve Seat ID, in	1.187

the engine is using its size to produce work independent of speed. The imep is particularly useful for comparing fuel performance in an engine because it represents the effectiveness of the chemical-thermal-mechanical energy transformation process of the engine.

When the indicated mean effective pressure is defined on the basis of the compression and expansion strokes, it can be calculated from the instantaneous cylinder pressure occurring in those portions of the cycle, using the following equation:

$$\text{imep} = (\text{mean pressure}_{\text{expansion}}) - (\text{mean pressure}_{\text{compression}}) = \frac{\int p_e \, d\theta}{\Delta \theta_e} - \frac{\int p_c \, d\theta}{\Delta \theta_c}$$

where:

p_e = instantaneous cylinder pressure during expansion portion of cycle

p_c = instantaneous cylinder pressure during intake portion of cycle

θ = time

The imep values can be generated by electronic circuits that first integrate the pressure signal from the compression and expansion strokes and then make the appropriate subtraction. In this manner, the response of the circuit, proportional to imep, is available for immediate display. Calibration curves for the imep meter were prepared by use of static loading. Accuracy of the meter was determined by generating the imep values by a manual technique and comparing with those obtained simultaneously with the meter. Accuracy was generally within 3% at selected conditions of operation.

Misfire, the failure of the flame to propagate, is characterized by low cylinder pressure during the power stroke relative to nonmisfiring power strokes. As a consequence, the imep is also low for those power strokes during which misfire occurs; therefore the imep meter response to misfire is a sudden spike toward zero imep levels. The imep meter response can be used in this manner to detect engine misfire, and the criterion selected for identification of misfire (three spikes per minute) was high enough to discriminate between random misfire that may occur near the misfire limit and consistent misfire. The lean misfire limit is then defined as the leanest fuel-air mixture at which exactly three misfires occur per minute.

EXHAUST GAS ANALYSIS - Measurement techniques applied in this study are described below.

Hydrocarbon and Methanol - Hydrocarbons and methanol appearing in the exhaust gases were detected with a Beckman 108A flame ionization analyzer*. The response of this instrument is proportional to hydrocarbon or methanol concentrations in the exhaust stream. One minor difficulty in its application is the effect of oxygen in the sample on the response of the unit (14). Studies (15, 16) have shown that the adoption of fuel and flow rates other than those specified by the manufacturer will reduce the oxygen error to 1% for oxygen concentration ranges of 0-24%. The operating conditions suggested in Ref. 17 which have been adopted for use in this study are: fuel mixture, 40% H_2 + 60% H_e; fuel rate, 122 cm^3/min; diffusion air rate, 220 cm^3/min; sample rate, 6 cm^3/min.

Ideally, the response of the hydrogen flame detector is proportional to the rate of introduction of carbon into the flame, independent of the hydrocarbon type. The proportionality factor is 1.0 for an aliphatic carbon atom. The effective carbon number of a particular type of carbon atom is defined as the ratio between the instrument response caused by an atom of the given type and the instrument response caused by an aliphatic carbon atom. Instrument operation with the adopted operating conditions provides a uniform response to all hydrocarbons, independent of molecular structure, and with an effective response of approximately 1 per carbon atom. The effective response from a molecule of methanol was determined to be 0.85.

The flame detector, because of its uniform response, provides estimates of the total concentration of hydrocarbon and alcohol in the exhaust. The presence of either component is considered undesirable; therefore, the total concentration more closely represents the potential contaminant burden. As a consequence, the relative components of hydrocarbon and alcohol appearing in the exhaust were not determined, and the measured concentrations are simply referred to as hydrocarbons.

Carbon Monoxide and Carbon Dioxide - Concentrations of CO and CO_2 in the exhaust stream were each determined with Beckman infrared analyzers, Model IR315A.* Calibration curves supplied by the manufacturer were verified before the instruments were used for data collection.

Oxygen - Concentrations of oxygen in the exhaust stream were determined with a Beckman process oxygen analyzer, Model 778*, which used the polarographic technique. The direct reading of the instrument was calibrated using oxygen in room air. However, the instrument response in the range of oxygen concentrations expected in exhaust gases was verified by chromatograph analysis.

Sampling System - The collection of CO, CO_2, O_2 and hydrocarbon concentration data was simplified by operating these instruments simultaneously on a sample stream and making permanent records of the instrument responses.

The design of the sampling system was similar to those required to certify vehicle exhaust emissions for U.S. Health, Education, and Welfare standards (17). Fig. 2 is a schematic diagram of the sample and analysis system.

---

*Beckman Instruments, Inc., Scientific and Process Instruments Div., 2500 Harbor Blvd., Fullerton, Calif.

*Beckman Instruments, Inc., Scientific and Process Instruments Div., 2500 Harbor Blvd., Fullerton, Calif.

Nitrogen Oxides - Nitric oxide was determined colorimetrically as nitrogen dioxide, using the reagent system of Saltzman (18). Nitric oxide conversion to nitrogen dioxide was accomplished via a 24 h oxidation period in which the exhaust sample, air, and reagent were contacted in a sealed flask. The transmittance of the developed reagent was measured at 550 mμ.

Aldehydes - Aldehydes were determined by the MBTH method (19), which provides for addition of exhaust samples to a flask containing an MBTH solution. After conversion, ferric chloride was added to form the reagent solution. The transmittance of the developed reagent was measured at 668 mμ.

RESULTS AND DISCUSSION

COMPRESSION RATIO SELECTION - The data collected at wide open throttle and maximum power spark timing to determine the compression ratio that yields incipient knock are given in Fig. 3. The compression ratio for incipient knock was influenced by engine speed, fuel type, and equivalence ratio. However, a ratio of 7.5:1 gave knock-free operation for all combinations of fuel and engine speeds considered in the experimental design. As a result, a compression ratio of 7.5:1 was used for all subsequent portions of the study.

LEAN MISFIRE LIMITS - The data collected to define the lean equipment misfire limits in terms of equivalence ratio and airflow are given in Fig. 4. The figure contains a curve for methanol and isooctane at each speed, 1000 and 1800 rpm. The engine was operated at 7.5:1 compression ratio and maximum power spark timing for all points on these curves.

Fig. 4 shows that the lean limit was influenced by fuel type, airflow rate, and engine speed. Methanol had a considerably lower lean limit than isooctane at equivalent airflow rates and engine speeds. The lean limits for isooctane at 1000 rpm are essentially equivalent to those previously reported for gasoline (12), which were obtained with a single-cylinder CFE engine at 1000 rpm. Since engine lean limits for methanol could not be found in the literature, comparison of the methanol data was not possible. However, the lean limits obtained with methanol at 1000 rpm in this study are similar to those reported for methane and propane at 1000 rpm in Ref. 12.

For given airflow rates that is, throttle opening, these curves established the minimum equivalence ratio for misfire-free operation with methanol and isooctane at 1000 and 1800 rpm. These values formed the lowest levels of fuel and airflow rates in the experimental design, and increments in both parameters began at this point.

TREATMENT OF DATA FROM EXPERIMENTAL DESIGN - Values of imep, airflow, and concentrations of CO, CO_2, hydrocarbon, and hydrocarbon and/or methanol were obtained by converting instrument signals through appropriate calibration curves. Concentrations of aldehydes and nitrogen oxides were obtained from simple calculations, using calibrations and specific data generated from exhaust gas samples.

The quantity of exhaust gases generated by the production

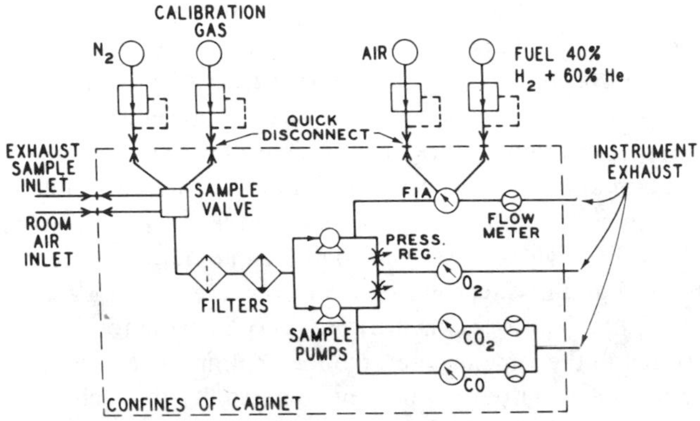

Fig. 2 - Flow schematic of exhaust gas analysis instruments

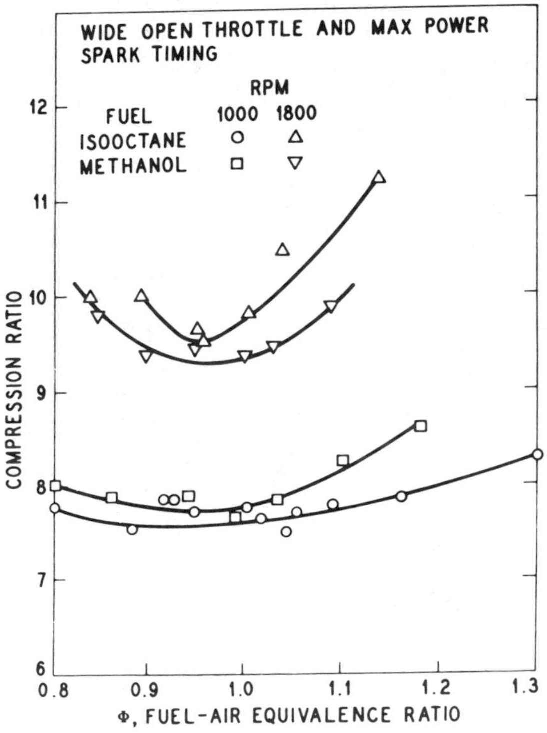

Fig. 3 - Compression ratio for incipient knock

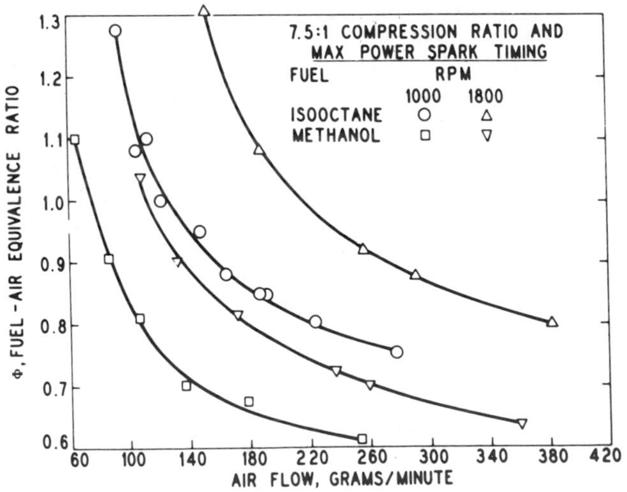

Fig. 4 - Airflow and equivalence ratio at incipient misfire

of a given quantity of shaft work can vary markedly, limiting the significance of concentration measurement. This ambiguity was avoided by referencing the emission measurements to the basic product of engine operation—indicated work. The specific units used are grams per indicated horsepower-hour. The equations needed to convert concentration data to absolute quantities were derived from mass balances (20).

PERFORMANCE MAPS - The comparison between exhaust emissions with methanol and those with isooctane requires consideration of engine operating characteristics, since emissions result from the use of specific quantities of fuel and air to produce engine power, imep. A map is a technique for showing the interrelation between fuel and airflow rates, the independent variables, and imep, the dependent variable.

Performance maps, also referred to as wedge charts (21), for each fuel and engine speed combination result from a series of graphical transformations of the raw data. First, the following three cross-plots were formed: fuel flow versus equivalence-ratio at values of the throttling parameter; airflow versus equivalence ratio at values of the throttling parameter; and imep versus equivalence ratio at values of the throttling parameter.

Second, from the cross-plot of imep versus equivalence ratio, uniform values of imep and equivalence ratio were selected; this fixed the throttling parameter for each set of imep and equivalence ratios.

Third, fuel and airflow rates for each imep value were extracted from the remaining cross-plots by using the common values of the throttling parameter and the equivalence ratio. Finally, the map was constructed by plotting lines of constant imep for values of airflow on the x axis and for fuel flow on the y axis, Fig. 5. Lines of constant stoichiometry have been superimposed on this figure.

The performance map in Fig. 5 describes the imep and relationship of fuel and airflow rates needed for operation of the engine with isooctane at maximum power spark timing and 1000 rpm. Examples of the usefulness of such a plot in describing engine performance follow.

The imep contours have three boundaries. The boundary, nearly parallel to the y-axis on the right-hand side of the figure, represents the conditions of wide open throttle operation. Maximum imep is obtained on the wide open throttle line. When traversing the wide open throttle boundary from ϕ = 0.8 to 1.3, the airflow rate decreases slightly as the additional (vaporized) fuel displaces air in the induction process.

A second boundary extends along the bottom of Fig. 5 from the wide open throttle conditions at ϕ = 0.8 to a third boundary, the line of ϕ = 1.3. The second boundary is properly called the equipment lean misfire limit, but is called the lean misfire limit in this study. The first and second boundaries are dictated by the engine, while the boundary at ϕ = 1.3 is a part of the experimental design.

The lean misfire limit is the locus of fuel and airflow rates that yield engine power with incipient misfire. Engine operation with fuel and airflow rates less than the misfire line is associated with a high frequency of misfire.

The equivalence ratio for lean misfire ranges from a low of 0.8 at wide open throttle to a high of 1.3 at lower airflows. Misfire limits at low values of equivalence ratios offer the advantage of extracting engine power at low fuel-flow rates. For example, when the 100 imep contour is traversed, starting at ϕ = 1.3 and going toward ϕ = 0.8, the minimum fuel-flow rate, 13.5 g/min, is reached at an air-flow rate of 231 g/min, or ϕ = 0.88. However, when the 50- or 6-imep contours are traversed in the same manner, the misfire limit is reached before the imep contour has had the opportunity to reach a minimum fuel consumption.

When the contours of another dependent variable—indicated specific fuel consumption (isfc) for example—are superimposed on the performance map, the relationships include those of the second dependent parameter, Fig. 6. Questions relating minimum indicated specific fuel consumption rela-

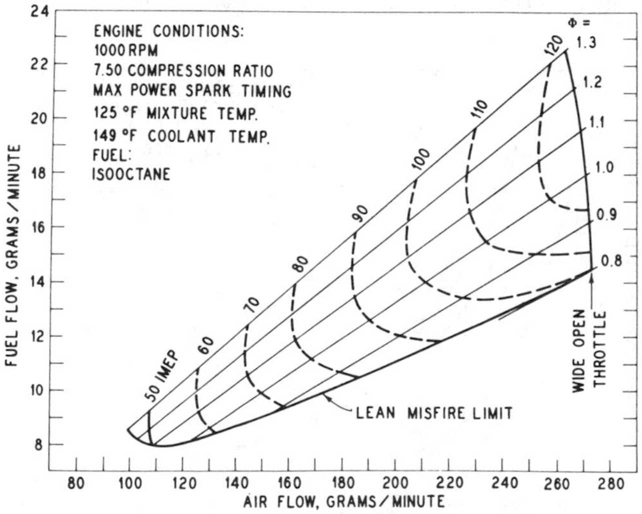

Fig. 5 - Interrelations of power, fuel and air consumption with isooctane at 1000 rpm

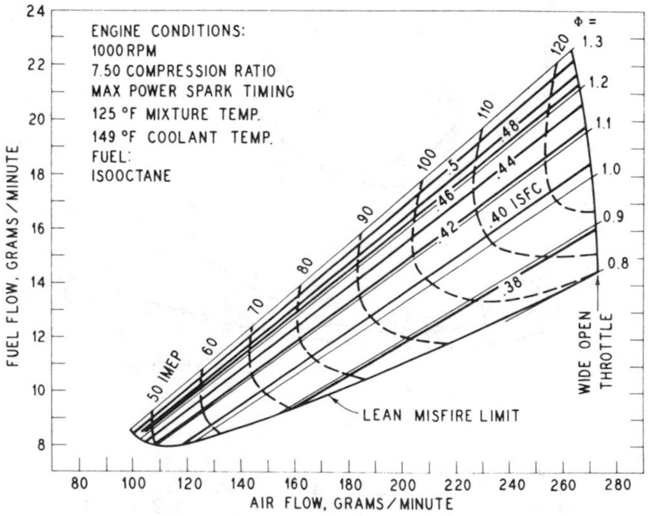

Fig. 6 - Interrelations of power, fuel and air consumption, and isfc with isooctane at 1000 rpm

Table 3 - Engine Performance and Exhaust Mass Data at 1000 rpm with Isooctane at MPSA

Pct Thro	Airf, g/min	Fuelf, g/min	Equil Ratio	IMEP, psi	SA, btdc	ISFC, lb/ihp-h	HC, g/ihp-h	CO, g/ihp-h	NO, g/ihp-h	HCHO, g/ihp-h	Run No.
41.5	112.3	8.30	1.116	54.1	35.0	0.431	4.16	67.1	2.78	0.046	81
41.5	112.3	8.83	1.187	53.4	32.5	0.464	4.34	111.4	1.33	0.044	82
41.5	112.1	9.30	1.253	53.4	32.5	0.489	4.38	145.0	0.73	0.044	83
49.4	136.4	8.73	0.966	62.8	37.5	0.390	2.62	1.2	9.91	0.108	75
49.4	136.4	9.07	1.004	64.5	35.0	0.395	2.66	8.4	9.02	0.090	76
49.5	135.9	8.98	0.998	64.0	35.0	0.394	2.74	5.0	9.07	0.075	129
49.4	136.1	9.54	1.058	65.6	32.5	0.408	2.95	32.6	4.15	0.073	77
49.4	136.1	9.94	1.103	65.7	30.0	0.425	3.31	59.7	2.85	0.056	78
49.4	135.5	10.61	1.182	65.6	30.0	0.454	3.80	99.1	1.31	0.054	79
49.4	135.6	11.22	1.249	64.9	30.0	0.485	4.30	150.7	0.61	0.055	80
60.2	165.2	10.12	0.925	74.6	35.0	0.381	2.04	0.3	11.75	0.120	61
60.2	165.1	10.48	0.958	76.7	35.0	0.383	2.05	0.2	11.37	0.095	62
60.2	165.2	10.90	0.996	78.2	32.5	0.391	2.12	12.0	8.69	0.097	63
60.2	164.1	11.70	1.077	79.7	30.0	0.412	2.34	44.5	4.64	0.050	64
60.2	164.3	12.40	1.140	79.9	27.5	0.436	2.67	84.1	2.07	0.040	65
60.2	163.8	13.33	1.229	79.5	27.5	0.471	3.04	128.9	0.72	0.043	66
90.2	246.5	13.17	0.807	99.0	40.0	0.373	1.24	0.6	13.54	0.159	67
90.2	245.9	14.10	0.866	103.8	35.0	0.381	1.21	0.5	13.89	0.140	68
90.2	244.4	14.75	0.911	108.7	32.5	0.381	1.21	1.8	13.34	0.102	69
90.2	244.4	15.55	0.961	112.2	30.0	0.389	1.35	1.7	11.59	0.080	70
90.2	244.0	16.22	1.004	115.7	30.0	0.393	1.56	8.4	10.25	0.087	71
90.2	243.5	17.25	1.070	116.4	27.5	0.416	1.94	50.3	4.44	0.074	72
90.2	243.5	18.90	1.172	115.8	25.0	0.458	2.22	105.7	1.34	0.054	73
90.2	242.8	20.35	1.266	115.0	27.5	0.497	2.65	157.7	0.55	0.058	74

Table 4 - Engine Performance and Exhaust Mass Data at 1800 rpm with Isooctane at MPSA

Pct Thro	Airf, g/min	Fuelf, g/min	Equil Ratio	IMEP, psi	SA, btdc	ISFC, lb/ihp-h	HC, g/ihp-h	CO g/ihp-h	NO, g/ihp-h	HCHO, g/ihp-h	Run No.
41.7	209.2	14.54	1.049	45.8	37.5	0.495	2.89	42.0	5.42	0.071	84
41.7	209.2	15.46	1.116	46.3	35.0	0.521	3.32	94.7	2.97	0.062	85
41.7	209.2	16.45	1.187	46.3	32.5	0.554	3.63	141.6	1.25	0.056	86
41.7	209.5	17.45	1.258	46.2	32.5	0.589	4.02	178.3	0.66	0.050	87
50.0	251.5	16.40	0.985	54.0	37.5	0.474	1.95	1.5	10.52	0.127	96
49.9	253.4	16.85	1.004	56.5	35.0	0.465	2.29	11.9	8.18	0.088	139
50.0	251.5	16.85	1.012	55.2	35.0	0.476	2.37	13.2	6.22	0.080	97
50.0	251.5	17.55	1.054	56.5	32.5	0.484	2.74	40.8	5.36	0.054	98
50.0	251.4	18.60	1.117	56.5	30.0	0.513	3.14	84.6	2.66	0.049	99
50.0	250.3	19.62	1.184	56.2	27.5	0.544	3.51	132.3	1.66	0.055	100
50.0	250.3	20.75	1.252	56.1	27.5	0.577	3.81	173.1	0.92	0.058	101
59.8	301.0	17.81	0.893	58.4	40.0	0.475	1.43	2.3	11.22	0.155	102
59.8	300.9	18.90	0.948	62.1	37.5	0.475	1.46	3.1	14.13	0.136	103
59.8	301.9	19.95	0.998	65.9	35.0	0.472	1.84	10.2	11.62	0.072	104
59.8	301.7	21.05	1.054	66.9	32.5	0.491	2.44	50.4	6.66	0.062	105
59.8	301.5	22.95	1.149	66.8	30.0	0.536	2.97	120.5	2.33	0.040	106
59.8	299.6	25.00	1.260	66.3	30.0	0.588	3.49	192.3	0.96	0.053	107
75.3	376.4	19.95	0.800	65.3	45.0	0.476	1.34	1.8	15.18	0.181	88
75.3	375.1	21.40	0.861	70.9	37.5	0.471	1.29	1.7	17.08	0.171	89
75.3	371.6	23.00	0.935	76.6	32.5	0.468	1.36	1.5	14.81	0.109	90
75.3	370.3	24.10	0.983	80.4	30.0	0.467	1.58	8.2	12.91	0.104	91
75.3	370.3	25.15	1.026	80.4	30.0	0.488	2.14	35.8	7.80	0.056	92
75.3	369.0	26.70	1.093	80.4	27.5	0.518	2.60	82.3	4.66	0.044	93
75.3	371.7	28.45	1.156	79.7	27.5	0.557	2.94	134.2	2.19	0.042	95
75.3	367.0	30.90	1.271	79.1	27.5	0.609	3.51	205.4	0.76	0.042	94

tive to airflow (that is, throttle), stoichiometry, or power levels can be answered by examining the plot in terms of the criterion in the question.

The scheme is used to illustrate the interrelations between fuel and airflow, the independent variables, and each dependent emission parameter, and between isfc and the common dependent parameter, engine power. Figs. 6 through 15 are performance maps for isooctane at 1000 and 1800 rpm, respectively, and Figs. 16 through 25 are the parallel maps for methanol. Tables 3 and 4 list the data for Figs. 6-15, and Table 5 gives data for comparable values for methanol. These performance maps establish the engine output, fuel consumption, and exhaust emission characteristics from one engine operated at maximum power spark timing for each fuel.

REDUCED PERFORMANCE MAPS - The characteristics shown on the performance maps, Figs. 6-25, suggest that the two fuels may have different emission and fuel consumption characteristics. Because the performance maps do not yield direct comparisons between the two fuels, reduced performance maps were prepared by extracting and replotting certain data from the complete performance map.

The reduced map takes the form of parameter contours plotted on a coordinate system of imep on the y-axis and

Table 5 - Engine Performance and Exhaust Mass Data at 1000 rpm with Methanol at MPSA

Pct Thro	Airf, g/min	Fuelf, g/min	Equil Ratio	IMEP, psi	SA, btdc	ISFC, lb/ihp-h	HC, g/ihp-h	CO, g/ihp-h	NO, g/ihp-h	HCHO, g/ihp-h	Run No.
29.8	74.6	12.21	1.056	38.7	37.5	0.885	0.67	18.3	1.94	0.194	22
29.8	74.4	13.31	1.154	39.8	35.0	0.939	0.86	62.6	0.81	0.187	23
29.8	73.9	14.46	1.262	38.7	32.5	1.049	1.04	111.6	0.23	0.187	24
39.5	102.4	13.10	0.825	45.0	42.5	0.817	0.49	3.7	2.29	0.282	11
39.7	101.9	13.68	0.866	48.3	37.5	0.795	0.44	2.4	4.70	0.246	111
39.4	101.5	14.30	0.909	49.5	37.5	0.811	0.42	3.3	5.12	0.185	1
39.5	101.9	14.38	0.910	50.5	40.0	0.799	0.41	3.3	5.20	0.229	12
39.7	100.8	15.18	0.971	51.5	30.0	0.827	0.39	2.3	5.48	0.191	112
39.4	101.5	15.70	0.998	52.3	32.0	0.842	0.38	4.6	4.86	0.179	2
39.5	101.2	15.80	1.007	52.0	32.5	0.853	0.40	6.3	5.01	0.182	13
39.7	99.7	16.16	1.045	52.3	27.5	0.867	0.45	31.9	2.51	0.179	113
39.5	100.4	17.20	1.105	53.0	30.0	0.911	0.63	50.1	2.13	0.197	14
39.4	98.5	17.54	1.149	52.7	30.0	0.934	0.75	63.9	1.09	0.181	3
39.5	98.2	18.65	1.225	52.3	30.0	1.001	0.86	101.8	0.55	0.183	15
39.4	96.9	19.17	1.276	51.8	30.0	1.039	0.87	109.5	0.28	0.188	4
50.0	128.4	14.90	0.748	50.2	42.5	0.833	0.51	2.7	1.20	0.268	5
49.8	128.2	15.98	0.804	55.2	40.0	0.812	0.43	0.5	4.46	0.231	114
50.0	126.9	16.88	0.858	56.9	35.0	0.833	0.39	2.8	7.78	0.274	6
49.8	126.6	17.69	0.901	61.6	33.0	0.806	0.36	0.5	7.70	0.211	115
50.0	125.5	18.70	0.961	62.8	30.0	0.836	0.34	2.6	7.57	0.224	7
49.8	126.4	19.45	0.993	65.3	30.0	0.836	0.33	2.2	7.49	0.207	116
49.5	126.1	19.55	1.000	66.3	25.0	0.828	0.34	5.5	7.06	0.188	134
50.0	125.2	20.13	1.037	64.9	27.5	0.870	0.38	20.2	5.56	0.163	8
50.0	123.9	22.77	1.185	65.2	25.0	0.980	0.69	88.3	1.16	0.175	9
50.0	118.9	24.83	1.347	62.8	27.5	1.110	0.89	153.1	0.23	0.194	10
59.6	153.3	17.28	0.727	60.3	37.5	0.804	0.43	2.1	1.84	0.257	16
60.1	153.1	18.27	0.770	64.9	35.0	0.790	0.37	1.9	5.26	0.235	108
59.6	151.0	19.50	0.833	68.4	30.0	0.800	0.33	2.6	8.30	0.268	17
60.1	150.1	20.65	0.887	72.5	30.0	0.799	0.29	1.7	9.58	0.193	109
59.6	149.6	21.35	0.921	73.5	27.5	0.815	0.29	2.3	10.09	0.219	18
60.1	148.5	22.20	0.964	76.7	27.5	0.812	0.28	1.6	8.83	0.204	110
59.6	147.1	23.52	1.031	77.4	25.0	0.853	0.31	21.9	5.92	0.169	19
59.6	145.6	25.18	1.115	77.4	25.0	0.913	0.38	59.6	2.55	0.179	20
59.6	139.9	26.95	1.243	77.1	22.5	0.981	0.64	110.8	0.48	0.184	21
90.0	229.1	22.50	0.633	78.1	45.0	0.809	0.47	1.7	1.81	0.367	25
90.3	228.7	25.35	0.715	90.6	37.5	0.785	0.36	0.6	6.16	0.312	117
90.0	225.5	26.10	0.747	93.0	32.5	0.788	0.33	2.0	7.77	0.247	26
90.3	225.6	28.18	0.806	99.0	30.0	0.799	0.30	0.5	11.56	0.225	118
90.0	220.9	29.38	0.858	103.7	27.5	0.795	0.27	1.7	12.39	0.201	27
90.3	222.0	31.45	0.914	108.7	27.5	0.812	0.25	0.5	13.08	0.180	119
90.0	216.0	33.13	0.989	113.0	25.0	0.823	0.25	4.0	10.48	0.163	28
90.3	217.3	35.15	1.043	115.0	24.0	0.858	0.28	30.1	5.54	0.208	120
90.0	207.8	36.77	1.141	112.5	25.0	0.917	0.42	74.5	1.72	0.159	29
90.0	202.3	39.75	1.267	107.3	25.0	1.040	0.66	124.8	0.67	0.159	30

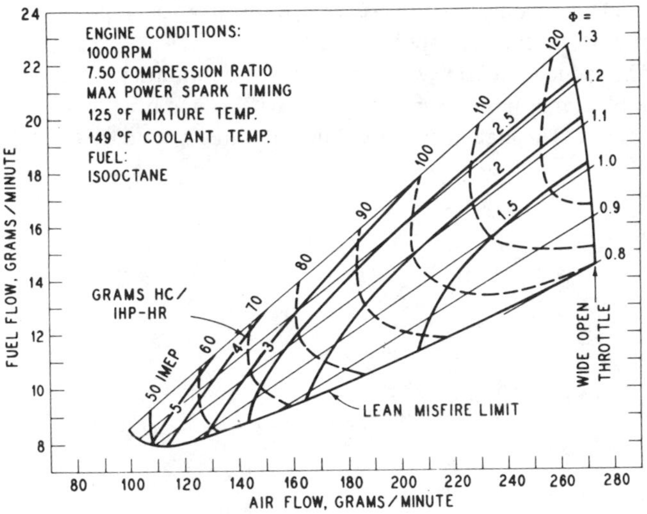

Fig. 7 - Interrelations of power, fuel and air consumption, and hydrocarbon with isooctane at 1000 rpm

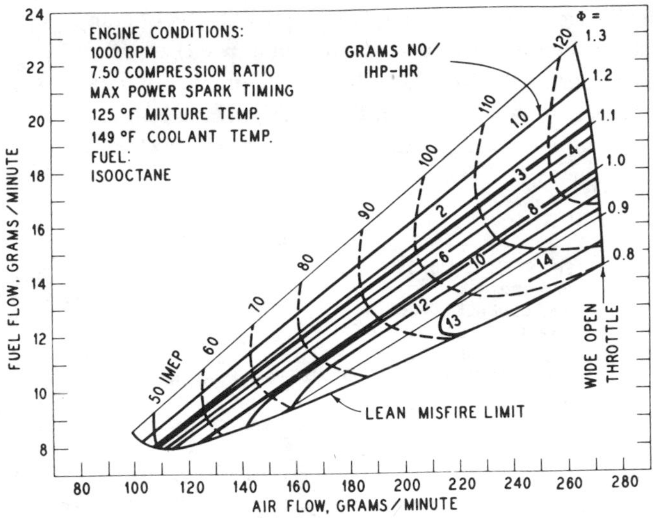

Fig. 9 - Interrelations of power, fuel and air consumption, and nitric oxide with isooctane at 1000 rpm

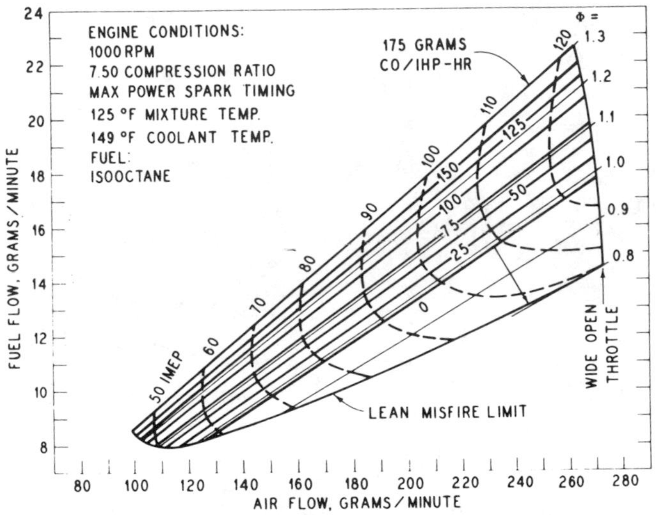

Fig. 8 - Interrelations of power, fuel and air consumption, and carbon monoxide with isooctane at 1000 rpm

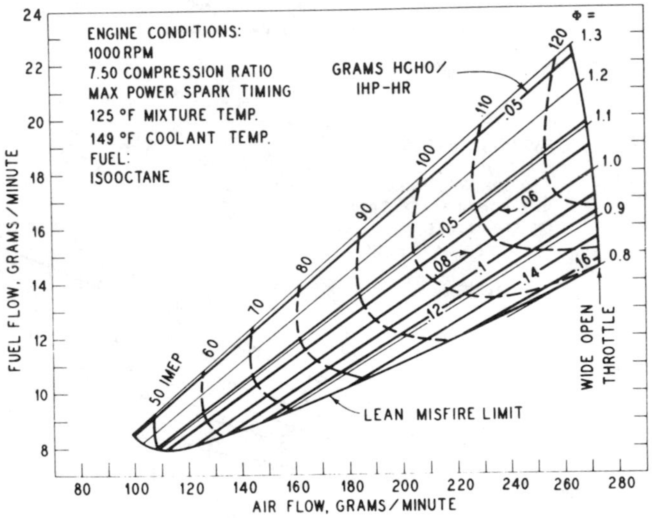

Fig. 10 - Interrelations of power, fuel and air consumption, and aldehydes with isooctane at 1000 rpm

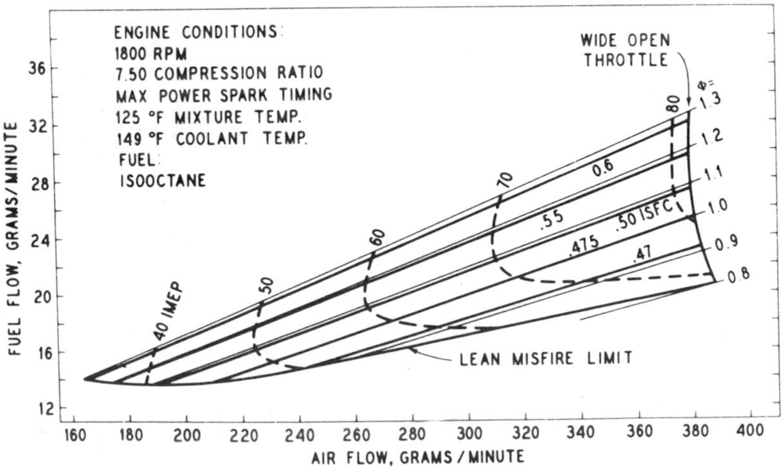

Fig. 11 - Interrelations of power, fuel and air consumption and with isfc isooctane at 1800 rpm

equivalence ratio on the x-axis. All possible direct comparisons between the two fuels (at the same speed) are made by superimposing the isooctane data on the methanol data. Reduced performance maps for hydrocarbon (Figs. 26 and 27), carbon monoxide (Figs. 28 and 29), nitric oxide (Figs. 30 and 31), and aldehydes, (Figs. 32 and 33) have been prepared for both engine speeds.

The reduced maps have the same three boundaries as the complete maps; that is, wide open throttle, lean misfire, and $\phi = 1.3$. The area included within the three boundaries of

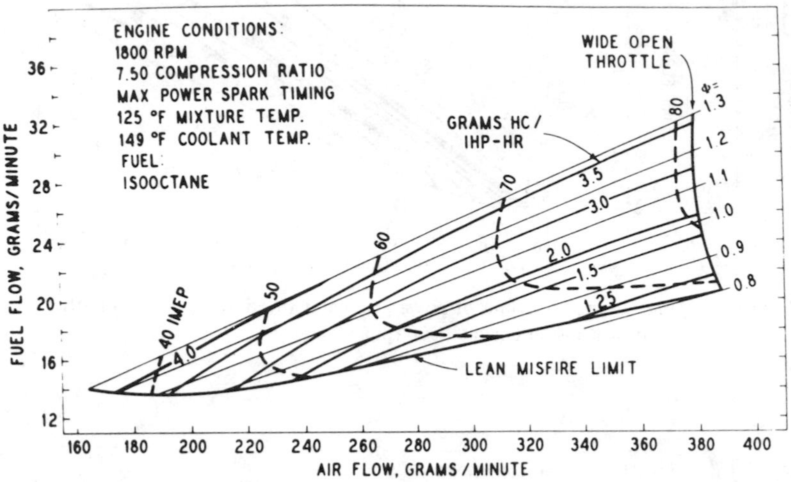

Fig. 12 - Interrelations of power, fuel and air consumption, and hydrocarbon with isooctane at 1800 rpm

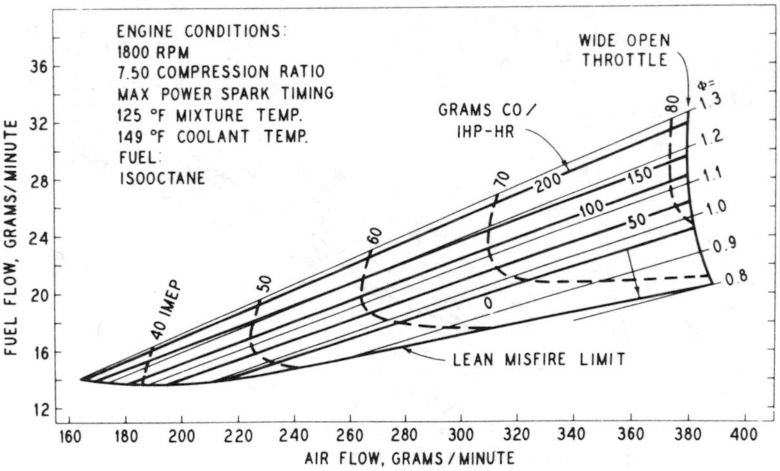

Fig. 13 - Interrelations of power, fuel and air consumption, and hydrocarbon with isooctane at 1800 rpm

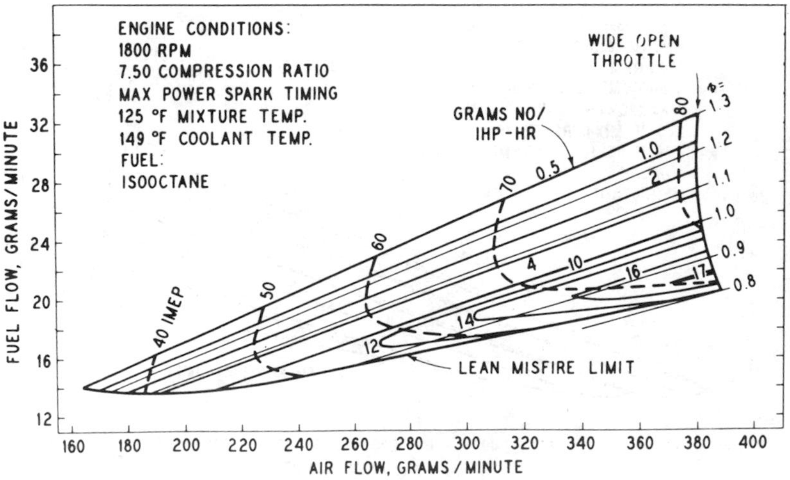

Fig. 14 - Interrelations of power, fuel and air consumption, and nitric oxide with isooctane at 1800 rpm

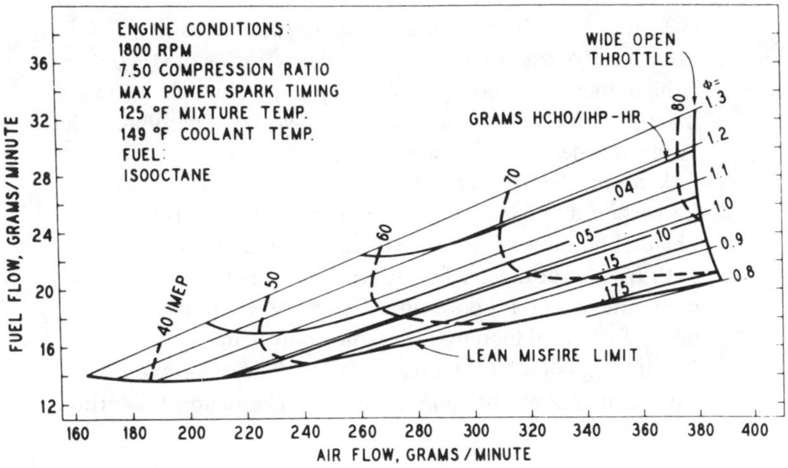

Fig. 15 - Interrelations of power, fuel and air consumption, and aldehydes with isooctane at 1800 rpm

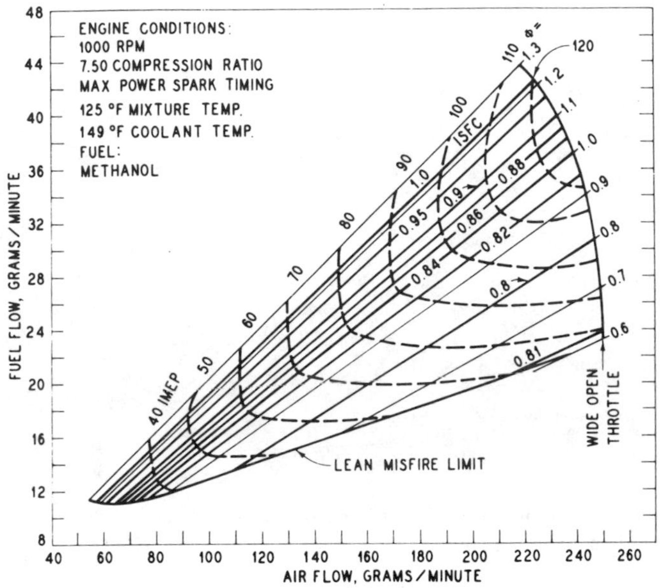

Fig. 16 - Interrelations of power, fuel and air consumption, and isfc with methanol at 1000 rpm

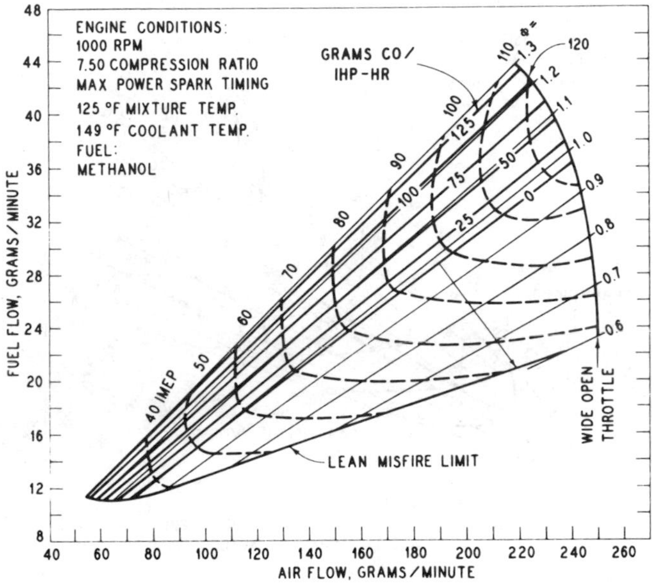

Fig. 18 - Interrelations of power, fuel and air consumption, and carbon monoxide with methanol at 1000 rpm

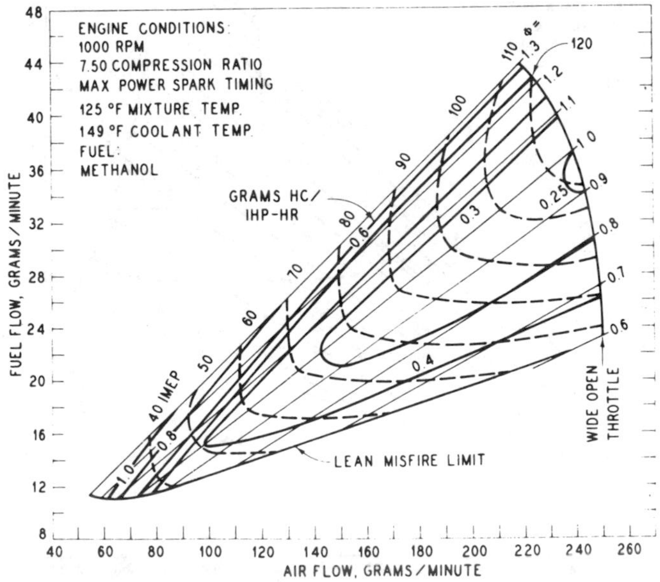

Fig. 17 - Interrelations of power, fuel and air consumption, and hydrocarbon with methanol at 1000 rpm

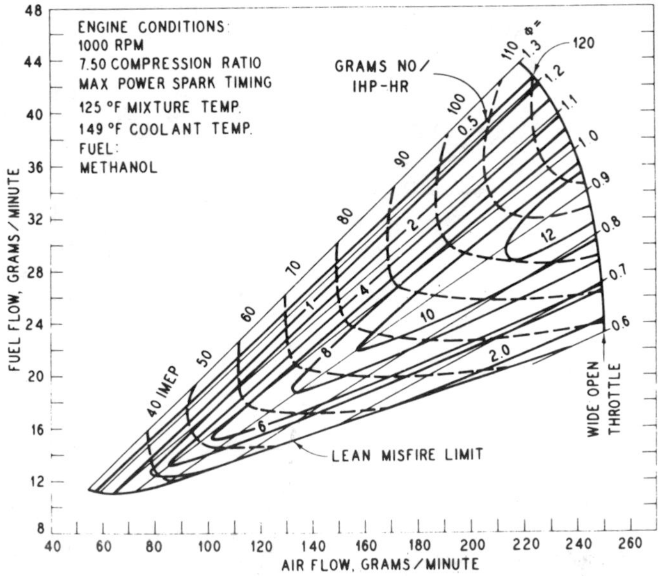

Fig. 19 - Interrelations of power, fuel and air consumption, and nitric oxide with methanol at 1000 rpm

the reduced map represents all possible combinations of imep and equivalence ratio that permit misfire-free operation at maximum power spark timing. The larger area associated with methanol occurs because the methanol lean limits are reached at lower equivalence ratios.

As shown by Figs. 26-33, the larger operating area associated with methanol provides the following advantages compared to isooctane: first, lower imep's are achieved at given equivalence ratios; and second, given imep's are achieved with lower equivalence ratios.

Reduced Map for Hydrocarbons - The reduced maps for hydrocarbon, Figs. 26 and 27, yield the following: Hydrocarbon contours inside the boundaries generally have the same trends for both fuels at both speeds, and low hydrocarbon emissions occur near equivalence ratios of 0.8-0.9 and at high imep's. At comparable imep's and equivalence ratios, the emission level with methanol is lower, by nearly an order of magnitude, than that with isooctane.

Reduced Map for Carbon Monoxide - The reduced maps for carbon monoxide, Figs. 28 and 29, yield the following: The carbon monoxide contours have the same trends for both fuels at both speeds, and the contours for zero carbon monoxide occur at approximately the same equivalence ratio; For equal increments of the equivalence ratio above 1.0, the emission level increases faster with isooctane than with methanol at both engine speeds. The region for carbon-

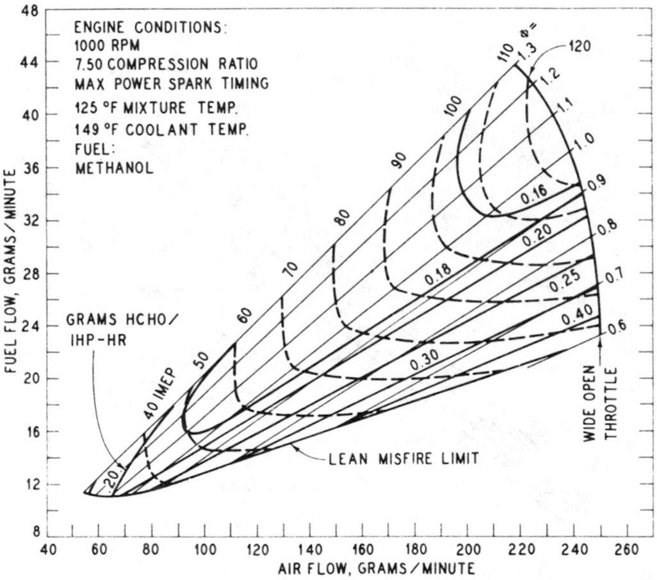

Fig. 20 - Interrelations of power, fuel and air consumption and aldehydes with methanol at 1000 rpm

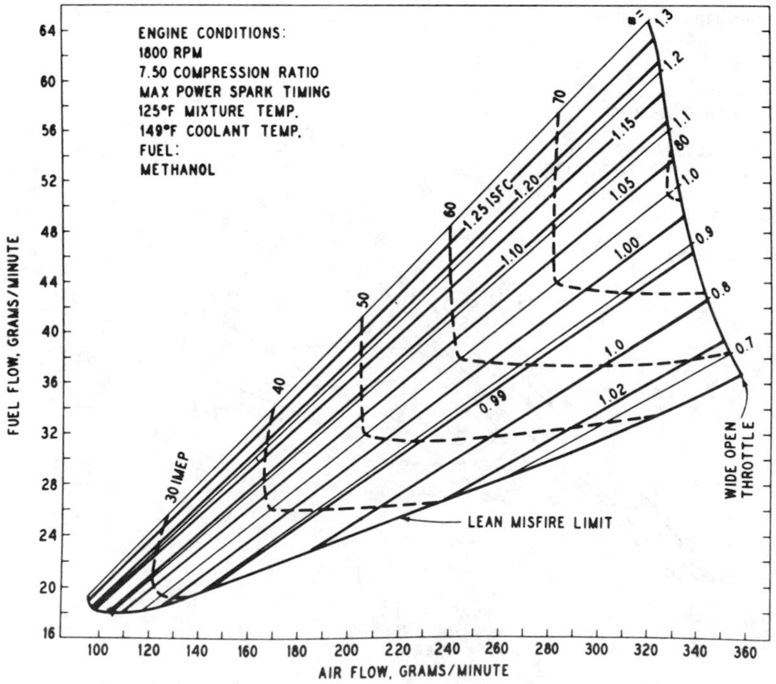

Fig. 21 - Interrelations of power, fuel and air consumption, and methanol with isfc at 1800 rpm

monoxide-free operation—that is, the area enclosed by the wide open throttle, lean misfire, and ϕ = 1.0 boundaries—for methanol is approximately double that for isooctane.

Reduced Map for Nitric Oxide - The reduced maps for nitric oxide, Figs. 30 and 31, yield the following: The contours generally show the same trends for both fuels at both speeds. However, the emission levels at comparable imep's and equivalence ratios are somewhat less with methanol compared to isooctane, especially at equivalence ratios above 1.0.

For equivalence ratios of 1.0 or less, the wider range of operation with methanol permits operation at given imep's with reduced nitric oxide emission by selecting lower equivalence ratios.

Reduced Map for Aldehydes - The reduced maps for aldehydes, Figs. 32 and 33, yield the following: The aldehyde contours for methanol tend to be influenced more by equivalence ratio than those for isooctane. This is true at both engine speeds. The emission levels with methanol are generally higher than those with isooctane at comparable imep's and equivalence ratios.

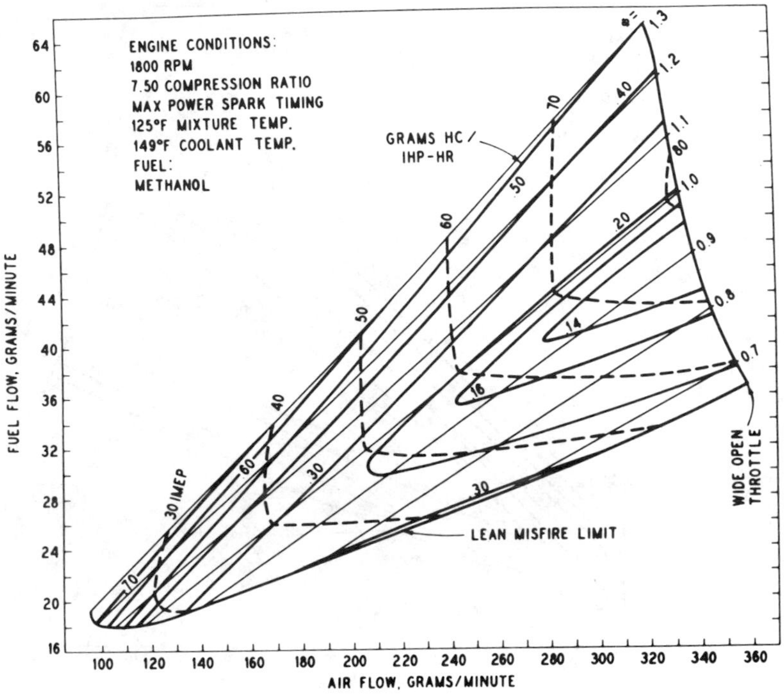

Fig. 22 - Interrelations of power, fuel and air consumption, and hydrocarbon with methanol at 1800 rpm

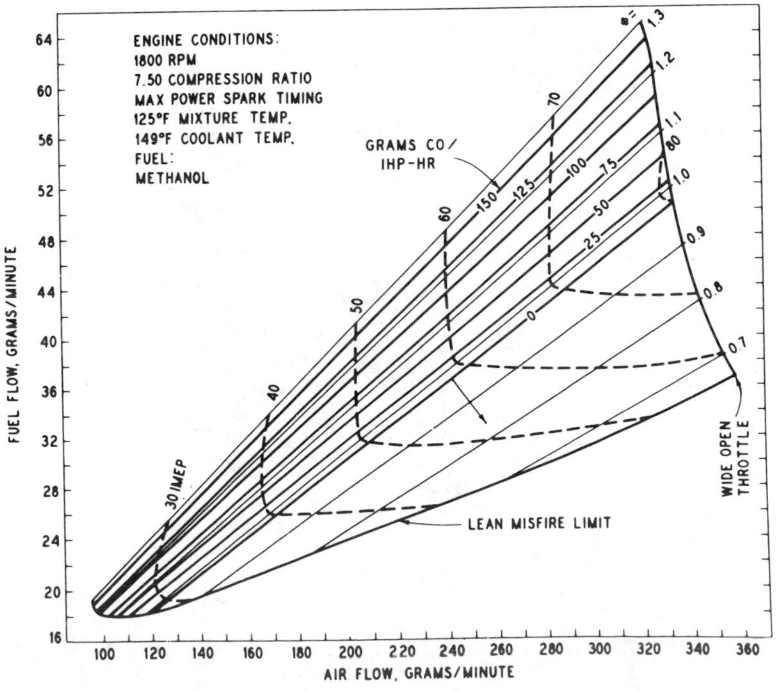

Fig. 23 - Interrelations of power, fuel and air consumption, and carbon monoxide with methanol at 1800 rpm

Reduced Map for Indicated Specific Fuel Consumption - Complete performance maps that include isfc, Figs. 6, 11, 16, and 21, show that isfc is dependent on equivalence ratio but not on changes in imep. The coordinate system used in the reduced performance maps, Figs. 26 through 33, is therefore not applicable. The coordinate system used for the isfc map is isfc itself on the y-axis and equivalence ratio on the x-axis. These data, Fig. 34, are obtained without cross-plotting.

The fuel consumption data show that the isfc for methanol averages 2.15 times that of isooctane for both speeds. This is very close to the ratio of constant volume heat of combustion for the two fuels, 2.14.

CONSIDERATION OF SPARK RETARDATION - The comparisons show that engine operation with methanol occurs with near-zero carbon monoxide emissions over a large power range. However, with isooctane, low imep's are achieved at the expense of operating at higher equivalence ratios where increases in the carbon monoxide emissions occur.

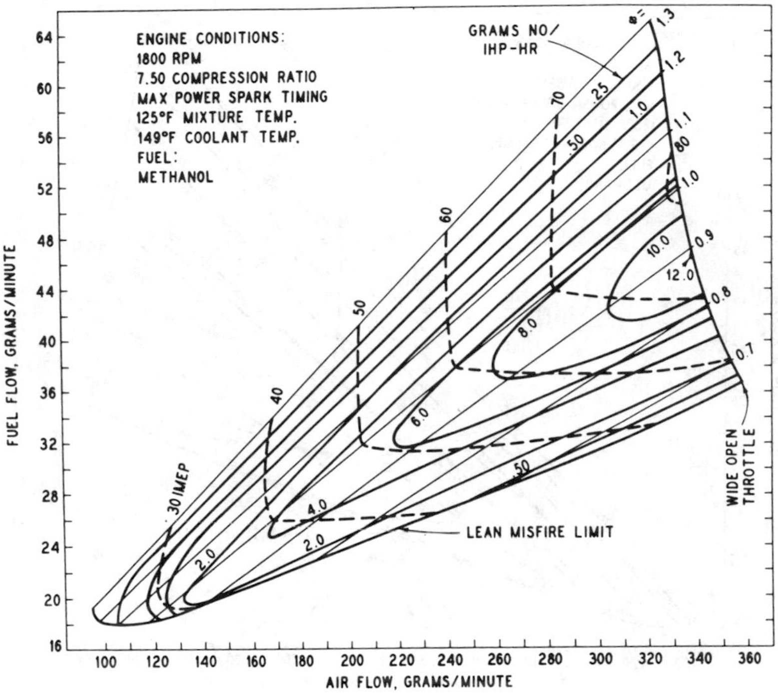

Fig. 24 - Interrelations of power, fuel and air consumption, and nitric oxide with methanol at 1800 rpm

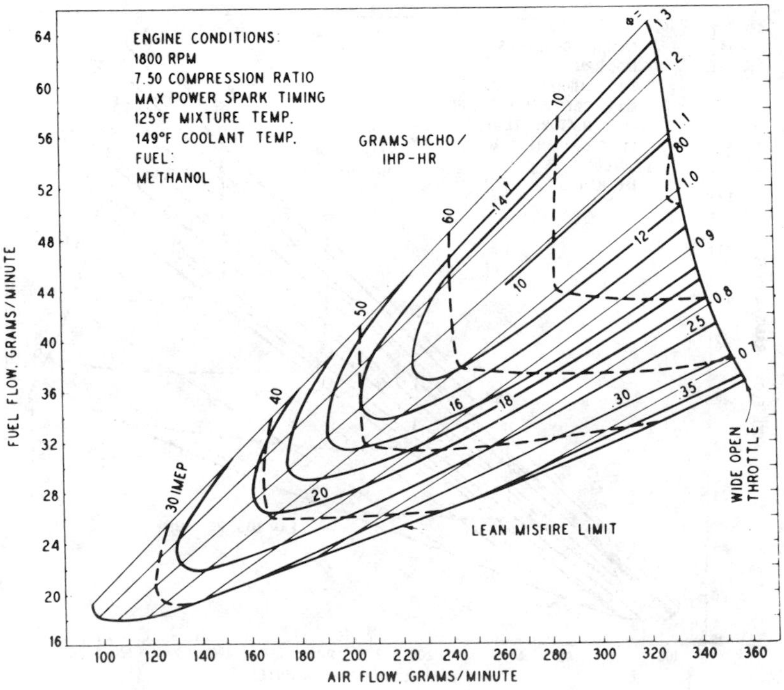

Fig. 25 - Interrelations of power, fuel and air consumption, and aldehydes with methanol at 1800 rpm

Low imep's accompanied by low carbon monoxide emissions can be achieved with isooctane by relaxing the restriction of maximum power spark timing. At the point where equivalence ratios are increased to achieve low imep's, spark retardation could be employed to achieve the low imep's and preserve low carbon monoxide emissions. Reference to the carbon monoxide contours for isooctane in Figs. 28 and 29 indicates that retardation is required to achieve imep's of less than 56 at 1000 rpm and of 46 at 1800 rpm while maintaining low carbon monoxide emissions.

Performance at Retarded Spark - The effect of spark retardation on all parameters referenced to those at maximum power spark timing at the cited imep's is shown in Figs. 35

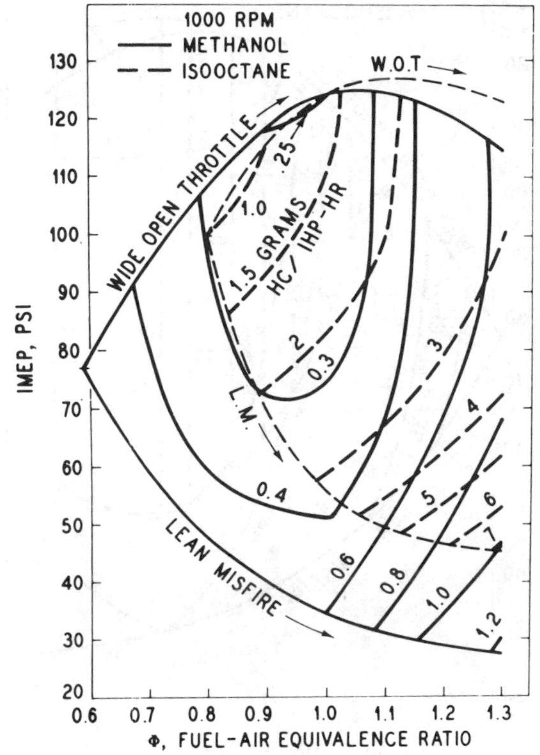

Fig. 26 - Interrelations of power, equivalence ratio, and hydrocarbon with isooctane and methanol at 1000 rpm

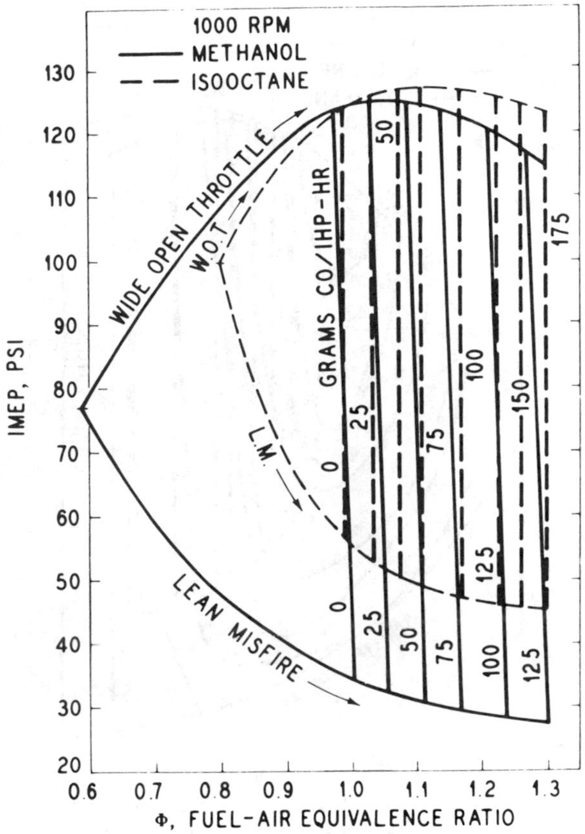

Fig. 28 - Interrelations of power, equivalence ratio, and carbon monoxide with isooctane and methanol at 1000 rpm

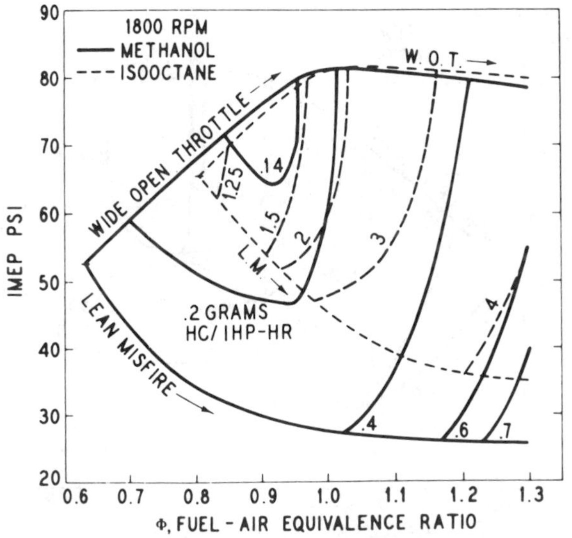

Fig. 27 - Interrelations of power, equivalence ratio, and hydrocarbon with isooctane and methanol at 1800 rpm

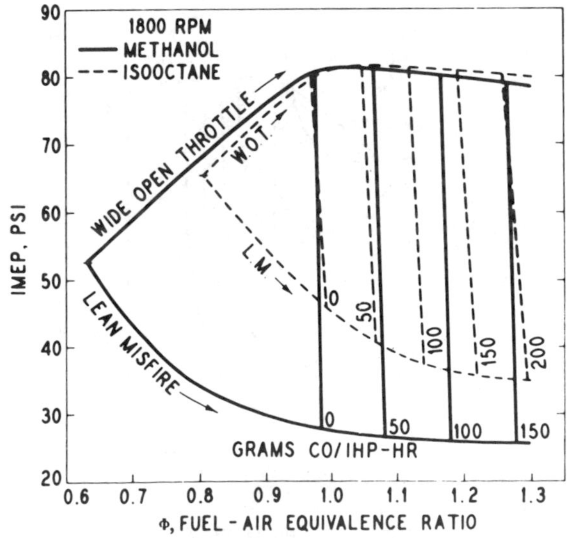

Fig. 29 - Interrelations of power, equivalence ratio, and carbon monoxide with isooctane and methanol at 1800 rpm

and 36 for 1000 and 1800 rpm, respectively; basic data are given in Tables 7 and 8. Similar data (Tables 9 and 10) were collected with methanol to provide comparisons between the fuels.

As the spark is retarded from maximum power spark timing, the imep's decrease almost in a linear manner from the maximum imep obtained. At the same time, the inefficiencies of the power produced are evidenced by the increasing indicated specific fuel consumption.

Emissions of nitric oxide decrease, while emissions of carbon monoxide and of aldehydes increase as the spark is retarded. The increases in the carbon monoxide ratios are

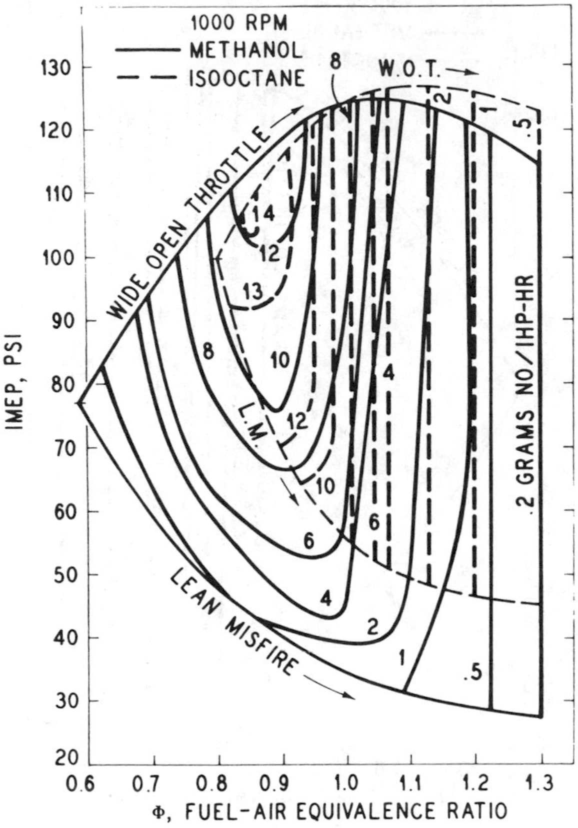

Fig. 30 - Interrelations of power, equivalence ratio, and nitric oxide with isooctane and methanol at 1000 rpm

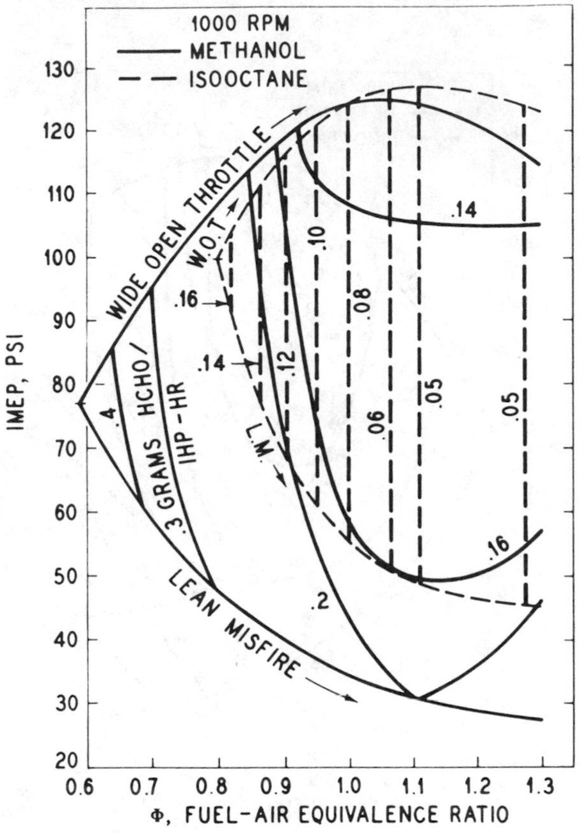

Fig. 32 - Interrelations of power, equivalence ratio, and aldehydes with isooctane and methanol at 1000 rpm

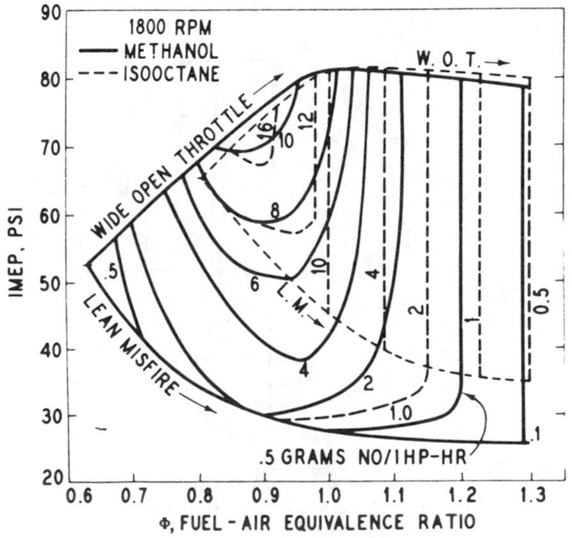

Fig. 31 - Interrelations of power, equivalence ratio, and nitric oxide with isooctane and methanol at 1800 rpm

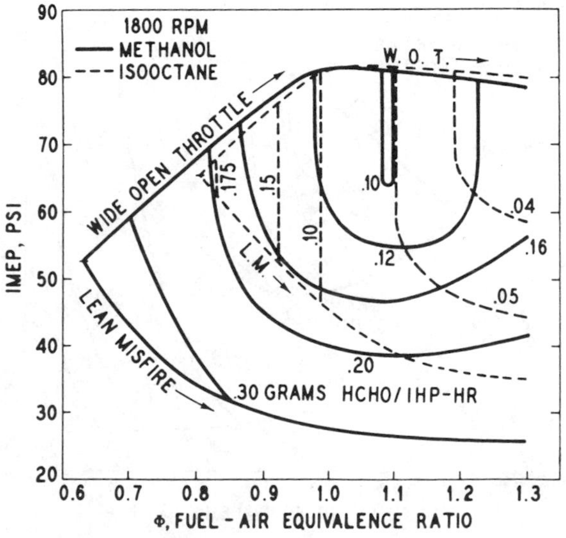

Fig. 33 - Interrelations of power, equivalence ratio, and aldehydes with isooctane and methanol at 1800 rpm

Table 6 - Engine Performance and Exhaust Mass Data at 1800 rpm with Methanol at MPSA

Pct Thro	Airf, g/min	Fuelf, g/min	Equil Ratio	IMEP, psi	SA, btdc	ISFC, lb/ihp-h	HC, g/ihp-h	CO, g/ihp-h	NO, g/ihp-h	HCHO, g/ihp-h	Run No.
30.5	139.4	20.40	0.944	30.8	37.5	1.033	0.27	2.0	2.46	0.272	56
30.5	137.9	22.10	1.034	32.6	35.0	1.057	0.37	25.1	2.29	0.249	57
30.5	137.5	23.00	1.079	33.5	32.5	1.070	0.42	47.4	1.44	0.240	58
30.5	137.0	24.30	1.144	33.6	30.0	1.128	0.48	79.2	0.66	0.240	59
30.5	136.5	26.35	1.245	33.3	30.0	1.234	0.69	132.1	0.20	0.261	60
40.2	192.1	23.40	0.786	35.8	37.5	1.019	0.32	3.4	1.12	0.385	50
40.2	191.8	25.65	0.863	38.9	32.5	1.028	0.24	3.1	2.86	0.212	51
40.2	190.2	27.75	0.941	42.7	27.5	1.013	0.20	2.8	5.02	0.137	52
40.2	188.2	29.80	1.021	45.2	27.5	1.028	0.32	23.8	3.95	0.108	53
40.2	185.5	32.55	1.132	45.2	25.0	1.123	0.44	78.7	1.33	0.163	54
40.2	180.9	35.60	1.269	43.9	25.0	1.264	0.65	150.9	0.13	0.172	55
49.6	236.7	27.00	0.736	42.1	40.0	1.000	0.35	0.7	0.45	0.450	38
49.6	234.3	29.80	0.820	45.2	32.5	1.028	0.26	1.7	2.41	0.343	39
49.4	230.6	32.17	0.900	51.5	27.5	0.974	0.17	2.0	6.36	0.209	121
49.6	228.7	33.15	0.935	52.7	25.0	0.981	0.17	1.4	6.85	0.163	40
49.4	227.6	33.95	0.962	54.2	25.0	0.977	0.17	2.9	6.43	0.158	122
49.9	229.4	35.75	1.005	56.5	25.0	0.987	0.22	12.7	5.06	0.105	148
49.4	227.3	35.50	1.007	55.5	23.5	0.997	0.23	15.6	5.07	0.151	123
49.4	226.2	36.80	1.049	56.5	24.0	1.016	0.31	40.9	3.82	0.145	124
49.6	225.2	36.90	1.057	55.9	25.0	1.029	0.32	42.0	2.90	0.134	41
49.6	217.6	39.80	1.180	54.0	25.0	1.149	0.46	105.1	0.98	0.144	42
59.3	280.4	31.20	0.718	47.7	37.5	1.020	0.28	0.8	0.76	0.362	43
59.3	275.6	34.75	0.813	53.4	32.5	1.015	0.21	3.3	5.18	0.219	44
59.3	269.0	38.50	0.923	59.6	27.5	1.007	0.15	2.9	8.42	0.164	45
59.3	268.3	39.95	0.960	62.1	25.0	1.003	0.15	2.8	7.47	0.139	46
59.3	267.8	41.30	0.995	64.0	25.0	1.006	0.20	8.5	6.37	0.116	47
59.3	258.5	45.00	1.123	65.1	25.0	1.078	0.35	71.6	2.45	0.098	48
59.3	254.9	47.80	1.210	64.0	25.0	1.164	0.44	117.9	0.91	0.121	49
75.7	358.8	35.80	0.644	54.0	42.5	1.034	0.23	0.9	0.42	0.465	31
75.7	352.5	40.40	0.739	62.1	32.5	1.014	0.18	1.8	4.00	0.280	32
75.3	344.6	42.65	0.798	67.2	27.5	0.990	0.16	0.7	7.59	0.311	125
75.7	344.6	44.60	0.835	70.0	27.5	0.993	0.14	3.2	11.13	0.200	33
75.3	338.7	45.90	0.874	73.4	25.0	0.975	0.13	2.1	11.67	0.209	128
75.3	335.3	47.80	0.920	76.0	25.0	0.981	0.12	2.0	11.50	0.196	126
75.7	338.2	49.30	0.940	79.7	25.0	0.964	0.13	4.1	10.75	0.145	34
75.3	331.8	51.70	1.005	80.4	23.0	1.003	0.17	15.7	7.75	0.132	127
75.7	331.8	53.50	1.040	79.7	22.5	1.047	0.22	40.4	4.90	0.099	35
75.7	329.4	58.20	1.140	79.7	22.5	1.139	0.33	92.1	1.40	0.102	36
75.7	324.2	62.30	1.239	78.5	25.0	1.237	0.42	138.4	0.54	0.135	37

Table 7 - Engine Performance and Exhaust Mass Data at 1000 rpm with Isooctane Retarded Spark

Pct Thro	Airf, g/min	Fuelf, g/min	Equil Ratio	IMEP, psi	SA, btdc	ISFC, lb/ihp-h	HC, g/ihp-h	CO, g/ihp-h	NO, g/ihp-h	HCHO, g/ihp-h	Run No.
49.5	135.9	8.98	0.998	64.0	35.0	0.394	2.77	5.1	9.17	0.102	129
49.5	135.5	8.98	1.001	58.7	25.0	0.429	2.91	6.6	7.71	0.105	130
49.5	136.1	8.98	0.996	48.1	15.0	0.524	3.15	8.1	4.80	0.118	131
49.5	136.4	8.98	0.994	34.9	5.0	0.722	3.02	10.0	2.80	0.139	132
49.5	136.4	8.98	0.994	25.1	−5.0	1.004	1.48	11.1	2.49	0.166	133

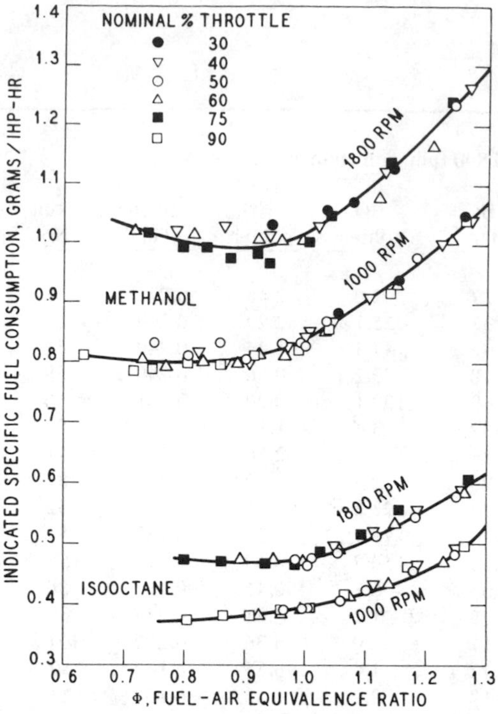

Fig. 34 - Isfc versus equivalence ratio for methanol and isooctane

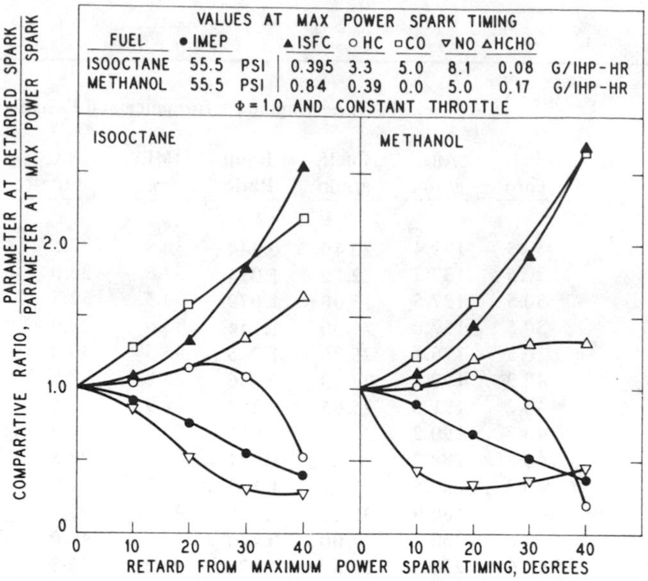

Fig. 35 - Effect of spark retard on power, fuel consumption, and emissions with isooctane and methanol at 1000 rpm

Table 8 - Engine Performance and Exhaust Mass Data at 1800 rpm with Isooctane-Retarded Spark

Pct Thro	Airf, g/min	Fuelf, g/min	Equil Ratio	IMEP, psi	SA, btdc	ISFC, lb/ihp-h	HC, g/ihp-h	CO, g/ihp-h	NO, g/ihp-h	HCHO, g/ihp-h	Run No.
49.9	253.4	16.85	1.004	56.5	35.0	0.465	2.30	12.0	8.23	0.098	139
49.9	255.2	16.85	0.997	50.8	25.0	0.517	2.34	13.1	5.30	0.096	140
49.9	254.9	16.85	0.998	39.5	15.0	0.665	1.90	15.9	3.14	0.101	141
49.9	254.9	16.85	0.998	27.0	5.0	0.973	0.89	20.9	2.44	0.102	142
49.9	254.9	16.85	0.998	15.7	-5.0	1.673	0.69	40.0	2.86	0.099	143

Table 9 - Engine Performance and Exhaust Mass Data at 1000 rpm with Methanol-Retarded Spark

Pct Thro	Airf, g/min	Fuelf, g/min	Equil Ratio	IMEP, psi	SA, btdc	ISFC, lb/ihp-h	HC, g/ihp-h	CO, g/ihp-h	NO, g/ihp-h	HCHO, g/ihp-h	Run No.
49.5	126.1	19.55	1.000	66.3	25.0	0.828	0.34	5.5	7.06	0.188	134
49.5	125.8	19.49	0.999	59.3	15.0	0.922	0.35	6.8	3.07	0.191	135
49.5	126.5	19.49	0.994	45.7	5.0	1.197	0.37	8.9	2.40	0.228	136
49.5	126.3	19.65	1.004	34.5	-5.0	1.598	0.31	11.6	2.59	0.248	137
49.5	126.5	19.65	1.002	24.8	15.0	2.224	0.07	14.6	3.17	0.251	138

Table 10 - Engine Performance and Exhaust Mass Data at 1800 rpm with Methanol-Retarded Spark

Pct Thro	Airf, g/min	Fuelf, g/min	Equil Ratio	IMEP, psi	SA, btdc	ISFC, lb/ihp-h	HC, g/ihp-h	CO, g/ihp-h	NO, g/ihp-h	HCHO, g/ihp-h	Run No.
49.9	229.4	35.75	1.005	56.5	25.0	0.987	0.22	12.7	5.07	0.106	148
49.9	230.0	35.75	1.003	49.6	15.0	1.124	0.22	12.4	2.89	0.113	147
49.9	230.8	35.75	0.999	37.7	5.0	1.478	0.21	16.3	1.30	0.122	146
49.9	230.8	35.75	0.999	26.6	-5.0	2.095	0.14	23.2	1.13	0.134	145
49.9	231.0	35.75	0.998	16.9	15.0	3.298	0.12	36.4	1.56	0.144	144

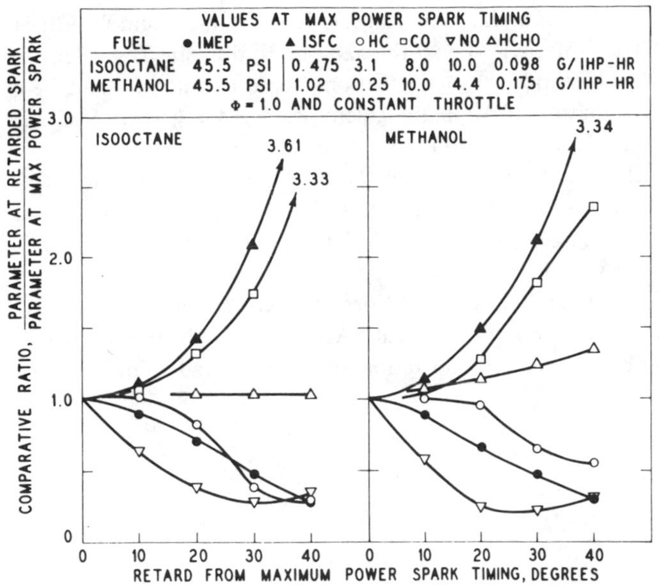

Fig. 36 - Effect of spark retard on power, fuel consumption, and emissions with isooctane and methanol at 1800 rpm

somewhat misleading because absolute values are relatively low. Emissions of hydrocarbon are unaffected by up to 15 degrees retard, after which they increase slightly and then decrease. These trends apply to both fuels and to both speeds.

CONCLUSIONS

This study has established the engine output, fuel consumption, and exhaust emission characteristics of a single-cylinder spark-ignition engine operated at maximum power spark timing with two fuels, isooctane and methanol.

Comparison of these characteristics for the two fuels yields the following conclusions, applicable to both engine speeds considered.

1. The lean misfire limits with methanol were approximately 0.2 equivalence ratios leaner than with isooctane.
2. The maximum engine output with methanol was approximately equal to that obtained with isooctane. The leaner limit associated with methanol permitted engine operation at considerable lower imep's than with isooctane.

For comparisons at the same power outputs and equivalence ratios:

1. Isfc with methanol was 2.15 times greater than with isooctane.
2. The quantity of unburned fuel in the exhaust with methanol was 0.1-0.3 times that with isooctane.
3. Zero carbon monoxide levels were achieved with both fuels at stoichiometric and learner mixtures. Carbon monoxide emissions with isooctane increased at a faster rate as equivalence ratio increased.
4. Emissions of nitric oxide with methanol were generally lower than with isooctane, (one-half at midrange imep), while the inverse was true for emissions of aldehydes.
5. The effect of spark retardation on performance at stoichiometric mixtures and a constant airflow (which yielded low imep's at maximum power spark timing) was the same for both fuels.

ACKNOWLEDGMENTS

The authors wish to thank Phillips Petroleum Co. who made available the test engine, the test fuels, the analytical instrumentation, and supporting facilities with which the experimental work was conducted. The senior author also wishes to express appreciation to Phillips for financial support through their Graduate Level Educational Assistance program, under which this study was made possible, and to affiliates of the University of Tulsa for their cooperation.

REFERENCES

1. "Alcohols and Hydrocarbons as Motor Fuels," SP-254. New York: Society of Automotive Engineers, June 1964.
2. Prepared Discussions and Author's Closures to Ref. 1. New York: Society of Automotive Engineers, June 1964.
3. Patent No. 1, 252, 198, Dec. 26, 1969. Patent Office, 25 Southampton Building, London.
4. H. R. Ricardo, "Report on the Empire Motor Fuels Committee." The Institute of Auto Engineers, XVIII, 1923-1924.
5. L. C. Lichty and E. J. Ziurys, "Engine Performance with Gasoline and Alcohol." Ind. Engr. Chem., Vol. 28, No. 9 (September 1956), pp. 1094-1101.
6. E. S. Starkman, H. K. Newhall, and R. D. Sutton, "Comparative Performance of Alcohol and Hydrocarbon Fuels." SP-254. New York: Society of Automotive Engineers, June 1964.
7. E. S. Starkman, R. F. Sawyer, R. Carr, G. Johnson, and L. Muzio, "Alternative Fuels for Control of Engine Emissions." Jrl. Air Pollution Control Assoc., Vol. 20, No. 2 (February 1970), pp. 87-92.
8. R. E. Fitch and J. D. Kilgore, "Investigation of a Substitute Fuel to Control Automotive Air Pollution." Consolidated Engineering Technology Corp., 188 Whisman Road, Mountain View, Calif., February 1970.
9. E. F. Obert, "Internal Combustion Engines." Scranton, Pa.: International Textbook Co., 1968.
10. "Reference Data for Hydrocarbons." Bulletin No. 521, Phillips Petroleum Co., Bartlesville, Oklahoma.
11. Lester C. Lichty, "Combustion Engine Processes." New York: McGraw-Hill, 1967.
12. R. C. Lee and D. E. Wimmer, "Exhaust Emission Abatement by Fuel Variations to Produce Lean Combustion." SAE Transactions, Vol. 77 (1968), paper 680769.
13. ASTM Standards–Petroleum Products, Part 18. American Society of Test and Materials, Philadelphia. (Issued annually.)
14. "Oxygen Effect in Flame Ionization Response to Hydrocarbons in Automotive Engine Exhaust." CRC Report No. 410, March 1967.
15. M. W. Jackson. "Analysis for Exhaust Gas Hydrocarbons–Nondispersive Infrared versus Flame Ionization."

Jrl. Air Pollution Control Assoc., Vol. 11 (1966), pp. 697-702.

16. D. Maxwell Teague, E. J. Lesniak, Jr., and E. H. Loeser, "A Recommended Flame Ionization Detector Procedure for Automotive Exhaust Hydrocarbons." Paper 700468, presented at SAE Mid-Year Meeting, Detroit, May 1970.

17. "Control of Air Pollution from New Motor Vehicles and New Motor Vehicle Engines." Federal Register, Vol. 33, No. 108 (June 1968).

18. B. E. Saltzman, "Modified Nitrogen Dioxide Reagent for Recording Air Analyzers." Anal. Chem., Vol. 32 (1960), p. 135.

19. E. Sawicki, T. R. Hauser, T. W. Stanley, and W. Elbert, "The 3-Methyl-2-Benzothioazolone Hydrazone Test–Sensitive New Methods for the Detection, Rapid Estimation, and Determination of Aliphatic Aldehydes." Anal. Chem., Vol. 33, No. 1 (January 1961), pp. 93-96.

20. G. D. Ebersole, "Power, Fuel Consumption, and Exhaust Emission Characteristics of an Internal Combustion Engine Using Isooctane and Methanol," Ph.D. Thesis, The University of Tulsa, 1971.

21. P. H. Schweitzer, "Correct Mixtures for Otto Engines." Paper 700884, presented at SAE Combined National Fuels & Lubricants & Transportation Meetings, Philadelphia, November 1970.

Comparison of Emission Indexes Within a Turbine Combustor Operated on Diesel Fuel or Methanol *

Clayton W. LaPointe
and Weston L. Schultz
Ford Motor Co.

WITH INCREASED EMPHASIS upon automotive emission control, interest in powerplants other than the conventional otto cycle reciprocating engine has been intensified. One of the most attractive alternative powerplants, both from an emissions point of view and in terms of development already received, is the regenerative gas turbine. Were it not for the emission of oxides of nitrogen (NO_x), some form of the turbine could in all likelihood eventually meet the stringent 1976 federal emissions requirements (1)* (Table 1) as presently specified for light-duty vehicles. For this reason, a large amount of effort (2-5) has been devoted toward the reduction of this exhaust component.

Early studies (3) revealed that alteration of conventional burner design parameters (for example, dilution port location) would not achieve the desired reduction in overall burner NO_x emission rate. Detailed chemical probing (4, 5) within a laboratory burner operating at atmospheric inlet pressure and temperature suggested that, as with the otto cycle engine (6), the NO formation process was a nonequilibrium, time-dependent process which could perhaps be described by the Zeldovich chain mechanism (7). There is some evidence (8, 9) that the latter is not the case under all conditions and the situation has not been completely resolved to date.

The Zeldovich NO formation mechanism does provide an analytical basis for all NO_x reduction strategies, however. These strategies state that if combustion temperature is lowered, or if reactant residence time is shortened, or if the reactant residence time is shortened, or if the reactants (O_2 and N_2) are separated during the peak temperature periods (or zones) in their flow history, the ultimate level of NO produced by a particular combustor will be lowered. Stratified (rich-lean) and premixed very lean (2, 10) combustors are examples of the physical implementation of these principles.

As well as varying the customary burner design parameters to determine their influence on NO_x production, it was natural at first to inquire what might happen to the NO_x emis-

*Numbers in parentheses designate References at end of paper.

*Paper 730669 presented at the National Powerplant Meeting, Chicago, Illinois, June 1973.

ABSTRACT

The emission index (grams of species per kilogram of fuel) field within a regenerative turbine combustor has been mapped using a water-cooled sampling probe. The probe employed a choked orifice to simultaneously determine the local temperature. Derived from measurements are: air-fuel ratio, combustion efficiency, average fuel velocity, and fuel distribution factor. Methods of averaging the discrete data are developed.

A comparison of the data obtained when the combustor was operated on each of two fuels revealed that the use of methanol leads to lower nitric oxide but higher carbon monoxide emission than does the use of diesel fuel.

SAE/PT-80/19/$02.50

Table 1 - Maximum Allowable Emissions Indexes for a Light Duty Vehicle in 1976*

Substance	1976 Federal Standard, g/mile	Emissions Index, g/kg fuel: Hydrocarbon Fuel	Emissions Index, g/kg fuel: Methanol
HC	0.41	1.42	0.62
CO	3.4	11.8	5.2
NO	0.26**	0.9	0.4

*These are based on a hydrocarbon fuel economy of 10 mpg.
**The NO_x standard is 0.41 g/mile.

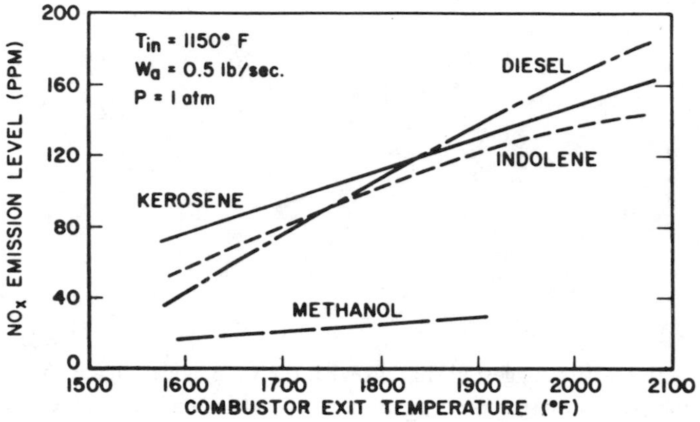

Fig. 1 - NO_x emission levels versus combustor exit temperature for several fuels

sion level when one liquid fuel was substituted for another, all other controllable conditions being the same. Since the Zeldovich thermal theory depends on the fuel-oxygen reaction only to supply the energy to start the N-O chain, it was conceivable that the type of fuel used might be of secondary importance to the production of NO. Experiments with hydrocarbon (HC) fuels in our laboratory (11) (Fig. 1) tended to support this conclusion. On the other hand, when alcohols were substituted for the HC fuels a marked, roughly fourfold reduction in NO_x emission occurred at all combustor exit temperatures.

This paper describes the results of an experiment which attempts to determine the causes of this phenomenon. It will be shown that the substitution of alcohol for HC fuel results in changes to all three Zeldovich parameters: temperature, time, and reactant availability. The diminished NO formation observed with the use of alcohol can therefore be reasonably explained without resort to chemical considerations other than those of the Zeldovich mechanism. A preliminary report of this investigation was presented at the General Motors-sponsored "Emissions from Continuous Combustion Systems" Symposium (11). The data presented therein have since been further analyzed and these constitute the major part of the present paper.

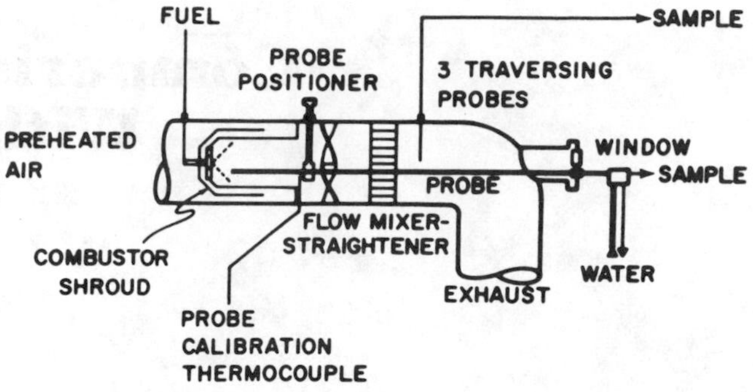

Fig. 2 - Combustor test rig

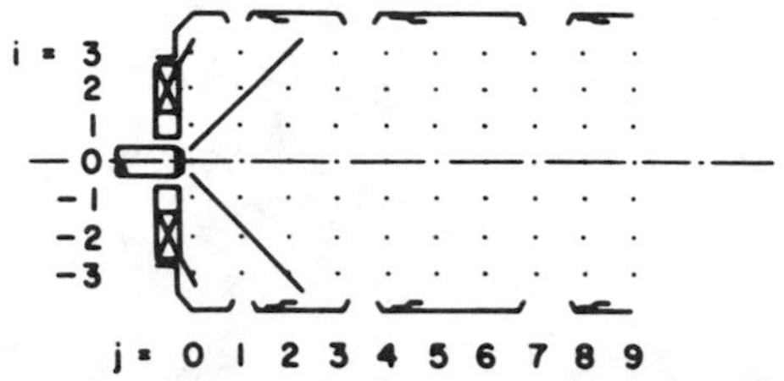

Fig. 3 - Combustor configuration and locations of sample extraction points. Radial coordinate index, i, has 0.75 in spacing; axial coordinate index, j, has 1 in spacing

EXPERIMENT

The experiment utilized a conventional turbine combustor mounted in a test stand (Fig. 2) fed by high pressure preheated air in order to simulate a regenerative turbine combustor environment. The combustor is shown in Fig. 3 and its dimensions are listed in Table 2. A specially designed water-cooled probe extracted samples of the combustion products at the points shown in Figs. 3 and 4. There are seven radial sampling locations spaced 0.75 in apart along a combustor diameter in each of 10 cross-sectional planes spaced 1 in apart along the combustor axis.

Advantage was taken of the elevated combustion pressure to choke the probe orifice and thereby obtain a measure of the gas temperature. Under these conditions, the sample flow rate is inversely proportional to the square root of the gas temperature upstream of the orifice (12). The orifice pressure ratio was monitored to guarantee that the orifice was indeed operating in a choked condition.

For purposes of comparison, two fuels, No. 2 diesel and methanol, were chosen. The combustor operating parameters were adjusted such that the inlet and outlet temperature as well as the airflow rate were identical when each fuel was used (Table 3). Gas samples were taken beginning with the exit plane and proceeded inward until probe clogging prevented further penetration. This occurred 2 in from the fuel nozzle with diesel oil and 3 in from the nozzle with methanol. The greater quantity of water produced in the combustion of methanol, while not actually clogging the probe, rendered unreliable those readings taken closer to the nozzle than this.

Details of the gas analysis instrumentation are given in Ref. 11.

Table 2 - Combustor Dimensions*

	Diameter	Location (from nozzle)
Primary dilution ports	0.5	1.2
Secondary dilution ports	0.7	3.75
Tertiary dilution ports	1.0	7.5

Swirler: 1.72 ID × 2.7 OD; 15 deg air entry angle with respect to frontal plane.

Nozzle: 90 deg hollow cone, orifice diameter = 0.025.

Length: 9.25; diameter: 5.75.

*All dimensions are expressed in inches.

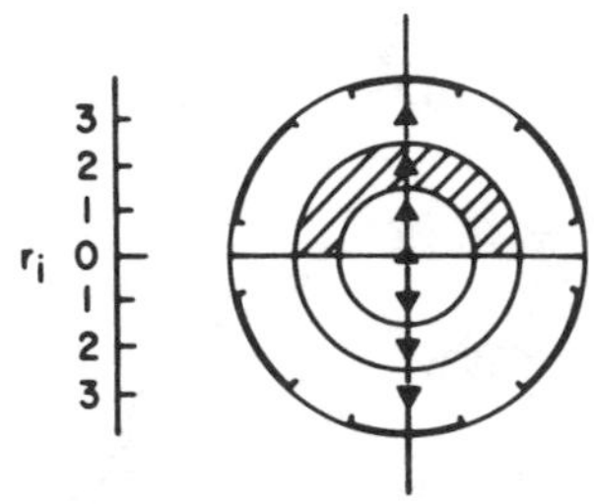

i	r_i (in.)	S_i (sq. in.)
3	2.25	7.5
2	1.50	3.51
1	0.75	1.775
0	0	0.442
-1	-0.75	1.775
-2	-1.50	3.51
-3	-2.25	7.5

Fig. 4 - Radial sampling locations and their associated area weights

EXPERIMENTAL RESULTS

From the individual gas sample analyses, the following gas properties were deduced:

1. Concentrations (mole fractions) of NO, NO_2, CO, CO_2, H_2O, O_2, N_2, HC.
2. Temperature.
3. Local air-fuel ratio (A/F).
4. Molecular weight.
5. Local emission indices for NO, CO and HC (for example, grams NO per kilogram fuel burned).

The method of analysis is given in Appendix A.

Shown in Fig. 5 are the radial profiles of NO mole fraction, A/F, and temperature for both fuels. Note that, in general, where NO concentration is low, A/F is high and temperature is low. This suggests that, for the combustor locations represented by these data, the NO might be chemically frozen and varying in concentration only because of dilution. This is indeed the case but cannot be concluded on the basis of the data in Fig. 5 alone.

Other characteristics apparent from these data are the lack of radial symmetry and the similarity of the profiles for the two fuels. Maximum profile distortion is evident at planes 3 and 7 which are dilution port locations.

The locally measured values of NO, CO and HC emission indices (EI) for the combustion of No. 2 diesel oil are shown in Fig. 6. From these it is clearly evident that NO is chemically frozen even before plane 3 is reached. The NO emission index

Table 3 - Experimental Conditions

	Fuel	
	Diesel No. 2	Methanol
Inlet temp., °F	1000	1000
Exit temp., °F	1400	1400
Airflow, lb/s	1.6	1.6
Fuel flow, lb/h	40.5	84
Fuel temp., °F	220	190
Lower heating valve, Btu/lb	18500	8600
Latent heat of vaporization, Btu/lb	155	503
Stoichiometric A/F	14.6	6.4
Overall A/F	141	68
Sauter mean diameter, μm	127	55.1
Injection velocity, fps	65	139
Drop Rn	106	99
Pressure, atm	3	3
Combustor pressure drop, %	3	3
Space rate, Btu/h-ft^3 atm	1.8×10^6	1.8×10^6

is practically constant everywhere and frozen at a value of approximately 6 g NO/kg fuel.

On the other hand, HC and CO both continue to oxidize throughout much of the combustor. The profiles of EI_{HC} and EI_{CO} also display wide variation with combustor radius. The situation is similar in the case of methanol combustion (Fig. 7). In this case, the NO emission index is everywhere relatively constant at a value of approximately 0.7 g NO/kg methanol.

Except for the NO emission index which is approximately constant, the overall emission of CO and HC is not sufficiently determined by the point values as presented above. What is required are mass-weighted averages of the radial values. In order to accomplish such an averaging process, strictly speaking, the local gas velocity must be known. This is an exceedingly difficult measurement in a turbine combustor environment and was not attempted in the present experiments. This deficiency may be largely overcome, however, if some loss of accuracy is tolerated. Appendix B presents the details of an averaging process which effectively mass-weights locally measured values to arrive at approximate cross-sectional averages. It will be shown that very little accuracy is sacrificed in this process. The averages thus obtained illustrate the combustor's chemical behavior much more clearly than do the foregoing radial mappings.

The diesel fuel cross-sectional average emission indices of NO, CO, and HC are shown in Fig. 8. Values for planes 0 and 1 are computed by means of a one-dimensional combustor model to be described later. From the figure, it is apparent that all of the NO has been formed prior to plane 2. The NO formation zone is thus a relatively small region compared to the reaction zones of HC and CO. It may be seen that both HC and CO undergo significant reaction up to plane 6 and continue to exhibit diminished activity throughout the rest of the combustor.

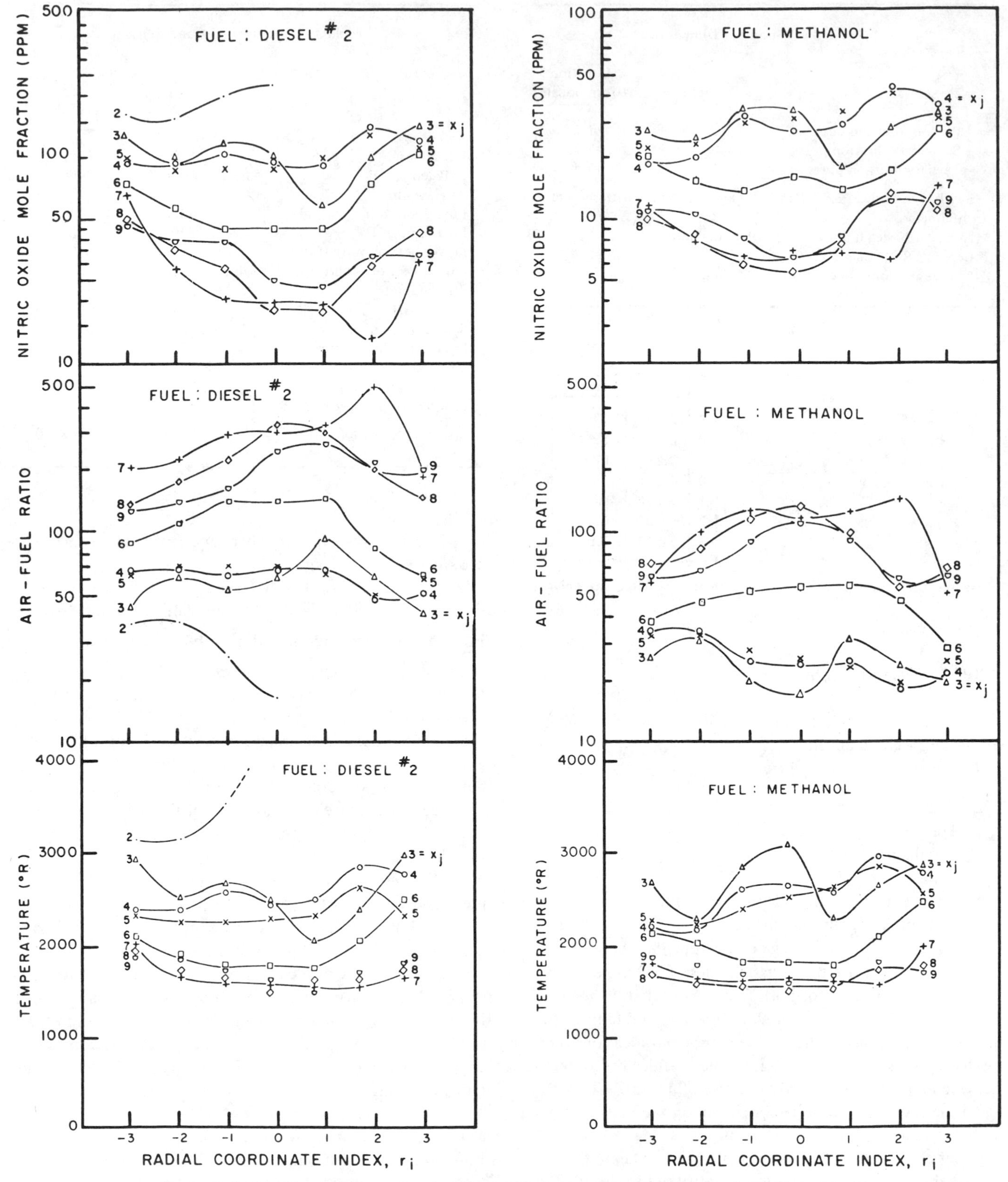

Fig. 5 - Radial profiles of NO concentration, local A/F, and temperature; No. 2 diesel oil and methanol

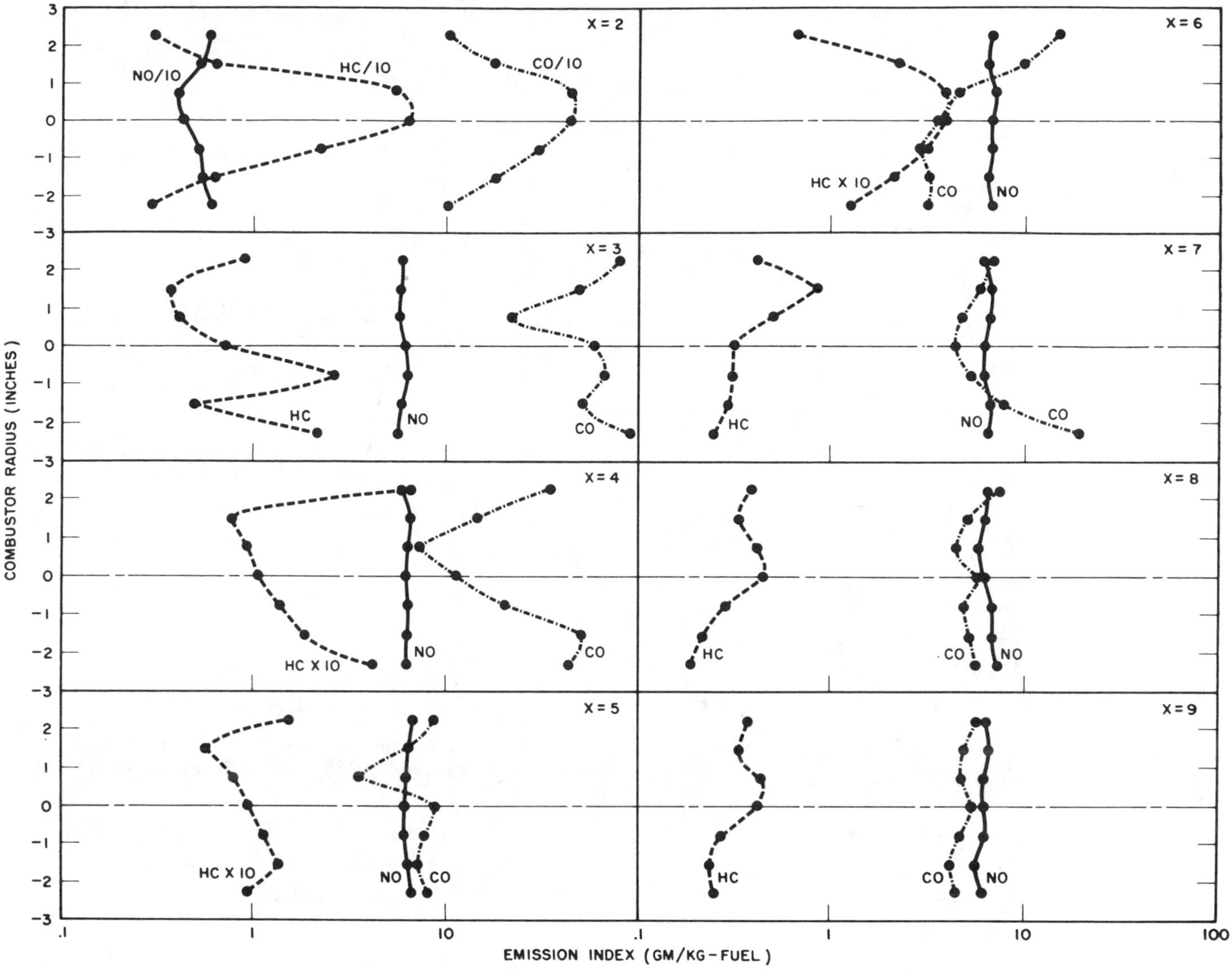

Fig. 6 - HC, CO, and NO emission index profiles versus combustor radius for planes 2-9; No. 2 diesel oil

Also indicated on the figure are the maximum permissible values of EI as listed in Table 1, assuming a diesel fuel economy of 10 mpg. At these combustor operating conditions, the exiting CO and HC levels are both well below their respective limits while NO exceeds its limit by a factor of seven.

The corresponding EI arising from methanol combustion are different in several respects (Fig. 9). The NO emission index is constant throughout the measured portion of the combustor at a value of approximately 0.7. This exceeds the corresponding methanol limit value of 0.4. Whereas during the combustion of diesel fuel, EI_{CO} and EI_{HC} monotonically decrease throughout the combustor as the exit is approached, these quantities now approach a minimum near the midway point and then begin to increase as the exit plane is approached. The limiting value of EI_{HC} is eventually exceeded and the limiting value of EI_{CO} is reached. Thus, in producing less NO, methanol produces unacceptable CO and HC.

A possible explanation for the observed minimums in CO and HC emission indices may lie in the known production of formaldehyde as an intermediate product of methanol combustion. The flame ionization detector (FID) would not respond to this component which upon decomposition would again yield products (HC and CO) which were detectable. In this way, both HC and CO may artificially be produced in the downstream portion of the combustor.

Table 4 contains a summary of the cross-sectional averages of several diesel fuel combustion parameters. Listed are the average mass fractions of several components, the average temperature, A/F, combustion efficiency (ETA1 and ETA2), fuel distribution factor, FD, and average fuel velocity, UF. The combustion efficiencies are computed on the basis of the amount of unburned fuel remaining or the amount of oxygen consumed (ETA1 and ETA2, respectively). They are given by

$$ETA1 = 1 - Y_f(x_j)\left[1 + \frac{A}{F}(x_j)\right] \quad (1)$$

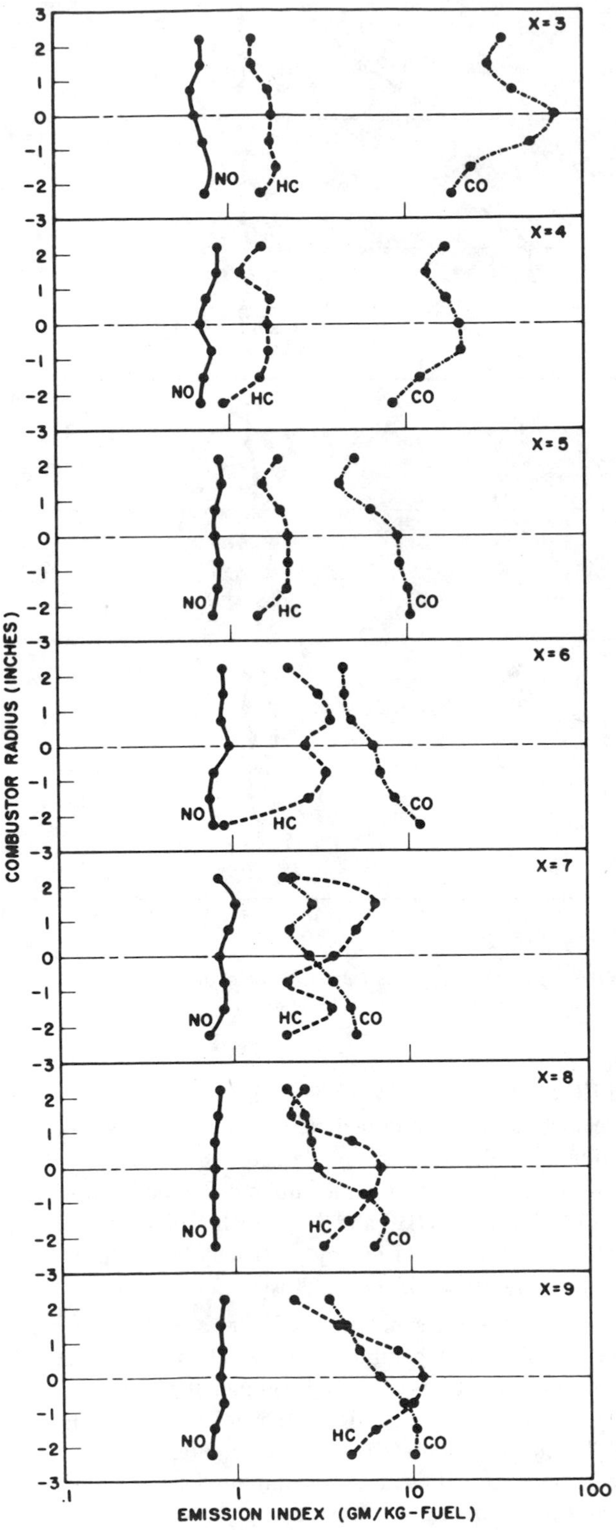

Fig. 7 - HC, CO, and NO emission index profiles versus combustor radius for planes 3-9; methanol

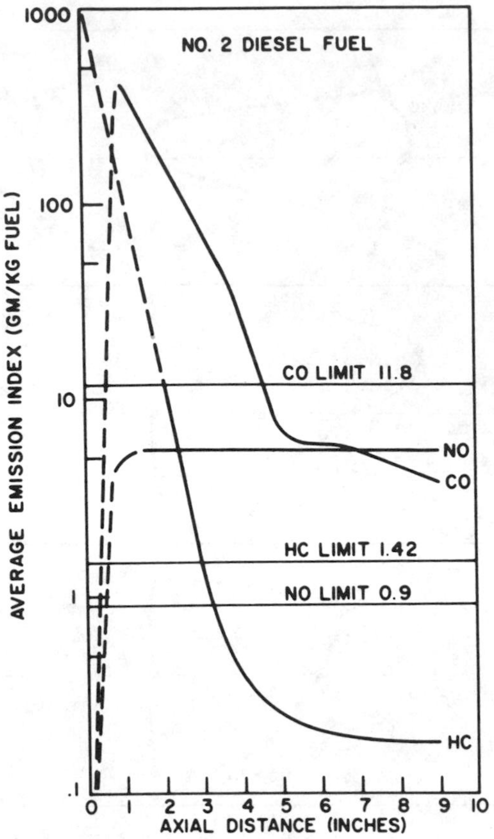

Fig. 8 - Average HC, CO, and NO emission index versus combustor axial distance. Respective permissible limits for 1976 are indicated. Fuel: No. 2 diesel oil

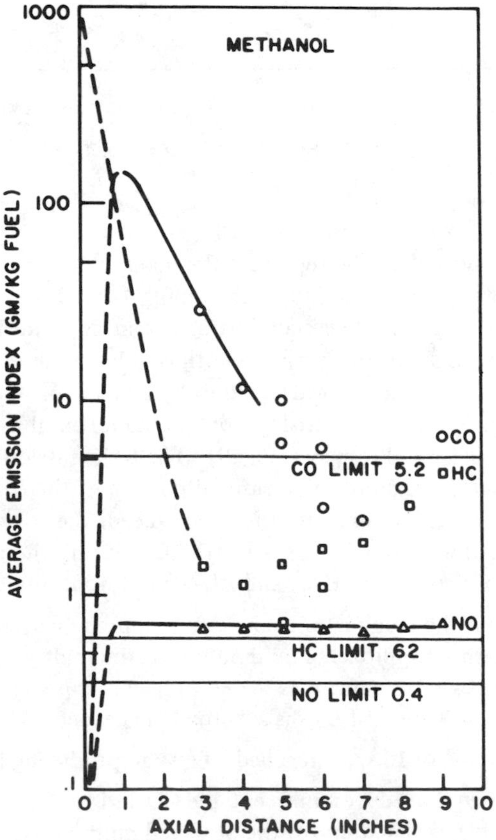

Fig. 9 - Average HC, CO, and NO emission index versus combustor axial distance. Respective permissible limits for 1976 are indicated. Fuel: methanol

Table 4 - Experimental Average Combustor Mass Fractions, Temperature, Combustion Efficiencies, A/F, Fuel Distribution Factor and Fuel Velocity Versus Combustor Axial Distance*

X	NO, GPMG	O_2, G%	CO_2, G%	CO, GPMG	H_2O, G%	N_2, G%	HC, GPMG	T, R	ETA1	ETA2	A/F	FD	VF, ft/s
2	166.6	13.4	8.3	5337.5	3.5	74.2	398.6	3246.5	0.986	0.927	33.21	88.31	58.80
3	115.0	16.4	5.9	1379.9	2.3	75.1	26.6	2738.4	0.999	0.968	50.65	123.36	75.23
4	103.9	17.3	5.2	539.1	2.0	75.4	6.4	2623.9	1.000	0.978	59.19	117.42	83.04
5	102.0	17.7	4.9	117.2	1.9	75.5	1.9	2391.5	1.000	0.983	63.38	115.11	80.24
6	73.9	19.3	3.5	85.0	1.4	75.8	1.8	2126.6	1.000	0.975	88.66	145.00	100.72
7	27.0	21.7	1.4	27.5	0.5	76.4	2.0	1686.0	0.999	0.929	231.05	138.95	208.93
8	39.1	21.1	1.9	35.2	0.7	76.2	1.9	1815.5	1.000	0.950	166.96	165.51	162.19
9	37.3	21.1	1.9	29.2	0.7	76.2	2.0	1806.5	1.000	0.951	165.41	135.77	161.05

*Fuel: No. 2 diesel oil.

Table 5 - Experimental Average Combustor Mass Fractions, Temperature, Combustion Efficiencies, A/F, Fuel Distribution Factor and Fuel Velocity Versus Combustor Axial Distance*

X	NO, GPMG	O_2, G%	CO_2, G%	CO, GPMG	H_2O, G%	N_2, G%	HC, GPMG	T, R	ETA1	ETA2	A/F	F/D	VF, ft/s
3	27.9	16.5	5.2	1218.8	4.4	73.7	59.5	2671.5	0.998	0.978	24.41	102.26	72.62
4	28.0	17.1	4.9	498.5	4.1	73.9	47.2	2528.4	0.999	0.983	26.78	106.63	75.31
5	29.2	17.3	4.7	253.6	3.9	74.0	61.0	2450.8	0.998	0.984	28.10	104.44	76.77
6	20.1	19.1	3.3	169.3	2.7	74.8	49.6	2210.7	0.998	0.975	40.55	129.86	98.54
5	26.7	17.4	4.6	416.6	3.9	74.1	28.6	2518.9	0.999	0.983	28.37	99.58	79.37
6	19.0	19.0	3.4	83.8	2.8	74.8	31.3	2222.8	0.999	0.978	39.32	133.86	95.60
7	10.5	21.1	1.8	43.9	1.4	75.7	33.8	1776.6	0.997	0.949	77.12	160.33	151.61
8	10.4	21.1	1.8	54.2	1.5	75.7	41.0	1688.3	0.997	0.949	76.19	160.50	144.29
9	11.1	20.9	1.9	99.1	1.6	75.6	63.8	1766.9	0.998	0.951	79.84	151.81	137.35

*Fuel: methanol.

$$ETA2 = \frac{A/F\,(x_j)}{(A/F)_s}\left\{1 - \frac{Y_{O_2}(x_j)}{0.232}\left[1 + \frac{F}{A}(x_j)\right]\right\} \qquad (2)$$

where the factor 0.232 is the mass fraction of O_2 in the incoming air. From Table 4 (and Table 5, which is the corresponding table for methanol), it is seen that the combustion efficiency is everywhere greater than 95% throughout the measurable portion of the combustor. Note that the dimensions GPMG and G% are mass dimensions as contrasted to the more familiar volume-based expression of concentration. They signify "grams per million grams" and "gram %" respectively. Note also that Table 5 contains repeated data for planes 5 and 6 obtained in consecutive experiments. The complete mapping of the combustor required more time than a normal working day.

Before proceeding with further discussion of these tables, the results of a one-dimensional combustor model prediction program (11) are presented in Tables 6 and 7 for comparison. Inputs to the predictor program are the burning rate (11) and air addition rate. For this comparison, the air addition rate was specified by the measured dilution rate of NO concentration since it has been shown that NO behaves like an inert tracer element. On the other hand, the dilution rate entering into the calculations of Tables 4 and 5 is based on the measured average A/F. Thus, the average velocities predicted in Tables 6 and 7 and measured in Tables 4 and 5 are based on independent inputs and should serve as a consistency check on the approximations entering into the averaging process outlined in Appendix B. The success of this approximation is shown in Figs. 10 and 11 which compare the measured fuel velocity, VF (Eq. B-17), to the predicted average combustor velocity for each fuel.

Defined in Appendix B is an average quantity designated as FD, the fuel distribution factor. In dealing with as many individual data and variables as is necessary in the present experiment, it is desirable to attempt to quantify certain overall combustor properties with a single parameter. The fuel distribution factor is such an attempt. As shown in Appendix B, FD assumes a dimensionless value depending on the data grid geometry and the distribution of fuel throughout the grid. Thus, for the grid of Fig. 4, a value FD <109 indicates a fuel

Table 6 - One-Dimensional Combustor Model Computer Prediction of Average Temperature, Velocity, Density, Cumulative Burning Rate, Cumulative Air Addition Rate, NO Equilibrium and Actual Concentration, and Equilibrium CO and CO_2 Concentrations Versus Combustor Distance and Residence Time*

Dist.	Time		Temp.	Speed,	Dens.,	Fuel	Total	NO		CO	CO_2
% of Total	Cum. ms	Cum. %	°R	ft/s	rho/rhoo	Burn, cum. %	Air, cum. %	Eq. ppm	Actual ppm	Eq. ppm	Eq. mol %
0.0	0.00	0.	1459	6.7	1.105	0.0	6.0	0.	0.	0.	0.00
0.5	0.47	4.	1938	9.3	0.825	6.7	6.2	71.	0.	0.	1.51
2.5	1.60	14.	3255	17.9	0.477	31.8	7.0	3188.	0.	24.	6.25
5.0	2.45	21.	4106	26.2	0.368	59.2	8.0	7316.	13.	2647.	9.77
7.5	3.09	26.	4427	32.3	0.333	81.5	9.0	6724.	156.	12967.	10.79
10.0	3.63	31.	4500	36.7	0.323	96.8	10.0	5753.	379.	20345.	10.81
11.5	3.85	33.	3395	53.4	0.434	100.0	20.2	3941.	199.	59.	6.87
12.5	3.99	34.	3382	53.6	0.435	100.0	20.3	3872.	197.	54.	6.82
15.0	4.34	37.	3351	54.1	0.439	100.0	20.7	3708.	193.	44.	6.70
17.5	4.69	40.	3229	56.2	0.456	100.0	22.4	3092.	179.	20.	6.21
20.0	5.02	43.	3182	57.1	0.462	100.0	23.1	2865.	174.	15.	6.03
22.5	5.33	46.	3063	59.6	0.481	100.0	25.1	2335.	161.	6.	5.56
25.0	5.63	48.	2925	62.9	0.503	100.0	27.8	1791.	145.	2.	5.03
27.5	5.92	51.	2809	66.1	0.524	100.0	30.5	1392.	133.	1.	4.60
30.0	6.19	53.	2724	68.8	0.540	100.0	32.8	1143.	124.	0.	4.28
32.5	6.46	55.	2676	70.5	0.550	100.0	34.2	1015.	118.	0.	4.10
35.0	6.71	57.	2616	72.8	0.562	100.0	36.2	870.	112.	0.	3.89
37.5	6.97	60.	2585	74.1	0.569	100.0	37.3	801.	109.	0.	3.77
40.0	7.22	62.	2561	75.2	0.575	100.0	38.2	749.	106.	0.	3.69
42.5	7.47	64.	2548	75.7	0.577	100.0	38.7	723.	105.	0.	3.64
45.0	7.71	66.	2540	76.1	0.579	100.0	39.0	707.	104.	0.	3.61
47.5	7.96	68.	2538	76.2	0.580	100.0	39.1	702.	104.	0.	3.60
50.0	8.20	70.	2534	76.4	0.581	100.0	39.2	695.	104.	0.	3.59
52.5	8.45	72.	2528	76.7	0.582	100.0	39.5	682.	103.	0.	3.57
55.0	8.69	74.	2518	77.1	0.584	100.0	39.9	663.	102.	0.	3.53
57.5	8.93	76.	2492	78.4	0.590	100.0	41.0	614.	99.	0.	3.44
60.0	9.16	78.	2448	80.7	0.601	100.0	43.0	537.	95.	0.	3.28
62.5	9.39	80.	2388	84.1	0.616	100.0	46.0	445.	89.	0.	3.07
65.0	9.60	82.	2311	89.3	0.637	100.0	50.5	343.	81.	0.	2.80
67.5	9.79	84.	2217	96.9	0.663	100.0	57.2	244.	71.	0.	2.48
70.0	9.97	85.	2131	105.7	0.690	100.0	65.0	173.	63.	0.	2.18
72.5	10.14	87.	2073	113.0	0.709	100.0	71.5	135.	57.	0.	1.99
75.0	10.30	88.	2030	119.4	0.724	100.0	77.2	111.	53.	0.	1.84
77.5	10.46	90.	1994	125.5	0.737	100.0	82.6	94.	50.	0.	1.72
80.0	10.60	91.	1972	129.8	0.746	100.0	86.5	85.	47.	0.	1.64
82.5	10.74	92.	1952	133.7	0.753	100.0	90.0	77.	46.	0.	1.58
85.0	10.88	93.	1940	136.5	0.758	100.0	92.5	72.	44.	0.	1.54
87.5	11.02	94.	1929	138.8	0.762	100.0	94.6	68.	43.	0.	1.50
90.0	11.15	96.	1923	140.4	0.765	100.0	96.0	66.	43.	0.	1.48
92.5	11.29	97.	1917	141.8	0.767	100.0	97.2	64.	42.	0.	1.46
95.0	11.42	98.	1912	143.2	0.769	100.0	98.5	62.	42.	0.	1.44
97.5	11.25	99.	1908	144.1	0.771	100.0	99.4	61.	41.	0.	1.43
100.0	11.68	100.	1905	144.8	0.772	100.0	100.0	60.	41.	0.	1.42

Steps = 200. Combustion profile.
*Fuel: No. 2 diesel oil.

flux density peaking near the combustor centerline and vice versa. The fuel distribution factor for both fuels is shown in Fig. 12. Note that the fuel, in both cases, originates near the centerline and migrates toward the combustor walls as a function of axial distance. No obvious differences in the dispersion of the two fuels is apparent from the figure.

DISCUSSION OF RESULTS

From consideration of the emission indices and other average combustor parameters, it is evident that diesel fuel and methanol both effectively complete NO production in the immediate vicinity of the fuel nozzle at regenerative turbine

Table 7 - One-Dimensional Combustor Model Computer Prediction of Average Temperature, Velocity, Density, Cumulative Burning Rate, Cumulative Air Addition Rate, NO Equilibrium and Actual Concentration, and Equilibrium CO and CO_2 Concentrations Versus Combustor Distance and Residence Time*

Dist. % of Total	Time Cum. ms	Time Cum, %	Temp. °R	Speed, ft/s	Dens., rho/rhoo	Fuel Burn, cum. %	Total Air, cum. %	NO Eq. ppm	NO Actual ppm	CO Eq. ppm	CO_2 Eq. mol %
0.0	0.00	0.	1459	8.2	1.029	0.0	6.0	0.	0.	0.	0.00
0.5	0.38	3.	2038	11.8	0.731	8.2	6.2	95.	0.	0.	1.50
2.5	1.26	12.	3434	22.5	0.423	38.6	7.0	3306.	0.	75.	6.22
5.0	1.96	18.	4148	31.1	0.341	70.4	8.0	4592.	11.	4624.	9.43
7.5	2.51	23.	4287	36.3	0.323	93.2	9.0	2416.	49.	17264.	9.89
9.0	2.81	26.	4288	38.4	0.322	100.0	9.6	2333.	64.	18125.	9.95
10.0	3.00	28.	4278	39.5	0.325	100.0	10.0	3079.	76.	13208.	10.10
12.5	3.36	31.	3268	55.3	0.439	100.0	20.3	2971.	42.	27.	6.23
15.0	3.69	34.	3241	55.8	0.443	100.0	20.7	2859	41.	23.	6.12
17.5	4.03	37.	3170	57.2	0.453	100.0	21.8	2567.	39.	14.	5.84
20.0	4.35	40.	3093	58.8	0.465	100.0	23.1	2263.	37.	8.	5.54
22.5	4.65	43.	2986	61.3	0.483	100.0	25.1	1875.	34.	3.	5.13
25.0	4.95	46.	2863	64.6	0.504	100.0	27.8	1471.	31.	1.	4.67
27.5	5.22	48.	2758	67.9	0.524	100.0	30.0	1171.	29.	1.	4.28
30.0	5.49	51.	2680	70.6	0.540	100.0	32.8	975.	27.	0.	4.00
32.5	5.75	53.	2635	72.3	0.550	100.0	34.2	873.	26.	0.	3.84
35.0	6.00	55.	2580	74.6	0.562	100.0	36.2	756.	24.	0.	3.65
37.5	6.25	57.	2552	75.9	0.568	100.0	37.3	699.	24.	0.	3.54
40.0	6.49	60.	2529	77.0	0.574	100.0	38.2	657.	23.	0.	3.46
42.5	6.74	62.	2517	77.5	0.576	100.0	38.7	635.	23.	0.	3.42
45.0	6.98	64.	2510	77.9	0.578	100.0	39.0	622.	23.	0.	3.40
47.5	7.22	66.	2507	78.0	0.579	100.0	39.1	618.	23.	0.	3.39
50.0	7.46	69.	2504	78.2	0.580	100.0	39.2	612.	23.	0.	3.38
52.5	7.70	71.	2498	78.5	0.581	100.0	39.5	602.	23.	0.	3.36
55.0	7.93	73.	2489	78.9	0.583	100.0	39.9	586.	22.	0.	3.32
57.5	8.17	75.	2465	80.2	0.589	100.0	41.0	545.	22.	0.	3.24
60.0	8.40	77.	2423	82.5	0.599	100.0	43.0	481.	21.	0.	3.10
62.5	8.61	79.	2367	86.0	0.614	100.0	46.0	402.	19.	0.	2.90
65.0	8.82	81.	2294	91.1	0.634	100.0	50.5	314.	18.	0.	2.65
67.5	9.01	83.	2205	98.8	0.661	100.0	57.2	227.	16.	0.	2.35
70.0	9.19	85.	2123	107.6	0.687	100.0	65.0	164.	14.	0.	2.08
72.5	9.35	86.	2067	114.9	0.706	100.0	71.5	129.	13.	0.	1.90
75.0	9.51	88.	2025	121.4	0.721	100.0	77.2	107.	12.	0.	1.76
77.5	9.66	89.	1991	127.4	0.734	100.0	82.6	91.	11.	0.	1.65
80.0	9.81	90.	1968	131.7	0.742	100.0	86.5	82.	11.	0.	1.58
82.5	9.95	92.	1950	135.6	0.750	100.0	90.0	75.	10.	0.	1.52
85.0	10.08	93.	1937	138.4	0.754	100.0	92.5	70.	10.	0.	1.48
87.5	10.22	94.	1927	140.8	0.759	100.0	94.6	67.	10.	0.	1.44
90.0	10.35	95.	1921	142.3	0.761	100.0	96.0	65.	10.	0.	1.42
92.5	10.48	96.	1915	143.7	0.763	100.0	97.2	63.	9.	0.	1.41
95.0	10.61	98.	1910	145.1	0.766	100.0	98.5	61.	9.	0.	1.39
97.5	10.74	99.	1906	146.1	0.767	100.0	99.4	60.	9.	0.	1.38
100.0	10.87	100.	1903	146.8	0.768	100.0	100.0	59.	9.	0.	1.37

Combustion profile.
*Fuel: methanol.

combustor conditions. Thus, eventual differences in the ultimate emission level must be attributable to details of the droplet burning process and associated heterogeneous chemistry. On the other hand, it is also apparent that considerable differences in HC and CO chemistry between the two fuels exist throughout a substantial portion of the combustor.

In view of the inaccessibility of the NO formation zone to direct measurement, it is necessary to resort to analytical means to characterize the burning zone. In doing so, trends rather than absolute value accuracy are obtained.

Consideration of the Zeldovich NO formation rate expression reveals one source of rate differential which may be ex-

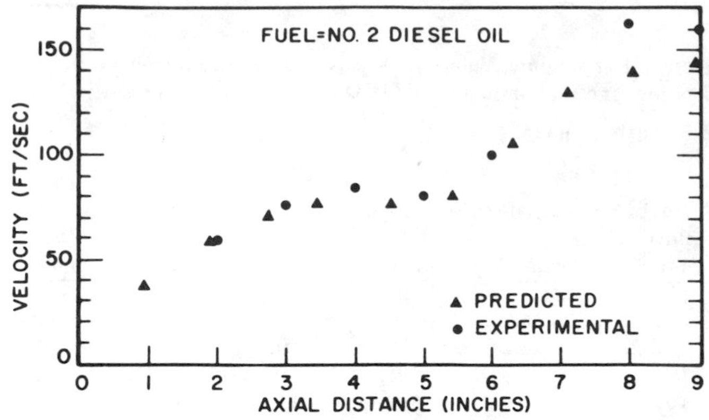

Fig. 10 - Predicted average mixture velocity and experimental "fuel" velocity versus combustor axial distance; fuel: No. 2 diesel oil

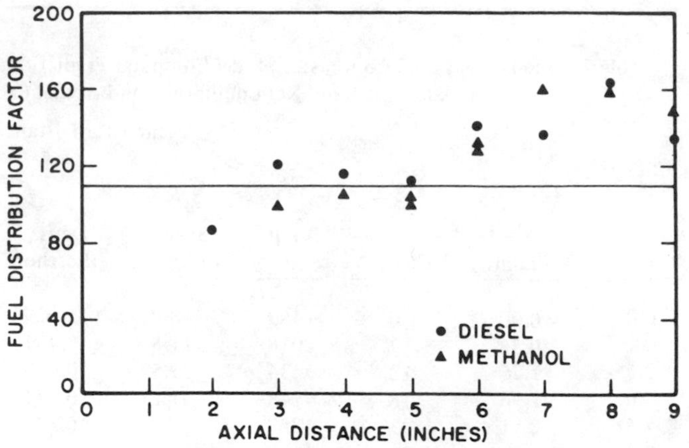

Fig. 12 - Measured fuel distribution factor versus combustor axial distance; fuels: No. 2 diesel oil and methanol

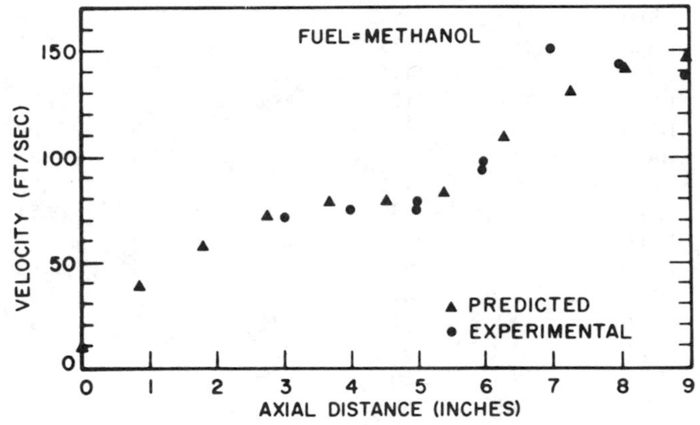

Fig. 11 - Predicted average mixture velocity and experimental "fuel" velocity versus combustor axial distance; fuel: methanol

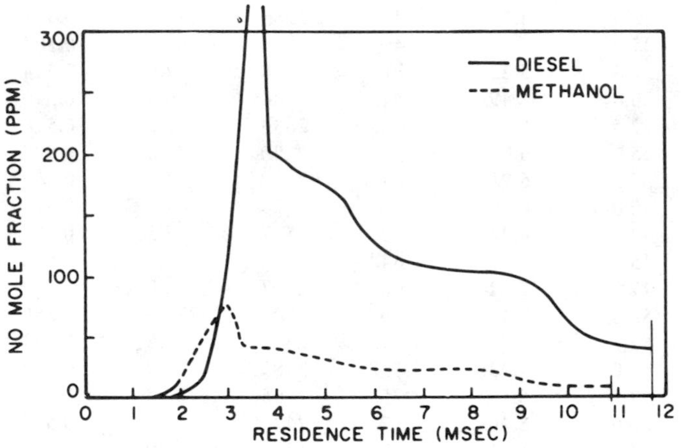

Fig. 13 - Average NO mole fraction versus residence time

pected to exist between the two fuel situations. It has to do with the reactant availability corresponding to each fuel. The rate is given by (7)

$$\frac{dX_{NO}}{dt} \sim P^{1/2} X_{O_2}^{1/2} X_{N_2} T^{-1} \left[1 - \left(\frac{X_{NO}}{X_{NO_e}} \right)^2 \right] \exp(-122000/T) \quad (3)$$

The average concentrations of O_2 and N_2 may be expressed in terms of the combustion efficiency, η, and the fuel equivalence ratio, ϕ.

$$X_{O_2}(t) = \frac{0.21}{1 + \dfrac{\phi(t)\, W_a}{(A/F)s\, W_f}} \left[1 - \eta(t)\, \phi(t)\right] \quad (4)$$

$$X_{N_2}(t) \cong \frac{0.79}{1 + \dfrac{\phi(t)\, W_a}{(A/F)s\, W_f}} \quad (5)$$

At stoichiometric conditions, the true NO mass formation rate ratio (methanol/diesel), neglecting the derivative of the airflow rate, would therefore be

$$\frac{d(\dot{m}_{NO})\text{ methanol}}{d(\dot{m}_{NO})\text{ diesel}} = 0.94 \quad (6)$$

In other words, the NO formation rate for methanol on a mass basis is 6% lower than that for diesel fuel due to the difference in O_2 and N_2 availability in the reaction zone. On a ppm (volume) basis, this effect amounts to a 16% difference in dX_{NO}/dt with methanol producing at the reduced rate.

With the combustion of methanol, the residence time per unit combustor length (for example, the distance to a quench location defined by the first row of dilution holes) is also less than the corresponding residence time for the products of diesel combustion. An estimate of this difference is provided in Fig. 13 which was derived from the previously mentioned one-dimensional predictor program. From the curves of NO mole fraction plotted as a function of time, it is seen that methanol products suffer dilution (quenching) 1/2 ms sooner than do the diesel products. One half millisecond is sufficient to account for a 25% difference in the NO produced by the two fuels. This may be estimated by extrapolating the methanol NO mole fraction curve to the quench time for diesel products.

The reason for this difference stems from the difference in stoichiometric A/F and heat of combustion between the two fuels (Table 3). Since the same fuel nozzle was used for both fuels, the methanol injection rate for equivalent heat addition was more than twice as great as the diesel fuel injection rate. This leads to a methanol droplet diameter (Sauter mean diameter) 43% smaller than a diesel fuel droplet (13). This, in turn, leads to a methanol evaporation rate, based on Godsave's law (14) corrected for convection (15), 2.9 times greater for methanol than for diesel. Thus, if the fuel reaction rate is diffusion controlled, the heat release rate for methanol will be greater than that for diesel. The methanol combustion products are therefore heated earlier in their flow history and accelerated to higher velocity with the result that they exit the NO formation region earlier than do the diesel products.

Finally, the peak temperature attained with methanol is slightly lower (200°F) than that attained with diesel fuel (Tables 6 and 7). Identical exit temperatures are ultimately attained because there is more mass which must be cooled by dilution in the case of methanol. The difference in peak temperature accounts for a major portion of the difference in NO production rates. From Eq. 3, a 4500°R peak diesel combustion temperature leads to 3.5 times the NO production rate than that which accrues with the 4300°R peak methanol combustion temperature.

Thus, on the basis of these considerations, it is possible to ascribe the difference in NO production rates of the two fuels to factors contained within the framework of the Zeldovich Thermal Theory. It is apparent that the slight difference in flame temperature accounts for most of the observed difference in NO produced by methanol and by diesel fuel. The different stoichiometric ratios also contribute to this result by affecting both the reactant concentrations and their residence times within the reaction zone.

NOMENCLATURE

A/F = Air-fuel ratio
(A/F)s = Stoichiometric air-fuel ratio
a = Hydrogen-carbon ratio of fuel; also subscript denoting "air"
b = Hydrogen-carbon ratio of unburned hydrocarbons
e = Subscript denoting "equilibrium"
F/A = Fuel-air ratio
f = Subscript denoting "fuel"
h = Static enthalpy
i = Radical coordinate index
j = Axial coordinate index
ℓ = Subscript denoting ℓth chemical species
$\dot{m}$ = Mass flow rate
n = Number of moles of a reactant divided by total number of moles of products
P = Pressure
R = Universal gas constant
r_i = ith radial coordinate
S = Area
t = Temperature, R
V = Velocity
W = Molecular weight
w_a = Airflow rate
X = Mole fraction
x = Number of carbon atoms in unburned hydrocarbon
x_j = jth axial coordinate
Y = Mass fraction
y = Number of hydrogen atoms in unburned hydrocarbon
z = Number of oxygen atoms in methanol combustion product
ΔH_c = Heat of combustion
η = Combustion completeness (efficiency)
ν = Moles of product
ν' = Moles reactant
ϵ = Number of carbon atoms in fuel
ϕ = Fuel-air equivalence ratio
ρ = Density

ACKNOWLEDGMENTS

The authors would like to acknowledge the assistance of Dr. T. E. Sharp who helped to develop the NO prediction computer program, N. Azelborn who programmed the data reduction analysis, and A. Kolb, N. L. Smith, G. W. Andrews, T. F. Amman, and M. D. Noldy for their assistance during the experimental phase of the study.

REFERENCES

1. Environmental Protection Agency, Rules and Regulations, Federal Register, Vol. 37, No. 221, Part II, Wednesday, Nov. 15, 1972.

2. W. R. Wade et al., "Low Emissions Combustion for the Regenerative Gas Turbine." Papers Nos. GT1173 and GT1273, Presented at the ASME Gas Turbine Conference, April 9, 1973.

3. W. R. Wade and W. Cornelius, "Emission Characteristics of Continuous Combustion Systems of Vehicular Powerplants–Gas Turbine, Steam, Stirling." Emissions from Continuous Combustion Systems. Eds. W. Cornelius and W. G. Agnew, New York: Plenium Press, 1972.

4. P. G. Parikh, R. F. Sawyer, and A. L. London, "Pollutants from Methane Fueled Gas Turbine Combustion." College of Engineering Rept. No. TS-70-15, Univ. of California, Berkeley, January 1971.

5. L. Caretto, R. E. Sawyer, and E. Starkman, "Formation of Nitric Oxide in Combustion Processes." College of Engineering Rept. No. TS-68-1, Univ. of California, Berkeley, March 1968.

6. P. Blumberg and J. T. Kummer, "Prediction of NO Formation in Spark-Ignited Engines–An Analysis of Method

of Control." Combustion Science and Technology, Vol. 4 (1971), pp. 73-95.

7. Ya. B. Zeldovich, P. Ya. Sadovnikov, and D. A. Frank-Kamenetskii, "Oxidation of Nitrogen in Combustion." Publishing House of the Academy of Sciences, USSR, Moscow, 1947. (Translated by M. Shelef, Scientific Research Staff, Ford Motor Co. Copies available from translator.)

8. C. T. Bowman and D. J. Seery, "Investigation of NO Formation Kinetics in Combustion Processes." Emissions from Continuous Combustion Systems. Eds. W. Cornelius and W. G. Agnew. New York: Plenium Press, 1972.

9. C. P. Fenimore, "Formation of Nitric Oxide in Premixed Hydrocarbon Flames." Thirteenth Symposium (International) on Combustion: The Combustion Institute, Pittsburgh, Pa. (1971), pp. 373-379.

10. EPA Gas Turbine Contractors' Coordination Meeting, Ann Arbor, Mich., December 1972. There are several corporations under contract to the Environmental Protection Agency to develop low emissions Brayton and Rankine cycle combustors. G. M. Thur of EPA is coordinator of gas turbine development.

11. C. W. LaPointe and W. L. Schultz, "Measurement of Nitric Oxide Formation within a Multi-fueled Turbine Combustor." Emissions from Continuous Combustion Systems. Eds. W. Cornelius and W. G. Agnew. New York: Plenium Press, 1972.

12. A. M. Keuthe and J. D. Schetzer, "Foundations of Aerodynamics." New York: John Wiley & Sons, 1959.

13. W. R. Marshall, "Atomization and Spray Drying." Chem. Eng. Prog. Series No. 2, Vol. 5, AIChE, New York, 1954.

14. S. S. Penner, "Chemistry Problems in Jet Propulsion." New York: Pergamon Press, 1957.

15. S. Way, "Combustion in the Turbojet Engine." Selected Combustion Problems II, AGARD, London: Butterworths, 1956.

16. A. G. Piken and C. H. Ruof, "Chemical Composition of Automobile Exhaust and A/F Ratio." Scientific Research Staff, Ford Motor Co., June 1968.

APPENDIX A

DATA REDUCTION–LOCAL VALUES

A gas sample is extracted at the point (r_i, x_j) within the combustor. Denoting measured mole fractions by an asterisk, the relations between measured and actual mole fractions are given by

$$3X^*_{HC} = X_{HC} \tag{A-1}$$

$$X^*_{NO} = X_{NO} \tag{A-2}$$

$$X^*_{O_2} = X_{O_2} \tag{A-3}$$

$$X^*_{CO_2} = X_{CO_2}/(1 - X_{H_2O}) \tag{A-4}$$

$$X^*_{CO} = X_{CO}/(1 - X_{H_2O} - X_{CO_2}) \tag{A-5}$$

The factor of 3 in the HC relation arises because the FID is calibrated using propane. CO_2 and CO measurements are taken on dried samples, and the CO sample has CO_2 removed by the use of ascarite. At the low CO concentrations of interest, the presence of CO_2 leads to an NDIR interference comparable to the indicated CO reading.

For a hydrocarbon fuel, the combustion reaction is assumed to be

$$\begin{aligned} \nu_1' C_\xi H_\eta + \nu_{O_2}' O_2 + 3.76\nu_{O_2}' N_2 &\longrightarrow \nu_2 C_x H_y \\ &+ \nu_{O_2} O_2 + \nu_{N_2} N_2 + \nu_{CO_2} CO_2 + \nu_{H_2O} H_2O \\ &+ \nu_{CO} CO + \nu_{NO} NO \end{aligned} \tag{A-6}$$

This relation ignores H_2 which may be present in amounts comparable to CO. In the past, 16 equilibrium of the water-gas shift reaction has been assumed and the H_2 computed from the concentrations of H_2O, CO_2, and CO. Since CO is always present in super-equilibrium amounts, as was seen in the section on experimental results, this equilibrum relation cannot be employed. It is theoretically possible to use the normalization condition, $\sum_\ell X_\ell = 1$, to solve for X_{H_2}, but this would entail differencing large numbers to obtain a number in the ppm range and is therefore not practical in view of the accuracy of the measurements involved. Neglecting H_2 introduces no serious error because it is a minor constituent over the major portion of the combustion chamber. The computation of the local A/F depends mainly on O_2 and CO_2 mole fractions.

Dividing Eq. A-6 by $\sum_{\ell} \upsilon_{\ell}$ expresses the products in terms of mole fractions.

$$n_1 \xi CH_a + n_{O_2} O_2 + 3.76\, n_{O_2} N_2 \longrightarrow xX_2 CH_b + X_{O_2} O_2 + X_{N_2} N_2 + X_{CO_2} CO_2 + X_{CO} CO + X_{H_2O} H_2O + X_{NO} NO \tag{A-7}$$

Here, the hydrogen-carbon atom ratios of the fuel and the unburned hydrocarbons have been designated a and b, respectively. For diesel fuel, a = 1.86 and b is assumed to be equal to a/2. The quantity xX_2 equals X_{HC} in Eq. A-1.

From the conservation of atomic species, H_2O and N_2 may now be computed:

$$X_{H_2O} = \frac{a}{2}\left[\left(1 - \frac{b}{a}\right) xX_2 + X_{CO_2} + X_{CO}\right] \tag{A-8}$$

$$X_{N_2} = \left[1.88 \quad 2X_{O_2} + 2X_{CO_2} + X_{CO} + X_{H_2O} + 0.734\, X_{NO}\right] \tag{A-9}$$

Substitution of Eqs. A-1-A-5 into these relations expresses X_{H_2O} and X_{N_2} in terms of measured quantities. This completes the chemical description of the gas mixture at a point for the case where a hydrocarbon mixture characterized by the ratio, a, reacts with air. As a check on accuracy, the sum, $\sum_{\ell} X_{\ell}$, was computed at each point. In general, it remained within 5% of unity for all points. The local molecular weight was computed from

$$W(r_i, X_j) = \sum_{\ell} W_{\ell} X_{\ell} (r_i, x_j) \tag{A-10}$$

and found to vary between 29 and 30 for all combustor locations which could be reliably sampled. (In the immediate vicinity of the fuel nozzle, there were indications that W exceeded 30.) For purposes of this computation, the hydrocarbon's molecular weight, xW_{HC}, was assumed to be 86 (hexane).

The local A/F is computed from Eq. A-6 and the atom conservation equations

$$\frac{A}{F}(r_i, x_j) = \frac{138 n_{O_2}}{\xi n_1 (12 + a)} = \frac{36.7}{12 + a}\left(\frac{X_{N_2} + X_{NO}/2}{xX_2 + X_{CO_2} + X_{CO}}\right) \tag{A-11}$$

In the case of methanol combustion, a similar treatment may be made assuming the following combustion reaction:

$$\nu_1' CH_3OH + \nu_{O_2}' O_2 + 3.76 \nu_{O_2}' N_2 \dashrightarrow \nu_2 C_x H_y O_z + \nu_{O_2} O_2 + \nu_{CO_2} CO_2 + \nu_{H_2O} H_2O + \nu_{CO} CO + \nu_{N_2} N_2 + \nu_{NO} NO \tag{A-12}$$

Here, for lack of chromatographic analysis, the unburned fuel is assumed to be methanol. The response of the FID to methanol is reduced by 25%, such that now,

$$4X_{HC}^{*} = X_{HC} \tag{A-13}$$

while the other measurements remain unchanged. Water and nitrogen are computed from:

$$X_{H_2O} = 2X_{CO_2} + 2X_{CO} + \left(x - \frac{y}{4}\right) X_2 \tag{A-14}$$

and

$$X_{N_2} = 1.88\left[3X_{CO_2} + 2X_{CO} + 2X_{O_2} + 0.734\, X_{NO} + \left(z - \frac{y}{4}\right) X_2\right] \tag{A-15}$$

For pure methanol, $x = z = \frac{y}{4}$, so that X_2 does not appear in these computations.

The local A/F for methanol combustion is given by

$$\frac{A}{F}(r_i, x_j) = 1.148\left(\frac{X_{N_2} + X_{NO}/2}{X_{CO_2} + X_{CO}}\right) \tag{A-16}$$

The emission index, EI_{ℓ}, of the exhaust component, ℓ, is defined as the mass of ℓ per unit mass of fuel burned. It is a convenient figure of merit by which to evaluate a particular combustor's emission performance and can be related to an absolute emission level in grams per mile given a fuel economy value in miles per gallon. It may be computed, on a local basis, from the foregoing measurements as follows:

$$EI_{\ell}(r_i, x_j) = \frac{1000\, W_{\ell}}{W} X_{\ell} (1 + A/F) \tag{A-17}$$

Here, EI_{ℓ} has units of grams ℓ per kilogram fuel.

These are point values. They are representative of conditions in a small volume surrounding the sampling point. The change in any quantity from point to point may be due to either chemical reaction or dilution and may not be determined unless the history of the fluid element passing through the sampling volume is known. Since the origin of all the fluid passing through the incremental sampling volume is unknown, one must resort to averages taken across the combustor to separate dilution from chemical effects and this may only be done if the average dilution profile is known as a function of axial distance. The latter may be estimated from cold flow measurement or obtained from the decrease in the cross-sectional mass-average value of a chemically inert tracer gas. NO satisfies this criterion throughout much of the combustor. It may also be estimated from the cross-sectional average A/F derived in Appendix B which deals with the relationship between point values and averages taken in a cross-sectional plane.

APPENDIX B

DATA REDUCTION–CROSS-SECTIONAL AVERAGES

Associated with each data point (r_i, x_j) is an area, S_i, in the x_j plane, over which the measured values are assumed constant. For the data network employed here, the magnitude of S is completely specified by the radial coordinate alone. Through S_i, the net mass flux in the x_j direction is given by

$$\dot{m}(r_i, x_j) = \rho(r_i, x_j)\, V(r_i, x_j)\, S_i \qquad \text{(B-1)}$$

The net mass flux of species, ℓ, through S_i is similarly

$$\dot{m}_\ell(r_i, x_j) = \rho_\ell(r_i, x_j)\, V(r_i, x_j)\, S_i \qquad \text{(B-2)}$$

where the diffusion velocity of species, ℓ, has been neglected compared to the mixture mass-average velocity, $V(r_i, x_j)$.

The local mass fraction of species, ℓ, is by definition

$$Y_\ell(r_i, x_j) = \frac{\dot{m}_\ell(r_i, x_j)}{\dot{m}(r_i, x_j)} = \frac{W_\ell X_\ell(r_i, x_j)}{W(r_i, x_j)} \qquad \text{(B-3)}$$

The total mass flux of species, ℓ, through the entire cross section is the sum over all the S_i. Introducing the perfect gas law and neglecting the small pressure change with r_i, this becomes

$$\dot{m}_\ell(x_j) = \frac{P}{R}\sum_i \frac{W_\ell X_\ell(r_i, x_j)}{T(r_i, x_j)}\, S_i V(r_i, x_j) \qquad \text{(B-4)}$$

The total mass flux of all species is obtained by summing this relation over the index, ℓ.

$$\dot{m}(x_j) = \frac{P}{R}\sum_i \frac{W(r_i, x_j)}{T(r_i, x_j)}\, S_i V(r_i, x_j) \qquad \text{(B-5)}$$

Velocity is not measured, but we may, without loss of generality, define a cross-sectional average velocity

$$V(x_j) = \frac{R\dot{m}_\ell(x_j)}{P\sum_i \frac{W_\ell X_\ell(r_i, x_j) S_i}{T(r_i, x_j)}} \qquad \text{(B-6)}$$

and a similar one in terms of the total mixture mass flux. The cross-sectional average mass fraction of species, ℓ, is now defined as the ratio of Eqs. B-4 and B-5.

$$Y_\ell(x_j) = \frac{\sum_i \frac{W_\ell X_\ell(r_i, x_j) S_i}{T(r_i, x_j)}}{\sum_i \frac{W(r_i, x_j) S_i}{T(r_i, x_j)}} \qquad \text{(B-7)}$$

Here, it has been assumed that $V = V_\ell$. This is an approximation which should be increasingly accurate as x_j increases.

This is the desired expression. It relates the average mass fraction to the locally measured mole fractions and the measured local temperature. It is essentially a mass-weighting of the individual point values to yield an average mass fraction which can then be used in conjunction with an independently obtained $m(x_j)$ to calculate

$$\dot{m}_\ell(x_j) = Y_\ell(x_j)\, \dot{m}(x_j) \qquad \text{(B-8)}$$

The average cross-sectional temperature may be similarly related to the locally measured temperatures. In this case, one considers the flux of enthalpy through S_i, sums over S_i to get the total enthalpy flux, assumes that mass and energy convect with the same average velocity and arrives at an expression for cross-sectional average enthalpy in terms of local temperatures

$$h(x_j) = \frac{\sum_i \frac{W(r_i, x_j) S_i h(r_i, x_j)}{T(r_i, x_j)}}{\sum_i \frac{W(r_i, x_j) S_i}{T(r_i, x_j)}} \quad \text{(B-9)}$$

In the present data reduction program, the specific heat is assumed constant so that this expression reduces to

$$T(x_j) = \frac{\sum_i W(r_i, x_j) S_i}{\sum_i \frac{W(r_i, x_j) S_i}{T(r_i, x_j)}} \quad \text{(B-10)}$$

One definition of cross-sectional average A/F is

$$1 + \frac{A}{F}(x_j) = \frac{\dot{m}(x_j)}{\dot{m}_f(x_o)} \quad \text{(B-11)}$$

where it has been assumed that all the fuel enters the combustor at the plane x_o. In this form, A/F (x_j) may not be computed from the individually measured local A/F ratios without knowledge of the fuel distribution across the plane x_j. This may be seen by defining the local A/F in terms of the individual fluxes passing through S_i and summing as with $Y_\ell(x_j)$. There results the following expression:

$$1 + \frac{A}{F}(x_j) = \frac{\sum_i \frac{W(r_i, x_j) S_i}{T(r_i, x_j)}}{\sum_i \frac{\dot{m}_f(x_o)}{\dot{m}_f(r_i, x_o)}}$$

$$\sum_i \left\{ \frac{T(r_i, x_j)}{W(r_i, x_j) S_i} \left[1 + \frac{A}{F}(r_i, x_j) \right] \right\} \quad \text{(B-12)}$$

Everything in this expression relating A/F (x_j) to A/F (r_i, x_j) is known except the sum in the denominator which may be termed the fuel distribution factor, FD. It is the sum of the total fuel flow divided by the individual fuel fluxes passing through each S_i. For uniform fuel flux density $(\rho_f V_f)$ distribution, FD = 109 in the present data grid system. For parabolic fuel flux density distribution (maximum density near the centerline), FD = 49. The fuel distribution factor is greater than 109 and approaches infinity if the centerline flux density approaches zero (the fuel is concentrated near the walls).

Using the basic definition of A/F, the fuel distribution factor may be calculated by substitution in Eq. B-12. That is,

$$\frac{A}{F}(x_j) = \frac{\dot{m}_a(x_j)}{\dot{m}_f(x_o)} \quad \text{(B-13)}$$

Expressing $\dot{m}_a(x_j)$ and $\dot{m}_f(x_o)$ in terms of sums of individual fluxes, employing the perfect gas law and noting that the mass fractions of air and of fuel are

$$Y_a = \frac{1}{1 + \frac{F}{A}} \quad \text{(B-14)}$$

and

$$Y_f = \frac{1}{1 + \frac{A}{F}} \quad \text{(B-15)}$$

respectively, one obtains for the cross-sectional average A/F:

$$\frac{A}{F}(x_j) = \frac{\sum_i \frac{WVS}{T\left[1 + \frac{F}{A}\right]}}{\sum_i \frac{WVS}{\left[1 + \frac{A}{F}\right]}} \quad \text{(B-16)}$$

where all variables within the summation have locally measured point values.

As before, an average velocity may be defined in terms of the above summations and their respective overall fluxes. Thus, for example, the mass average fuel velocity may be written

$$v_f(x_j) = \dot{m}_f(x_o) \left\{ \frac{P}{R} \sum_i \left[\frac{WS}{T\left(1 + \frac{A}{F}\right)} \right] \right\}^{-1} \quad \text{(B-17)}$$

and an analogous expression for $V_a(x_j)$. If it is assumed as an approximation that $V_f(x_j) = V_a(x_j)$, then A/F (x_j) may be calculated from the sum of the locally measured A/F (V_i, x_j) as given by

$$\frac{A}{F}(x_j) = \frac{\sum_i \frac{WS}{T(1 + F/A)}}{\sum_i \frac{WS}{T(1 + A/F)}} \quad \text{(B-18)}$$

Thus A/F (x_j) and FD (x_j) may both be obtained from locally measured values using Eqs. B-12 and B-18 subject to the approximation that the defined average velocities are roughly equal.

Methanol as a Gasoline Extender - Fuel Economy, Emissions, and High Temperature Driveability*

Eric E. Wigg and Robert S. Lunt

The possibility of using alcohols as motor fuels, either in the "pure" form or in gasoline blends, has been considered from time to time for over forty years[(1)]*. The current energy shortage and high price of crude oil has generated renewed interest in this use for alcohols, particularly methanol, since it can be produced from coal[(2)] at costs which have been estimated to be roughly comparable to those for other coal-derived synthetic fuels[(3)]. In the past, the product-quality disadvantages associated with methanol-gasoline blends; phase separation, increased volatility, and corrosion problems, have always vastly outweighed any potential advantages. However, in today's climate of energy shortages and environmental awareness, the idea is being reevaluated. Experimental results have been published[(4)] which suggest that significant benefits in fuel economy, emissions, and performance are possible with methanol-gasoline blends. These blends have also been found to have increased octane quality compared to similar blends without methanol[(5)].

Firm data dealing with fuel economy and emissions from vehicles using methanol-gasoline blends are very limited. And in general, product quality aspects as they relate to today's vehicles have not been evaluated. The study described here was designed to provide data on fuel economy and emissions using a cross-section of vehicle types with known carburetion characteristics, tested under controlled conditions. Experiments were also conducted to establish the magnitude of the fuel volatility problem as it relates to high temperature driveability. Tests relating to phase separation in the presence of water were also carried out.

It should be noted that the use of "pure" methanol as a motor fuel is a completely separate issue and beyond the scope of the paper. It can be stated, however, that because methanol and gasoline cannot be used interchangeably, due to widely different carburetion requirements, any use of "pure" methanol would be limited to fleet operation, at least for the foreseeable future.

*Numbers in parentheses designate references which appear at end of paper.

*Paper 741008 presented at the Automobile Engineering Meeting, Toronto, Canada, October 1974.

ABSTRACT

Methanol's potential as a gasoline extender has been evaluated, with data being obtained in the areas of fuel economy, exhaust emissions, and driveability. The results of tests with three cars, having carburetion spanning the range normally encountered in the existing car population, showed that methanol's effect on fuel economy and emissions could be directly related to its leaning effect on carburetion. The data suggest that any benefits in these two areas would only be significant for older, rich-operating cars.

A 13-car driveability study indicated that the large increase in fuel volatility which occurs with the addition of methanol to gasoline could pose serious problems. A marked increase in vapor locking tendency was observed when no front-end volatility adjustments were made to the methanol blends. Stretchiness, a lack of expected response to throttle movement, was also found with the methanol blends. This operational characteristic, being related to excessively lean operation, was more pronounced with the newer cars tested.

Phase separation is also a potential problem with methanol-gasoline blends. Data are presented which show the effect of including higher molecular weight alcohols along with the methanol. Phase separation still occurred in the presence of less than 1% water.

Taken as a whole, the data suggest that, if it becomes available in large quantities, the use of methanol in applications other than in motor gasoline would be preferred.

SAE/PT-80/19/$02.50

EXPERIMENTAL

Fuels

A number of test fuels, all unleaded, with and without methanol, were used in the program. These can be divided into two groups: those used in the fuel economy and emissions tests, and those used in the volatility studies.

Three fuels were used to study the effect of methanol on fuel economy and emissions. A typical unleaded fuel containing no methanol was used as a base gasoline. One methanol blend was prepared by adding 15% methanol, by volume, directly to this base blend. The other methanol blend (also 15% methanol) was adjusted to give the same RVP as the base blend. This required backing out all the butane and half the pentanes from the base blend before adding the methanol. It can be seen from the fuel specifications given in Table 1, that the Reid vapor pressure of the non-matched RVP methanol blend is much higher than that of the base blend. 15 percent was chosen as the blending concentration since this is generally considered to be the maximum tolerable for typical gasolines from a solubility point of view.

Tabel 1 – Fuel Properties
Fuel Economy and Emissions Testing

	Base Blend	Matched RVP MeOH Blend	Non-Matched RVP MeOH Blend
Reid Vapor Pressure (RVP)	11.9	11.7	16.0
% Distilled at °C			
70 (158°F)	30	50	55
100 (212°F)	48	53	59
150 (302°F)	87	87	89
Gravity (g/cc)	0.76	0.78	0.76
% MeOH by volume	0	15.0	15.0

Eight fuels were used in the experiments to determine methanol's effect on driveability. Testing was carried out at 21°C (70°F) and 38°C (100°F), with a separate set of four fuels used at each temperature, consisting of a base hydrocarbon fuel and three fuels containing 15% methanol. The volatility of one of the methanol blends was adjusted to give the same RVP as the base blend. The second methanol blend was matched with the base blend according to vapor lock index (VLI)* and the third methanol blend had no volatility adjustments, being prepared by adding the methanol directly to the appropriate base blend. A further description of these fuels is given in the section of the paper dealing with volatility effects.

The complete fuel inspections and compositions for all the test fuels are given in Appendix 1.

*VLI = RVP + 0.13% distilled @ 70°C (158°F)

Vehicles

As with fuels, the vehicles are most conveniently described according to the two phases of the test program. The fuel economy and emissions tests were carried out with three vehicles, having carburetion spanning the range normally encountered in the field. Carburetion was viewed as the most important vehicle characteristic, since theoretical considerations suggest that methanol's influence on fuel economy should be primarily due to its leaning effect on carburetion. 1967, 1973 and a catalyst-equipped 1973 vehicle were considered representative of pre-emission control, current and near future vehicle systems respectively. The vehicle characteristics are given in Appendix 2.

The driveability studies included tests on 13 cars selected from the 1967 through 1974 model years. There was no attempt to match the vehicles chosen with current vehicle population. However, the inclusion of cars with and without exhaust emission controls provided information on how these two broad classes performed with methanol-containing fuels. A complete listing of vehicle characteristics is given in Appendix 2.

Test Procedures

The fuel economy and emissions tests were carried out using the 1975 Federal Test Procedure (FTP)[6]. Data were obtained for each of the three portions of this test. Tests with a given vehicle were run back-to-back, with each fuel being run at least twice. The reproducibility of the fuel economy data, expressed as percent deviation from the average in any set of replicate runs, was about 1%.

The vapor lock test procedure was an adaptation of the CRC track test procedure[7] for use on a chassis dynamometer. The driving schedule, shown in Table 2, simulates both bumper-to-bumper heavy city driving and interstate operation. Vehicle temperatures were stabilized by driving at 55 mph, after which a base acceleration time was obtained for the fully warmed up vehicle. Accelerations were then made after a 20 minute heat soak and also after a 10 minute idle period. Hot starting characteristics were determined by measuring the cranking time. The hot start was made in accord with the instructions in the manual for each vehicle. Stretchiness, the lack of anticipated response to throttle movement, was measured by a change in acceleration time for a 32-80 km/hr. (20-50 mph) acceleration under fixed manifold pressure.

To obtain meaningful data in this type of study, care must be taken to obtain realistic fuel system operating temperatures. The location and speed of the cooling fan in the controlled temperature facility was adjusted to give fuel system temperatures in line with those reported in a recent study[8]. It was necessary to use a fuel supply other than the car tank to facilitate the many fuel changes which had to be made during the course of the fuel volatility testing. An auxiliary fuel can, placed at about the same level as the fuel tank, was connected to the car fuel line at the outlet from the vehicle tank. The can was vented into the line to the vehicle's carbon canister. If present on the car, the vapor return line was also connected to the can.

Table 2 – High Temperature Driveability Procedure for Controlled Temperature Room

Step No.	
1.	**Warm-Up** – 1st run of day – 15 min. at 89 km/hr (55 mph)
2.	**Charge Test Fuel**
3.	**Temp. Stabilization** – 5 min. at 89 km/hr
4.	**Base Acceleration Time** – 24-113 km/hr full throttle (15-70 mph) 2 accels
5.	**Temp. Stabilization** – 5 min. at 89 km/hr
6.	**Idle** – 1 min. drive
7.	**Shutdown** – 20 min.
8.	**Startup** – Depress throttle ⅓ way and hold. Record cranking current.
9.	**Idle** – 1 min. drive
10.	**Acceleration Time** – 24-113 km/hr full throttle (15-70 mph) 2 accels.
11.	**Temp. Stabilization** – 5 min. at 89 km/hr (55 mph)
12.	**Idle** – 10 min. in drive – brake on
13.	**Acceleration Time** – repeat of 10.
14.	**Stretchiness** – 32-80 km/hr (20-50 mph) at 8, 10 and 12 inches man. vac.
15.	**Charge next test fuel** and go to Step 3.

Analytical Methods

Mass emissions of CO, HC, and NO_x were determined using conventional instrumentation (non-dispersive IR for CO, flame ionization detection for HC, and chemiluminescent reaction with ozone for NO_x). Total aldehydes were measured using the 3-methyl-2-benzothiazolone hydrazone hydrochloride (MBTH) method. Aldehyde composition was determined using a gas chromatographic procedure on the dinitrophenylhydrazine (DNPH) derivatives. The methods for aldehydes are described in detail in reference 9. Individual hydrocarbon concentrations were determined by the gas chromatographic procedure, also outlined in reference 9.

RESULTS AND DISCUSSION
FUEL ECONOMY AND EMISSIONS

Fuel Economy

The fuel economy and emissions study was carried out in two phases. The first phase involved a back-to-back comparison of the base blend with the matched RVP methanol blend for all three cars. A minimum of two runs for each fuel-vehicle combination was completed, the tests with a given vehicle being run in a block, with random fuel selection. The second phase involved a similar series of tests where the base blend was compared with the high RVP methanol blend, the blend prepared by adding the methanol directly to the base blend with no volatility adjustments.

The data obtained in the two phases are given in Table 3. The effect of methanol addition is shown in the two right hand columns of the table on a volume basis as well as an energy basis. Methanol prices are normally quoted in terms of cost per unit energy, so the latter value may be of primary interest. From the customer's point of view, however, the volume figures are probably more relevant.

Table 3 – Fuel Economy During the 1975 Federal Test Procedure

Fuel	Car	Fuel Economy, Km/Liter: Base	Fuel Economy, Km/Liter: 15% MeOH	Effect of MeOH: Km/Liter	Effect of MeOH: Km/Unit Energy
Matched RVP	1967	6.05(14.3)*	6.09(14.4)	+1%	+8%
	1973	4.74(11.2)	4.44(10.5)	−6%	+1%
	Catalyst	4.86(11.5)	4.61(10.9)	−5%	+2%
Non-matched RVP	1967	6.05(14.3)	5.71(13.5)	−6%	+1%
	1973	5.12(12.1)	4.82(11.4)	−6%	+1%
	Catalyst	4.86(11.5)	4.57(10.8)	−6%	+1%

*Miles/gal. data shown in parentheses

In order to understand the effect of methanol on fuel economy, it is useful to examine first the data from the tests with matched RVP fuels. These data are free from any effects due to volatility differences, which previous studies[10] have shown can have a significant bearing on CO and HC emissions and possibly on fuel economy as well.

It can be seen that, on a volume basis, the 1967 car gave about the same fuel economy with the two fuels, while the 1973 car and the catalyst-equipped car showed a significant debit. Since methanol's energy content per unit volume is about half that of gasoline, expressing fuel economy on an energy basis shows an improvement for the 1967 car and very little change for the other two cars.

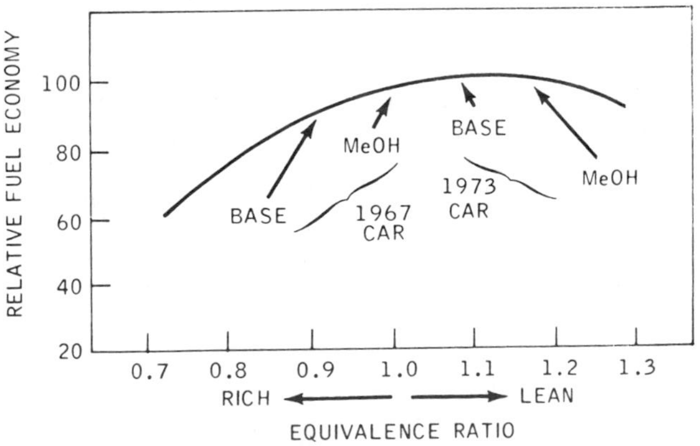

Fig 1 - Relative fuel economy as a function of carburetion

These fuel economy data agree well with the fuel economy changes which would be expected from methanol's leaning effect on carburetion. Figure 1 shows the relationship between fuel economy and equivalence ratio for a typical gasoline engine[11]. Equivalence ratio here is defined as the weight ratio of air to fuel, with the stoichiometrically correct ratio set at 1.0. Fuel-rich carburetion gives equivalence ratios less than 1.0, with the opposite being true for fuel-lean carburetion. The addition of 15% methanol to gasoline results in an increase in the equivalence ratio of about 0.1 units.

It can be seen from Figure 1 that maximum fuel economy from an equivalence ratio point of view, occurs on the fuel-lean side of the stoichiometric point (~1.1). Since methanol has a leaning effect on carburetion, a car operating at or on the lean side of this maximum with gasoline should experience a decrease in fuel economy, on an energy basis, with the addition of methanol to the gasoline. Conversely, a car operating on gasoline below about 1.05 equivalence ratio, e.g. fuel rich, should benefit as a result of methanol's leaning effect.

The operating equivalence ratios for the 1967 and 1973 cars were calculated from raw exhaust concentration data obtained at various steady state conditions. Similar measurements were not made with the catalyst car because of air injection at the exhaust valves. The data for the 1967 and 1973 cars are shown in Table 4. The 1967 car, under most conditions, was net rich, while the 1973 car was net lean. The catalyst-equipped car had carburetion adjusted slightly richer than the standard 1973 car tested (0.056" main carburetor jets in place of 0.054"). Its operating equivalence ratio would probably be slightly net lean.

Table 4 – Equivalence Ratio Data with the Base Fuel

Speed, km/hr.	1967 Car	1973 Car
Idle	0.94	1.04
32 (20)*	0.90	1.05
48 (30)	0.88	1.08
64 (40)	0.95	1.08
81 (50)	1.01	1.05

*Numbers in parentheses in miles/hr.

An examination of the fuel economy vs. equivalence ratio relationship shown in Figure 1, in the light of the operating equivalence ratios of the three cars, indicates that the addition of methanol, with its resultant increase in equivalence ratio, should yield an increase in fuel economy, on an energy basis, for the 1967 car and very little change for the 1973 car. The fuel economy data given in Table 3 are in excellent agreement with the predicted changes, suggesting that the effect of methanol on fuel economy is primarily due to its influence on the operating air-fuel ratio. The data, therefore, imply that fuel economy benefits would not be expected for vehicles carburetted net lean with gasoline. This would rule out significant benefits for most 1968 and later vehicles.

The results from the second set of experiments, where the base blend was compared to the methanol blend prepared by adding 15% methanol directly to the base blend, shows one interesting difference. The 1967 car no longer experiences a significant fuel economy benefit with the methanol blend. The high volatility of this fuel apparently led to carburetor enrichment during warmed up operation, counterbalancing the benefits realized from methanol's leaning effect. As will be discussed in the emissions section, CO data support this conclusion. A similar effect was not apparent in the case of the 1973 car and the catalyst-equipped car. The extent of this volatility effect in normally rich-operating cars cannot be estimated from data on one car. It can be stated, however, that benefits beyond those observed with the matched RVP fuels would not be expected for higher RVP methanol blends.

Mass Emissions of CO, HC, and NO_X

As was the case with fuel economy, methanol's effect on emissions can be explained on the basis of equivalence ratio considerations. Figure 2 shows the general relationships which the various pollutants follow as equivalence ratio is changed[12]. It would be expected, then, based on the equivalence ratio data given earlier, that the leaning effect of methanol should result in a large decrease in CO emissions for the 1967 car, operating at about 0.9 equivalence ratio on the base fuel, and a lesser effect for the 1973 car, operating between 1.0 and 1.1. The CO emissions from the catalyst car would not be expected to be significantly affected. As for hydrocarbons, a significant drop in HC emissions from the 1967 car would be expected with the methanol blend, but very little change in the case of the other two cars. The NO_x curve in Figure 2 predicts an increase in NO_x emissions for the 1967 car and very little change for the other two cars.

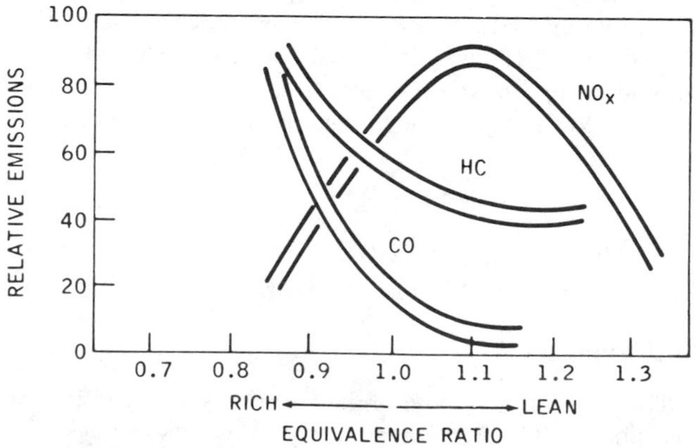

Fig 2 - Relationship between equivalence ratio and exhaust emissions

The mass emissions data are displayed in Figures 3, 4 and 5. Generally speaking, the matched RVP fuels gave data in good agreement with the predicted changes associated with methanol's leaning effect on carburetion. The only apparent anomoly was the observed decrease in NO_x emissions for the 1973 and catalyst cars. It is possible that this decrease is due to the high latent heat of vaporization of methanol, which could result in lower peak flame temperatures in the combustion chamber.

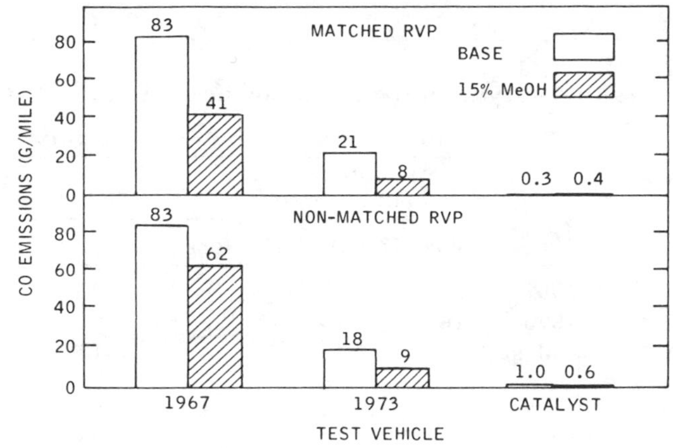

Fig 3 - Effect of methanol on 1975 FTP CO emissions

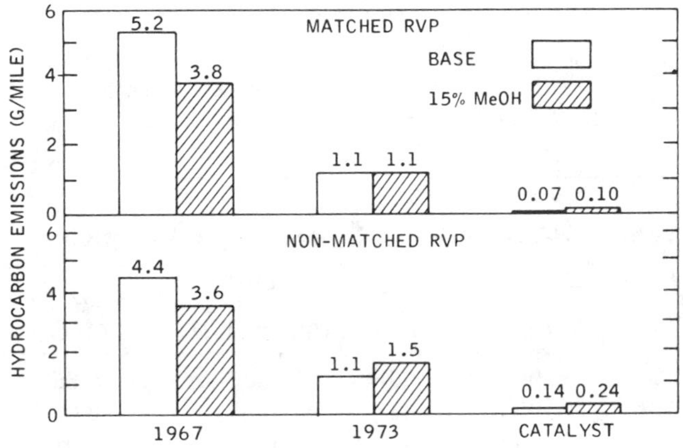

Fig 4 - Effect of methanol on 1975 FTP hydrocarbon emissions

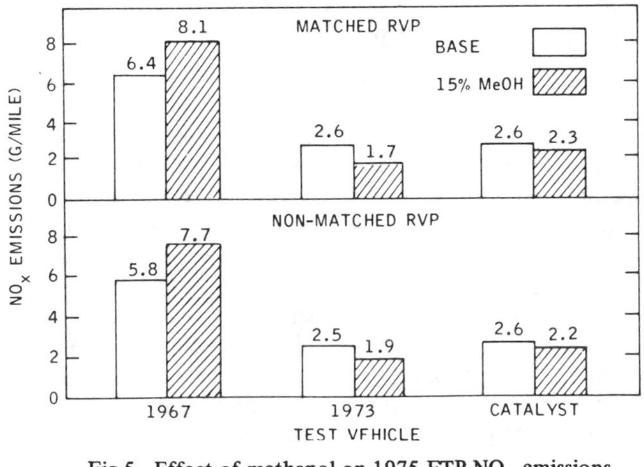

Fig 5 - Effect of methanol on 1975 FTP NO_x emissions

As was the case with fuel economy, adding methanol to the base blend, with no volatility adjustments, partially nullified the benefits observed with the 1967 car. As can be seen in Figures 3 and 4, the decrease in CO and HC emissions found due to methanol was only about half that realized with the matched RVP blend. An examination of the data from the three segments of the 1975 Federal Test Procedure shows that this was largely due to higher emissions during warmed up operation. Table 5 shows the percentage decrease in CO and HC emissions with methanol for the two blends, broken down according to test segment. The first segment is the cold start portion and segments 2 and 3 involve warmed up operation.

Table 5 – Effect of Methanol on CO and HC Emissions from the 1967 car for the three parts of the 1975 FTP

	Percentage Decrease Due to Methanol			
	CO		HC	
Test Segment	RVP = 12*	RVP = 16	RVP = 12	RVP = 16
1	41	39	27	29
2	52	19	27	13
3	57	27	25	8

*RVP matched with base blend RVP

It can be seen that for both CO and HC, the cold start data were very little affected by the difference in fuel volatility. However, the high RVP fuel gave much less improvement than the fuel with matched RVP during warmed up operation. These data strongly suggest that the high front-end volatility is causing carburetor enrichment during warmed up operation.

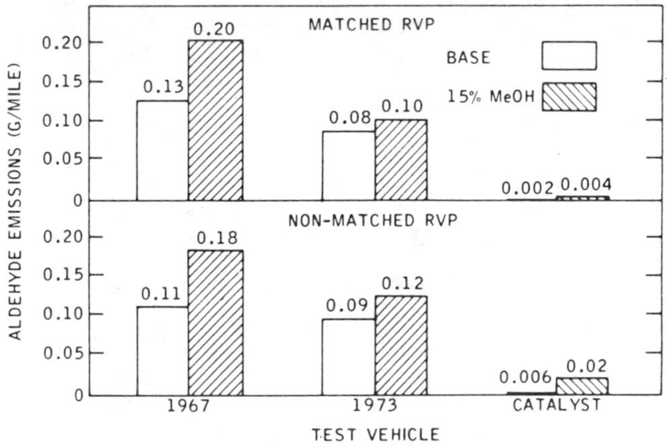

Fig 6 - Effect of methanol on 1975 FTP aldehyde emissions

Aldehyde Emissions

Aldehyde emissions from current gasoline-powered vehicles account for about 5-10% of the total organic emissions. These partially oxidized species, as a class, are considered to be about as photochemically reactive as olefins[13]. Formaldehyde, the predominant aldehyde compound in auto exhaust, is an eye irritant.

The addition of methanol to gasoline would be expected to lead to an increase in formaldehyde emissions. Methanol, being a partially oxidized species itself, can readily undergo reaction to give formaldehyde. The total aldehyde emissions data, as measured by MBTH, are presented in Figure 6. The data show, on a percentage basis, a significant increase in aldehyde emissions for all cars. There was no marked difference between the behavior of the non-matched RVP methanol blend and the matched RVP blend except possibly in the case of the catalyst car, where the observed increase in aldehyde emissions was considerably greater with the high RVP blend.

Aldehyde compositional analyses showed the observed increase in aldehyde emissions with the methanol blends could be accounted for on the basis of an increase in formaldehyde emissions. This is consistent with the observation made earlier that formaldehyde is a likely partial oxidation product of methanol. It is not clear whether this observed increase in formaldehyde emissions would lead to an environmentally significant problem if the use of methanol blends were to become wide-spread.

Methanol Emissions

Unburned methanol emissions were determined in the test series with matched RVP fuels. Methanol emissions would only be of concern if they approached high levels, since methanol is relatively unreactive in the atmosphere[14]. Using a methanol response factor of 0.85 for the flame ionization detector[15], it was found that unburned methanol concentrations in the exhaust were on the same order of magnitude as the formaldehyde emissions. Table 6 shows the methanol data expressed as ppmV and compares them with formaldehyde and total hydrocarbon ppmV concentrations. These levels would not appear to pose any significant problems.

Table 6 – Methanol, Formaldehyde, and Total Hydrocarbon Emissions During 1975 FTP – Matched RVP Fuel

	Raw Exhaust Concentration (ppm V)		
Car	MeOH	HCHO	Total HC
1967	72	44	500
1973	11	22	170
Catalyst	0.8	2.4	24

Hydrocarbon Composition

Individual hydrocarbon concentrations were determined on exhaust samples collected during the 1975 FTP The olefin and aromatic class hydrocarbon data are given in Table 7. These data refer to the composition of the exhaust hydrocarbon fraction and, as such, do not reflect the large differences in mass emissions from car to car. It can be seen that the presence of methanol in the blend leads to relatively small changes in aromatic and olefin content of the exhaust hydrocarbon, with no consistent trends apparent.

The hydrocarbon compositional data were used to calculate total exhaust hydrocarbon reactivity for the various tests. A composite reactivity scale[16] was used, whereby each individual hydrocarbon was given a reactivity value. Total reactivity was calculated by summing the individual "concentration x reactivity factor" terms with hydrocarbon concentrations expressed as mole %. The values are shown in Table 8.

Table 7 – Hydrocarbon Compositional Data

		Mole % Hydrocarbon Composition			
		Aromatics minus benzene*		Olefins	
	Car	Base	15% MeOH	Base	15% MeOH
	1967	29	35	18	20
Matched	1973	19	24	22	20
RVP	Catalyst	15	18	10	13
Non-	1967	32	29	17	20
matched	1973	22	25	21	22
RVP	Catalyst	10	20	11	14

*Benzene not included because of its low photochemical reactivity

Table 8 – Total Exhaust Hydrocarbon Reactivity

		Total Reactivity		Change Due
	Car	Base	15% MeOH	to MeOH
Matched	1967	2.95	2.29	−0.66
RVP	1973	0.64	0.61	−0.03
	Catalyst	0.03	0.04	+0.01
Non-	1967	2.50	2.17	−0.33
matched	1973	0.66	0.90	+0.24
RVP	Catalyst	0.04	0.12	+0.08

The values in Table 8 reflect the mass emissions from the various cars, as well as the reactivity of these emissions. Because of the relatively small effect of methanol on hydrocarbon composition, the major factor influencing total reactivity is the mass emissions factor. Significant decreases in reactive emissions were only observed for the 1967 car, with the effect being greatest for the matched RVP fuel. The combination of a small increase in reactivity and an increase in hydrocarbon mass emissions in the case of the 1973 car, using the non-matched RVP methanol blend, resulted in a significant increase in total hydrocarbon reactivity.

A consideration of the emissions data as a whole suggests that adding methanol to gasoline will provide meaningful benefits only in the case of older, rich-operating cars. These benefits, which involve substantial reductions in CO and hydrocarbon emissions are obtained at the expense of increased NO_x and formaldehyde emissions. The observed effects of methanol on emissions are in good agreement with theoretical expectations.

PRODUCT QUALITY CONSIDERATIONS

Volatility

Methanol has a marked effect on the volatility of methanol-gasoline blends. The introduction of small quantities of polar methanol into non-polar gasoline results in large increases in fuel vapor pressure. This vapor pressure effect is demonstrated in Figure 7 where it can be seen that adding as little as 2 percent methanol gives a 3 psi increase in RVP, despite the fact that the volatility of pure methanol is much less than that of typical gasolines. When added to gasoline, the methanol behaves more nearly as one would predict from its molecular weight, the non-polar gasoline breaking up the hydrogen bonds which exist in the pure state.

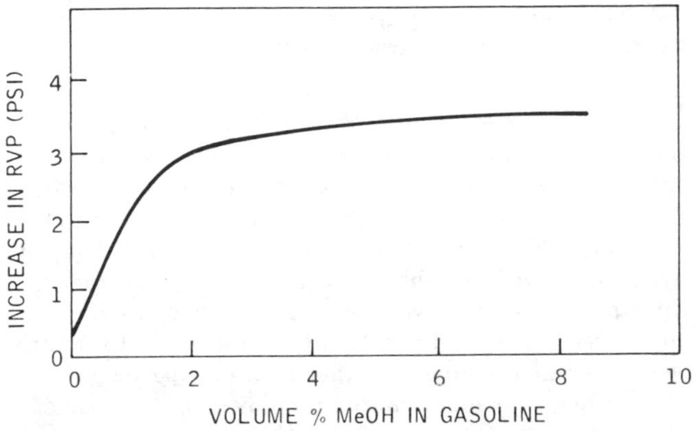

Fig 7 - Effect of methanol on gasoline vapor pressure

In addition to its effect on RVP, methanol markedly changes the shape of the fuel distillation curve. Figure 8 shows a curve for a 15% methanol-gasoline blend, compared to the same fuel without methanol. It can be seen that the presence of the methanol leads to a much greater fraction distilled at 70°C (158°F).

A fuel's tendency to give high temperature driveability problems, such as vapor lock, has been satisfactorily predicted by an expression which involves both parameters discussed above, RVP, and percent distilled at 70°C:

$$VLI^{*} = RVP + 0.13\% \text{ dist. @ } 70°C$$

*vapor lock index

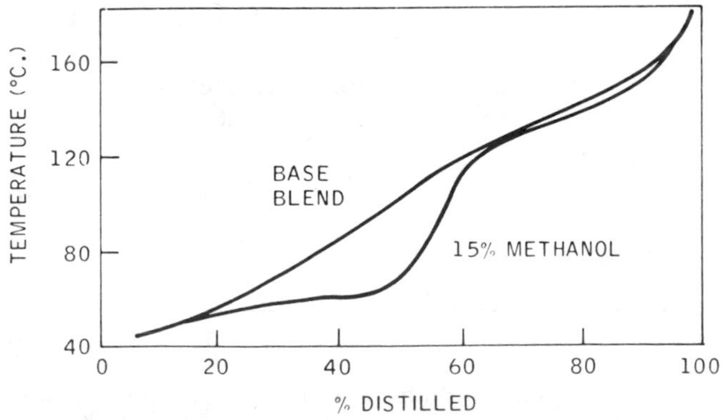

Fig 8 - Effect of methanol on fuel distillation characteristics

The higher the VLI, the greater the high temperature driveability problems. From the foregoing discussion, it can be seen that adding methanol to gasoline will give a substantial increase in VLI.

The experiments carried out in this area were designed to determine if the VLI equation, which had been developed for gasolines, could also be applied to methanol gasoline blends. Tests were made at 21°C (70°F) and at 38°C (100°F), with a separate set of four fuels being blended for use at each test temperature. The base fuel of each set was a typical full boiling gasoline, having VLI adjusted to provide only borderline protection for the test temperature in question. Based on past experience, this fuel would be expected to give vapor lock problems in 10 to 30% of the cars tested. The other three fuels all contained 15% methanol. One was blended to give the same VLI as the base blend, which required removal of some of the front end hydrocarbon components. The second fuel was prepared by adding the methanol directly to the base blend which gave very high VLI. The third methanol blend represented a compromise between the matched VLI and very high VLI of the first two blends. This fuel was blended to give the same RVP as the base blend. Relevant fuel data are shown in Table 9.

A comparison of car performance on the base fuel with that found with the VLI-matched blend should indicate whether the VLI equation is valid for methanol blends. Tests with the other two fuels gives a measure of the severity of any problems which could be encountered if the vapor pressure of these blends were not controlled.

Table 9 – Fuel Properties – Driveability Testing

21°C (70°F) Test Blends

	(1) Base Blend	(2) Matched VLI Blend	(3) Matched RVP Blend	(4) No Constraints
Reid Vapor Pressure	13.6	11.4	13.3	15.7
Vapor Lock Index (VLI)	18.9	18.7	19.7	23.7
% MeOH by volume	0	15.0	15.0	15.0

38°C (100°F) Test Blends

	(5) Base Blend	(6) Matched VLI Blend	(7) Matched RVP Blend	(8) No Constraints
Reid Vapor Pressure (RVP)	10.4	8.3	10.7	12.8
Vapor Lock Index (VLI)	13.9	13.7	17.9	19.6
% MeOH by volume	0	15.0	15.0	15.0

The major malfunction found during the tests was acceleration vapor lock, which was defined as (a) hesitations or bucking which resulted in at least 25% longer acceleration times, or (b) stalling during accelerations. Difficult hot starting and rough hot idle were only of low frequency in the test fleet. Those vehicles which had problems starting were the oldest vehicles tested and these occurrences may have been more age-related than fuel-related.

The acceleration vapor lock data are presented in Figures 9 and 10. Figure 9 shows the number of actual problems encountered with the various fuels. Vapor lock performance was measured three times for each of the 13 cars, making a total of 42 occurrences possible, if difficulties were encountered in each test. Within the experimental limitations of the test, the VLI-matched

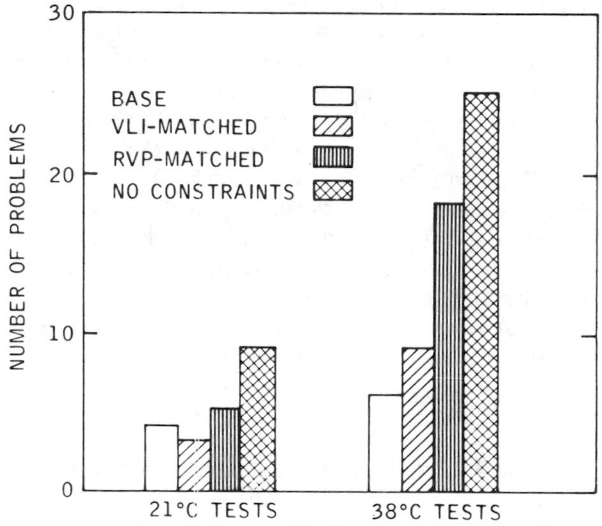

Fig 9 - Number of vapor lock problems encountered as a function of fuel at two test temperatures

methanol blends behaved about the same as the base blend, indicating that the VLI equation is applicable to methanol blends. This means that the other two methanol blends should lead to more problems than found with the base blend, which was observed, particularly at the higher test temperature.

Figure 10 shows the number of cars which showed vapor lock in the various tests. As would be predicted, the increase in problems, as VLI increased, involved a greater number of cars.

The fact that more problems were encountered at 38°C than at 21°C implies that the VLI target of 13.7 used for the 38°C blends was somewhat more restrictive than the 18.3 target used for the 21°C blends.

The only other warm vehicle difficulty which was encountered was stretchiness. Stretchiness is defined as a lack of anticipated response to throttle movement, a condition normally caused by excessively lean carburetion. Methanol, due to its leaning effect on carburetion, tends to compound any problems that may be present in late model cars which are adjusted to operate net lean on gasoline.

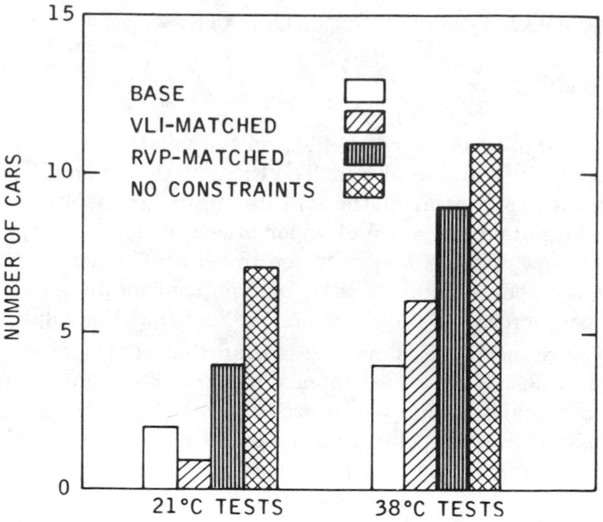

Fig 10 - Number of cars experiencing vapor lock problems as a function of fuel at two test temperatures

A measure of stretchiness with the methanol blends was obtained by observing the increase in part throttle acceleration times at constant manifold pressure, over the times found with the base blend. Accelerations from 32 to 80 km/hr. (20 to 50 mph) were used. The data are summarized in Table 10, where the results from the 1968 and newer cars were averaged to give one set of numbers (1968+) and the data from the three 1967 cars were averaged to give the other set. As expected, stretchiness was much more pronounced in the case of the newer cars. One of the cars, a 1973 model was unable to reach the final speed of the test (80 km/hr.), at the manifold pressure setting used in the tests, with 4 of the 6 methanol blends. The data from such tests were not included in the average values shown in Table 10. It is also apparent from the data in Table 10, that increased volatility lessens the stretchiness problem. This may be due to a richening of carburetion as was observed in the fuel economy and emissions tests with the high volatility methanol fuel. A complete tabulation of the stretchiness data is given in Appendix 3.

Table 10 – Stretchiness with Methanol Blends

		% Increase in Accel. Time Relative to Base Fuel Time		
	Car Year	Matched VLI	Matched RVP	No Constraints
21°C (70°F)	1968+	53	39	22
Tests	1967	14	12	10
38°C (100°F)	1968 +	23	7	6
Tests	1967	11	0	1

The data presented above indicate that some adjustment to the front-end composition of the gasoline would be required if methanol blends were to be expected to give reasonable vehicle performance characteristics. Any adjustment which involved removing front-end hydrocarbon components from the gasoline pool would detract from methanol's usefulness as a gasoline extender since the removal of the hydrocarbons would counterbalance the increased energy availability provided by the methanol. To further compound the problem, methanol's low energy content per unit volume (half that of gasoline's) means that the addition of 15% methanol only adds about 7% to the available energy on a hydrocarbon equivalent basis. Thus, if an adjustment of more than 7% to the hydrocarbon front-end is required, a net loss in available energy for motor gasoline would result.

Taking the matched-RVP methanol blend prepared for the 21°C runs as an example, it can be seen from the data in Appendix 1, that the sum of butane and pentanes for this fuel was 19% by volume, compared to 34% by volume for the base fuel. The 15% volume decrease due to butane and pentanes removal cancels the 15% increase due to methanol and leads to a net *loss* in available energy of about 7%. The displaced butane and pentanes would still be available for other uses. However, by the same token, the methanol could be used in other applications as well, without the potential problems associated with its use as a motor fuel.

This question of available energy to the gasoline pool is highly complex, depending upon refinery design, as well as upon the degree of vapor lock protection desired. To more clearly define the energy content of the gasoline pool with methanol-gasoline blends, extensive study of overall refinery operations would be required. Nevertheless, for 15% methanol blending concentrations, the total increase in available energy to the gasoline pool could not exceed about 7%. Any volatility adjustments would lead to a decrease in that percentage.

Phase Separation

Another potential problem with the use of methanol in motor gasoline is associated with its tendency to undergo phase separation in the presence of very small concentrations of water. Methanol, because of its polar character, has only limited solubility in non-polar gasoline, even under anhydrous conditions. The actual solubility is dependent on the hydrocarbon composition of the fuel, with higher aromatic content fuels giving higher methanol solubility. Generally speaking, 15% methanol by volume should result in no solubility problems if anhydrous conditions are maintained. However, for typical gasolines phase separation occurs in the presence of less than 1% water.[17].

Technology is available[18] for the manufacture of methanol containing up to about 15% higher molecular weight alcohols. To determine the benefit, if any, which would be obtained if higher molecular weight alcohols were present in the fuel along with the methanol, blends containing i-propanol, n-butanol, n-pentanol, and a mixture of these three were prepared. The alcohol fraction of these blends contained 15% of the higher molecular weight alcohol(s).

The degree of phase separation was measured by adding 2 cc of water to 300 cc of the blend, shaking, and determining the volume of the water-alcohol layer. Temperatures were maintained at about 10°C (50°F) during these tests to simulate reasonably severe conditions.

The results are shown in Table 11. It can be seen that, although the higher molecular weight alcohols did stabilize the blend to a certain extent, separation still occurred.

Table 11—Effect of Higher Molecular Weight Alcohols on the stability of a Methanol-Gasoline Blend

Composition of Alcohol Fraction	Volume Alcohol in Blend	Volume Separated*
Methanol only	45 cc	42 cc
15% i-propanol bal. methanol	52	34
15% n-butanol bal. methanol	52	28
15% n-pentanol bal. methanol	52	27
5% i-propanol 5% n-butanol 5% n-pentanol bal. methanol	52	28

*Includes most of the 2 cc water added to the 300 cc blend.

Blending the methanol and gasoline at the service station pump would be one way of preventing phase separation in the fuel distribution system. This would still require that extreme care be taken to keep the methanol anhydrous. If water contamination occurred, phase separation would take place in the car's fuel tank. A car adjusted to operate on gasoline would not function on the methanol-water layer which would separate out at the bottom of the tank.

Stabilizing methanol-gasoline blends with the addition of a third component may be possible. However, recent experiments[19] suggest that the treat levels required would be too high to be practical.

SUMMARY AND CONCLUSIONS

The results from this study suggest the following:

(1) Fuel economy and emissions benefits which result from the use of methanol-gasoline blends would probably only be significant for older rich-operating cars. On an energy basis, an eight percent benefit was found for the 1967 car tested. The other two cars in this part of the program representing current and future models, showed no significant benefit. The data are in excellent agreement with predictions based on methanol's leaning effect on carburetion.

(2) Methanol's effect on emissions also follows the trend expected from equivalence ratio considerations. Substantial reductions in CO and HC emissions were only observed for the rich-operating 1967 car, and these were accompanied by an increase in NO_x emissions.

(3) If no adjustments are made to allow for the large increase in fuel volatility found with the addition of methanol, the menthanol-related fuel economy and emissions benefit may diminish. The data suggest that carburetor enrichment during warmed up operation counteracts the leaning effect of the methanol.

(4) Formaldehyde emissions increase with the use of methanol-gasoline blends. It is not clear whether the 25-50% increases found in the program would be environmentally significant.

(5) The presence of methanol in the fuel has very little effect on exhaust hydrocarbon photochemical reactivity. Hydrocarbon mass emissions were the dominant factor in determining total reactivity.

(6) Expressions used to estimate vapor locking tendency with gasolines can also be used for the methanol-gasoline blends. A significant increase in vapor lock problems would be expected if methanol were added directly to current gasolines because of methanol's large effect on front-end volatility.

(7) Removal of butane and pentanes to provide reasonable volatility characteristics for methanol-gasoline blends would detract from methanol's role as a gasoline extender. A net loss in energy available to the gasoline pool could actually occur if current volatility specifications were met.

(8) Stretchiness (a lack of expected response to throttle movement) is a problem with methanol-gasoline blends. This problem, being related to excessively lean carburetion, was much more pronounced with the newer cars tested.

(9) Phase separation is probably the most critical product quality factor associated with the use of methanol-gasoline blends. Tests indicated that the presence of higher molecular weight alcohols in the methanol blends, accounting for 15% of the alcohol fraction, reduced the severity of the problem but did not eliminate it.

Taken as a whole, the results of this program suggest that fuel-related uses for methanol outside the motor gasoline area are to be preferred. Its use in electric utility gas turbines, for example, would free the petroleum-derived fuels currently burned in these installations for other applications. Such uses for methanol would provide an overall increase in the available energy from liquid fuels, without introducing the undesirable side effects associated with its use as a motor gasoline blending component.

REFERENCES

(1) D. A. Howes, "The Use of Synthetic Methanol as a Motor Fuel", Jour. Inst. Petr. Tech. *19,* 301 (1933). S. J. W. Pleeth, "Alcohol, a Fuel for Internal Combustion Engines", Chapman & Hall, London (1949).

(2) R. G. Jackson, "The Role of Methanol as a Clean Fuel", SAE Paper #740642 (October, 1973).

(3) J. C. Gillis, J. B. Pangborn and J. G. Fore, "Synthetic Fuels for Automotive Transportation", paper presented at the spring meeting of the Combustion Institute, Madison (March, 1974).

(4) T. B. Reed and R. M. Lerner, "Methanol: A Versatile Fuel for Immediate Use", Science *182,* 1299 (1973).

(5) P. Breisacher and R. Nichols, "Fuel Modification: Methanol Instead of Lead as the Octane Booster for Gasoline", paper presented at the spring meeting of Combustion Institute, Madison (March, 1974).

(6) Environmental Protection Agency, Part II, "New Motor Vehicles and New Motor Vehicle Engines, Control of Air Pollution", Federal Register, *37,* No. 221 (November 15, 1972).

(7) Coordinating Research Council Inc., "1966 CRC Vapor Lock Tests," New York (1966).

(8) Scott Research Laboratory Report #2602 (October, 1969).

(9) E. E. Wigg, R. J. Campion, and W. L. Petersen, "The Effect of Fuel Hydrocarbon Composition on Exhaust Hydrocarbon and Oxygenate Emissions", SAE Paper #720251 (January, 1972).

(10) P. J. Clarke, "The Effect of Gasoline Volatility on Exhaust Emissions", SAE Paper #720932 (1972).

(11) C. F. Taylor, "The Internal Combustion Engine in Theory and Practice", M.I.T. Press (1965) p. 438.

(12) T. A. Huls, *et al.,* "Spark Ignition Engine Operation and Design for Minimum Exhaust Emission", SAE Paper #660405 (June, 1966).

(13) B. Dimitriades and T. C. Wesson, "Reactivities of Exhaust Aldehydes", Bureau of Mines RI #7527 (April, 1968).

(14) M. F. Brunelle, J. E. Dickinson and W. J. Hamming, "Effectiveness of Organic Solvents in Photochemical Smog", Los Angeles Air Pollution Control District Solvent Project, Final Report (July, 1966).

(15) G. D. Ebersole and F. S. Manning, "Engine Performance and Exhaust Emissions: Methanol vs. Isooctane", SAE Paper #720692 (1972).

(16) A. P. Altshuller, "An Evaluation of Techniques for the Determination of the Photochemical Reactivity of Organic Emissions", J. Air. Poll. Cont. Assoc. *16,* 257 (1966).

(17) Committee for Air and Water Conservation, API, "Use of Alcohol in Motor Gasoline – A Review", Report #4082 (August, 1971).

(18) T. D. Wentworth, "Outlook Bright for Methyl-Fuel", Environ. Sci. Technol. *7,* 1002 (1973).

(19) J. C. Ingamells and R. H. Lindquist, Chevron Research Company, "Methanol as a Motor Fuel", to be published.

APPENDIX 1 - Fuel Compositions and Inspections

Fuel Code	Base	Matched RVP	Non-Matched RVP	(1)	(2)	(3)	(4)	(5)	(6)	(7)	(8)
	Fuel Economy & Emissions			**Fuel Volatility Studies**							
Components/Vol. %											
Reformate	58	56	49	46	45	46	39	48	50	45	40
Heavy Cat. Naphtha	0	0	0	0	0	11	0	18	25	2	16
Light Cat. Naphtha	22	22	19	20	20	9	17	14	0	20	12
Cat C_5	13	7	11	25	19	13	21	14	10	18	12
Butane	7	0	6	9	1	6	8	6	0	0	5
Methanol	0	15	15	0	15	15	15	0	15	15	15
RVP	11.9	11.7	16.0	13.6	11.4	13.3	15.7	10.4	8.3	10.7	12.8
ASTM D-86 Dist.											
IBP	88°F	96°F	88°F	88°F	93°F	94°F	80°F	85°F	112°F	96°F	92°F
D+L 5%	102	110	90	96	107	106	91	103	123	109	102
10%	115	118	103	104	111	114	99	117	127	113	110
20%	133	127	117	118	117	125	110	140	133	119	120
30%	157	134	127	136	124	133	119	166	138	126	129
40%	188	139	135	157	130	137	127	197	148	132	135
50%	218	158	138	184	135	174	134	229	243	147	139
60%	245	242	221	218	208	244	140	257	268	217	235
70%	267	263	252	252	247	271	234	283	293	256	269
80%	285	285	277	283	281	299	273	307	315	287	299
90%	310	309	305	313	313	330	310	335	342	319	330
95%	333	326	333	339	337	352	333	360	367	345	358
FBP	372	384	371	383°F	383°F	398°F	385°F	408°F	408°F	390°F	402°F
RON	98.3	101.7	101.3	96.2	99.4	99.3	99.3	99.3	98.9	96.0	99.1
MON	86.8	87.5	87.0	85.3	85.6	86.1	86.2	85.7	85.9	84.9	85.7
API Gravity @ 60°F	54.9	50.2	54.5	60.3	54.4	52.7	58.2	53.8	48.9	53.9	53.0

APPENDIX 2 - Vehicle Characteristics

No.	Model Year	Displacement	Cylinders	Air Cond.
		Fuel Economy and Emission Tests		
1	1973*	351	8	No
2	1973	351	8	No
3	1967	289	8	Yes
		Volatility Studies		
1	1974	350	8	Yes
2	1974	318	8	Yes
3	1974	351	8	Yes
4	1974	140	4	Yes
5	1973	225	6	No
6	1973	200	6	No
7	1972	455	8	Yes
8	1972	400	8	Yes
9	1971	351	8	Yes
10	1968	283	8	No
11	1967	289	8	No
12	1967	289	8	Yes
13	1967	230	6	No

*Equipped with monolithic noble metal oxidation catalysts. Carburetor main jets to 0.056" from 0.054". Air injection at the exhaust valves.

APPENDIX 3 - Stretchiness

Percent Increase in Acceleration Time Relative to the Base Gasoline

	21°C			38°C		
Fuel	**(2)**	**(3)**	**(4)**	**(6)**	**(7)**	**(8)**
Vehicle						
1	49	9	29	26	6	18
2	59	30	14	12	0	0
3	56	50	14	26	18	∞*
4	154+	108	39	25	5	0
5	21	21	22	12	4	1
6	∞*	∞*	27	40	∞*	∞*
7	19	19	8	26	23	25
8	29	43	19	37	7	4
9	46	35	23	12	4	-5
10	46	39	25	16	2	13
11	6	6	11	33	0	1
12	9	9	2	0	0	0
13	28	21	16	0	0	1
Average	43.5	32.5	19.2	20.4	5.8	5.3
Range	6-∞	6-∞	2-39	0-40	0-∞	-5-∞

*Data ignored when calculating average. ∞ value resulted when car failed to reach final test speed of 80 km/hr.

Single-Cylinder Engine Evaluation of Methanol - Improved Energy Economy and Reduced NO_x*

W. J. Most and J. P. Longwell
Corporate Research Labs
Exxon Research and Engineering Co.

INTRODUCTION

The increasingly stringent restrictions imposed on the pollutant efflux from automotive engines coupled with a marked rise in the price of gasoline has increased the incentive for an improved combination of lower exhaust emissions and higher fuel economy. The objective of this work was to determine the degree to which the combustion characteristics and other attributes of methanol might be utilized to achieve this goal. The use of alcohols as automotive fuels is, of course, not unique to this work, having been promoted and investigated to a varying degree for more than 50 years. References 1 and 2 provide a history and technical evaluation up to 1964. More recent evaluations of alcohol fuels have been carried out by many investigators including Ebersole and Manning(3),* Pefley et al.(4) and Adelman et. al.(5). Indeed, many of the elements of this work can be found in these references. However, optimum advantage of methanol's lean combustion characteristics by equivalence ratio variation (fuel throttling) and of the high octane value of methanol/water blends has not been effected to date.

As background, it is worthwhile reviewing the impact of the variables considered in the experimental work presented below. The performance, pollutant efflux, and economy of an automotive engine are complex functions of the engine design, speed, spark timing, inlet manifold pressure, fuel-air ratio, pollutant suppression or clean-up systems, and fuel. Of these many variables, engine compression ratio has a significant and well-documented effect on fuel economy. Reference 6, for example, illustrates that between a compression ratio of 8, typical of current engines, and compression ratios of 11 to 12 an improvement of approximately 20 percent is possible.

A survey of the operational characteristics of automotive engines also indicates that there is potential for significant gains, in both fuel economy and pollution levels, by operating such engines on very fuel-lean mixtures(7, 8, 9). Lean combustion has the potential advantage of improving the cycle thermal efficiency for a fixed compression ratio by decreasing the peak flame temperature. The effect of reducing the peak

*Numbers in parentheses designate References at end of paper.

*Paper 750119 presented at the Automotive Engineering Congress and Exposition, Detroit, Michigan, February 1975.

ABSTRACT

Comparative testing of pure methanol, methanol/water blends and isooctane in single-cylinder engines has demonstrated that through proper utilization of methanol's fuel-lean combustion characteristics it may be possible to reach CO emissions of the order of 0.1 percent and NO_x emission levels of less than 100 ppm in the raw (undiluted) exhaust. Exhaust treatment to remove unburned methanol and partial oxidation products might be required. Concomitant with decreased emissions are specific energy consumption improvements estimated to be in the range of 26 to 45 percent better than achievable with current gasolines and the associated low compression ratio engines and emission control systems. These energy consumption improvements are obtained by virtue of efficient lean operation and by utilizing the high octane values of methanol/water blends at high compression ratios. Despite these potential end-use technical advantages for methanol, its large scale use as an automotive fuel is precluded for at least one to two decades because of inadequate supply, the need for immense capital expenditures to increase supply and the need for special engine and fuel control designs.

SAE/PT-80/19/$02.50

temperature is generally attributed to the real-world effects of decreasing the specific heat capacity of the combustion products, suppressing dissociation and lowering the heat losses to the engine[e.g. 7]. Concomitant with the potential efficiency improvements of lean operation is a reduction in engine power. Fuel-lean conditions simply provide less chemical energy for conversion to mechanical horsepower. However, the peak power of an engine is used only over a small fraction of the operating envelope. Thus combining lean combustion with fuel throttling, as opposed to the current practice of air throttling, has the added potential advantage of reducing the energy expanded in pumping air into the engine.

Varying the equivalence ratio* has a profound impact on exhaust emissions by virtue of the corresponding effects on the peak flame temperature achieved in the cycle, the rate of flame propagation through the mixture, and the concentration of species, particularly the critical oxygen concentration. As the equivalence ratio is reduced from rich to slightly lean, the relative oxygen concentration is increased to the degree necessary to convert a larger fraction of the hydrocarbons and carbon monoxide to water and carbon dioxide. For the same reason the oxides of nitrogen (NO_x) increase to a peak value at an optimum combination of temperature and oxygen availability. Further decreasing the equivalence ratio lowers the peak flame temperature, resulting in a rapid decrease of NO_x due to the exponential dependence of NO_x production on temperature.

Based on these arguments it is apparent that there is potential for significant gains, in both fuel economy and pollution levels, by operating automotive engines at high compression ratios and on very lean mixtures. Unfortunately, additives to improve gasoline octane values which are both economically feasible and environmentally acceptable are not yet available. Furthermore, the current generation of gasolines cannot be used to effect the potential advantages of lean combustion because of the rapid decrease in flame speed as the mixture is leaned out from stoichiometric conditions and because of the proximity of the lean combustion limit to the stoichiometric condition[10]. As a consequence, attempts to burn homogeneous charges of gasoline and air near the lean combustion limit in spark-ignition engines result in erratic combustion, poor drivability and a rapid increase in unburned hydrocarbon emissions[11].

Methanol is a fuel which, in contrast to gasolines and other hydrocarbon fuels, has relatively good lean combustion characteristics. (The properties of methanol and, for contrast, those of isooctane are given in Table 1.) The wider flammability limits and higher-than-gasoline flame speeds under lean conditions have been demonstrated in a variety of combustion situations [10]. Relative to current unleaded gasolines, pure methanol's basic octane ratings are slightly higher by the motor method and significantly higher by the research method. In addition, methanol has no tendency to smoke or to form carbonaceous engine deposits. Thus it can be theorized that a smaller margin for increase in the road octane requirement would be needed, allowing an engine to be designed for the full octane value of the fuel. On the negative side, the high latent heat of vaporization and single boiling point characteristics of methanol may complicate the task of achieving a correct fuel vapor-liquid ratio without encountering vapor lock problems. The most frequently cited disadvantage of methanol as an automotive fuel is its low volumetric energy density. Operating at the same energy efficiency as with gasoline, a methanol-fueled vehicle would require roughly twice the fuel tank volume or have half the range.

Table 1—Properties of Methanol and Isooctane

Property	Methanol (CH_3OH)	Isooctane (C_8H_{18})
Molecular Weight	32	114
Density at 60°F.—		
g/cc	0.791	0.692
lb/gal	6.62	5.79
Boiling Temperature, °F.	148	211
Vapor Pressure at 100°F., psia	4.55	1.72
Autoignition Temperature, °F.	878	837
Flammable Limits, % by volume in air at STP	7.3-36	1.1-6.0
Specific Heat of Liquid Btu/lb °F. at 60F and 1 atm.	0.599	0.489
Heat of Vaporization at boiling point and 1 atm., Btu/lb	473	117
Lower Heating Value, Btu/lb	8644	19,065
Stoichiometric Air/Fuel Ratio	6.46	15.1
Octane Ratings—		
Motor	87.4	100
Research	109.6	100

Thus it appears that methanol has potential for effecting to some undefined degree the advantages outlined above, albeit with certain limitations. With these considerations in mind, an experimental evaluation of pure methanol's potential as an automotive fuel was undertaken. This work does not consider blends of methanol and gasoline or methanol and any other hydrocarbon fuel.

EXPERIMENTAL EQUIPMENT AND PROCEDURES

A portion of the experimental work was carried out using a standard CFR single-cylinder engine. This is

*The equivalence ratio is defined as the ratio of the actual fuel-air ratio to the stoichiometric value. Values of the equivalence ratio, Φ, less than one signify fuel-lean conditions.

the variable-compression engine used in octane rating of fuels. Detailed specifications are given in Reference 12. The engine features a cylindrical combustion chamber with the valves mounted in the head. The point and spark plug gaps, ignition source, valve clearance, oil and coolant temperatures, and oil pressure were maintained in accord with the specifications of Reference 12. A standard three-bowl octane-rating carburetor was used, with the exception of replacing the standard 9/16 in. dia. venturi with a 3/4 in. dia. venturi. Power absorption and speed control to 965 rpm was accomplished with the standard motor-generator set. Brake output power and friction power absorbed by the engine were determined by measuring the electrical power to or from the motor-generator.

The second engine used in this work was the single-cylinder CLR oil test engine(13). The wedge-shape combustion chamber of the CLR engine is more similar to modern design than that of the CFR engine. The compression ratio of this engine is 8.1. The ignition system, valve clearance, etc. were maintained at standard conditions. Brake and friction horsepower were determined with a water-cooled dynamometer.

For both engines the inlet airflow was measured with Meriam laminar flow elements. The inlet air humidity, inlet air temperature and subsequent air-fuel mixture temperature were left uncontrolled, being allowed to adjust to the levels dictated by ambient conditions, fuel type and flow rate. The fuel flow rates were determined with a burette and a stopwatch. All equivalence ratios reported herein are based on these measurements of the input flows.

The exhaust emission and analysis system provided continuous monitoring of oxides of nitrogen, oxygen, carbon monoxide, carbon dioxide, and unburned hydrocarbons using the instruments listed in Table 2. This system was calibrated at the beginning and end of each test series with gases of known composition.

Table 2—Exhaust Emissions Instrumentation

Oxides of Nitrogen	Beckman Model IR315	Infrared Analyzer
Oxygen	Beckman Model 741	Oxygen Analyzer
Carbon Monoxide	Beckman Model 864	Infrared Analyzer
Carbon Dioxide	Beckman Model 15A HB	Infrared Analyzer
Unburned Hydrocarbons	Beckman Model 109-A	Hydrocarbon Analyzer

The exhaust gas was passed through a series of traps and filters upstream of the instruments. Thus all of the exhaust gas compositions reported herein are on a dry basis.

CFR Engine Results

The first series of tests were carried out in the CFR engine described above. Initially the engine was set for a compression ratio of 7.82 and a spark advance of 18°BTDC. These values were selected on the basis of the ASTM octane rating "book" values for standard knock intensity with isooctane(12). Note, however, that as described above other octane rating specifications as to inlet and mixture temperature, inlet humidity, etc. were not rigorously followed. Each of these parameters is known to have an effect on the engine performance and emissions. However, these effects are sufficiently small that general trends are not altered.

The results of these tests conducted with methanol and with isooctane (as a baseline case) are illustrated in Figures 1 and 2. The data in Figure 1 and where appropriate throughout this paper are given as indicated specific energy consumption (ISEC) in terms of Btu's consumed per hour per indicated horsepower

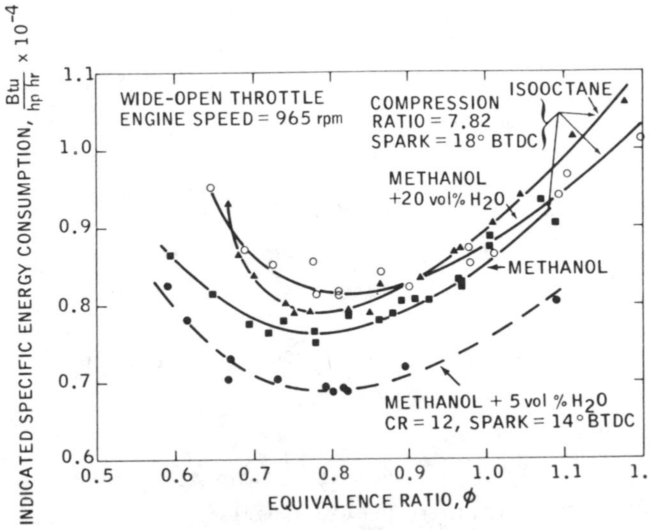

FIGURE 1—EFFECTS OF FUEL TYPE AND EQUIVALENCE RATIO ON INDICATED SPECIFIC ENERGY CONSUMPTION IN A SINGLE-CYLINDER CFR ENGINE

produced rather than the more familiar specific fuel consumption in terms of pounds of fuel per hour per indicated horsepower produced. (The indicated horsepower is the net brake output of the engine plus the friction horsepower of the engine.) The use of specific energy consumption provides a more direct indication of the energy cost of producing a unit of automotive power and normalizes the energy density (Btu/lb.) differences between fuels. Comparison of the methanol and isooctane data in Figure 1 indicates that at these fixed engine conditions methanol has an advantage over isooctane in terms of indicated specific energy consumption. Furthermore, it is obvious from Figure 1 that methanol burns efficiently at leaner equivalence ratios than does isooctane. This improvement in lean operation suggests that some of the benefits of fuel throttling may be realized.

Figure 2 compares the NO_x emissions of isooctane and methanol as functions of equivalence ratio. Under the conditions tested methanol actually produced more NO_x than did isooctane for equivalence ratios greater than 0.9. At leaner conditions both fuels display the anticipated trend of rapidly decreasing NO_x with decreasing equivalence ratio. However, because of its superior lean combustion characteristics much lower NO_x levels can be achieved with methanol than with isooctane. It is important from a practical viewpoint to note that NO_x levels on the order of 100 ppm can be achieved at equivalence ratios well above the lean combustion limits. Thus, it is possible that the equivalence ratio corresponding, for example, to 100 ppm

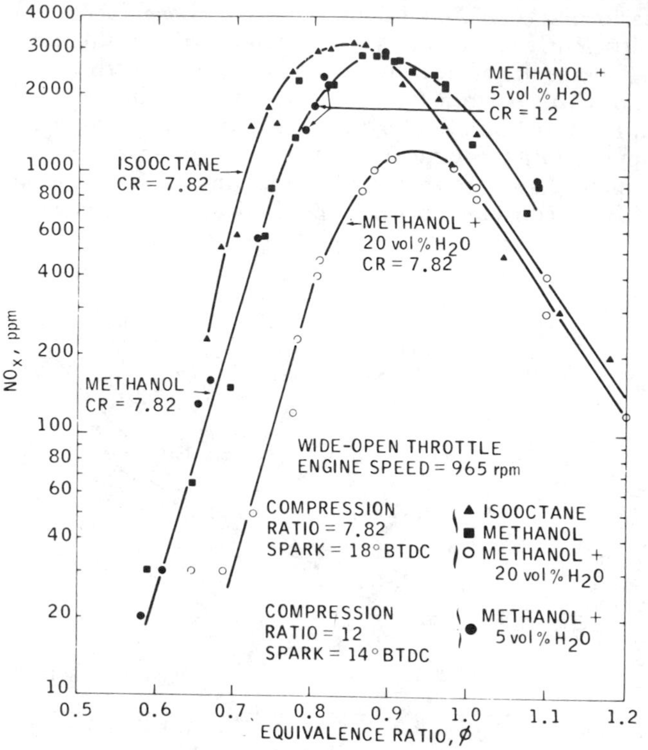

FIGURE 2—EFFECTS OF FUEL TYPE AND EQUIVALENCE RATIO ON NO_x IN A SINGLE-CYLINDER CFR ENGINE

of NO_x could be an operating point in a multi-cylinder engine without the normal maldistribution of fuel between cylinders resulting in one or more of the cylinders misfiring because of too lean a mixture. Conversely, an isooctane fueled multi-cylinder engine could not be run satisfactorily at the minimum NO_x level of 230 ppm ($\Phi \simeq 0.66$) shown in Figure 2 because this condition is very near the lean limit even in the single-cylinder engine. The equivalence ratio at which an increase in fuel consumption begins to occur in a multi-cylinder engine is a complex function of the engine design (particularly the carburetion and inlet manifolding) and the operating conditions. No degradation in mileage was suffered down to an equivalence ratio of approximately 0.75 (A/F = 19-20) in Hansel's work[11] with a 1969 318 CID V-8 Plymouth engine and Indolene fuel. These data are for steady-state 50 mph cruise. The isooctane data in Figure 1 show a minimum ISEC in the same equivalence ratio range indicating no large difference between single and multi-cylinder engines under steady-state conditions. However, enriched mixtures would probably be required for rapid acceleration because such conditions are likely to result in more severe cylinder-to-cylinder variation in equivalence ratio. Thus it is the dynamics of each carburetor/inlet manifold under rapidly varying conditions that would limit the minimum equivalence ratio for acceptable vehicle operation. Such limitations cannot be calculated and require experimental evaluation.

Addition of water to any fuel has long been known to suppress the formation of the oxides of nitrogen. The effect on NO_x emissions of adding 20 volume percent water to methanol at a compression ratio of 7.82 is shown in Figure 2. Figure 3 shows the NO_x emissions data for a family of methanol/water blends at a compression ratio of 12. The knock regions indicated in Figure 3 (and Figure 4 below) for pure methanol and isooctane indicate the minimum equivalence ratio at

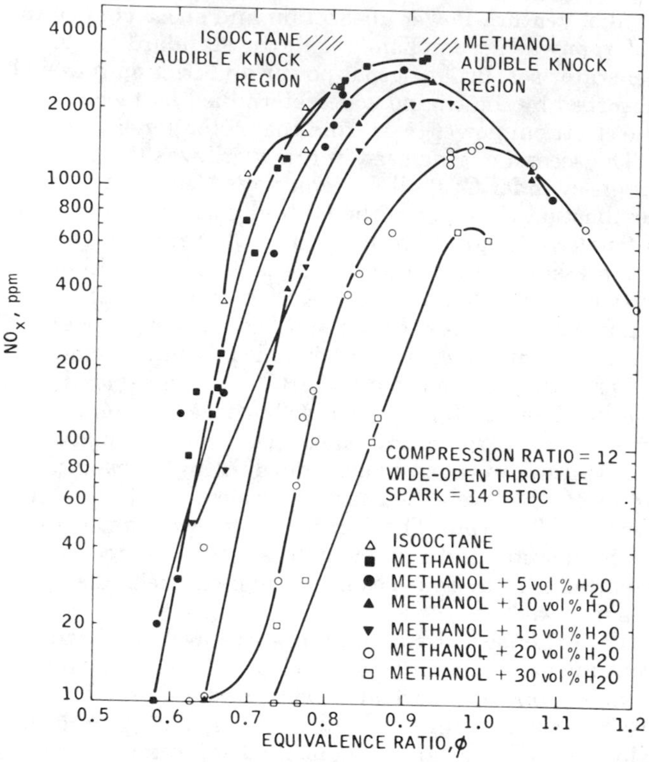

FIGURE 3—EFFECT OF WATER ADDITION TO METHANOL ON NO_x IN A CFR ENGINE

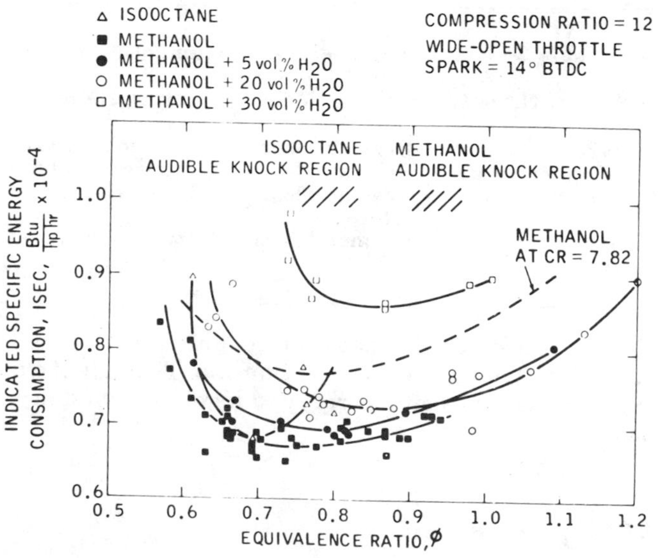

FIGURE 4—EFFECT OF WATER ADDITION TO METHANOL ON ISEC IN A CFR ENGINE

which audible knock was first detected. The audible knock level increased in intensity as the equivalence ratio was increased up to the condition at which the test series was terminated. The effect of water addition on the specific energy consumption at fixed engine conditions is illustrated by the data for methanol/20 vol. percent water in Figure 1 (CR = 7.82) and Figure 4 (CR = 12). In Figure 1 it can be seen that the performance with this fraction of water is slightly better or slightly worse than the baseline isooctane, depending on the equivalence ratio at which the comparison is made. Thus it is feasible to effect substantial NO_x reduction by adding water to methanol with little or no energy consumption penalty relative to traditional hydrocarbon fuels. Relative to pure methanol this NO_x reduction can be achieved at richer equivalence ratios, thereby permitting operation at higher power levels for the same NO_x emission level. This comparison does show an energy consumption penalty however. At the respective minima of the specific energy consumption curves for methanol ($\phi \simeq 0.78$) and methanol/20 percent water ($\phi \simeq 0.82$) of Figure 1, pure methanol has a 6-7 percent advantage. From Figure 2 it can be seen that this energy consumption penalty results in a 66 percent reduction in NO_x. This appears to be a more advantageous compromise than can be achieved with existing NO_x suppression systems for gasoline-fueled engines where an 0 to 9 percent economy debit is anticipated[16]. This is discussed more completely later in the paper.

In addition to suppressing NO_x formation, addition of water to methanol increases the octane rating of the fuel. The knock rating characteristics of such blends are given in Table 3.

Table 3—Methanol/Water Octane Ratings

Fuel	Research Octane No.	Motor Octane No.
Methanol	109.6	87.4
Methanol + 5 Vol. % Water	110.0	89.5
Methanol + 10 Vol. % Water	114.0	92.8

These octane ratings were determined in a fully qualified and standardized laboratory in accord with the specifications of Reference 12. The only exception to these specifications was that for the motor method ratings a second fuel-air mixture heater was installed between the carburetor and the engine. This second heater was necessary in order to maintain the specified mixture temperature (300°F.) at the engine inlet.

The energy efficiency improvements available by utilizing the increased octane value of the methanol/water fuel at higher compression ratios are illustrated in Figure 1 by the data for the methanol/5 percent water blend at a compression ratio of 12. Note, in Figure 2, that this improvement is accomplished without suffering an increase in the level of NO_x pollution. It was not possible to operate the engine under these test conditions with pure methanol near the full-power stoichiometric-mixture condition because of knock limitations. Addition of 5 percent water to the methanol completely suppressed any audible indication of knock at those test conditions.

It should be recognized that the energy consumption comparison shown in Figure 1 is only illustrative in nature. This is because it effectively compares a situation where the full potential of the methanol is not utilized (CR = 7.82) to a situation where the full value is realized with the methanol/5 percent water fuel (CR = 12). Similarly, the increases in energy consumption with increasing water content shown in both Figures 1 and 4 at fixed engine conditions can be interpreted in a misleading manner because the full octane value of the blends were not utilized. Furthermore, the high-compression-ratio data of Figure 1 were obtained by allowing the engine output to increase as the compression ratio increased. Maintaining constant engine performance by decreasing engine displacement or rear axle ratio would result in larger fuel economy gains[6]. In any event, the data do demonstrate that energy efficiency improvements similar in magnitude to those obtainable with hydrocarbon fuels can be achieved with methanol/water blends.

A justifiable criticism of the work presented above is that the test conditions were restricted in range and were not optimized for any of the fuels. This prompted the work described below.

CLR Engine Results

The additional parameters of maximum torque spark advance, engine speed (rpm) and inlet manifold pressure were studied in a single-cylinder CLR engine. Isooctane was again used as the baseline fuel for comparison with pure methanol. Both fuels were tested at three engine speeds (1000, 2000 and 3000 rpm). At each engine speed three inlet manifold pressures (29, 25 and 20 in. Hg absolute) were tested. At each selected equivalence ratio the spark advance was adjusted to give maximum brake torque. For each engine condition and fuel a sufficient range of equivalence ratios were tested to define the variation of emissions and power. The results of these tests in terms of indicated specific energy consumption vs. indicated horsepower are shown in Figures 5, 6 and 7 and in terms of NO_x emissions vs. equivalence ratio in Figures 8, 9 and 10.

Consider as examples the curves in Figure 6 for 29 in. Hg. manifold pressure. As explained earlier in this paper, as the equivalence ratio is reduced from rich to lean, the efficiency of the system, being inversely proportional to the specific energy consumption, is seen to improve. This, however, is accomplished at the sacrifice of engine power. Continuing to lean out the engine ultimately results in conditions for which the combustion becomes erratic, resulting in rapid increases in specific fuel consumption and in unburned hydrocarbon emissions.

Although there are many interesting features of the data curves presented in Figures 5, 6 and 7, the essential point is that for all conditions tested the methanol fuel results in better fuel economy than does the isooctane. This advantage becomes more pronounced at the lower manifold pressures. An automotive engine is operated under part throttle conditions most of the time, going to full throttle (high manifold pressure) only for rapid acceleration.

It should be noted that because these tests were conducted at a compression ratio of 8.1 they do not reflect the additional energy efficiency potential of methanol or methanol/water mixtures by virtue of their high octane values. Such potential advantage would, at best, be small relative to isooctane but would be substantial relative to the current unleaded gasolines.

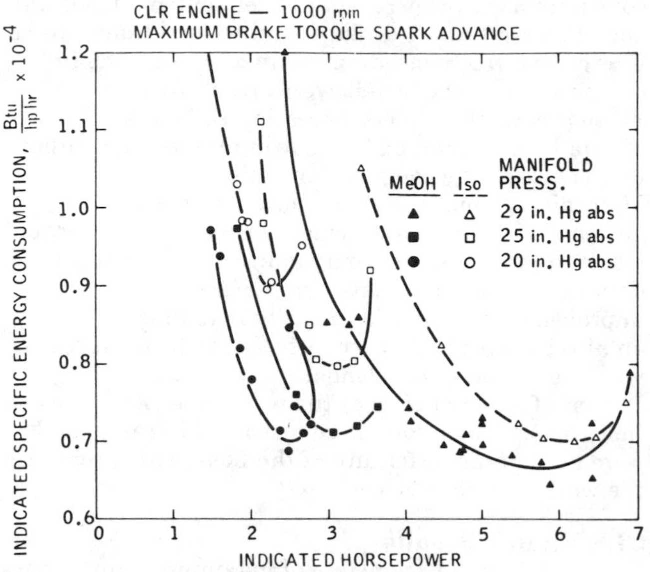

FIGURE 5—EFFECTS OF FUEL TYPE, EQUIVALENCE RATIO AND INLET MANIFOLD PRESSURE ON SPECIFIC ENERGY CONSUMPTION

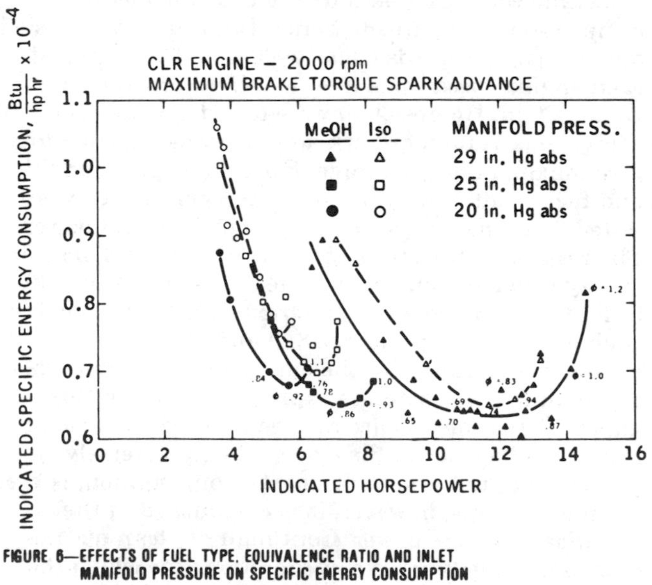

FIGURE 6—EFFECTS OF FUEL TYPE, EQUIVALENCE RATIO AND INLET MANIFOLD PRESSURE ON SPECIFIC ENERGY CONSUMPTION

The NO_x emission data are presented in Figures 8, 9 and 10. Examination of these curves indicates that, in the range of equivalence ratios within which automotive engines are normally operated ($0.9 < \Phi < 1.1$), the relative advantage between the two fuels varies as a function of engine condition. If the engine were restricted to this range, there would be no significant incentive for methanol from a NO_x emissions point of view. However, methanol results in smooth, stable engine operation at much leaner conditions than is possible with gasoline. This results in achievement of the very low NO_x levels shown in Figures 8, 9 and 10.

Some approximate calculations emphasize the significance of the low NO_x levels shown in these

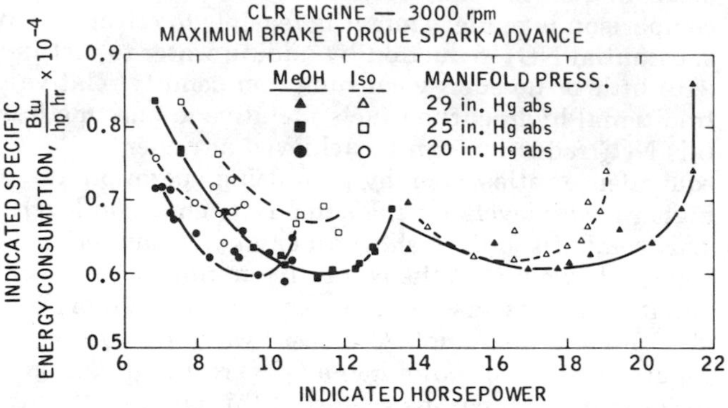

FIGURE 7—EFFECTS OF FUEL TYPE, EQUIVALENCE RATIO AND INLET MANIFOLD PRESSURE ON SPECIFIC ENERGY CONSUMPTION

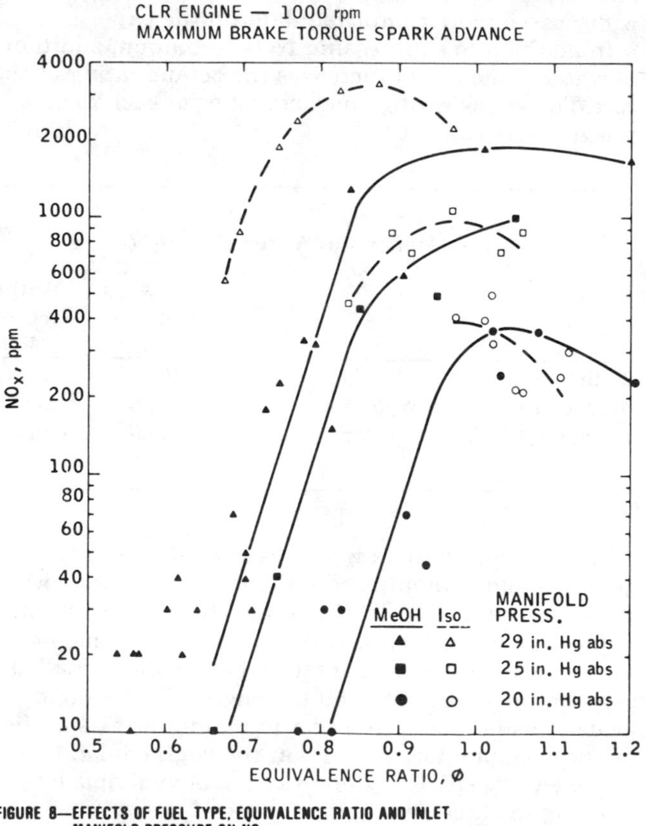

FIGURE 8—EFFECTS OF FUEL TYPE, EQUIVALENCE RATIO AND INLET MANIFOLD PRESSURE ON NO_x

figures. The stoichiometric ratio of air to methanol is 6.5. Thus, the total exhaust weighs 7.5 times the weight of methanol consumed. Methanol weighs approximately 3000 g/gal. (6.6 lb./gal.). The 1978 Federal NO_x standard is 0.4 g/mile averaged over the extensive variation of driving conditions associated with the

Federal Driving Cycle. The conversion from grams to parts per million (ppm) is 10^6. Thus, one can write:

$$\text{ppm} = \frac{\text{Federal Standard (g/mi)} \times \text{mpg} \times 10^6 \text{ (ppm)}}{[1 + \text{Air-Fuel Ratio}] \times \text{Fuel Density (g/ gal.)}} \quad (1)$$

$$= \frac{(0.4)\,(10^6)\,(\text{mpg})}{(7.5)\,(3000)} = 17.8 \times (\text{mpg})$$

Due to the energy density difference between methanol and gasoline, a methanol-fueled vehicle averaging 8 mpg over the Federal Driving Cycle would be equivalent to a gasoline-fueled vehicle averaging approximately 18 mpg. At 8 mpg of methanol, the 0.4 g/mi of NO_X standard would be satisfied by an average level of 140 ppm of NO_X. The data in Figures 8, 9 and 10 indicate that this level could be met by a methanol-fueled engine for most of the part power

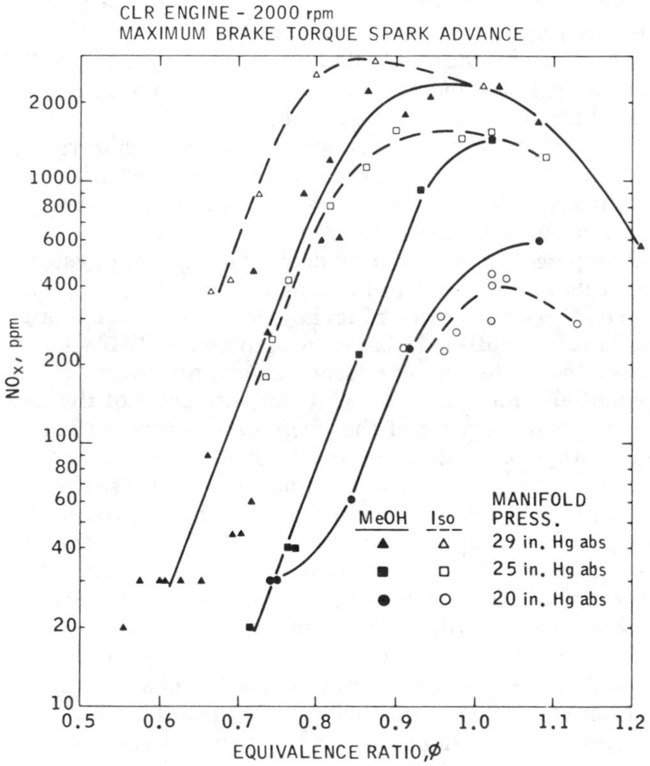

FIGURE 9—EFFECTS OF FUEL TYPE, EQUIVALENCE RATIO AND INLET MANIFOLD PRESSURE ON NO_X

conditions tested. Furthermore, the lean equivalence ratios required to reduce the NO_X levels to the 100-140 ppm range are at or near the optimum efficiency points of the curves in Figures 5, 6 and 7.

This is illustrated in detail by comparing Figures 6 and 9. In Figure 6 the equivalence ratios have been noted for some of the methanol data points. The optimum efficiency (minimum specific energy consumption) points translate, in Figure 9, to NO_X levels in the 200-300 ppm range or roughly 0.6 to 0.8 g/mi. Further leaning out of the engine to the extent of sacrificing 1-2 percent in efficiency would reduce the NO_X emissions to the order of 100 ppm. Referring back to the methanol data obtained with the CFR engine at a compression ratio of 7.82 (Figures 1 and 2), the comparison between minimum ISEC and the ISEC at 140 ppm NO_X is less favorable. For these tests where the spark timing was not optimized, the 140 ppm

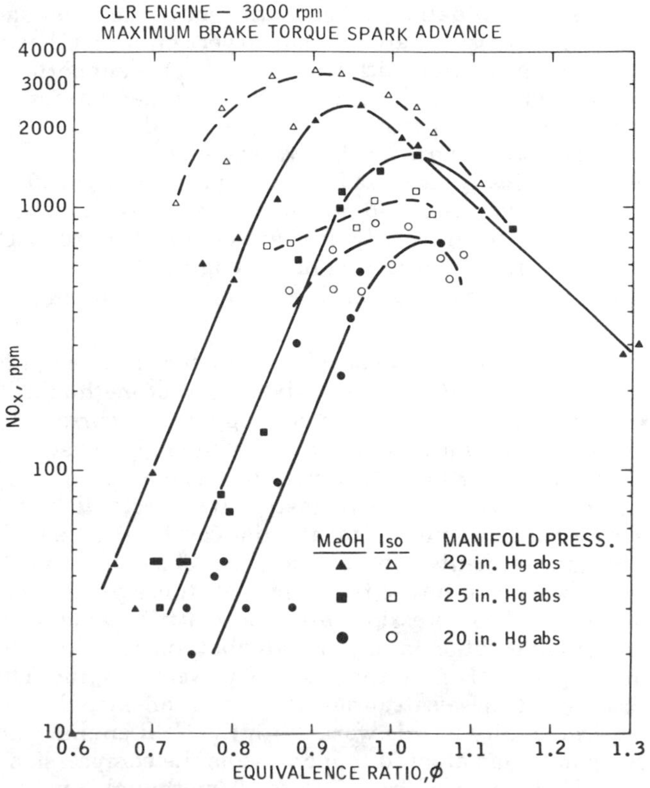

FIGURE 10—EFFECTS OF FUEL TYPE, EQUIVALENCE RATIO AND INLET MANIFOLD PRESSURE ON NO_X

NO_X level was achieved at a sacrifice of approximately 4 percent from the minimum ISEC equivalence ratio. In this case the ISEC for methanol at 140 ppm NO_X is approximately equal to the minimum ISEC for isooctane at a NO_X level on the order of 2000 ppm.

The Federal Driving Cycle involves various accelerations, decelerations and cruise modes of vehicle operation. The above calculations represent a gross averaging, particularly with regard to the high power acceleration portions of the cycle. Considerably more sophisticated and detailed calculations are required in order to give a complete evaluation. However, it is very encouraging that these rough calculations indicate that satisfaction of the most stringent proposed NO_X standard may be feasible with a methanol-fueled system.

The other two principal automotive pollutants for which standards have been written are carbon monoxide and unburned hydrocarbons (HC). For equivalence ratios less than approximately 0.9 - 0.95 the CO concentrations in the undiluted exhaust gas were always below 0.1 to 0.2 percent. (This is true for isooctane, methanol and the methanol/water blends). Thus, as is well known, fuel-lean operation practically eliminates carbon monoxide emissions.

In the case of methanol, the term hydrocarbon is a misnomer in that it is extended to include the alcohol fuel and other partially oxygenated exhaust products such as formaldehyde. The measurements of unburned hydrocarbons were less than fully successful due to difficulties in sampling the water soluble partially oxygenated species present in the exhaust when methanol is burned. Although inconclusive, preliminary qualitative evaluation indicates that unburned fuel and other HC emissions are no more severe on a ppm basis with methanol than with isooctane. This is supported by Reference 3. However, the nature of other species present is believed to be primarily aldehydes[3, 4, 5 and 14]. Such compounds are on the average more photochemically reactive than the species present in the exhaust from gasoline-powered engines[15]. Although warranting further study, it is anticipated that the HC emissions from a methanol-fueled engine can be handled with existing oxidation catalyst technology as a last resort.

It is of interest to pause at this juncture to attempt to add up the energy economy advantages of methanol relative to isooctane and, by analogy, to the current unleaded gasolines and the associated engine systems. This task, unfortunately, involves considerable uncertainty, but it remains instructive to establish at least a range of improvements that can be anticipated. The improvements between the two fuel systems can be broken down into three categories: those attributable to increased compression ratio, those attributable to fuel-lean operation, and those attributable to elimination of NO_x controls from the gasoline-powered engine. The first two of these categories have been considered experimentally in this work. With the CFR engine where the power was allowed to increase as the compression ratio was increased from 7.82 to 12, methanol/5 vol. percent water was demonstrated to have a 16 percent ISEC advantage over isooctane at stoichiometric conditions and a 13 percent advantage at the minimum energy consumption point for each fuel (Figure 1). The work of Reference 6 indicates that for a constant performance system and constant speed driving this compression ratio effect would increase to a 23 percent advantage for methanol. Allowing for stop-and-go driving[6] would reduce this advantage by 3-7 percentage points. This results in a range of estimated economy improvements from 16 to 20 percent due to increasing the compression ratio from 8 to 12. This seems reasonable on the basis of the experimental data.

The effects of lean operation have been shown by the CLR engine data to vary from 2 percent at high speed and high inlet manifold pressure to 22 percent at low speed and low manifold pressure (Figures 5, 6 and 7). From this range a value of 10 percent advantage for methanol was arbitrarily selected as representative of an urban driving cycle.

Note that although arbitrary, this is a conservative estimate in that it compares the best that can be achieved with lean operation with isooctane to the best that can be achieved with methanol. A larger advantage for methanol can be justified by comparison with isooctane at $\phi \simeq 1$, a value more typical of traditional automotive operation. Ignoring any increase in equivalence ratio which might be required for multi-cylinder engine operation is justified on the grounds that the relative difference between methanol and isooctane observed in the single-cylinder engine tests would remain the same.

The estimates of the economy debits associated with the current and anticipated pollution controls to meet the 1978 Federal emission standards (CO = 3.4 g/mi., HC = 0.41 g/mi. and NO_x = 0.40 g/mi.) are based on Reference 16. This Reference provides fuel debit projections which are among the least severe of those available in the literature[e.g., 17, 18]. The baseline developed from Reference 16 represents the compromise in tuning between economy and performance typical of a pre-emission control 4000 lb. vehicle.

For the purposes of this exercise, these fuel economy debits are assigned entirely to NO_x control on the grounds that the more severe penalties of required catalyst temperatures or exhaust gas recycle are attributable to this pollutant. Exhaust treatment for HC and aldehyde emissions would probably be required for a methanol system but is assumed to involve no fuel debit. This is justified on the basis of the high hydrocarbon conversion efficiencies achieved by both oxidation catalysts and thermal systems at the normal exhaust temperatures available from an engine tuned for good fuel economy. Reference 16 considers both catalytic systems operating at compression ratios of 8 and 9 and a lead-tolerant thermal system with exhaust gas recycle operating at a compression ratio of 10. The catalytic systems are projected to entail a 0 to 9 percent fuel economy debit. The thermal system projects to a debit of 18 to 25 percent.

The three categories of fuel system energy economy advantages outlined above are summed in Table 4. These estimates involve large uncertainty, with potential errors and omissions on both sides of the ledger. Not the least of these potential errors is the long-range projection of two fuel systems which are currently far from optimized and undergoing rapid evolution. There will inevitably be both compromises in translating these laboratory results for methanol to a commercial vehicle and improvements in the gasoline-fueled systems. Even relaxation of the NO_x emission standard, for example, to the EPA-recommended level of 2.0 g/mi. through 1981, would reduce, but not eliminate, methanol's advantage. Despite these uncertainties, the magnitude of the estimated advantages provides a strong driving force for evaluation of methanol in multi-cylinder engines.

Table 4—Methanol Energy Economy Advantages

Pre-emission Control, 4000 lb. Vehicle Baseline Projected to 1978 Emission Standards

	Catalytic System CR = 8	Thermal System CR = 10
CR to 12	16—20	8—10
Lean Operation	10	10
Pollution Control	0—9	18—25
Methanol Advantage	+26—38%	36—45%

Engineering Development Problems

It is believed that the data presented herein demonstrate the potential of utilizing the unique combustion properties of methanol as an automotive fuel to achieve both substantial improvements in end-use energy efficiency and reduction in NO_X and CO pollution. It is fully recognized that there are of course many significant engineering problems between these single-cylinder engine tests and an operational road vehicle. The data dictate that optimum utilization of methanol requires a system capable of operating at the lean equivalence ratios required to achieve efficient part-power operation and low NO_X and CO levels. However, in order to maintain the peak power potential for rapid acceleration, the system must also provide enriched operation ($\phi \simeq 1$) at wide open throttle and high engine speeds. Such a variable air-fuel ratio carburetor would have to be designed and developed before the acceleration capabilities and other drivability characteristics of a methanol-fueled vehicle could be assessed.

Other engineering problem areas include development of a system to achieve and maintain the proper degree of manifold vaporization of a single boiling point material such as methanol. Cold starting and carburetor vapor lock are also potential problem areas. It might, for a practical system, be necessary to add a "front-end" component such as butane or dimethyl ether to the methanol. It is also probable that problems involving solvency and corrosion would require solutions.

This is not an all-inclusive listing of the engineering problems associated with development of a methanol-powered vehicle. However, the essential point is that no problems have been identified which are outside of the scope of existing engineering practice and technology. Furthermore, there are a number of development approaches which could actually enhance the advantages of methanol. For example, advanced ignition systems have been demonstrated to expand the effective flammability limits of gasoline[19]. The same techniques seem applicable to a methanol-fueled system.

LOGISTICS OF COMMERCIAL APPLICATION

The single-cylinder engine data presented in this paper suggest that methanol may have certain end-use technical advantages as an automotive fuel relative to current gasolines and engine systems. Application of these advantages to the real world, however, requires solutions to the normal questions of commerce, viz. cost, availability of appropriate end-use equipment and supplies in appropriate volume.

The cost of producing methanol has been estimated by a variety of sources to lie in the range of 1. to 1.5 times the current cost of gasoline on an energy unit basis[20, 21, and 22]. This exensive range is, in large part, attributable to differences in assumed geographic and raw material sources. Although sensitive to decisions of the oil producing nations, imported methanol does not at the present time appear competitive with oil or LNG imports. Indeed, using existing technology no synthetic fuel appears cost competitive. If, however, it should become a matter of federal policy to reduce USA dependence on external fuel sources, the question becomes a matter of selecting the most appropriate synthetic liquid fuel to be derived from domestic sources. A recent study[23] projects a methanol to gasoline cost ratio of 1.14 to 1.09 (heat content basis) from the early 1980's through 2000 when both fuels are derived from coal. On an overall source-to-end-use energy basis, methanol would appear to provide more useful liquid energy, but slightly less total liquid and gas energy, than the coal/synthetic crude oil path. However, these conclusions are very sensitive to assumptions regarding relative end-use conversion efficiencies between the fuels considered. If the end-use energy efficiency advantages of pure methanol and methanol-water blends as demonstrated in single-cylinder engines were to be realized in multi-cylinder engines, methanol would be economically competitive with gasoline even under the more pessimistic projections and could offer a significant pollution advantage.

Two hurdles, however, remain to be passed. The first of these concerns the development of appropriate end-use equipment. As discussed in the previous section of this paper, there are many significant engineering problems to be solved before optimum use of methanol and methanol-water blends could be realized. The automotive industry would have to develop a new family of engines to accommodate the new fuel type. Initially, at least, such new engines could be more costly and assert their own effect on comparative economics relative to existing and anticipated gasoline/engine systems. Failure to achieve any one of the three categories of advantages identified for methanol (Table 4) would significantly decrease the incentive for developing such a methanol/engine system.

Finally, the question of supply must be considered. Our analysis indicates that supply is the principal limitation to large-scale use of methanol as an automotive fuel. The Atomic Energy Commission[22] projects that with "sufficient effort" it would be possible by 1980 to produce methanol to the extent of only 3% of the gasoline energy market projected for that time. At that level, even if this entire supply of methanol were to be allocated to specially equipped fleet use in areas with high pollution problems, the environmental impact would be minimal. (Fleet use would be necessary because pure methanol would not be interchangeable within the general automotive population). Thus it appears that, even following a most-favorable scenario, methanol could not contribute significantly to either the national energy or environmental situations within the next decade.

Although this supply situation appears to preclude large scale automotive use of methanol for some time, a longer range approach to selection of the most advantageous synthetic fuel may prove to be more favorable. As the world petroleum reserves continue to be depleted, an increasing fraction of the automotive energy supply must inevitably be assumed by other resources. Wise selection of the best fuel from the many options available (synthetic crude oil, synthetic natural gas, hydrogen, methane, methanol, etc.) requires, in addition to studies of the energy and capital investment costs of manufacturing and distributing these fuels, that the end-use efficiency and environmental impact be more clearly defined. This is an urgent task, as might be deduced from the supply situation described above, because of the substantial lead times involved.

CONCLUSIONS

The following conclusions can be drawn on the basis of the single-cylinder data presented in this paper.

1. Methanol provides an improved specific energy consumption relative to isooctane and, by analogy, to current unleaded gasolines.
2. Methanol's superior lean combustion characteristics provide a potential for high efficiency concomitant with very low CO and NO_X emission levels, possibly even to the extent of eliminating the need for exhaust gas treatment for these pollutants. However, treatment for removal of unburned methanol and its partial oxidation products might be necessary.
3. Methanol combined with small amounts of water permits operation at significantly higher compression ratios, resulting in improved specific energy consumption with no NO_X penalty.
4. Methanol combined with large amounts of water (20 percent) permits operation at higher power density consistent with significantly decreased NO_X levels.
5. Multi-cylinder engine studies are required to determine the degree to which methanol's potential advantages can be realized in practice, but no fundamental limitations have been identified.

Despite these technical advantages for methanol, large-scale use as an automotive fuel will be limited for at least the next decade by lack of adequate supply. This situation does not negate the need for near-term determination of the relative end-use advantages of the several alternative fuels, including methanol, because of the long lead times between commitment to and large-scale production of any alternative fuel.

ACKNOWLEDGEMENTS

This paper represents the contributions of many people. In particular, the assistance of G. F. Holderied with the experimental work and the instructive consultations with M. J. Kittler and L. S. Bernstein on carburetion and other experimental matters are acknowledged.

REFERENCES

(1) J. A. Bolt, "A Survey of Alcohol as a Motor Fuel." SAE SP-254, June 1964.

(2) E. S. Starkman, H. K. Newhall and R. D. Sutton, "Comparative Performance of Alcohol and Hydrocarbon Fuels." SAE SP-254, June 1964.

(3) G. D. Ebersole and F. S. Manning, "Engine Performance and Exhaust Emissions: Methanol vs. Isooctane." SAE Paper 720692.

(4) R. K. Pefley, M. A. Saad, M. A. Sweeney, and J. D. Kilgroe, "Performance and Emission Characteristics Using Blends of Methanol and Dissociated Methanol as an Automotive Fuel." Paper 719008, Proceedings of the Sixth Intersociety Energy Conversion Engineering Conference, 1971.

(5) H. G. Adelman, D. G. Andrews and R. S. Devoto, "Exhaust Emissions from a Methanol-Fueled Automobile." SAE Paper 720693, 1972.

(6) E. S. Corner and A. R. Cunningham, "Value of High Octane Number Unleaded Gasolines in the U.S." Presented before the Division of Water, Air and Waste Chemistry, American Chemical Society Meeting, Los Angeles, California, March 1971.

(7) J. A. Bolt and D. H. Holkeboer, "Lean Fuel/Air Mixture for High-Compression Spark-Ignited Engines." Paper 380D presented at SAE Summer Meeting, 1961.

(8) R. C. Lee and D. B. Wimmer, "Exhaust Emission Abatement by Fuel Variations to Produce Lean Combustion." SAE Paper 680769, 1968.

(9) D. A. Hirschler, W. E. Adams and F. J. Marsee, "Lean Mixtures, Low Emissions and Energy Conservation." NPRA Annual Meeting, April 1973.

(10) E. S. Starkman, F. M. Strange and T. J. Dahm, "Flame Speeds and Pressure Rise Rates in Spark Ignition Engines." Presented at SAE International West Coast Meeting, Vancouver, B.C., August 1959.

(11) J. G. Hansel, "Lean Automotive Engine Operation-Hydrocarbon Exhaust Emissions and Combustion Characteristics." Paper 710164 presented at SAE Automotive Engineering Congress, Detroit, Michigan, January 1971.

(12) "ASTM Manual for Rating Motor, Diesel and Aviation Fuels." Published annually.

(13) Laboratory Equipment Corp., Mooresville, Indiana.

(14) E. E. Wigg and R. S. Lunt, "Methanol as a Gasoline Extender-Fuel Economy, Emissions and High Temperature Drivability." SAE Paper 741008, presented at Automobile Engineering Meeting, Toronto, Canada, October 1974.

(15) B. Dimitriades and T. C. Nesson, "Reactivities of Exhaust Aldehydes." Bureau of Mines, RI °7527, April 1968.

(16) L. E. Furlong, E. L. Holt and L. S. Bernstein, "Emission Control and Fuel Economy." American Chemical Society Meeting, Los Angeles, California, April 1974. Also Chemical Technology, January 1975.

(17) C. LaPointe, "Factors Affecting Vehicle Fuel Economy." Presented at SAE Combined National Farm, Construction, Industrial Machinery and Fuels and Lubricants Meeting, Milwaukee, Wisconsin, September 1973.

(18) E. N. Cantwell, "Exhaust Controls Use Energy." Hydrocarbon Processing, July 1974, pp. 94-103.

(19) T. W. Ryan, S. S. Lestz and W. E. Meyer, "Extention of the Lean Misfire Limit and Reduction of Exhaust Emissions of an SI Engine by Modification of the Ignition and Intake Systems." Paper 740105 presented at SAE Automotive Engineering Congress, Detroit, Michigan, February 1974.

(20) W. D. Harris and R. R. Davison, "Methanol from Coal Can Be Competitive with Gasoline." The Oil and Gas Journal, December 17, 1973, pp. 70-72.

(21) T. B. Reed and R. M. Lerner, "Methanol: A Versatile Fuel for Immediate Use." Science, Vol. 182, Number 4119, December 28, 1973, pp. 1299-1304.

(22) H. Jaffe, et. al., "Methanol from Coal for the Automotive Industry." Atomic Energy Commission, February 1974.

(23) F. H. Kant, et. al., "Feasibility Study of Alternative Fuels for Automotive Transportation; Vol. II—Technical Section." Report EPA-460/3-74-009-b, June 1974.

Exhaust Emissions, Fuel Economy, and Driveability of Vehicles Fueled with Alcohol-Gasoline Blends*

N. D. Brinkman and N. E. Gallopoulos
Fuels and Lubricants Dept.

M. W. Jackson
Environmental Science Dept.
Research Labs.
General Motors Corp.

ALCOHOLS HAVE long been considered potential automotive fuels and have been used in racing because they increase engine power and reduce fire hazards (1,2).* More recently, alcohols, especially methanol, have attracted attention because of the potential to relieve fuel shortages in the U. S. and to decrease its dependence on foreign petroleum (3,4,5,6,7). Reed and Lerner (3) suggested that the use of methanol-gasoline blends could ease fuel shortages very quickly, not only because of the greater amount of liquid fuel that would be available, but also because vehicle fuel economy would be increased due to intrinsic properties of methanol. Some authors have also suggested that the use of alcohols, either neat or in gasoline blends, would decrease exhaust emissions (3,8,9).

Before the recent resurgence of interest in alcohols as automotive fuels, one of the authors of this paper investigated the effects of blending ethanol with gasoline on exhaust hydrocarbon and nitrogen oxide emissions (10) from a single-cylinder engine. The present study extends the previous work. Vehicles, rather than single-cylinder engines were used, and both methanol and ethanol were investigated. Methanol was investigated more intensively than ethanol because in our judgement methanol derived from coal has the greater potential to increase the fuel supply in the U. S. Others (3,4,5,6) also agree with this assessment. From an engineering

* Numbers in parentheses designate References at end of paper.

*Paper 750120 presented at the Automotive Engineering Congress and Exposition, Detroit, Michigan, February 1975.

ABSTRACT

Current national interest in alternative fuels has placed considerable emphasis on alcohols, mainly methanol and its blends with gasoline. Vehicle studies with methanol-gasoline and ethanol-gasoline blends showed that adding alcohol to gasoline without carburetor modifications decreased carbon monoxide emissions, volume-based fuel economy, driveability, and performance. Depending on the carburetor's air-fuel ratio characteristics, hydrocarbon and nitrogen oxide emissions and road octane are either increased, decreased, or not affected.

These effects can be explained on the basis of changes in stoichiometry, energy content, combustion temperatures, and detonation resistance caused by the addition of alcohol to gasoline.

SAE/PT-80/19/$02.50

viewpoint, results of studies with methanol could also be used to predict the behavior of ethanol under similar circumstances.

If alcohols were to relieve automotive fuel shortages significantly and immediately, they would have to be used in vehicles already on the road. Neat methanol and ethanol with their low stoichiometric air-fuel ratios and high heats of vaporization cannot be used in vehicles designed to burn gasoline without extensive modifications to the engine. Therefore, only blends of alcohol with gasoline were studied. Another reason for studying blends was that at the time this work was initiated, comprehensive studies with blends had not been reported. Since then, the publications of Wigg (11,12) have filled some of the void.

The research reported in this paper had several specific objectives. First, to measure the changes in vehicle exhaust emissions, fuel economy, driveability, performance, and octane response. Second, to understand the reasons for any changes. Third, to estimate the potential advantages and disadvantages of fueling the existing car population with alcohol-gasoline blends. To attain these objectives, the research was conducted in two simultaneous stages. In the first stage, two recent model year cars were used for detailed studies in which the car was closely controlled. In the second stage, several used vehicles representing a fraction of the current vehicle population were tested in the "as received" condition.

This study does not address certain aspects of alcohol-gasoline blends germane to their use as automotive fuels. Aldehyde emissions, although significant with alcohols, were not measured. The limited solubility of alcohol in gasoline, especially at low temperatures, and the separation of the blends due to water contamination are not discussed in all their aspects. However, the literature is already sufficient in this area (1,2,11,12, 13). Also, conventional gasolines were blended with the alcohols without compensation for the increase in vapor pressure of the blend by selectively removing light hydrocarbons from the gasoline before blending. As compared to conventional gasoline, the high vapor pressure of the blend would aggravate vapor-lock and increase evaporative emissions. These aspects and the potential energy penalty associated with light hydrocarbon removal have been discussed by Wigg (11,12).

DETAILED STUDIES WITH TWO VEHICLES

TEST CARS AND FUELS - To investigate how and why alcohols affect vehicle operation, the two vehicles described in Table 1 were used. Both cars were equipped with automatic transmissions and had accumulated approximately 4 000 km in normal driving using a commercial unleaded gasoline. Each car was adjusted to manufacturer's specifications before each test. For some tests, vehicle X was modified as will be discussed later.

The gasolines and alcohol-gasoline blends studied are described in Table 2. Gasolines P and R were commercial unleaded gasolines. Clear Indolene is the gasoline specified for exhaust emissions testing (14). RU-86 is an octane reference gasoline similar to the 1973 FBRU-86 blend of the Coordinating Research Council (CRC) reference fuels (15).

Both the methanol and ethanol were commercial grade products. The methanol was approximately 99.9 percent pure, whereas the ethanol was denatured with 5 percent methanol.

The concentration of alcohol in gasoline is expressed in terms of volume percent of the total mixture. Thus, 10 percent methanol indicates a mixture which was obtained by blending 90 parts of gasoline with 10 parts of methanol, by volume. Concentrations were selected to allow the use of the maximum amount of alcohol compatible with driveability requirements. Concentrations lower than 10 percent alcohol were not considered because of their small impact on fuel supply and potential water sensitivity problems (1,2).

To determine the effect of alcohol on emissions, fuel economy, driveability, performance, and octane requirement, comparative tests were run. One set of tests was run with the base gasoline, and another set with the alcohol-base gasoline blend. Each test was run at least twice. For each set of repeat tests, the ratio of each test value to the mean was calculated. Using these ratios from all tests - i.e., tests with gasoline, various alcohol-gasoline blends, and various vehicle modifications - a standard deviation was calculated. This standard deviation multiplied by 100 yields the pooled coefficient of variation.

The experimental approach required frequent fuel changes. Therefore, all tests were run with the test fuels in auxiliary containers connected to the car's fuel pump.

EXHAUST EMISSIONS -

Experimental - The exhaust hydrocarbon (HC), carbon monoxide (CO), and nitrogen oxide (NO_x) emissions from each car were measured using the 1975 Federal Test Procedure (FTP) (14), which was modified as described in Appendix A. One of the modifications was to separate the FTP into 23 cycles

Table 1. Vehicles for Detailed Studies

Code	Model Year	Body Style	Engine Type	Engine Displacement, litres	Carburetor Barrels
X	1973	B	V-8	7.5	4
Z	1974	B	V-8	7.5	4

in the manner described by Weirs and Scheffler (16). In this scheme, cycles 1-5 represent the first "bag" of the FTP, and cycles 6-18 and 19-23 the second and third "bags," respectively. The modified method for calculating HC emissions with alcohol-gasoline blends is also described in Appendix A.

As required by the FTP, the test car was conditioned the day before the test.

The pooled coefficient of variation was 9, 7, and 5 percent for HC, CO, and NO_x, respectively.

Results and Discussion - Exhaust emissions during the FTP from car X using the commercial unleaded gasoline P were 0.93, 15.8, and 1.2 g/km for HC, CO, and NO_x, respectively. The effects of adding either 10 percent methanol or 10 percent ethanol to gasoline P are shown in Figure 1. Only CO and NO_x were significantly affected at the 95 percent confidence level. Contrary to expectations, ethanol affected NO_x emissions more than methanol, but the difference was not statistically significant at the 95 percent confidence level.

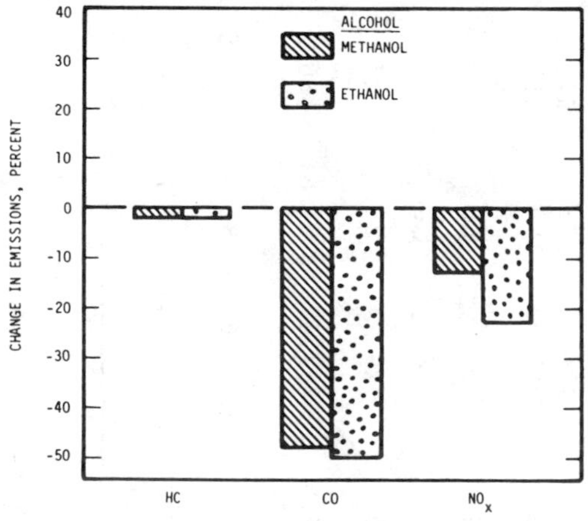

Fig. 1 Change in emissions caused by adding 10 percent alcohol to gasoline P, 1975 FTP, car X with production carburetor

The stoichiometric air-fuel ratios for alcohol combustion are lower than those for gasoline (Table 3); therefore, addition of alcohol to gasoline is similar to reducing the amount of fuel in the intake charge of an engine carbureted for gasoline combustion only. This leaning of the intake charge is expected to affect exhaust emissions according to the well established trends (17) illustrated in Figure 2. Thus, the effects of alcohol on exhaust emissions shown in Figure 1 were at least partially due to stoichiometry changes.

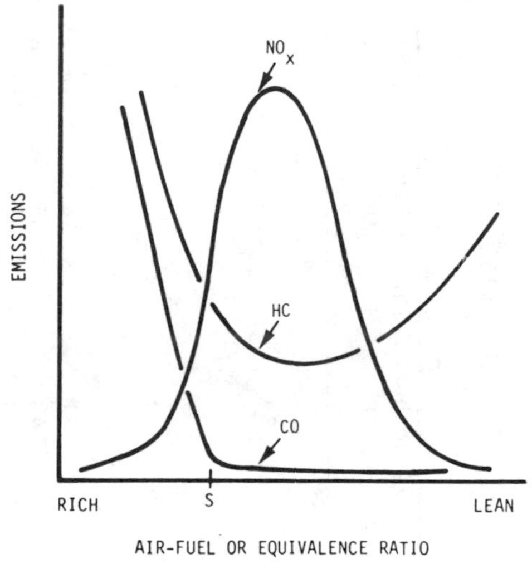

Fig. 2 Typical relationship between exhaust emissions and stoichiometry

To further explore how alcohol addition to gasoline affects exhaust emissions, tests were run in which stoichiometry was varied not only by varying the concentration and type of alcohol blended with gasoline, but also by modifying the carburetor of car X to obtain air-fuel ratios nominally 6 and 15 percent richer than that of the production carburetor (Appendix B). These two degrees of enrichening were achieved by varying the severity of the carburetor alterations as described in Appendix B, and were selected to compensate for the leaning effect which would be expected from the addition to gasoline P of 10 and 25 percent methanol, respectively.

The effect of 10 percent methanol on emissions is shown in Figure 3 for both carburetor modifications. With the nominally 6 percent richer carburetor, the results were similar to those shown in

Table 2. Properties of Test Fuels (ASTM Methods)

Properties	Fuels								
	Gasoline P	Clear Indolene*	Gasoline R	Gasoline RU-86	10% Methanol in P	25% Methanol in P	10% Ethanol in P	10% Methanol in Clear Indolene	10% Methanol in R
Reid Vapor Pressure, kPa	71	62	72	53	83	85	72	82	89
Distillation (Temperature, °C, Corresponding to Percentages of Fuel Evaporated):									
% Evaporated									
Initial	29	32	29	34	31	34	35	34	32
10	49	56	48	57	43	45	49	46	44
20	63	72	63	70	48	51	57	52	49
30	77	87	76	81	53	56	63	56	53
40	96	99	92	91	67	59	69	84	64
50	114	107	102	101	100	61	101	102	96
60	132	113	115	110	122	63	126	110	111
70	148	120	123	120	143	122	146	118	122
80	159	132	138	134	157	151	159	129	137
90	173	161	156	155	169	168	172	156	157
95	187	179	171	171	183	182	188	176	172
End Point	222	209	199	197	208	205	200	205	200
Paraffins, vol. %	62	65	66	73					
Olefins, vol. %	3	6	8	10					
Aromatics, vol. %	35	29	26	17					
Research Octane Number	96.0	97.6	91.6	86.2	98.8	102.5	99.0	100.6	96.0
Motor Octane Number	85.0	87.6	82.9	79.8	86.5	87.3	86.6	88.1	85.1
Specific Gravity	0.757	0.746	0.736	0.731	0.758	0.763	0.760	0.750	0.742

* Indolene HO III

Note: All gasolines were unleaded.

Table 3. Stoichiometry and Energy Content of Test Fuels

Fuel	Stoichiometric Air-Fuel Ratio	Lower Heating Value MJ/kg
Gasoline P	14.3	42.75
Clear Indolene	14.4	42.99
Gasoline R	14.6	43.13
Gasoline RU-86	14.6	43.51
Methanol	6.45	20.11
Ethanol	8.97	26.89

Figure 1 for the production carburetor; HC, CO, and NO_x decreased when methanol was added to gasoline. With the nominally 15 percent richer carburetor, methanol again decreased CO, but it increased HC slightly, and NO_x dramatically. As expected, the response of exhaust emissions to alcohol addition depends very strongly on the vehicle's carburetion. Contrary to expectations based on stoichiometry changes, CO decreased more with the production carburetor than with either of the modified carburetors. As will be shown later, this inconsistency may be due to data scatter in the lean region.

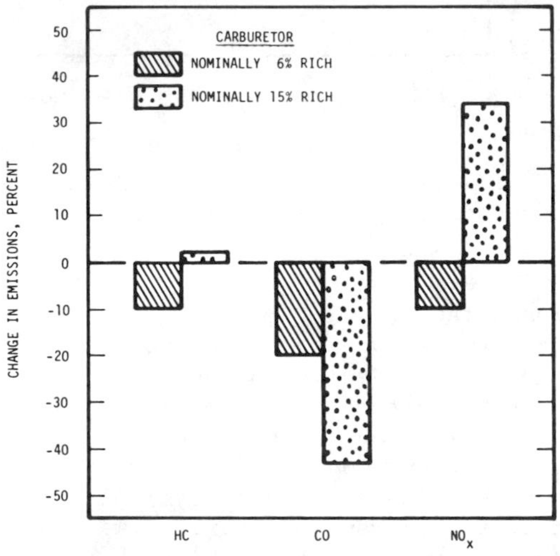

Fig. 3 Change in emissions caused by adding 10 percent methanol to gasoline P, 1975 FTP, car X with modified carburetors

To better evaluate the effects of stoichiometry on exhaust emissions, the FTP results obtained with all carburetors and all alcohol types and concentrations were plotted against equivalence ratio as shown in Figures 4, 5, and 6 for HC, CO, and NO_x, respectively. The equivalence ratio is the stoichiometric air-fuel ratio of the fuel divided by the air-fuel ratio during engine operation. Appendix B describes the determination of the average air-fuel ratio during the FTP with each of the three carburetors. These air-fuel ratios and the stoichiometric air-fuel ratios in Table 3 were used to calculate the equivalence ratios shown in Table 4.

Where data are plotted versus equivalence ratio, as in Figure 4, it is possible to identify the carburetor with which the data were obtained. Either with gasoline only, or with the 10 percent methanol-gasoline blend, each carburetor is represented by points at three different equivalence ratios: points at the smallest equivalence ratio were obtained with the production carburetor; points at the largest equivalence ratio were obtained with the 15 percent richer carburetor; and points at intermediate equivalence ratios were obtained with the 6 percent richer carburetor. With the 10 percent ethanol-gasoline blend, points are shown at one equivalence ratio only, because only the production carburetor was used. All the points for the 25 percent methanol-gasoline blend are also at one equivalence ratio, the one corresponding to the 15 percent richer carburetor.

Figure 4 shows the expected trend in HC emissions with equivalence ratio: in this range of equivalence ratios, the leaner the carburetion, the lower the HC emissions. However, the data follow an additional pattern. All the points obtained with gasoline and the three carburetors fall on one curve, whereas the data obtained with 10 percent methanol in gasoline fall on another curve at a higher level of HC emissions. This apparent increase in HC emissions was not observed with the 25 percent methanol blend. This could be experimental error, since in the lean region the gasoline and methanol curves appear to converge. Also, the low response of the HC analyzer to methanol referred to in Appendix A may be involved. The one point with ethanol lies between the gasoline and 10 percent methanol-gasoline curves.

Several factors may be responsible for the increase in HC emissions when comparing

Table 4. Air-Fuel and Equivalence Ratios for the Fuel-Carburetor Combinations Tested

Carburetor	Average Air-Fuel Ratio, Appendix B	Equivalence Ratio: Gasoline P	10% Methanol in Gasoline P	25% Methanol in Gasoline P	10% Ethanol in Gasoline P
Production	15.7	0.91	0.86	-	0.88
6% Richer	14.9	0.96	0.91	-	-
15% Richer	13.4	1.06	1.01	0.92	-

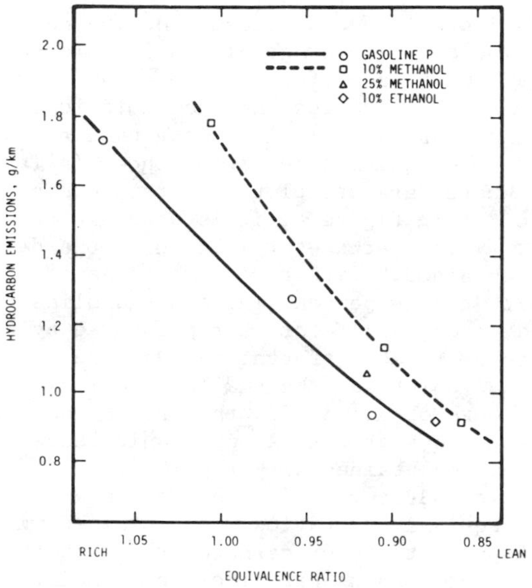

Fig. 4 Effects of alcohol and carburetion of hydorcarbon emissions, 1975 FTP, car X

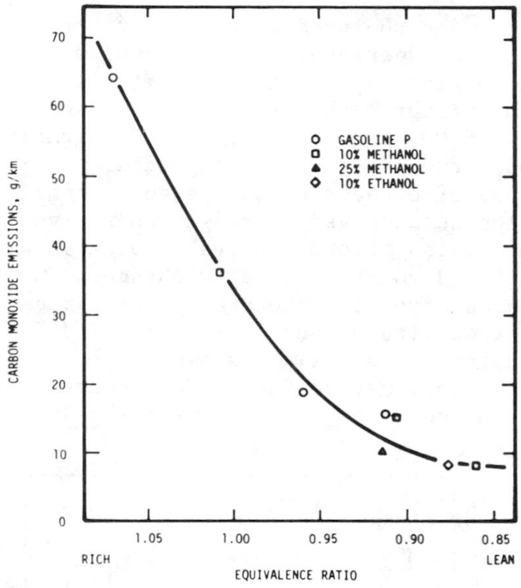

Fig. 5 Effects of alcohol and carburetion on carbon monoxide emissions, 1975 FTP, car X

methanol-gasoline blends to gasoline at the same equivalence ratio. One factor could be the increased mass of unburned fuel in the quench layer and crevices of the combustion chamber which are prime sources of exhaust HC emissions (18,19). The mass of fuel in these areas would be expected to be greater with methanol-gasoline blends than with gasoline alone, because at a given equivalence ratio the blends have a smaller air-fuel ratio than gasoline. This smaller air-fuel ratio indicates the presence of a greater amount of fuel per unit of air.

At stoichiometric conditions, the air-fuel mixture of the 10 percent methanol-gasoline blend contains 6 percent more fuel than that of gasoline. All other things being equal, this could cause a maximum increase of 6 percent in exhaust HC. As can be seen from Figure 4, the increase at the stoichiometric equivalence ratio (1.0) was much greater, about 24 percent. Thus, the increase in HC emissions must be due to additional factors.

Another factor may be the volatility of the methanol-gasoline blends which is greater than that of gasoline (Table 2). Fuel volatility can influence ease of engine starting and the amount of fuel evaporated during soak periods. During cycle 1 of the FTP (cold start), engine starting with gasoline could require more cranking than with the more volatile alcohol-gasoline blend; however, the higher heat of vaporization of the blend could counteract the volatility effect. Increased engine cranking could increase the amount of fuel in the engine. Some of this extra fuel could then pass through the engine unburned, thus increasing HC emissions.

During cycle 19 of the FTP, more fuel would be introduced into the engine with the alcohol-gasoline blend than with gasoline alone, because the blend would form more vapor during the hot-soak period. Again, this extra fuel could increase exhaust HC emissions.

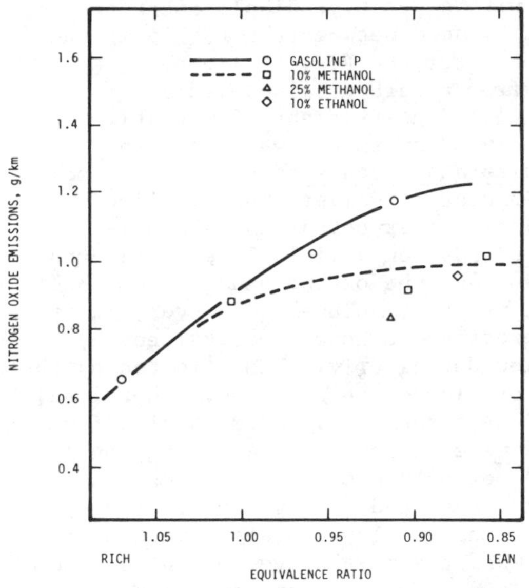

Fig. 6 Effects of alcohol and carburetion on nitrogen oxide emissions, 1975 FTP, car X

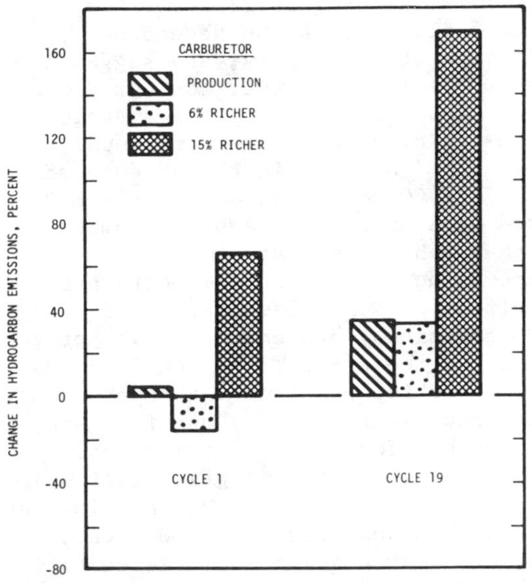

Fig. 7 Change in hydrocarbon emissions caused by adding 10 percent methanol to gasoline P, cycles 1 and 19 of 1975 FTP, car X with various carburetors

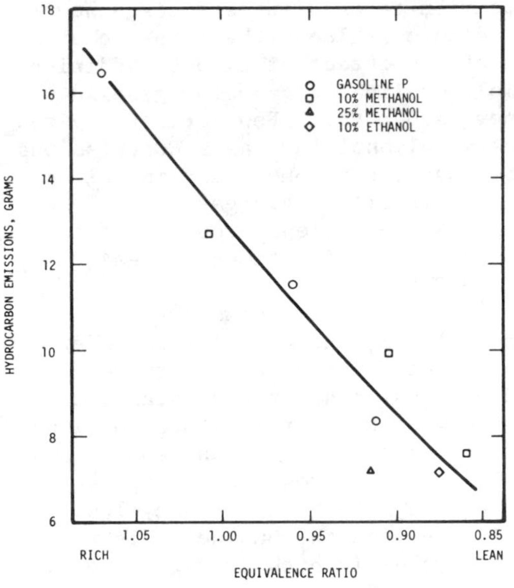

Fig. 8 Effects of alcohol and carburetion on hydrocarbon emissions, cycles 2-18 and 20-23 of 1975 FTP, car X

Experience suggests that starting difficulties could be irregular both in frequency and intensity. Therefore, they could magnify experimental error for results from either cycle 1 or cycle 19, but would not necessarily shift the results in one direction. Conversely, high volatility fuels should consistently generate more vapor during hot soaking than low volatility fuels. Consequently, HC emissions during cycle 19 should always increase as fuel volatility increases.

Figure 7 shows the change in HC emissions during cycles 1 and 19 of the FTP caused by the addition of 10 percent methanol to gasoline. These results generally follow the trends predicted on the basis of volatility considerations.

To further test the hypothesis that volatility effects were responsible for the two curves shown in Figure 4, HC emissions during cycles 1 and 19 were subtracted from the total FTP values and the results (HC emissions during cycles 2-18 and 20-23) were plotted versus equivalence ratio in Figure 8. All points now cluster about a single curve, suggesting that the effects of alcohol addition to gasoline can be explained by stoichiometry changes alone, but only when car operation is in regimes where fuel volatility has no effect. However, the 25 percent methanol point still deviates considerably from the rest of the data. Wigg (11,12) found similar effects of volatility on HC emissions. His and these results agree with respect to the importance of the hot-soak period, but Wigg also shows an effect of volatility in the middle portion of the FTP, cycles 6-18.

The effects, unrelated to stoichiometry, invoked to explain the higher HC emissions with methanol at a given equivalence ratio for the entire FTP do not apply to CO emissions. The data in Figure 5, which shows CO emissions during the entire FTP as a function of equivalence ratio, fall on a single curve which closely resembles the theoretical curve illustrated in Figure 2.

Figure 5 illustrates the dependence of CO emissions on stoichiometry, and suggests two other conclusions. First, for car X the assumptions used for computing the average equivalence ratio, as described previously, were reasonable. Second, in the absence of other data, the CO emissions during the FTP can be used as a gauge of the stoichiometry under which a vehicle operates.

Figure 6 shows that stoichiometry has the expected effect on NO_x emissions. As in the case of HC emissions, stoichiometry is not the only reason for the effect of alcohol on NO_x emissions. This follows from the apparent two curves shown in Figure 6: one for gasoline, and another for 10 percent methanol-gasoline blends. The reasons given earlier to explain the alcohol effects on HC emissions are not applicable to NO_x emissions. However, it is possible that at a given equivalence ratio, methanol and ethanol reduce NO_x emissions because their combustion temperatures are lower than those of gasoline (9,20). Wigg (11,12) has given similar results and explanations.

These vehicle studies have shown that alcohol addition to gasoline affects exhaust emissions because it leans the intake charge. Thus, much of the effect of alcohol addition can be simply duplicated by using gasoline and a leaner carburetor. However, at a given stoichiometry, alcohol increases HC emissions probably because the alcohol-gasoline blend has a higher volatility than gasoline alone, and decreases NO_x emissions probably because alcohol lowers combustion temperatures.

FUEL ECONOMY -

Experimental - Fuel economy was evaluated during the FTP, and on the road under the following conditions: constant speed, level road cruises; and three driving cycles commonly used within General Motors to simulate business district, suburban, and highway driving. Table 5 compares the average speed and stop frequency of the driving cycles. The FTP falls between the business district and suburban cycles in both average speed and stop frequency.

Car X, equipped with each of the three carburetors described previously, was used in the measurements. Fuel consumption was measured with a positive-displacement piston-type meter placed between the fuel pump and carburetor. For the FTP, fuel economy was computed from a carbon balance (21).

The pooled coefficient of variation for fuel economy measurements was 2 percent.

Results and Discussion - Since alcohols have lower heating values than gasoline (Table 3), it is expected that alcohol-gasoline blends would yield lower fuel economies than gasoline on the usual volume basis (distance per unit volume). However, it is also interesting to know if actual energy utilization during driving is affected by the presence of alcohol in gasoline. Therefore, in addition to the usual volume basis, fuel economies were expressed on an energy basis (distance per unit energy). Fuel economy results were converted from volume to energy basis by using the lower heating values of the fuels and their relative concentrations in each blend. Heat effects due to mixing were neglected because calculations showed they were insignificant.

Constant speed, level road load fuel economies with gasoline and 10 percent alcohol-gasoline blends are shown in Figure 9 on a volume basis. Car X was equipped with the production carburetor for these measurements. At low speeds, volume-based fuel economy was almost the same with either gasoline or the blends. However, at speeds greater than 48 km/h, 10 percent alcohol caused a loss in fuel economy. As expected from differences in heating value, the loss was greater with methanol than with ethanol.

Figure 10 shows the same data as Figure 9, but on an energy basis. If the effects of alcohol on fuel economy were solely due to differences in energy content, this replotting should have caused all the data to cluster about a single curve. Since this obviously did not happen, there must be other factors which influence fuel economy. One of these factors is probably related to the changes in the stoichiometry of the intake charge caused by the addition of methanol. It is well known that fuel economy is affected by stoichiometry (22).

To isolate the effect of stoichiometry, all the energy-based fuel economy data obtained with gasoline and alcohol-gasoline

Table 5. Comparison of Fuel Economy Cycles

	Average Speed, km/h	Stops/km
1975 FTP	31	1.5
Business District	26	2.4
Suburban	39	1.0
Highway	76	0.2

blends using all three carburetors were plotted versus the equivalence ratio obtained from measurements on the road at each of the test speeds. Results for five of the test speeds are shown in Figure 11, which displays

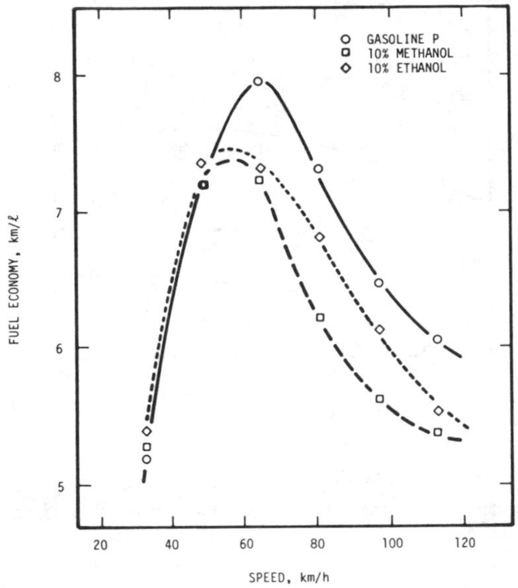

Fig. 9 Effect of adding 10 percent alcohol to gasoline P on level road load fuel economy (volume basis), car X with production carburetor

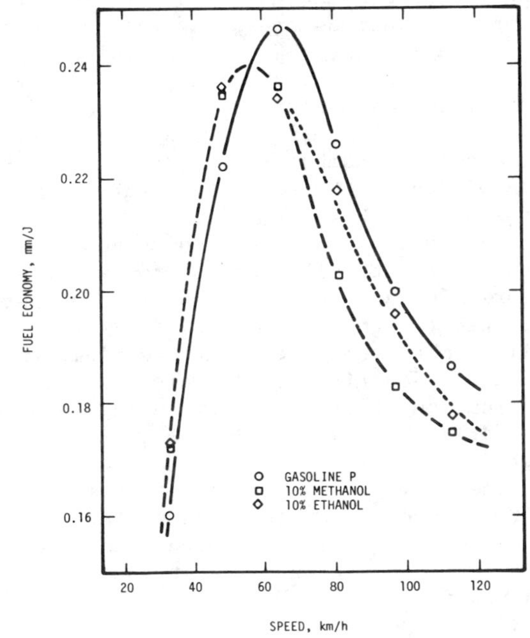

Fig. 10 Effect of adding 10 percent alcohol to gasoline P on level road load fuel economy (energy basis), car X with production carburetor

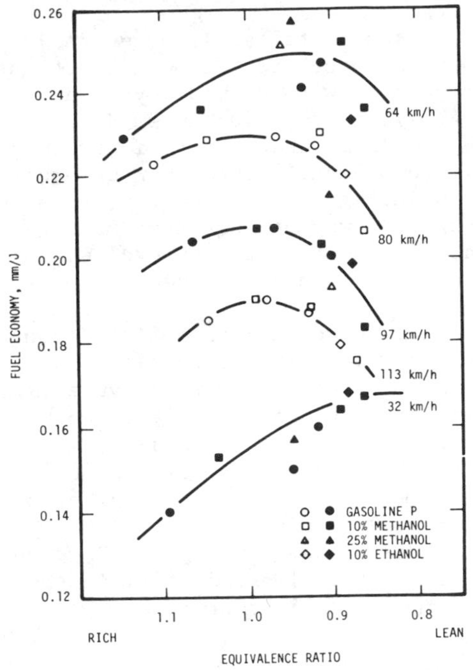

Fig. 11 Effects of alcohol and carburetion on level road load fuel economy (energy basis), car X

the classical pattern of fuel economy curves. Fuel economy decreases as the equivalence ratio either decreases or increases from the value at which fuel economy is maximum. However, this maximum fuel economy occurred at progressively richer equivalence ratios as car speed increased from 32 km/h to 80 km/h.

Figure 11 shows that when both energy content and stoichiometry differences are considered, the effects of alcohol on fuel economy can be explained. However, the data with 25 percent methanol at 80, 97, and 113 km/h disagree with this explanation. The reason for this discrepancy is not known. Adding alcohol to gasoline decreases volume-based fuel economy in direct proportion to the difference in heating value between alcohol and gasoline. However, the decrease can be augmented, reduced, and even canceled by the leaning of the intake charge caused by alcohol addition. This leaning effect will either decrease or increase fuel economy when basic engine carburetion is either on the lean or rich side of that for maximum economy, respectively.

The fuel economy results from the various driving cycles were treated in the same manner as that used for the constant speed data. First, the effects of alcohol on volume-based fuel economy were examined. Typical results are shown in Figure 12 for all four cycles and three carburetors, but with 10 percent methanol only. Again, the addition of methanol to gasoline generally decreased volume-based fuel economy, but the effect varied both with driving cycle and carburetion. To rationalize these differences, all the data were converted to energy-based fuel economy and plotted versus equivalence ratio in Figure 13. However, this equivalence ratio was not obtained from measurements during the driving cycles, but instead was computed as described in Appendix B.

The trends shown in Figure 11 for constant speed fuel economy are repeated in Figure 13 for the various driving cycles. The explanations given previously in conjunction with Figure 11 apply equally well to Figure 13. The effects of alcohol on fuel economy are explained by its effects on the energy content and stoichiometry of the alcohol-gasoline blend. In Figure 13, even the 25 percent methanol blends fall on the curves described by all other points.

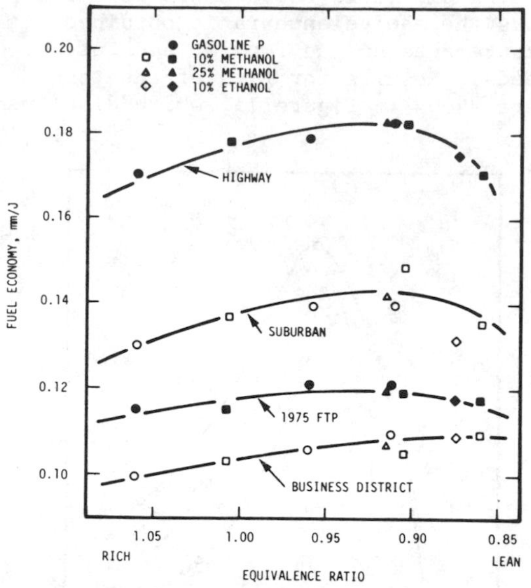

Fig. 13 Effects of alcohol and carburetion on fuel economy (energy basis), various driving cycles, car X

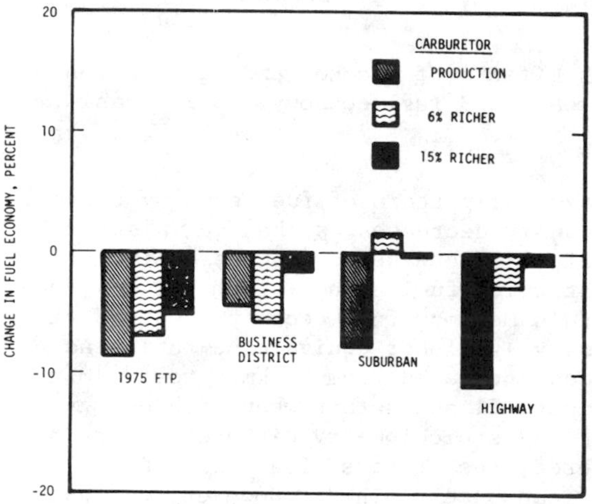

Fig. 12 Change in fuel economy (volume basis), caused by adding 10 percent methanol to gasoline P, various driving cycles, car X with various carburetors

The fuel economy results reported in this study agree with those previously published by Wigg (11,12) and help explain why older cars which were carbureted rich could show increases in fuel economy when using alcohol-gasoline blends rather than gasoline.

DRIVEABILITY -

Experimental - Car driveability was evaluated both on the road and on a chassis dynamometer located in a "cold" room. Measurements on the road were performed according to the 1972 CRC Intermediate Temperature Procedure (23) at an average temperature of 14°C (ranged from 10 to 17°C). For the measurements in the "cold" room (at temperatures from -34°C to 21°C), the CRC procedure was modified. Appendix C describes the driveability procedures.

The pooled coefficient of variation for all the paired driveability determinations was 7 percent for tests on the road, and 15 percent for tests in the "cold" room.

During each test, the time required to start the engine and the number of stalls were recorded. Various operating malfunctions such as hesitation, stumble, surge, idle roughness, and backfire were rated subjectively by the driver on a scale of trace, moderate, or heavy. To obtain a numerical value for driveability, a certain number of demerits was assigned to each of the subjective ratings as shown in Table 6. However, since all malfunctions are not of the same importance, the demerits were first multiplied by the weighting factors also shown in Table 6. The numbers thus obtained (weighted demerits) for each driveability maneuver were summed to yield the total weighted demerits (TWD), which were used as an indication of driveability during the test. As driveability deteriorates, the number of TWD increases.

Because it is difficult to subjectively rate driving problems on the dynamometer, continuous records of engine speed, manifold vacuum, throttle position, and car speed were taken as previously done by Stebar and Everett (24). Appendix C describes the method for determining TWD from these measurements.

The driveability measurements on the road were performed with car X (Table 1). All three previously described carburetors were used because stoichiometry affects

Table 6. Driveability Demerits and Weighting Factors

Demerits:

Trace	=	1
Moderate	=	2
Heavy	=	4
Start	=	[starting time (s) -2]

Weighting Factors:

Idle Roughness, start	=	1
Surge	=	4
Backfire, Stumble, Hesitation	=	6
Stall (idle)	=	8
Stall (maneuvering)	=	32

Calculation:

Total Weighted Demerits = Σ (Demerits x Wt Factor)

driveability (25). Gasoline P was the base test fuel. It was blended with methanol at 10 and 25 volume percent, and with ethanol at 10 volume percent for the road driveability tests.

The driveability measurements on the chassis dynamometer were performed with car Z described in Table 1. The base fuel was clear Indolene (Table 2). Methanol was blended with this fuel at concentrations of 10, 15, and 20 percent by volume.

Results and Discussion - With car X on the road, the total weighted demerits (TWD) with gasoline P were 88, 72, and 44 with the production, 6 percent richer, and 15 percent richer carburetors, respectively. Thus, enrichening the production carburetor, which was calibrated for lean operation (see Table 4), improved driveability.

As shown in Figure 14, adding alcohol to gasoline P significantly deteriorated vehicle driveability with each of the three carburetors. The alcohol blends caused more severe and/or frequent stumbles, surges, and stalls than gasoline alone. Normally, the richer the carburetor, the less the deterioration expected from the addition of methanol to gasoline. The results in Figure 14 do not conform to this trend. This is probably due to the amplification of differences between TWD of low numerical value when these differences are expressed as percentages.

Figure 14 also shows that the effect of alcohol on driveability varies with alcohol type and concentration and with carburetor stoichiometry. To determine whether changes in stoichiometry were solely responsible for the driveability changes, all driveability

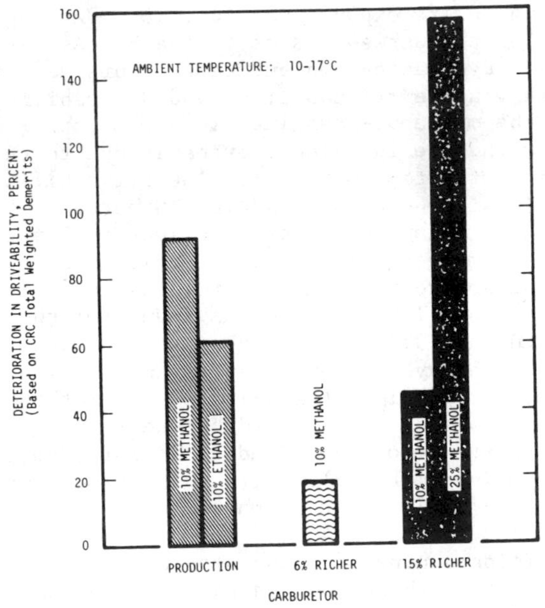

Fig. 14 Deterioration in driveability on the road caused by adding alcohol to gasoline P, car X with various carburetors

data (TWD) collected with car X on the road were plotted versus equivalence ratio in Figure 15. These equivalence ratios were obtained as described in Appendix B. With the exception of the data obtained with 25 percent methanol, all points fall remarkably close to a single curve. Consequently, driveability was a function of stoichiometry only, and as expected, driveability deteriorated as equivalence ratio decreased (leaned). The effect of stoichiometry on driveability was small at equivalence ratios as lean as 0.90, but it became extremely important at leaner ratios. Thus, using methanol-gasoline blends with a carburetor calibrated for lean operation with gasoline (as most carburetors are today) would severely deteriorate driveability at intermediate temperatures. On the contrary, with carburetors calibrated for rich operation (many pre-emission-control carburetors) using methanol-gasoline blends would not significantly affect driveability. The single curve obtained from all the data points in Figure 15 also suggests that the increase in volatility caused by the addition of alcohol to gasoline did not influence driveability at intermediate temperatures. This is contrary to what would be expected for gasoline alone (23).

The effect of ambient temperature on driveability was evaluated with car Z on a chassis dynamometer in a "cold" room. Total weighted demerits (TWD) obtained with the test gasoline and a 10 percent methanol-gasoline blend are shown as a function of ambient temperature in Figure 16. The curves shown in the figure are exponential functions developed from least squares fits of the data. As expected, decreasing ambient temperature deteriorated vehicle driveability, and driveability with the methanol-gasoline blend was poorer than that with gasoline alone, probably due to the leaning effect of methanol. The driveability deterioration caused by adding 10 percent methanol to Indolene was equivalent to lowering the ambient temperature from 20°C to -7°C.

An apparent deviation from the general trends shown in Figure 16 was found for the test run at -34°C. The engine stalled when the "driveaway" maneuver was attempted with gasoline, but it did not stall with the methanol-gasoline blend. This could be due to the higher vapor pressure of the blend as compared to that of gasoline (Table 2). With this 10 percent methanol-Indolene blend, phase separation did not occur even at -34°C.

To determine if phase separation would occur with methanol concentrations greater than 10 percent, additional tests were run at -29°C. With 15 percent methanol, no unusual problems occurred, but TWD increased from 168 with gasoline to 307 with the 15 percent blend. With 20 percent methanol, no "driveaway" could be achieved. The engine could be forced to idle by pumping the accelerator pedal, but attempts to "driveaway" resulted in 16 stalls and the test was aborted. Inspection of a sample removed from the bottom of the fuel can showed that two phases were present. Similar inspections of the 10 and 15 percent methanol blends showed only one phase. Apparently, clear Indolene in a moisture free fuel system can tolerate up to 15 percent methanol at temperatures as low as -29°C. However, other gasolines may not be as tolerant, especially if the concentration of aromatic hydrocarbons is small.

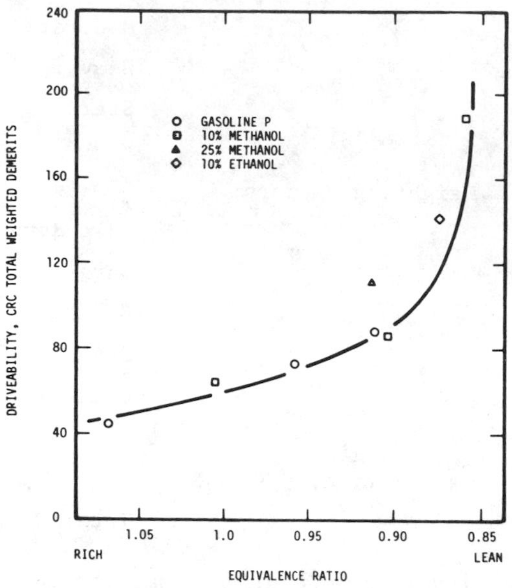

Fig. 15 Effects of alcohol and carburetion on driveability on the road, car X

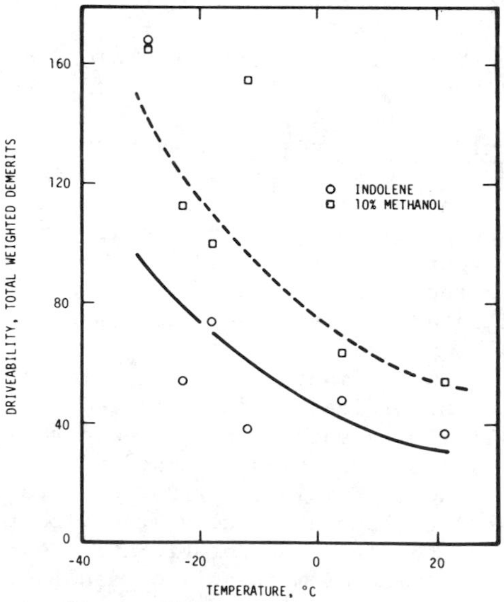

Fig. 16 Effects of temperature and methanol on driveability on chassis dynamometer, car Z

PERFORMANCE -

Experimental - The effect of alcohol addition to gasoline on car performance was evaluated by measuring the times required to perform wide-open-throttle (WOT) and part-throttle (PT) accelerations. The WOT accelerations were from 0 to 129 km/h, while the transmission shifted gears normally. The time required to reach the intermediate speed of 97 km/h and the final speed of 129 km/h was recorded. All PT accelerations were made in high gear, starting at 40 km/h and ending at 97 km/h. The three PT conditions were characterized by constant intake manifold vacuums of 20, 27, and 40 kPa.

Tests were run with car X equipped with the production or modified carburetors. Gasoline P, and varying concentrations of alcohol in gasoline P were used. The pooled coefficient of variation determined from triplicate runs was 1 and 3 percent for WOT and PT accelerations, respectively.

Results and Discussion - Figure 17 shows that with the production carburetor, 10 percent methanol in gasoline P had a negligible effect on WOT acceleration, but deteriorated car performance during PT accelerations by increasing acceleration time by as much as 51 percent. The effects of methanol with the two richer carburetors were similar, but not as large. With the production carburetor, the effects of ethanol were smaller than those of methanol. With gasoline only, PT acceleration time decreased as the carburetor was changed from production to the 6 and 15 percent richer versions. These observations are consistent with general knowledge (26) about the effects of stoichiometry on engine performance, and suggest that the results can be placed on a common basis by using the equivalence ratio computed from the measured air-fuel ratios shown in Appendix B.

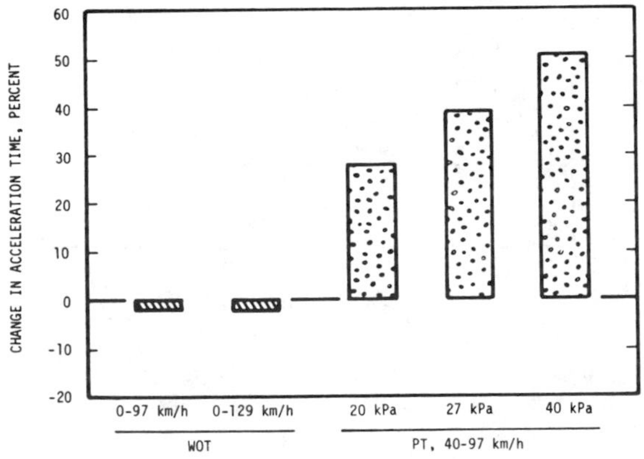

Fig. 17 Change in performance caused by adding 10 percent methanol to gasoline P, car X with production carburetor

Figures 18 and 19 show the acceleration times obtained with all the fuels and carburetors as a function of equivalence ratio. At WOT conditions (Figure 18), the effect of stoichiometry on acceleration time is very small, probably because the carburetor power enrichment circuit is fully operative and keeps the equivalence ratio on the rich side of best power, where the effects of stoichiometry on engine performance (brake mean effective pressure) are small (26).

At each PT condition, all data points fall on a single curve, supporting the as-

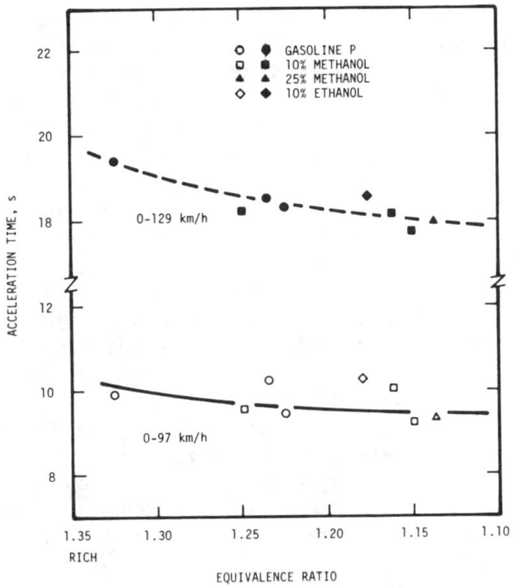

Fig. 18 Effects of alcohol and carburetion on wide-open throttle performance, car X

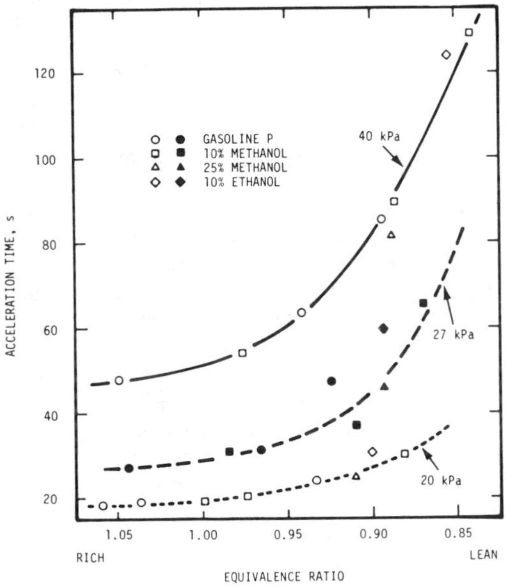

Fig. 19 Effects of alcohol and carburetion on part-throttle performance, car X

sumption that changes in acceleration time

were due to stoichiometry changes. The effects of stoichiometry on acceleration time are very pronounced, especially at equivalence ratios leaner than stoichiometric. This is attributed to the significant decrease in brake mean effective pressure caused by progressively leaning the equivalence ratio from its best power value (21). The effect of intake manifold vacuum on acceleration time shown in Figure 19 is also consistent with its effect on brake mean effective pressure, and reflects the influence of throttling losses (26).

Thus, alcohol addition to gasoline can either improve or deteriorate car performance depending on the initial carburetion of the car. For recent model-year cars, the effect would be adverse because they are carbureted lean, and further leaning by alcohol addition would decrease mean effective pressure.

OCTANE NUMBER -

Experimental - Octane numbers of the test fuels were measured in the laboratory using the usual ASTM procedures for Research and Motor octane (27), and on the road using the CRC Modified Borderline technique (28). As required by the CRC technique, the road octane measurements were performed at a variety of engine speeds. Car X equipped with the production carburetor was used for these tests. With this car, maximum knock occurred at WOT conditions.

Results and Discussion - The much publicized increase in octane number caused by blending alcohol with gasoline is demonstrated by the data in Table 2. Some of the same data were used for Figure 20, which shows the increase in octane number caused by adding 10 percent methanol to gasoline as a function of either the Research, or the Motor octane number of the base gasoline. Three interesting features are shown in Figure 20. First, the Research octane number increases considerably more than the Motor octane number when 10 percent methanol is added to gasoline. Second, as the octane number (either Research or Motor) of the base gasoline increases, the octane increase due to alcohol becomes progressively smaller. Third, the effect of 10 percent ethanol is comparable to that of 10 percent methanol.

To further understand the effects of alcohol on octane number, methanol-gasoline blends of various concentrations were studied. Figure 21 shows the results. With increasing methanol concentration, the Research octane number increased rapidly and almost linearly, but the Motor octane number increased slowly and only at the lower methanol concentrations. The resulting increase in fuel sensitivity (the difference between Research and Motor octane) would suggest that the blends with high methanol concentration would be no better in satisfying car octane requirements on the road

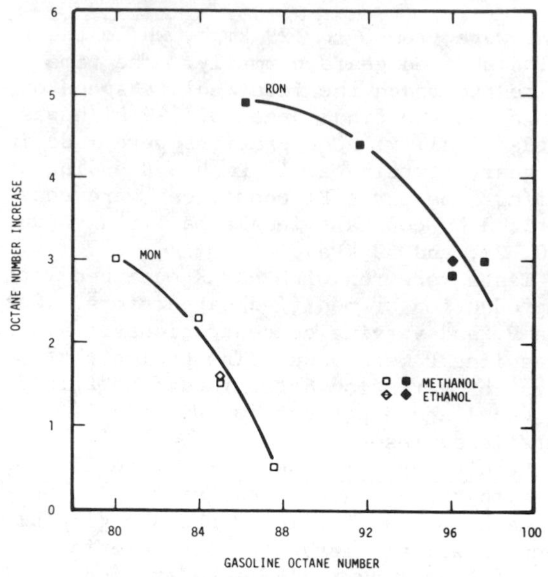

Fig. 20 Increase in research and motor octane number caused by adding 10 percent alcohol to various gasolines

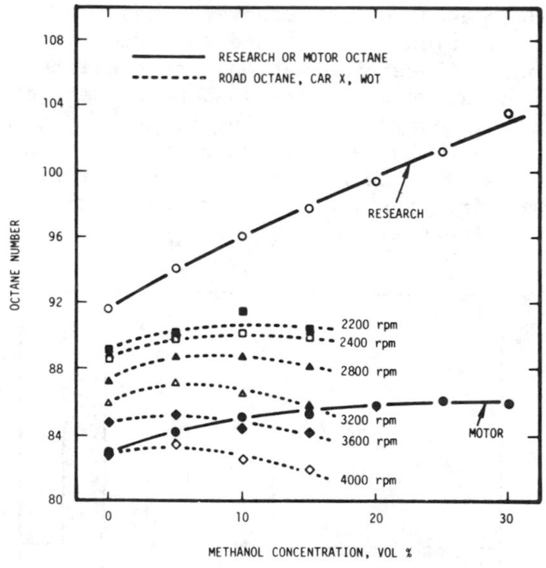

Fig. 21 Effect of adding methanol to gasoline R on research, motor, and road octane numbers

than the low methanol concentration blends. Also, since recent model-year cars require fuels of high Motor octane number rather than high Research number, the road octane number of the alcohol-gasoline blends would be only slightly higher than that of the base gasoline.

These expectations were confirmed by the road octane data also shown in Figure 21. As a matter of fact, at engine speeds greater than 2800 rpm, adding more than 5 percent methanol reduced the road octane number of the blend. This reduction was severe enough at 3600 and 4000 rpm to actually decrease the road octane of the blend to values smaller than those of the base gasoline. Thus, it appears that for alcohol-gasoline blends, the ASTM octane numbers do not adequately predict the performance of the fuel in cars on the road.

This apparent anomaly can be explained by considering that octane requirement is maximum at an equivalence ratio on the rich side of stoichiometric (1.1), and decreases at either richer or leaner ratios (22). The maximum octane requirement occurs at the same equivalence ratio as does the maximum brake mean effective pressure.

With gasoline P, car X had its highest octane requirement at 2400 rpm and wide-open-throttle. However, the measured equivalence ratio (Appendix B) at that condition was too rich (1.22) for this octane requirement to be maximum. Adding up to 15 percent methanol decreased the equivalence ratio towards values closer to 1.1 where octane requirement is maximum. Therefore in car X, methanol had two counteracting effects: it increased engine octane requirements by leaning the intake charge, whereas it increased fuel octane rating by improving resistance to detonation. This counteraction does not occur in the ASTM octane tests because air-fuel ratio is adjusted to obtain maximum knock for each fuel tested. Consequently, it is not surprising that the ASTM octane numbers did not adequately predict the anti-knock performance of the alcohol-gasoline blends in road tests.

The data show that the road octane ratings of alcohol-gasoline blends cannot be adequately predicted by ASTM Research and Motor octane numbers. Unfortunately, the customary use of Research octane number as the preliminary yardstick of a fuel's octane quality has created the misconception that alcohol-gasoline blends will automatically satisfy cars with high octane requirements, and thus permit the production of cars with improved fuel economy.

STUDIES WITH A REPRESENTATIVE VEHICLE POPULATION

The detailed studies reported in the previous section were intended to provide an understanding of the effects of alcohol addition to gasoline on vehicle exhaust emissions, fuel economy, driveability, performance, and road octane. With this understanding and suitable assumptions, the effect of blending alcohol and gasoline on entire vehicle populations could be predicted. However, automobiles are very complex machines, and the prediction process is perilous. Furthermore, the results of the detailed study were obtained with two recent model year vehicles only, and one can credibly argue against extrapolating the data to populations composed of both new and old vehicles of uncertain mechanical condition and varying mechanical designs and emission controls. Therefore, to better answer the practical question, "What happens to the exhaust emissions, fuel economy, and driveability of cars on the road if an alcohol-gasoline blend suddenly becomes the only available automotive fuel?" it was decided to obtain experimental information with a group of vehicles representing a large portion of the roughly one hundred million cars in the U. S. This question is not totally rhetorical, as such proposals have been considered in plans to alleviate liquid fuel shortages in the U. S.

A 10 percent methanol-gasoline blend was used for three reasons. First, the results of the detailed study indicated that 10 percent was a reasonable concentration. Second, methanol appears to be the most promising alcohol for automotive fuel use (3,4,5,6). Third, a 10 percent blend is a good compromise between the known miscibility and water sensitivity problems of methanol-gasoline blends (1,2,3,13).

TEST CARS AND FUELS - The 14 test cars are briefly described in Table 7. One car was a 1966 and another was a 1968 model with 135 500 and 121 000 accumulated kilometres, respectively. The remaining 12 cars were selected in matched pairs covering a wide range in engine displacements, and had been driven from 5 000 to 60 000 kilometres. One car of each pair was either a 1971 or 1972 model and the other was either a 1973 or 1974 model. The paired cars were divided in pre-1973 and post-1973 groups because 1973 was the year in which exhaust gas recirculation for control of nitrogen oxides was first used nationwide. Three of the cars were company-owned and used by employes for business transportation, and the other eleven cars were owned by employes. All of the cars had been driven in normal service.

The test fuels were clear Indolene and the 10 percent methanol-Indolene blend described in Table 2. As in the previously described tests, each car was modified with a special fuel line to permit operation from cans of test fuel during car conditioning and the emission, fuel economy, and driveability tests.

EXHAUST EMISSIONS -

Experimental - The test procedure and equipment were described in a previous section. For some of the tests, however, a heated flame ionization analyzer, Beckman 402, was used in addition to the unheated Beckman 108 analyzer to measure exhaust HC. This was to ascertain that no methanol was lost between the constant volume sampler and the detector. The results obtained did not show any loss in methanol due to condensation or adsorption.

Table 7. Vehicles for Representative Population Studies

Code	Model Year	Body Style	Engine Type	Engine Displacement, litres	Carburetor Barrels	Accumulated Kilometres
A	1972	B	V-8	7.5	4	35 000
B	1974	B	V-8	7.5	4	5 000
C	1972	C	V-8	7.7	4	60 000
D	1973	C	V-8	7.7	4	32 000
E	1968	B	V-8	5.4	4	117 000
F	1972	B	V-8	5.7	4	66 000
G	1973	B	V-8	5.7	4	19 000
H	1971	X	L-6	4.1	2	55 000
I	1974	X	L-6	4.1	2	5 000
J	1966	B	V-8	7.0	2	135 000
K	1971	B	V-8	6.6	2	51 000
L	1974	B	V-8	6.6	2	5 000
M	1971	H	L-4	2.3	1	60 000
N	1974	H	L-4	2.3	1	14 000

The pooled coefficient of variation was 6, 8, and 4 percent for HC, CO, and NO_x, respectively.

Results and Discussion - The HC, CO, and NO_x emissions of the 14 cars fueled with Indolene varied from 1.14 to 2.81 g/km, 6.65 to 48.34 g/km, and 1.16 to 3.80 g/km, respectively.

The percent change in exhaust hydrocarbon emissions caused by adding 10 percent methanol to Indolene, is shown in Figure 22 for each of the 14 test cars. The change in HC emissions varied from an increase of 41 percent to a decrease of 26 percent, and averaged an increase of 1 percent. The 1 percent average in-increase was not significant at the 95 percent confidence level and corresponded to an increase of only 0.02 g/km. Although the results are scattered, HC emissions from the 1966-8 and 1971-2 model cars generally decreased when methanol was added to Indolene, and HC emissions from the 1973-4 model cars generally increased.

To determine whether these changes were due to changes in stoichiometry caused by methanol addition to gasoline, the air-fuel ratio of each car must be known. In the absence of such information, the CO emissions during the FTP tests run with gasoline were used as a measure of the basic stoichiometry of each car. The validity of this approach was discussed previously in conjunction with CO emissions from car X. CO values only from cycles 2-18 and 20-23 were used as an indication of basic car carburetion. This was done to avoid any confusion which might have been introduced due to erratic or faulty choke operation, especially with old cars.

A previous section has illustrated that HC emissions during cycles 1 and 19 were dependent on factors other than stoichiometry. Therefore, the HC emissions during these two cycles were separated from the rest of the FTP. The change in HC emissions during cycles 1 and 19 caused by methanol for each of the test cars is shown in Figures 23 and 24, respectively. As with car X, the changes during cycle 1 are not unidirectional and probably reflect the irregular response of engine starting time to changes in stoichiometry and to the mechanical condition of each vehicle. Figure 24 shows that, with three exceptions, methanol increased HC emissions during cycle 19. As previously mentioned, this is attributed to increases in the volatility of the blend and the resulting greater fuel concentration in the intake system during the hot-soak period preceding cycle 19.

The decisions to eliminate cycles 1 and 19 from the CO emissions to obtain a better measure of basic car stoichiometry, and from the HC emissions to isolate the effect of methanol on HC due to stoichiometry changes, are supported by the data obtained with the 14 cars. A regression analysis of the change in HC emissions against CO emissions with Indolene yielded the following indices of determination (correlation coefficients squared) for a straight line relationship: 0.23 for HC

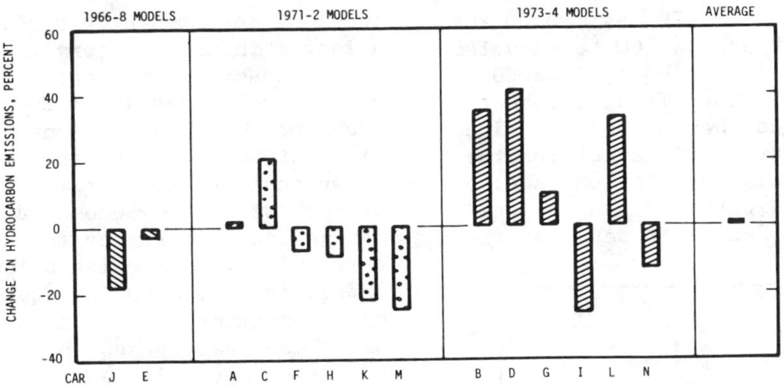

Fig. 22 Change in exhause hydrocarbon emissions caused by adding 10 percent methanol to indolene, 1975 FTP

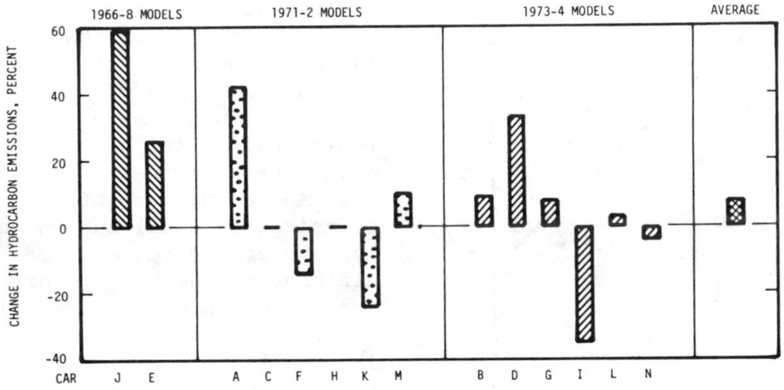

Fig. 23 Change in exhause hydrocarbon emissions caused by adding 10 percent methanol to indolene, cycle I of 1975 FTP

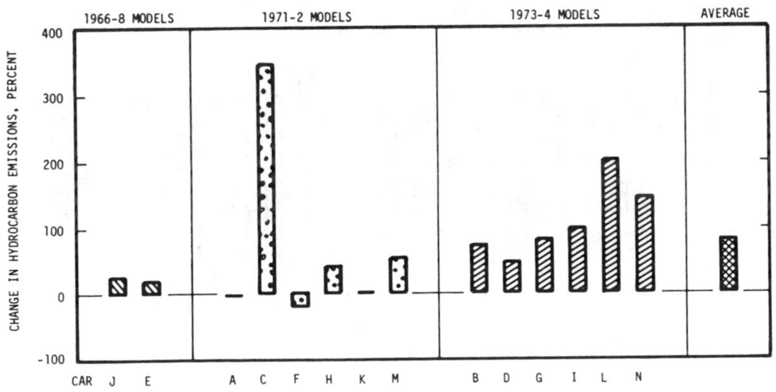

Fig. 24 Change in exhause hydrocarbon emissions caused by adding 10 percent methanol to indolene, cycle 19 of 1975 FTP

against CO from the entire FTP test; 0.35 for HC from the entire FTP against CO from cycles 2-18 and 20-23; and 0.63 for HC against CO from cycles 2-18 and 20-23. Thus, the best correlation is obtained when HC and CO values from cycles 1 and 19 are eliminated from the emissions results. This correlation is also shown graphically in Figure 25. The points in Figure 25 are somewhat scattered, but the

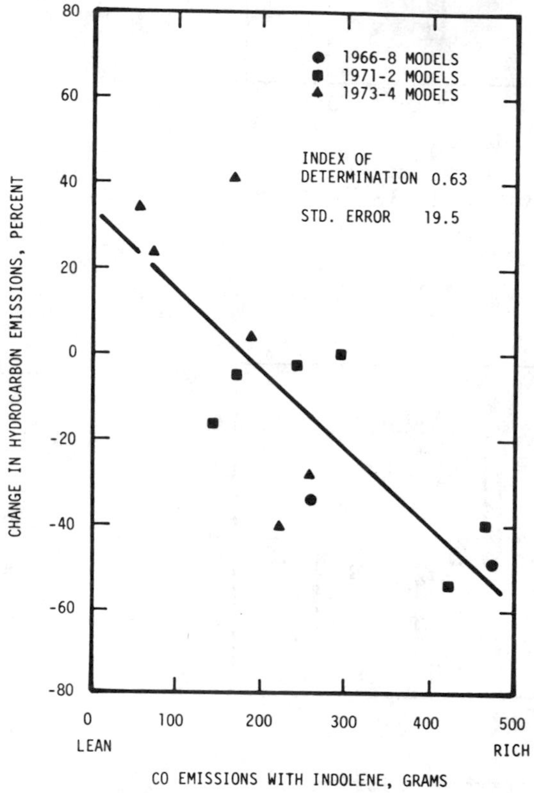

Fig. 25 Change in exhaust hydrocarbon emissions caused by adding 10 percent methanol to indolene correlated with carbon monoxide emissions with indolene, cycles 2-18 and 20-23 of 1975 FTP

index of determination of 0.63 is fairly good. A larger index of determination was not obtained for at least two reasons. First, HC emissions were correlated against CO mass emissions instead of CO concentrations. The CO mass emissions are affected by the variation in exhaust flowrates resulting from variation in vehicle inertia weights. Second, even if the observed changes in HC were due to changes in stoichiometry only, the correlation between percent change in HC and CO (or equivalence ratio) is only approximated by a straight line. This is a mathematical consequence of the shape of the HC versus equivalence ratio plot. Plotting percent change in HC for discreet increments of equivalence ratio from data such as those in Figure 2 yields a curve which can be approximated by a straight line.

Figure 25 shows that for cars with high CO emissions, adding methanol to Indolene decreased HC emissions. For cars with low CO emissions, adding methanol to Indolene increased HC emissions. These effects agree with those expected from changes in stoichiometry caused by addition of methanol to gasoline. They also explain why experiments with old cars only, which were usually carbureted rich, would lead to the conclusion that methanol always reduces HC emissions.

Adding methanol to Indolene decreased CO emissions with all but one car (Figure 26). The average decrease of 38 percent was significant at the 95 percent confidence level, and corresponded to a decrease in CO emissions of 9.82 g/km. Except for car B, the change in CO emissions was as generally expected from the leaning effect of methanol. The exception, car B, had the lowest CO emissions, 6.65 g/km, of the 14 cars tested. The 10 percent increase in CO emissions for car B with the methanol-Indolene blend was caused by an increase of 10.2 grams in CO emissions during cycles 1 and 19. This finding is consistent with previous assertions that changes in CO

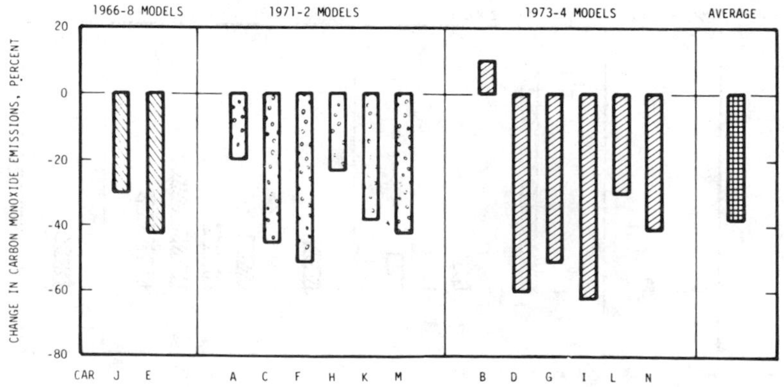

Fig. 26 Change in exhaust carbon monoxide emissions caused by adding 10 percent methanol to indolene, 1975 FTP

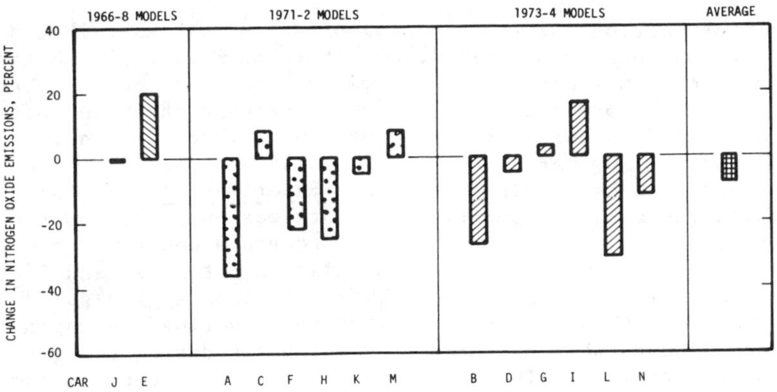

Fig. 27 Change in exhaust nitrogen oxide emissions caused by adding 10 percent methanol to indolene, 1975 FTP

emissions during cycles 1 and 19 are not exclusively related to stoichiometric effects. Therefore, these cycles should be eliminated from the results in efforts to correlate emissions with stoichiometry.

The percent change in CO emissions during cycles 2-18 and 20-23 caused by the addition of methanol did not correlate well with the CO emissions with Indolene during cycles 2-18 and 20-23; for a straight line relationship, the index of determination was only 0.10. This is again a mathematical consequence of the shape of the CO versus equivalence ratio curve (Figure 2). Unlike the case for HC, the percent change in CO for incremental changes in equivalence ratio plotted versus equivalence ratio is not approximated by a straight line. It has a minimum value at an equivalence ratio of 1.0, and higher values at either richer or leaner ratios. The previous comments about exhaust flowrates apply to this data also.

The change in NO_x emissions caused by adding methanol to Indolene (Figure 27) varied from an increase of 20 percent to a decrease of 36 percent, and averaged a decrease of 8 percent. The 8 percent average decrease was not significant at the 95 percent confidence level, and corresponded to a decrease in NO_x emissions of 0.18 g/km. The results also suggest that the effects of methanol were not influenced by the presence of exhaust gas recirculation in 1973 and 1974 model year cars.

The effect of methanol on NO_x emissions was generally consistent with expectations based on stoichiometry changes. Figure 28 shows the change in NO_x emissions as a function of CO emissions during cycles 2-18 and 20-23. Cycles 1 and 19 were omitted again because of the potential for erratic choke operation and startin difficulties with older cars. The cars with high CO emissions during cycles 2-18 and 20-23 were probably operating richer than the air-fuel ratio required for maximum NO_x emissions, and adding methanol to Indolene leaned the equivalence ratio and increased NO_x emissions. If the cars with low CO emissions were operating at an air-fuel ratio equal to or leaner than that required for maximum NO_x emissions, adding methanol to Indolene, which leans the equivalence ratio, would have decreased NO_x emissions. Also, the low combustion temperature of methanol would decrease NO_x emissions at all equivalence

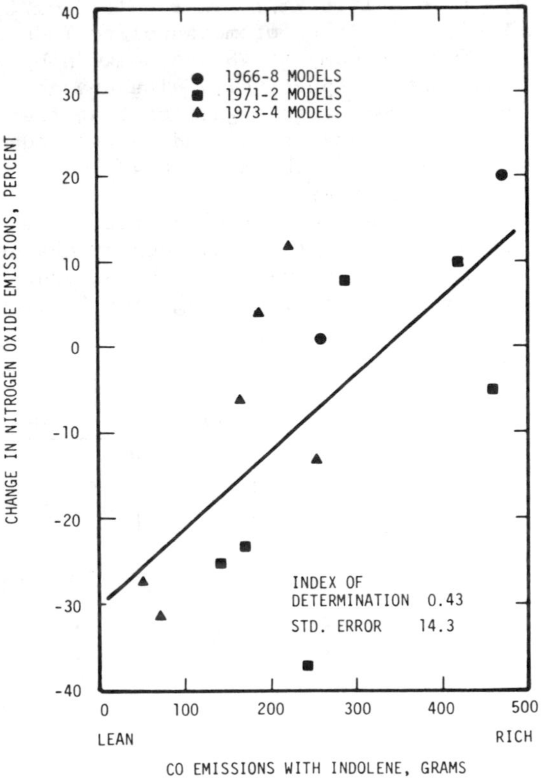

Fig. 28 Change in exhaust nitrogen oxide emissions caused by adding 10 percent methanol to indolene correlated with carbon monoxide emissions with indolene, cycles 2-18 and 20-23 of 1975 FTP

ratios. The index of determination for the results shown in Figure 28 was only 0.43 for a straight line relationship. This poor correlation is partially due to the previously mentioned flowrate effects, and to the fact that plotting percent change in NO_x versus equivalence ratio from data such as in Figure 2 yields a curve which can only be approximated by a straight line.

FUEL ECONOMY -

Experimental - Fuel economy was determined from measured fuel flow as described previously All measurements with the 14 cars were performe on the chassis dynamometer during the FTP, and at two steady-state conditions.

For all these paired tests, the pooled coefficient of variation was 1 percent.

Results and Discussion - During the 1975 FTP, fuel economy with Indolene ranged from 3.7 to 7.0 km/ℓ. The change in volume-based fuel economy due to the addition of methanol to Indolene is shown in Figure 29. This change varied from an increase of 6 percent to a decrease of 9 percent, and averaged a decrease of 3 percent. The 3 percent average decrease was significant at the 95 percent confidence level, and corresponded to a decrease in fuel economy of 0.13 km/ℓ. Although not illustrated, the fuel economy ranges with Indolene at 48 and 64 km/h were 6.0-15.8 km/ℓ and 6.4-15.2 km/ℓ, respectively. Addition of methanol to Indolene decreased fuel economy at 48 and 64 km/h by 5 and 4 percent, respectively. The changes at 48 and 64 km/h were also significant at the 95 percent confidence level, and corresponded to decreases in fuel economy of 0.43 and 0.38 km/ℓ, respectively.

Following the procedures and rationale described previously, the fuel economy changes, on an energy basis, caused by methanol addition to gasoline were plotted against CO emissions during cycles 2-18 and 20-23 of the FTP as shown in Figure 30. Methanol decreased volume-based fuel economy because it has a lower energy content and stoichiometric air-fuel ratio than gasoline.

DRIVEABILITY -

Experimental - The driveability of the 14 cars was evaluated on the road using the CRC procedure described previously. However, hesitation, stumble, and idle roughness were evaluated from engine speed and vacuum measurements as described in Appendix C. Ambient temperatures during these tests varied from -1 to 27°C. However, in comparing gasoline with the 10 percent methanol-gasoline blend in each car, the maximum temperature difference was 10°C. Gasoline R was the base fuel.

Since paired tests were not run, a pooled coefficient of variation could not be determined.

Results and Discussion - The driveability demerits with gasoline R ranged from a low of 18 to a high of 143. The percent change in driveability demerits caused by adding methanol to gasoline R is shown in Figure 31. The change in demerits varied from an increase of 427 percent to a decrease of 12 percent and averaged an increase of 104 percent. The 104 percent increase in demerits was significant at the 95 percent confidence level and corresponded to an increase of 69 demerits.

This deterioration in driveability was expected based on the effects of stoichiometry on driveability illustrated in Figure 15. A direct comparison of driveability demerits and stoichiometry for these cars was not possible. The air-fuel ratios of these cars were not measured, and the CO values from the FTP cannot be used since the FTP represents a different driving schedule than that used for driveability measurements.

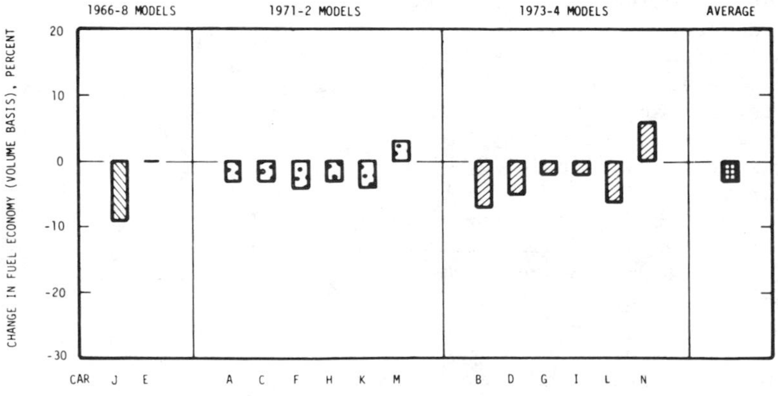

Fig. 29 Change in fuel economy (volume basis) caused by adding 10 percent methanol to indolene, 1975 FTP

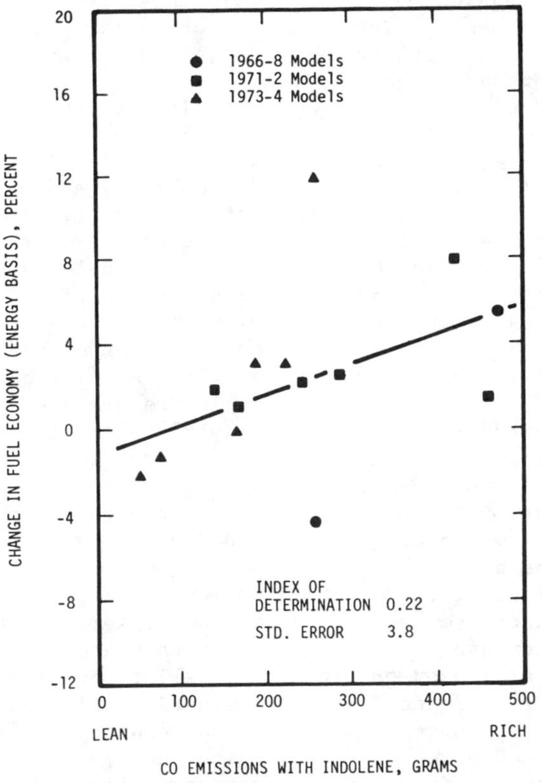

Fig. 30 Change in fuel economy (energy basis) during 1975 FTP caused by adding 10 percent methanol to indolene correlated with carbon monoxide emissions during cycles 2-18 and 20-23 of 1975 FTP with indolene

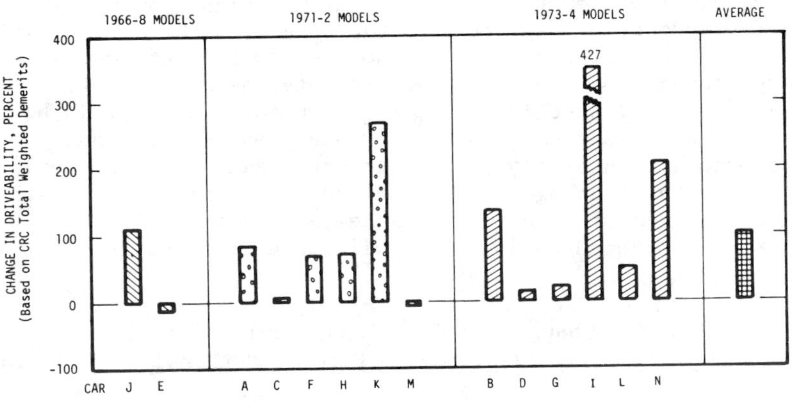

Fig. 31 Change in driveability (CRC total weighted demerits) caused by adding 10 percent methanol to gasoline

ALCOHOL-GASOLINE BLENDS - A TECHNICAL PERSPECTIVE

Findings in this paper suggest that the use of alcohol-gasoline blends in passenger cars would neither offer the tremendous advantages claimed by some, nor cause the insurmountable difficulties feared by others. Alcohol-gasoline blends behave according to well established principles. When stoichiometry, energy content, volatility, and combustion temperatures are considered, gasoline and alcohol-gasoline blends can be evaluated and compared on a fair and rational basis. In such a comparison, blending alcohol with gasoline has both advantages and disadvantages.

An estimate of the effect of using a 10 percent methanol-gasoline blend in all U. S. cars on the road through the 1974 model year can be obtained by arithmetically averaging the results obtained in this study. The estimate will be gross because the sample size is small, it represents a small fraction of the car population, and the arithmetic average does not account for the varying numbers of each model in the total population. Use of sales-weighted averages, however, cannot be justified.

Based on the average computed from the results with the 14 cars, adding 10 percent methanol to gasoline would cause the changes, at the 95 percent confidence level, shown in Table 8. Hydrocarbon emissions would not be affected, CO emissions would be decreased, NO_x emissions would not be affected, fuel economy (volume basis) would decrease, and driveability problems would increase.

The changes in emissions are based on quantitative observations and require no further comment. Some of the other changes, however, require some interpretation. In most instances, driveability and performance would be poor enough to merely annoy many drivers. In some cases, however, driveability could be unacceptable.
Fuel economy, on a volume basis, would decrease, but the only palpable effect would be the decrease in driving distance per tank of fuel. However, assuming adequate raw materials and production for methanol, the amount of liquid fuel available for automotive use would be increased - a very important advantage in terms of national energy reserves. Also the small octane advantage of alcohol-gasoline blends might alleviate knocking in some cars.

For 1975 model year cars with catalytic converter systems, the results of using methanol-gasoline blends instead of gasoline are expected to be similar to those for pre-1975 cars. Possibly, the carbon monoxide reduction may not be as substantial. Wigg (11,12) has presented data which agree with this.

If methanol-gasoline blends were marketed, future cars could be designed to accomodate the new fuel. To obtain acceptable and safe driveability and performance, richer carburetion would be provided. The effects of such a move on hydrocarbon and nitrogen oxide emissions would depend on the ultimate equivalence ratio of the new vehicles in accordance with the curves in Figure 2. The advantage of methanol for carbon monoxide emissions would be eliminated. The use of larger gas tanks could restore car driving distance to that possible with gasoline alone. Again, the liquid fuel supply for automotive use would be augmented.

These considerations of cars divorced from the fuel distribution system suggest that use of methanol to extend gasoline supplies is a reasonably sound approach. However, cars are fueled at service stations which are supplied via a vast and complex distribution system. The well recognized miscibility and water contamination problems with methanol-gasoline blends could generate difficulties in the distribution system. In addition, if it became necessary to remove volatile hydrocarbons from the methanol-gasoline blends to minimize vapor lock, the role of methanol as a gasoline extender would be diminished (11,12). Thus, when automobiles and the fuel distribution system are considered together, methanol's potential as a gasoline extender appears limited. Perhaps Wigg's (11,12) suggestion to augment automobile fuel supplies by substituting neat methanol in stationary power sources for the currently used petroleum products has merit.

Whether methanol is blended with gasoline or used neat in stationary powerplants, its impact on liquid fuel supplies would be immediate, but limited. If methanol is to be the fuel of the future as some believe (4,5,6), it will have to be burned neat in both stationary

Table 8. Changes in Emissions, Fuel Economy, and Driveability Caused by Adding 10 Percent Methanol to Gasoline

	95 Percent Confidence Limits for 14-Car Study	
	Change	Percent Change
HC	Not significant	Not significant
CO	-7.1 to -12.5 g/km	-27 to -48
NO_x	Not significant	Not significant
FTP Fuel Economy	-0.05 to -0.24 km/ℓ	-1 to -5
Cruise Fuel Economy	-0.26 to -0.70 km/ℓ	-3 to -8
Driveability	22 to 115 TWD	34 to 175

and mobile powerplants. The latter requirement presents an entirely new set of problems and opportunities for research.

SUMMARY AND CONCLUSIONS

The effect of blending alcohol - most frequently methanol - with gasoline on vehicle exhaust emissions, fuel economy, driveability, and performance was investigated with new and old cars. Adding methanol to gasoline (without altering the car's carburetor) decreased carbon monoxide emissions, volume-based fuel economy, driveability, and performance; however, it decreased, increased, or did not affect hydrocarbon and nitrogen oxide emissions. Ethanol and methanol behaved similarly. Alcohol addition to gasoline has these effects because the alcohol-gasoline blend has a lower stoichiometric air-fuel ratio, lower energy content, higher volatility, and lower combustion temperature than gasoline alone.

Considering air-fuel ratio effects alone, adding methanol to gasoline has the following effects. For cars carbureted "rich": exhaust hydrocarbons and carbon monoxide decrease; exhaust nitrogen oxides and energy-based fuel economy increase; and performance and driveability are not affected. For cars carbureted "lean": carbon monoxide also decreases; hydrocarbons, nitrogen oxides, and energy-based fuel economy increase, decrease, or are not affected, depending on the value of the equivalence ratio of the car-fuel combination with respect to the equivalence ratio at which each factor is at its minimum or maximum value.

At the same equivalence ratio, adding methanol to gasoline: increases hydrocarbon emissions because the volatility of the blend is greater than that of the parent gasoline (this increases the amount of fuel that can pass through the engine unburned); decreases nitrogen oxide emissions probably because it reduces combustion temperatures; and reduces volume-based fuel economy in proportion to methanol concentration, because methanol has a lower energy content than gasoline.

The Research and Motor octane numbers of methanol-gasoline blends do not predict the road octane number of the blend. This is partially due to the influence of equivalence ratio on vehicle octane requirements.

If the present U. S. car population were suddenly required to use 10 percent methanol-gasoline blends, then: hydrocarbon and nitrogen oxides emissions would not be affected, but carbon monoxide emissions would decrease; miles driven per tank of fuel would decrease; driveability and performance would deteriorate to an annoying or even unacceptable level; and knocking would be slightly decreased. Since carbon monoxide emissions with gasoline alone can also be reduced by mechanically leaning the carburetor, the only real advantage of using methanol-gasoline blends is that automotive fuel supplies in the U. S. would be augmented. This advantage is counterbalanced by the poor driveability of the methanol-gasoline blends, and by potential problems of separation of the blend's components in the marketing-distribution system of automotive fuels. Therefore, all practical alternative uses for methanol should be explored before deciding to blend it with gasoline for use in automobiles.

ACKNOWLEDGMENT

We are deeply indebted to the following colleagues who so capably performed the experimental work: C. G. Mitsopoulos, D. H. Coleman, R. G. Garvey, E. J. Grates, D. A. Hansen, E. G. Malzahn, W. E. Mili, and R. F. Milz of the Fuels and Lubricants Department; J. R. Collins, and G. J. Morris of the Environmental Science Department.

REFERENCES

1. J. A. Bolt, "A Survey of Alcohol as a Motor Fuel," First Part of SP-254, presented at SAE Summer Meeting, Chicago, Illinois, June 1964.

2. "Use of Alcohol in Motor Gasoline -- A Review," American Petroleum Institute, Washington, D. C., Publication 4082, 1971.

3. T. B. Reed and R. M. Lerner, "Methanol: A Versatile Fuel for Immediate Use," Science, Vol. 182, December 1973, pp. 1299-1304.

4. R. G. Jackson, "The Role of Methanol as a Clean Fuel," Paper 740642, presented at SAE Mid-Continent Section, October 1973.

5. W. D. Harris and R. R. Davison, "Methanol from Coal Can be Competitive with Gasoline," Oil and Gas Journal, Vol. 71, Dec. 17, 1973, pp. 70-72.

6. G. A. Mills and B. M. Harney, "Methanol -- The 'New Fuel' from Coal," Chemtech, January 1974, p. 26.

7. W. A. Scheller, "Agricultural Alcohol in Automotive Fuel -- Nebraska Gasohol," presented at the Eighth National Conference on Wheat Utilization Research, Denver, Colorado, October 1973.

8. H. G. Aldeman, D. G. Andrews, and R. S. DeVoto, "Exhaust Emissions from a Methanol-Fueled Automobile," Paper 720693, presented at SAE National West Coast Meeting, San Francisco, California, August 1972.

9. W. E. Bernhardt and W. Lee, "Combustion of Methyl Alcohol in Spark-Ignition Engines," Paper 136, presented at the 15th International Symposium on Combustion, Tokyo, Japan, August 1974.

10. M. W. Jackson, "Exhaust Hydrocarbon and Nitrogen Oxide Concentrations with an Ethyl Alcohol-Gasoline Fuel," Fourth Part of SP-254, presented at SAE Summer Meeting, Chicago, Illinois, June 1964.

11. E. E. Wigg and R. S. Lunt, "Methanol as a Gasoline Extender -- Fuel Economy, Emissions, and High Temperature Driveability," Paper 741008, presented at SAE Automobile Engineering Meeting, Toronto, Canada, October 1974.

12. E. E. Wigg, "Methanol as a Gasoline Extender: A Critique," Science, Vol. 186, November 29, 1974, pp. 785-790.

13. T. B. Reed, R. M. Lerner, E. D. Hinkley, and R. E. Fahey, "Improved Performance of Internal Combustion Engines Using 5-30% Methanol in Gasoline," Paper 749104, presented at the Ninth Intersociety Energy Conversion Engineering Conference, San Francisco, California, August 1974.

14. Federal Register, Vol. 37, No. 221, Part II, November 15, 1972, p. 24250.

15. "1973 Octane Number Requirement Survey," Coordinating Research Council (CRC), New York, Report No. 467, May 1974.

16. W. W. Wiers and C. E. Scheffler, "Carbon Dioxide (CO_2) Tracer Technique for Modal Mass Exhaust Emission Measurement," Paper 720126, presented at SAE Automotive Engineering Congress, Detroit, Michigan, January 1972.

17. T. A. Huls, P. S. Myers, and O. A. Uyehara, "Spark-Ignition Engine Operation and Design for Minimum Exhaust Emission," published in PT-12, Vehicle Emissions, Part II, New York: Society of Automotive Engineers, Inc., pp. 71-91.

18. W. A. Daniel and J. T. Wentworth, "Exhaust Gas Hydrocarbons -- Genesis and Exodus," published in PT-6, Vehicle Emissions, Part I, New York: Society of Automotive Engineers, Inc., 1964, pp. 192-205.

19. J. T. Wentworth, "The Piston Crevice Volume Effect on Exhaust Hydrocarbon Emission," Combustion Science and Technology, Vol. 4, October 1971, pp. 97-100.

20. E. S. Starkman, R. F. Sawyer, R. Carr, G. Johnson, and L. Muzio, "Alternative Fuels for Control of Engine Emission," Journal of the Air Pollution Control Association, Vol. 20, February 1970, pp. 87-92.

21. Federal Register, Vol. 39, No. 200, Part II, October 15, 1974, p. 36890.

22. D. F. Caris, B. J. Mitchell, A. D. McDuffie, and F. A. Wyczalek, "Mechanical Octanes for Higher Efficiency," SAE Transacions, Vol. 64, 1956, pp. 76-96.

23. "Driveability at Intermediate Temperatures -- 1972 Paso Robles," Coordinating Research Council (CRC), New York, to be published.

24. R. F. Stebar and R. L. Everett, "New Emphasis on Fuel Volatility -- Effects on Vehicle Warmup with Quick-Release Chokes," Paper 720934, presented at the National Fuels and Lubricants Meeting, Tulsa, Oklahoma, October-November 1972.

25. R. L. Everett, "Measuring Vehicle Driveability," Paper 710137, presented at SAE Automotive Engineering Congress, Detroit, Michigan, January 1971.

26. A. R. Rogowski, "Elements of Internal-Combustion Engines," New York: McGraw-Hill, 1953, pp. 129-139.

27. "ASTM Manual for Rating Motor, Diesel, and Aviation Fuels," Philadelphia: American Society for Testing and Materials, 1973, pp. 15-63.

28. "Modified Borderline Technique," Coordinating Research Council (CRC), New York, Designation F-27-70, June 1970.

29. S. H. Mick and J. B. Clark, Jr., "Weighing Automotive Exhaust Emissions," Paper 690523, presented at SAE Mid-Year Meeting, Chicago, Illinois, May 1969.

Appendix A

MODIFICATIONS TO THE 1975 FEDERAL TEST PROCEDURE (FTP) FOR EXHAUST EMISSIONS MEASUREMENTS

Instead of using the FTP specified constant volume sampler (CVS) bag measurements, exhaust emissions were determined from continuous measurements, which provide equivalent results (29) as well as more detailed individual mode and cycle information. Hydrocarbon, CO, CO_2, and NO_x concentrations were measured in the

CVS diluted exhaust gas stream. Hydrocarbons and CO were also measured in the CVS dilution air. Beckman model 108 flame ionization analyzers were used to measure HC in either the diluted exhaust or the dilution air. Beckman model 315A nondispersive infrared analyzers were used to measure CO and CO_2 in the CVS diluted exhaust gas, and a Bendix model 8501-5B nondispersive infrared analyzer was used to measure CO in the CVS dilution air. A Thermo-Electron model 10A chemiluminescent analyzer, equipped with a stainless steel converter heated to 725°C was used to measure NO_x in the diluted exhaust gas. The signals from each analyzer were scanned every one-half second by a General Automation model SPC-16/65 computer and integrated by an IBM model 370-135 computer to obtain mass emissions.

The FTP was further modified. Each emission test was started with the test car equipped with a purged evaporative emission control system charcoal canister. The purge lines from the carburetor or air cleaner were connected to the purged charcoal canister, and the line from the fuel tank remained connected to the car's original charcoal canister. This procedure may have masked some of the difference between the test fuels, since evaporative emissions from the fuels were probably not the same, and a different amount of fuel vapor should have been purged from the canister for each test fuel. The selected procedure, however, did improve the repeatability of the exhaust emission tests at the expense of obtaining information about any interaction between evaporative and exhaust emissions.

Hydrocarbon mass emissions were calculated in the usual manner from the product: (exhaust volume) times (hydrocarbon concentration fraction) times (hydrocarbon density). Exhaust hydrocarbon density was calculated from the ideal gas law and equaled 1.177 times the molecular weight of the exhaust hydrocarbon CH_xO_y species. Similar to the assumption made in the FTP, it was assumed that the exhaust hydrocarbon CH_xO_y species was the same as the fuel CH_xO_y species. Thus, instead of the $CH_{1.85}$ species and a density of 16.33 grams per cubic foot as specified in the FTP for Indolene, an exhaust hydrocarbon species of $CH_{1.71}$ and a hydrocarbon density of 16.17 grams per cubic foot was used for fuel P, and a hydrocarbon species of $CH_{1.82}O_{0.05}$ and a density of 17.24 was used for the 10 percent methanol-gasoline P blend. It was also necessary to change the species and density for Indolene because recent batches of Indolene have a lower hydrogen-carbon ratio than the ones used for establishing the values in the FTP. Therefore, a species of $CH_{1.78}$ and a density of 16.25 were used for Indolene, and a species of $CH_{1.89}O_{0.05}$ and a density of 17.33 were used for the methanol-Indolene blend.

Alcohol was not measured in the exhaust, and no compensation was made for the lower response of the flame ionization detector to alcohol. Therefore, some error may exist in the reported values for HC emissions from tests with alcohol-gasoline blends. Also, the HC analyzer used for the tests (Beckman 108) was not heated, and methanol could have been lost due to condensation and adsorption. The temperature of the diluted exhaust stream, however, was above the dew point. Also, some tests with a heated analyzer (Beckman 402) did not show any loss of methanol due to condensation or adsorption.

Appendix B

CARBURETOR MODIFICATIONS AND AIR-FUEL RATIO MEASUREMENTS

To obtain air-fuel ratios nominally 6 and 15 percent richer than the production carburetor of car X, the following carburetor alterations were performed: the main metering jets were enlarged; the part-throttle screw was adjusted; and the secondary metering rod hangers were changed to affect wide-open-throttle air-fuel ratio. In addition to these modifications, the idle mixture screws of the 6 and 15 percent richer carburetors were adjusted to give idle CO emissions with 10 and 25 percent methanol which were equivalent to the idle CO emissions obtained with gasoline and the production carburetor.

Table B-1 shows the air-fuel ratios obtained from measurements with the vehicle on the road equipped with either the production carburetor, or with each of the modified carburetors. Air-fuel ratios were computed from Orsat analyses of exhaust gas samples obtained during: constant speed, level road load cruising; constant vacuum part-throttle conditions; and wide-open-throttle (WOT) conditions. Constant vacuum part-throttle air-fuel ratios in Table B-1 are averages of results obtained at 1600, 2000, and 2400 rpm; and WOT air-fuel ratios are averages of results obtained at 2200, 2400, 2800, 3200, 3600, and 4000 rpm.

Air-fuel ratios for the production carburetor were measured either with gasoline P, or with 10 percent methanol in gasoline P. Air-fuel ratios with the two fuels were practically identical. This agrees with previously published observations (1,16) and with the expectation that methanol addition to gasoline would not significantly change fuel flow in the carburetor.

Air-fuel ratios with the two modified

carburetors were measured with blends of 10 and 25 percent methanol in gasoline P for the 6 percent and 15 percent richer carburetors, respectively. Air-fuel ratios for these carburetors were not measured with gasoline only, because a difference from the blends was not expected.

Table B-1 also shows the average of all the measurements both at cruising and at part-throttle, constant-vacuum conditions. This average was computed because in our judgement it best represents the overall air-fuel ratio of the vehicle during the FTP, cyclic fuel economy, and driveability tests. The wide-open-throttle data were excluded from these averages because they would have biased them towards the rich side. This would have misrepresented the results since wide-open-throttle conditions were infrequently encountered during these tests. This average can also be used to classify the carburetor as rich or lean with respect to either other carburetors, or the stoichiometry of the fuel. Thus, dividing the stoichiometric air-fuel ratio of a given fuel by this average air-fuel ratio yields the equivalence ratio for each fuel as shown in Table 4.

Appendix C

DRIVEABILITY

Driveability measurements on the road were performed according to the 1972 Coordinating Research Council (CRC) Intermediate Temperature Procedure (23). For the measurements on the chassis dynamometer in the "cold" room, the CRC procedure was modified.

The CRC procedure consists of a cold start (after an overnight soak) followed by 2.6 km of driving through various maneuvers such as light-throttle acceleration, cruise, detent acceleration,* full-throttle acceleration, crowd acceleration,** and idle. During each test, the time required to start the engine and the number of stalls are recorded. Various operating malfunctions such as hesitation, stumble, surge, idle roughness, and backfire

---

* Acceleration at a constant throttle position just prior to downshift.

** Acceleration at a constant vacuum just prior to power enrichment.

Appendix Table B-1

Air-Fuel Ratios Obtained with the Production and Modified Carburetors

Test Conditions	Air-Fuel Ratio: Production Carburetor	Nominally 6% Richer Carburetor	Nominally 15% Richer Carburetor
Level Road Load Cruise Speed, km/h			
32	15.6	15.1	13.0
48	15.5	14.9	12.8
64	15.7	15.2	12.8
80	16.0	15.2	13.2
97	16.1	15.2	14.0
113	15.9	15.0	14.0
Part-Throttle Constant-Vacuum, kPa			
20	15.3	13.8	13.5
27	15.5	14.8	13.7
40	16.0	15.2	13.8
Average of Cruise and Part-Throttle	15.7	14.9	13.4
Percent Change From Production		5.1	14.6
Wide-Open-Throttle	11.6	11.7	10.8

are rated subjectively by the driver on a scale of trace, moderate, or heavy. As described in the main text, these ratings are converted to total weighted demerits (TWD).

Two major modifications of the CRC procedure were required for the "cold" room tests. To avoid wheel slip on the dynamometer rolls at cold temperatures, wide-open-throttle accelerations were eliminated. Because it is difficult to subjectively rate driving problems on the dynamometer, engine speed, manifold vacuum, throttle position, and car speed were continuously recorded as previously done by Stebar and Everett (24). However, the driving procedure and method of rating hesitation, stumble, and idle quality slightly differed from those of Stebar and Everett. Their driving procedure did not closely follow the CRC procedure, and they used the recorder traces only as aides in assigning demerits to hesitation, stumble, and idle quality. A more quantitative approach was used in this work.

Figure C-1 shows engine speed versus time traces during normal conditions, and during either hesitation or stumble. The intensity of each malfunction was gauged by measuring the time "t" as shown in Figure C-1. These times "t" were then converted to ratings according to the schedule shown in Table C-1.

Figure C-2 shows a typical trace of intake manifold vacuum at idle versus time. Idle quality was measured in terms of the maximum vacuum fluctuation observed - the quantity "v" shown in Figure C-2. Table C-1 shows the ratings assigned to various values of "v".

The ratings in Table C-1 were converted to total weighted demerits as described in the main part of the report.

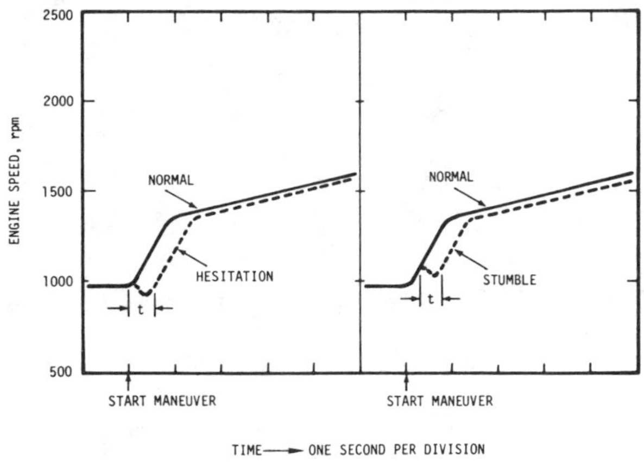

Fig. C-1 Method for rating hesitation and stumble intensity, light throttle acceleration from 0 to 40 km/h

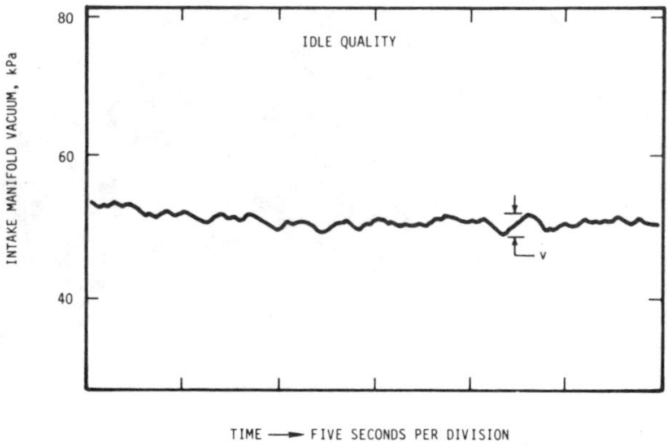

Fig. C-2 Method for rating idle quality

Appendix Table C-1

Conversion of Chart Measurements to Ratings

Hesitation and Stumble, "t", s	Idle Quality, "v", kPa	Rating
0.0 - 0.1	0.0 - 1.3	Clear
0.2 - 0.3	1.4 - 2.0	Trace
0.4 - 0.7	2.1 - 3.7	Moderate
0.8 and up	3.8 and up	Heavy

Methanol as a Motor Fuel or a Gasoline Blending Component*

J. C. Ingamells and R. H. Lindquist
Chevron Research Co.

THE IDEA OF USING ALCOHOL AS A MOTOR FUEL is as old as the automobile itself. Many studies have been made to assess the economics of making large quantities of ethyl alcohol from corn and grain for automotive use. More recently, the use of methanol as a motor fuel or as a gasoline blending component has been advocated in several articles (1, 2, 3).* Indeed, methanol is used in racing because of its special characteristics. However, the technique of starting these racers does not lend itself to the family car owner. They must be started with high speed electric starters or by pushing to get cylinder temperatures elevated for combustion.

Proposals to provide an independent energy supply for the U.S. in the 1980's include building nineteen 16,000—ton/day methanol plants which would burn 500,000 tons of coal per day to generate the equivalent of 2 million barrels of methanol (4). This volume of methanol corresponds to 15–20% of the motor gasoline

*Numbers in parentheses designate references at end of paper.

*Paper 750123 presented at the Automotive Engineering Congress and Exposition, Detroit, Michigan, February 1975.

ABSTRACT

Laboratory and road tests showed methanol to be an effective octane booster. Adding 10% methanol to unleaded gasoline raised the Road octane 2–3 numbers. However, significant deterioration in driveability tests occurred because of methanol's "leaning" effect. The water sensitivity of methanol/gasoline requires a separate fuel distribution system. Fuel storage in a vehicle must be protected from water absorption. Corrosion and degradation problems occur in the vehicle fuel system where methanol/gasoline mixtures contact lead, magnesium, aluminum, and some plastics. Methanol burned more efficiently under lean conditions than gasoline. However, the cold start problems require a separate starting fuel. Methanol is not a useful fuel additive for existing unmodified cars. Methanol could be used effectively in special vehicles designed to handle the corrosion, water absorption, and vaporization characteristics. The cost of manufacture and distribution in a separate system that overcomes the water sensitivity problem will determine the extent of methanol's use as a vehicular fuel.

SAE/PT-80/19/$02.50

requirements in the early 1980's. Before massive funds are invested in plants for making methanol from coal, we must consider the effects methanol will have on the fuel distribution system, the automobile fuel system, and the driveability of cars designed for 100% gasoline fuel. Also, the advantage or burden to the motoring public should be considered relative to other alternatives for enhancing the domestic U.S. gasoline supply.

A three–year test program was conducted at Chevron Research Company to examine all aspects of the use of methanol as a motor fuel or as a gasoline blending component. The results of the study are presented here.

Fuel Blending Properties of Methanol

It is well known that alcohols and other oxygenated hydrocarbons have unusual blending properties with hydrocarbons, and methanol is the extreme example since it has the greatest hydrogen bonding character (CH_3OH). One of the resulting differences between methanol and gasoline is that methanol can dissolve unlimited amounts of water, while gasoline can dissolve only about 200 ppm of water. Other major differences are in heating value and heat of vaporization. As shown on Table I, methanol has a heat of combustion half that of gasoline; but its latent heat of evaporation is nearly four times that of gasoline. This means that if cars were to run on straight methanol, they would need double–sized fuel tanks for equivalent driving range and much more heat on the intake system to vaporize the fuel.

Methanol has unusual effects on volatility when mixed with gasoline. In a distillation test, it causes the blend to distill much more rapidly at temperatures below methanol's boiling point, 149°F (65°C). (See Figure 1.) At higher temperatures, the effect becomes small, however. Methanol shows a positive deviation from Raoult's law for molar vapor pressure additivity when added to gasoline. The vapor pressure of the blend is higher than either component.

A problem associated with methanol as a motor fuel is its air/fuel ratio (A/F) requirements. The theoretical A/F ratio for complete combustion is 6.4 lb/lb of fuel; whereas, gasoline requires an A/F ratio of about 14.5. For methanol/gasoline blends, the theoretical A/F ratios fall between the two extremes, e.g., 13.6 for a 10% methanol blend – about 6% richer than required for gasoline alone.

TABLE I

PROPERTIES OF METHANOL AND GASOLINE

	Methanol	Gasoline
Density		
Lb/Gal.	6.6	5.9 to 6.4
g/cc	0.79	0.71 to 0.77
Lower Heating Value		
Btu/Gal.	57,000	111,000 to 116,000
kg–Cal/l	3,800	7,400 to 7,700
Latent Heat of Evaporation		
Btu/Gal.	3,320	ca. 900
kg–Cal/l	220	60
Boiling Point		
°F	149	90 to 430
°C	65	32 to 221
Reid Vapor Pressure		
psi	5	6 to 15
kPa	34	41 to 103
Stoichiometric Air/Fuel Ratio, Lb/Lb	6.4	ca. 14.5
Research Octane No.	ca. 112	91 to 100
Motor Octane No.	ca. 90	82 to 92

FIGURE 1

EFFECT OF METHANOL ON FRONT–END VOLATILITY

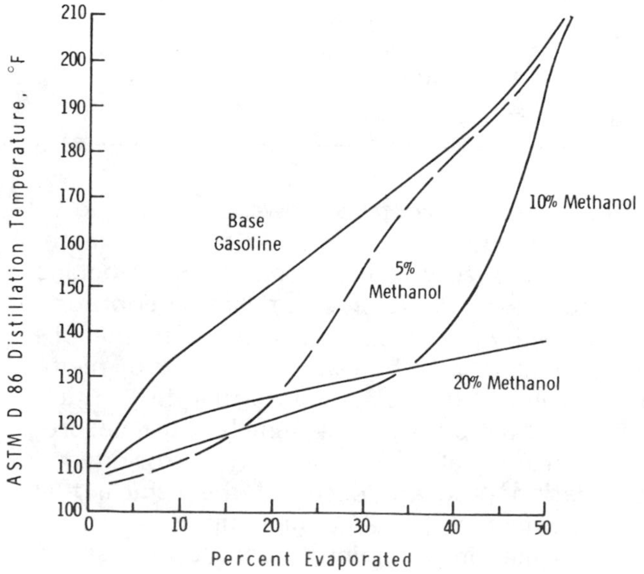

Gasoline Octane Boosting

Methanol's most desirable feature as a motor fuel is its high octane quality. As measured by the Research octane method, methanol rates 106–115 octane numbers (O.N.), depending on the investigator (5). Similarly, the Motor octane ratings are 88–92 O.N. As Table I shows, methanol is equal to the best gasoline on the basis of the Motor test and much better judging by the Research method.

Even more impressive is methanol's ability to "boost" gasoline's octane quality. Table II shows octane data obtained with three different unleaded gasolines. The Research O.N. was increased 4.2 to 4.6 O.N. by adding only 10% methanol. This gives the extremely high octane blending value of 135 O.N., on the average. (A blending value is calculated from the measured octane data using a simple linear equation.) The Motor octanes increased 1.6 to 2.5 O.N., indicating an average blending value of 102.5. Chassis dynamometer tests were run with a fleet of ten 1967–1971 cars to provide Road octane data on the same fuels. The results fell between the Research and Motor data, with boosts of 2.4 to 3.15 O.N.

TABLE II

OCTANE BOOSTING WITH METHANOL

Base Fuel (Unleaded)	Testing Method*	No Methanol		5% Methanol		10% Methanol	
		Octane No	Blending Value	Octane No	Blending Value	Octane No	Blending Value
A	Research	90.9		92.7	126.9	95.1	132.9
	Motor	82.0		82.1	84.0	83.8	100.0
	Road	88.9		90.2	114.9	91.7	116.9
B	Research	91.8		93.9	133.8	96.1	134.8
	Motor	83.5		84.0	93.5	85.1	99.5
	Road	89.6		90.9	114.6	92.0	112.8
C	Research	90.7		93.0	136.7	95.3	136.7
	Motor	82.1		83.6	112.1	84.6	107.1
	Road	89.2		91.2	128.2	92.35	120.5

*Road octane data obtained with a fleet of ten 1967–1971 cars

The Road octane data were obtained with cars now four to eight years old. With newer cars, Road octanes are known to correlate more closely with Motor octanes, and thus their response to methanol may be smaller, i.e., 1.6 to 2.5 O.N. with 10% methanol. Also, the A/F leaning effect of methanol may be different in newer cars because newer cars already operate in leaner ranges of A/F's.

Power Output

Methanol has been the fuel of choice in automobile racing for several decades because of its high power output. The heat of combustion of equal volumes of stoichiometric air/methanol and air/gasoline mixtures are nearly identical. However, more power is obtained with methanol because its higher latent heat of vaporization cools the air entering the engine much more than gasoline, and this increases the air density and the mass flow. The gain in power output with methanol is as much as 10% if very rich mixtures are used. With normal mixtures, as used in present–day cars, the benefits are smaller.

Cold Starting Tests

Tests with a 1971 compact Make D car and a 1970 full–size Make B car showed that cold starting is a severe problem with 100% methanol. Below about 60°F, the cars would not start unless a volatile material, such as ether, was sprayed into the intake air. In order to study the cold start problem, a series of tests were run at 40°F with the Make D car. Although this car had intake port fuel injection, and was modified to run on methanol, it would not start after "soaking" at ambient temperatures below about 60°F. Butane, isopentane, and methyl ether were tested as starting additives. As shown on Table III, at least 10% butane or 20% isopentane was required for acceptable starting. This solution for the problem might not be

TABLE III

COLD STARTING* 1971 COMPACT MAKE D

Fuel	Lean, %	Starting Time, Sec.	No. of Restarts
Gasoline	0.0	1.6	1
Gasoline	10.0	2.0	0
Methanol	14.6	No Start	–
Methanol + 4% Butane	15.8	No Start	–
Methanol + 10% Butane	15.7	2.0, 20.0	1, 0
Methanol + 4% Methyl Ether	13.3	No Start	–
Methanol + 20% Isopentane	17.3	7.0, 2.6	4, 0

*Starts made after soaking at least four hours at 40°F. Modified cold start valve used with methanol blends to provide about three times the normal fuel volume during starts.

practical for large--scale use of methanol as a motor fuel because these light hydrocarbons may be in short supply. Methyl ether at 4% did not allow starting and, again, higher concentrations would not be practical.

A good solution for cars equipped with the typical electronic fuel injection system would be to provide a very small tank of gasoline connected to the "cold start" valve, which sprays fuel into the intake air during starting. Introducing gasoline instead of methanol would allow easy starting. With carbureted engines, a liquefied petroleum gas starting system could be used. A solenoid valve activated by the starting system would supply the gas during starting.

Experience in our other test programs indicated that cold starting is not a problem with methanol/ gasoline blends containing up to 40% methanol.

Vapor Lock

Another problem of concern is vapor lock, which is caused by gasoline boiling in the fuel pump inlet, thereby limiting the flow to the carburetor. It is believed that straight methanol will not cause vapor lock, but low concentrations of methanol have drastic effects on front–end gasoline volatility. In a typical gasoline, 10% methanol causes a 2.5 psi increase in Reid vapor pressure (ASTM D 323) and a 22°F decrease in 10% point (ASTM D 86), as shown on Figure 1. With this blend, it is estimated that vapor lock would occur at 15°F to 20°F lower ambient temperatures, relative to the base gasoline. This conclusion is based on the assumption that methanol/gasoline blends behave the same as normal gasolines when volatility is considered. This assumption was verified by Messrs. Wigg and Lunt in vapor lock tests with 13 cars (6).

Vapor lock would be a problem only if methanol were added to gasoline without regard to volatility; for example, it could be blended into production gasoline at service stations with blending pumps. Volatility could be adjusted at the refinery for later addition of methanol by omitting light blending stocks. However, as Messrs. Wigg and Lunt point out, this would result in inefficient petroleum usage unless markets could be found for the light stocks.

TABLE IV

COLD START DRIVEABILITY COMMUTING TESTS[1]

1971 Car Make	Car Size	Engine	Driveability Demerits: No Methanol[2]	Driveability Demerits: 10% Methanol	Driveability Demerits: Increase
A	Intermediate	V–8	12.3	24.0	11.7
A	Full	V–8	6.0	15.5	9.5
B	Intermediate	V–8	3.8	4.3	0.5
B	Full	V–8	8.5	15.3	6.8
C	Compact	6	4.8	10.8	6.0
C	Intermediate	V–8	2.5	3.8	1.3
Average	–	–	6.3	12.3	6.0

[1]40°F to 70°F Ambient Temperatures
[2]Commercial Unleaded Gasoline

Cold Start Driveability Tests

In recent years, cars have been modified and new equipment has been installed to reduce exhaust emissions of unburned fuel, carbon monoxide, and nitrogen oxides. These changes have led to driveability problems in the newer cars. Starting with the 1971 models, cars have tended to stall, to hesitate on opening the throttle, and to surge under some conditions. One of the main reasons for this is that A/F ratios have been greatly leaned. Thus, adding methanol to gasoline is expected to have an adverse effect on driveability because such blends require richer mixtures for comparable performance.

A test program was conducted with six 1971 cars to study the effect of 10% methanol on fuel economy and driveability. The cars were popular models from the three major manufacturers, with all but subcompact sizes represented. Six drivers drove the cars in commute service between their homes and the Richmond Chevron Research Laboratory. During the test, ambient temperatures ranged from about 40°F to 70°F. An unleaded gasoline and the same gasoline containing 10% methanol were alternately tested in each car.

During the test, the drivers recorded daily observations of driveability malfunctions. Table IV shows average demerits for the base gasoline and the 10% methanol blend. Driveability demerits are based on the number of malfunction occurrences and the severity of each occurrence during one commute trip. The maximum acceptable demerit level is about 5 for this test.

Every car showed a higher demerit level for the methanol fuel, although driveability was still acceptable in two cars. Three cars went from reasonable driveability to poor driveability with the addition of methanol, and one car deteriorated from poor to very poor.

Driveability tests were also conducted with three of the commute cars on an all–weather chassis dynamometer at 25°F. Table V shows that the methanol effects were even larger under these conditions. Average demerits were tripled with the addition of 10% methanol.

TABLE V

COLD START DRIVEABILITY CHASSIS DYNAMOMETER TESTS[1]

1971 Car Make	Car Size	Driveability Demerits		
		No Methanol[2]	10% Methanol	Increase
A	Full	7	11	4
B	Intermediate	3	12	9
C	Intermediate	4	22	18
Average	–	4.7	15.0	10.3

[1] Tests Run at 25°F

[2] Unleaded Laboratory Test Gasoline

The methanol/gasoline blend used in these test programs were not corrected for the volatility effect of methanol. If this had been done, the relative performance of the methanol blend would have been directionally worse since front–end volatility does affect cold start driveability.

These tests showed that many drivers of 1971 cars would easily detect the adverse effects of adding 10% methanol. Older cars probably would be affected very little by the addition of methanol, but newer cars should show large effects because their driveability is already generally worse (7).

Chassis dynamometer testing was also done with the Make D compact car at 40°F. The electronic fuel injection control system was modified to allow changing A/F's with a selector switch. Tests were run with methanol, gasoline, and a 40%/60% methanol/gasoline blend. The car had to be started by squirting ether into the air intake when straight methanol was tested. Figure 2 shows driveability demerits (not including starting problems) versus "percent lean," which is relative to the stoichiometric A/F of each fuel. Driveability deteriorated rapidly as the A/F's were leaned with straight gasoline as the fuel. However, the addition of 40% methanol delayed the onset of poor driveability to A/F's 15% to 20% lean, and straight methanol operated satisfactorily with A/F's more than 20% lean.

FIGURE 2

COLD START DRIVEABILITY VERSUS RELATIVE AIR/FUEL RATIO 1971 COMPACT MAKE D

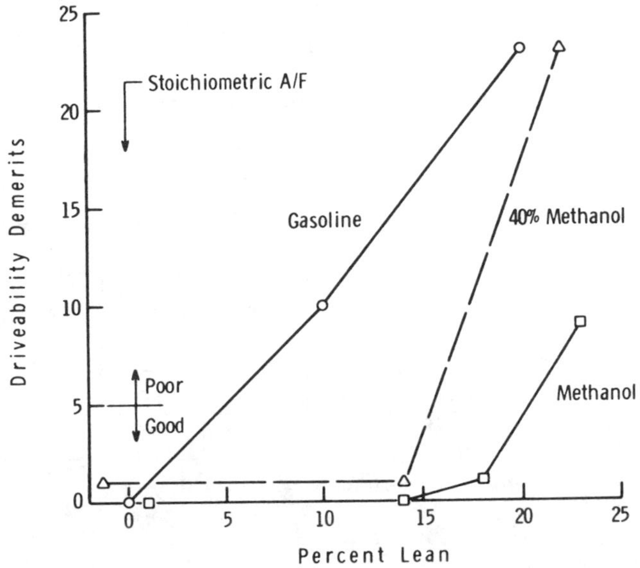

These results show that the use of methanol can be beneficial if A/F's are properly selected. With high concentrations of methanol, leaner mixtures can be used without adversely affecting driveability; and this should give lower exhaust emissions levels. An emissions study made with this car is discussed in the "Exhaust Emissions Tests" section.

Fuel Economy Tests

During the six–car driveability test, fuel consumption was measured with accurate fuel meters installed in the cars. Fuel economy was determined twice a day for each car over a six–week period. The average fuel economy data are given in Table VI. Meaningful data could not be obtained with one of the cars, medium–size car Make A, because the 10% methanol blend caused periodic stoppage of the fuel meter due to deterioration of plastic parts in the meter.

Fuel economy was 0.1–0.8 mpg less with the methanol fuel among the five cars. On a percentage basis, the loss was 1–5%, with a 3% average. The effect of adding 10% methanol to the base gasoline was a 5% loss in net heat of combustion. Because the measured losses in fuel economy were less than 5%, improvements in thermal efficiency with methanol are indicated. Table VII shows the data converted to miles per million Btu, which is a measure of the efficiency of converting chemical energy to work output. Ten percent methanol showed increases in efficiency ranging from −0.1% to 4.4%, with an overall average of 1.9%.

TABLE VI

COMMUTING FUEL ECONOMY

1971 Car Make	Car Size	Engine	Fuel Economy, mpg		
			No Methanol	10% Methanol	Increase, %
A	Intermediate	V–8	14.82	*	–
A	Full	V–8	12.99	12.44	–4.2
B	Intermediate	V–8	14.99	14.87	–0.8
B	Full	V–8	10.84	10.41	–4.0
C	Compact	6	20.37	19.85	–2.55
C	Intermediate	V–8	15.49	14.71	–5.0
Average	–	–	14.94 (5 Cars)	14.46	–3.2

*No data obtained because of deterioration of plastic material in fuel meter.

TABLE VII

COMMUTING FUEL ECONOMY ENERGY BASIS

1971 Car Make	Car Size	Engine	Fuel Economy, Miles/10^6 Btu		
			No Methanol	10% Methanol	Increase, %
A	Intermediate	V–8	130.0	*	–
A	Full	V–8	113.9	114.9	0.9
B	Intermediate	V–8	131.5	137.3	4.4
B	Full	V–8	95.1	96.1	1.05
C	Compact	6	178.7	183.3	2.6
C	Intermediate	V–8	135.9	135.8	–0.1
Average	–	–	131.05	133.5	1.9

*No data obtained because of deterioration of plastic material in fuel meter.

The 1971 compact Make D car was also used to evaluate methanol's fuel economy performance. Tests were run at 40 mph and 60 mph (road load) with gasoline, a 40%/60% methanol/gasoline blend, and with pure methanol. As Figure 3 shows, methanol gave about half the fuel economy of gasoline, and the blend fell between the two. This was due to the low heat content of methanol – just half that of the gasoline.

FIGURE 3
CRUISING FUEL ECONOMY
1971 COMPACT MAKE D
40/60 MPH AVERAGES

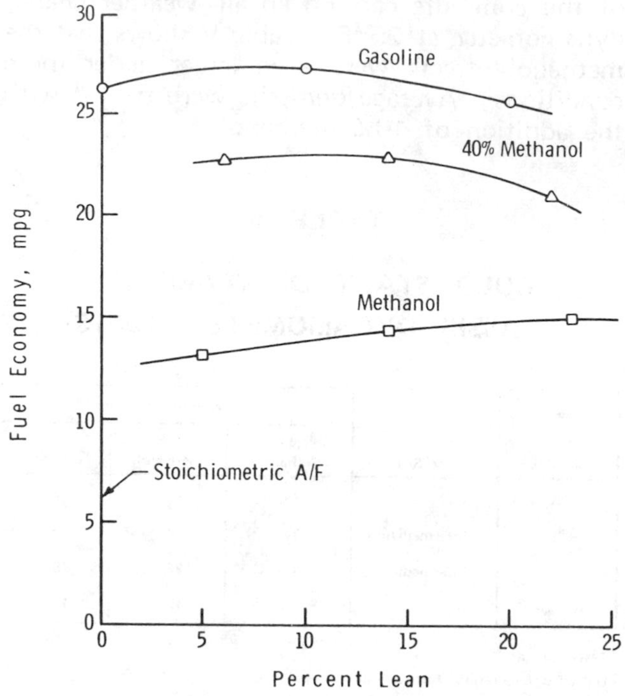

Figure 4 shows that the blend gave a 3–6% higher thermal efficiency than gasoline over a broad range of A/F's. Pure methanol, however, performed in an unusual manner. Its efficiency ranged from 94% to 123% relative to gasoline, as the A/F was leaned from 0% to 25%. The low efficiencies are believed to be due to methanol's large evaporative cooling effect causing lower combustion temperatures. Efficiency increased as A/F was increased because of methanol's high lean–mixture burning velocities. In addition, the thermodynamic properties of lean–mixture methanol combustion gases are inherently favorable.

These findings are supported by fuel economy data obtained during exhaust emissions testing with the Make D car. Tests were conducted with gasoline and with pure methanol using the 1972 Federal test procedure (8). Both hot and cold start tests showed results (Figure 5) similar to the steady speed data, although the general level of fuel economy was lower due to cyclic operation. Figure 6 shows the benefits for methanol on an energy basis. Again, methanol's efficiency was much better when operating lean. Methanol showed much better efficiency than gasoline. particularly after hot starts. Apparently, the engine had to be warmed up for best combustion efficiency.

FIGURE 4

CRUISING THERMAL EFFICIENCY
1971 COMPACT MAKE D
40/60 MPH AVERAGES

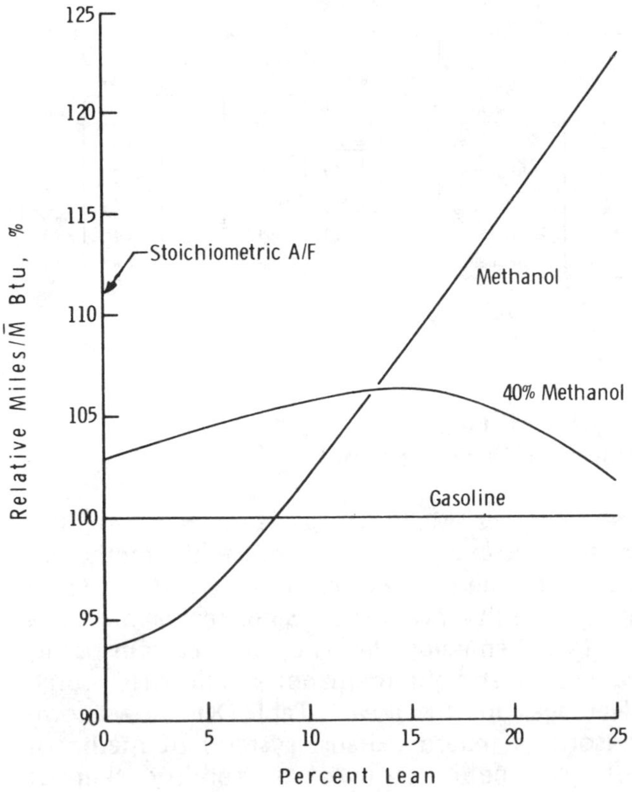

FIGURE 5

FUEL ECONOMY
1971 COMPACT MAKE D
EMISSIONS TEST CYCLE

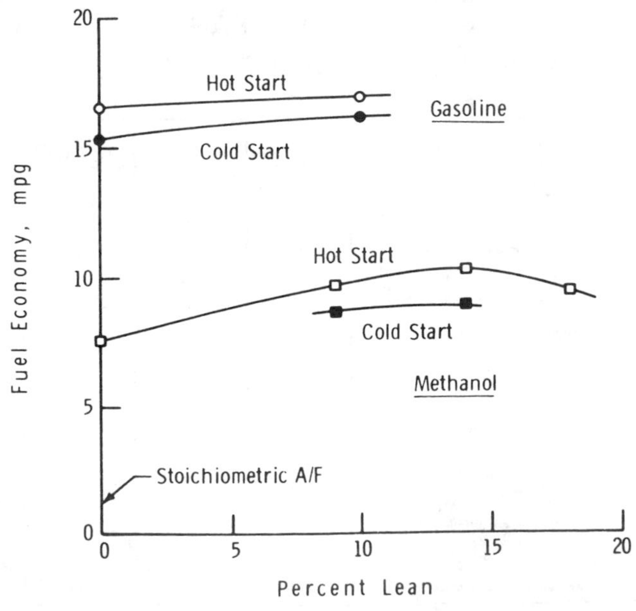

FIGURE 6

THERMAL EFFICIENCY
1971 COMPACT MAKE D
EMISSIONS TEST CYCLE

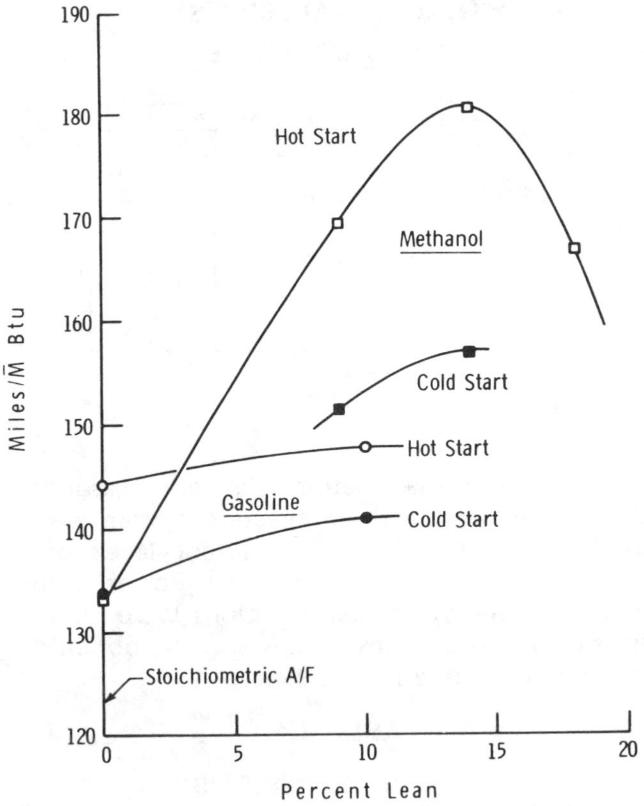

Exhaust Emissions Tests

To study the effect of methanol on exhaust emissions, two series of tests were run with gasoline and with pure methanol in the Make D car. In one series, the car was equipped with a catalytic reactor and air injection into the exhaust ports. In the other series, the standard exhaust system was used. Emissions were measured using the 1972 Federal test procedure which simulates a 23–minute city driving cycle on a chassis dynamometer. Aldehyde measurements were made in addition to the usual unburned fuel, carbon monoxide, and nitrogen oxides analyses, because one of the products of concern with methanol combustion is formaldehyde. The test results are shown on Tables VIII and IX in terms of grams of pollutant per vehicle mile.

Table VIII shows that methanol produced larger concentrations of unburned fuel and carbon monoxide than gasoline using the standard exhaust system. However, aldehydes were about the same; and nitrogen oxide levels were lower. The nitrogen oxide decrease is believed to be due to lower combustion temperatures with methanol because of its cooling ability.

TABLE VIII

EXHAUST EMISSIONS[1]
1971 COMPACT MAKE D

Air/Fuel Ratio	Start	Unburned Fuel		Carbon Monoxide		Nitrogen Oxides		Aldehydes[2]	
		Gasoline	Methanol	Gasoline	Methanol	Gasoline	Methanol	Gasoline	Methanol
Stoichiometric	Cold	3.2	–	32	–	7.7	–	0.13	–
9–10% Lean	Cold	2.7	5.5	10	26	8.5	4.2	0.18	0.15
14% Lean	Cold	–	4.2	–	14	–	3.1	–	0.22
Stoichiometric	Hot	2.8	5.3	29	58	6.4	5.0	0.11	–
9–10% Lean	Hot	2.5	3.1	10	20	7.1	4.7	0.13	0.12
14% Lean	Hot	–	0.3	–	8	–	2.0	–	0.19
18% Lean	Hot	–	3.1	–	6	–	3.7	–	0.23

[1] 1972 Federal test procedure. Results are in terms of grams per mile.
[2] MBTH Method (3–methyl–2–benzothiazolone hydrazone hydrochloride). Calculated as formaldehyde.

Methanol performed better relative to gasoline when the reactor and air injection system were installed (Table IX). All pollutant levels but unburned fuel were considerably lower with methanol; unburned fuel emissions were about the same. (Aldehydes data were not obtained with gasoline in this test series.)

TABLE IX

EXHAUST EMISSIONS[1]
1971 COMPACT MAKE D WITH CATALYTIC REACTOR AND AIR INJECTION

Air/Fuel Ratio	Start	Unburned Fuel		Carbon Monoxide		Nitrogen Oxides		Aldehydes[2]	
		Gasoline	Methanol	Gasoline	Methanol	Gasoline	Methanol	Gasoline	Methanol
Stoichiometric	Cold	0.9	1.15	8.0	5.3	7.6	3.5	–	0.09
9–10% Rich	Cold	1.3	1.2	39.8	5.3	4.0	1.5	–	–
Stoichiometric	Hot	0.3	0.3	3.6	1.2	7.8	2.5	–	0.06
9–10% Rich	Hot	0.4	0.4	16.8	3.3	4.2	1.4	–	0.04
23% Lean	Hot	–	0.7	–	0.3	–	3.1	–	0.23

[1] 1972 Federal test procedure. Results are in terms of grams per mile.
[2] MBTH Method (3–methyl–2–benzothiazolone hydrazone hydrochloride). Calculated as formaldehyde.

The reactor and air injection system greatly reduced unburned fuel and carbon monoxide concentrations with both gasoline and methanol (Table X). Although there was no significant effect of exhaust system on nitrogen oxides when the car ran on gasoline, nitrogen oxides were reduced 50% with methanol as the test fuel (stoichiometric A/F ratio, hot start). This is surprising because the exhaust system used is not expected to affect nitrogen oxides. Excess air should not be present in the reactor for the maximum reduction of nitrogen oxides. Further data should be obtained to verify this effect.

TABLE X

EFFECT OF CATALYTIC REACTOR AND AIR INJECTION ON EXHAUST EMISSIONS
1971 COMPACT MAKE D

Fuel	Air/Fuel Ratio	Grams/Mile					
		Unburned Fuel		Carbon Monoxide		Nitrogen Oxides	
		Without	With	Without	With	Without	With
		Cold Start Tests*					
Gasoline	Stoichiometric	3.2	0.9	32	8.0	7.7	7.6
		Hot Start Tests*					
Gasoline	Stoichiometric	2.8	0.3	29	3.6	6.4	7.8
Methanol	Stoichiometric	5.3	0.3	58	1.2	5.0	2.5

*Comparable tests with methanol were not run.

Benefits of Lean Operation With Methanol

The compact Make D car tests showed that leaner mixtures can be used with methanol, relative to the stoichiometric A/F, without causing driveability problems. Since leaner mixtures give lower emissions levels up to a certain point, the use of straight methanol should offer some advantages in this area. Table XI shows comparisons (standard exhaust system) of methanol with gasoline in terms of driveability, thermal efficiency, and exhaust emissions. A mixture ratio of 5% lean was selected for gasoline, because it corresponds to the maximum leanness for good driveability.

TABLE XI

BENEFITS OF LEAN OPERATION WITH METHANOL

	Gasoline, 5% Lean	Methanol, 14% Lean	Methanol, 18% Lean
Cold Start Tests			
Cold Start Driveability Demerits	5	0	1
Cycling Thermal Efficiency, Miles/10^6 Btu	138	157	–
Unburned Fuel, g/Mile	3.0	4.2	–
Carbon Monoxide, g/Mile	20	14	–
Nitrogen Oxides, g/Mile	8.3	3.1	–
Hot Start Tests			
Cruising Thermal Efficiency, Miles/10^6 Btu	236	252	263
Cycling Thermal Efficiency, Miles/10^6 Btu	146	180	167
Unburned Fuel, g/Mile	2.7	0.3	3.1
Carbon Monoxide, g/Mile	18	8	6
Nitrogen Oxides, g/Mile	6.9	2.0	3.7

At 14% lean, methanol showed better driveability, better thermal efficiency, and much lower carbon monoxide and nitrogen oxides levels compared to gasoline running 5% lean. Unburned fuel was much lower with methanol in the hot start tests, but it was higher in the cold start tests. Apparently, the high latent heat of vaporization of methanol adversely affected vaporization in the intake system when the engine was cold. Thermal efficiency with methanol was not drastically affected by cold starts, however.

The leanest mixture, 18%, showed trend reversals, indicating that 14% lean is about optimum for efficiency and emissions. The only improvements at 18% lean were in carbon monoxide emissions and cruising thermal efficiency (hot start).

Required Engine Modifications

Obviously, automobile engines must be modified to run on straight methanol because of its A/F ratio requirements and its air cooling ability. Even with low concentrations of methanol in gasoline, newer automobiles would need alterations because their A/F ratios are carefully tailored to give minimum exhaust emissions with reasonable driveability. Adding methanol would cause a deterioration of driveability in many cars and an increase in unburned fuel in the exhaust of some cars due to misfiring, because of its "leaning" effect.

For modern cars to run on gasoline containing 10% methanol, for example, carburetors would have to be recalibrated to provide an enrichment of about 6% under all operating conditions. This would not be difficult for carburetor manufacturers, but cars with modified carburetors would have to be certified by the Environmental Protection Agency for their exhaust emissions performance. In addition, these cars could not use conventional gasolines without affecting exhaust emissions. Other modern cars, of course, could not use this special gasoline.

The use of methanol alone in motor vehicles requires not only carburetor revisions, but also the intake system must be radically modified to provide several times the present heat input to the A/F charge. A 1970 model car with a six–cylinder engine was modified to run on methanol. It was not hard to recalibrate the carburetor, but several versions of a heated intake manifold were tried before the engine would run properly. The final solution was a thin–wall exhaust gas tube running the full length inside the intake manifold.

Water Sensitivity of Methanol/Gasoline Blends

Addition of small quantities of water to a methanol/gasoline single–phase blend produces a dramatic phase separation. A methanol–rich phase coalesces and settles to the bottom of the container; the gasoline–rich phase remains on the top. The occurrence of phase separation in a vehicular fuel system is intolerable since the methanol–rich phase has an A/F requirement approximately one–half of the blend.

Additionally, methanol and methanol/gasoline blends are hygroscopic. A dramatic demonstration can be made by swirling a 10% methanol/90% gasoline blend in an open beaker for a minute or two to maximize ambient air contact. The blend first becomes hazy and then the fine methanol/water droplets coalesce and separate into the bottom layer. A version of this experiment was used to determine the sensitivity of various methanol/gasoline blends to water contamination. Sealed vials of methanol/gasoline were titrated by water addition to the haze point. Within an hour

FIGURE 7

WATER TOLERANCE FOR METHANOL/GASOLINE MIXTURES

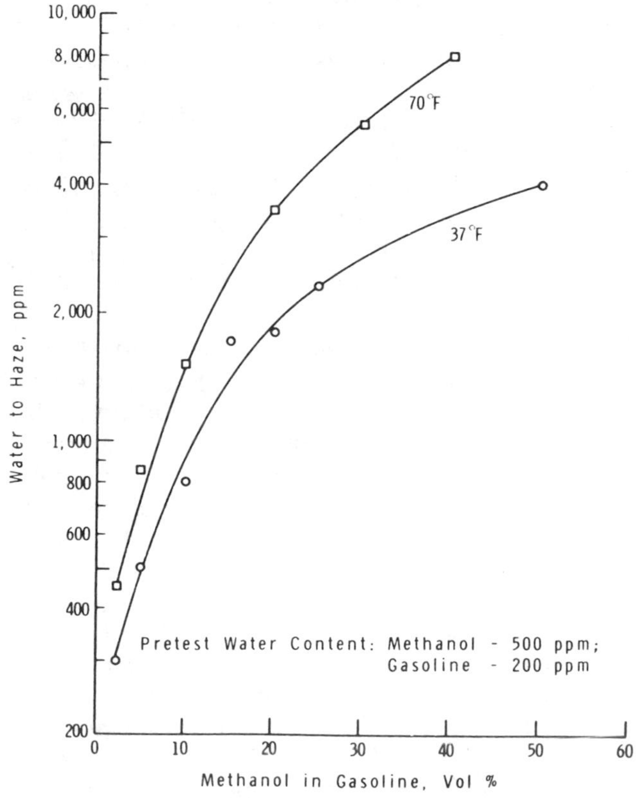

after titrating to this haze point, over 80% of the methanol has separated into the bottom phase. Thus, the haze point is the maximum amount of water than can be tolerated at a given temperature before separation. Figure 7 shows the water tolerance for methanol/gasoline mixtures at two temperatures. As the temperature is lowered, the blend becomes more sensitive to water addition.

A number of additives are used in the petroleum industry to solubilize small quantities of water in dry–cleaning fluids to act as ionic detergent carriers. We examined over 150 proprietary water–solubilizing additives and could find none that were effective in preventing phase separation at reasonable additive concentrations. The problem is that methanol appears to the additive as an ionic species like water. As a result, the additives involved form methanol–additive–hydrocarbon clusters, minimizing the water–solubilizing effect. The most effective additives at high concentration were the higher alcohols, particularly the branched alcohols such as isopropanol and tertiary butyl alcohol. For example, at 70°F the mixture of 5% tertiary butyl alcohol/15% methanol/80% gasoline tolerates 4000 ppm water, where a 15% methanol/85% gasoline blend will only tolerate 1500 ppm water. However, the cost of using higher alcohols at such concentrations negates large scale use of such additives for practical application.

Water can enter the methanol/gasoline blends in several ways during the normal distribution and consumption of the fuel. A significant proportion of the gasoline storage tanks have a water layer at the bottom. Sometimes this water layer is deliberate since water serves as a safety measure in case of tank leakage at the base of the tank. Water will leak out first, and the leak can be detected and stopped before significant quantities of gasoline flood the surrounding area.

Barge and tanker shipments of gasoline always result in contact with water from bilge bottom washdowns or rainy weather operation. In filling–station service tanks, water accumulates in the tank bottoms from normal breathing during diurnal temperature changes, water seepage through filling ports during rain and service station washing operations, and by ground water leakage into a corroded tank. The tanks are designed to accommodate 50 to 100 gallons of water without contaminating the fuel other than by dissolved water. Thus, gasoline as received in the motorist's tank is saturated with water in the range of 100–200 ppm. The main source of water contamination from the vehicle is by breathing of the fuel tank during diurnal temperature changes and intermittent operation when engine waste heat warms the fuel tank. The critical area in the vehicle is the carburetor bowl where normally 100–200 cc of fuel is in contact with the atmosphere for pressure balance, usually through the air cleaner as a filter.

In the fuel economy tests previously discussed, the cars were operated during a period of relative humidity ranging from 30–90% at moderate temperatures of 40–70°F. There were several cases of engine stall that were attributed to methanol separation in the carburetor. Whenever the carburetor bowl was examined during the test, one would usually see haze formation during exposure to ambient air. Because of the sensitivity to phase separation at low water concentrations, special fuel handling techniques would have to be used with methanol/gasoline mixtures. A separate distribution system similar to that used for aviation jet fuel would be required. The venting of vehicle fuel tanks and carburetor bowls would have to be via a drier to minimize moist air intrusion.

Deterioration of the Fuel System

The major problem with methanol/gasoline blends in a conventional vehicle fuel tank is corrosion of the terneplate lining by methanol. The terneplate coating consists of 75–90% lead and 10–25% tin. Methanol attacks this thin (1/2–1 mil thick) coating forming lead hydroxide. The lead hydroxide effectively plugs the fuel filter and clogs the carburetor jets or injector nozzles. Corrosion of the base steel soon results in the failure of the fuel tank. The possibility exists of developing a low cost corrosion inhibitor which when added to methanol/gasoline blends would prevent corrosion of the terneplate lining. We developed a simple test to screen a number of commercial corrosion inhibitors. Table XII shows the ineffectiveness (up to the 1000 ppm level) of nine classes of commercial corrosion inhibitors against attack of lead foil by methanol. The corrosion problem is amplified in those cars equipped with an in–line electric fuel pump. Since gasoline is an effective insulator, leakage current from the fuel pump's commutator to the grounded fuel tank is not a problem. However, when methanol is added to gasoline, we have measured currents of 40 microamperes between the fuel pump and the gas tank for 40% methanol/60% gasoline blends. In two days' exposure to this blend, we have seen a new fuel tank effectively stripped of the terneplate coating to the liquid level of the blend and severe corrosion develop in the exposed sheet metal.

TABLE XII

LEAD FOIL SOAK TESTS
20—ML FUEL
ADDITIVE AS INDICATED*

Class of Corrosion Inhibitors	Solubility in			Anticorrosion Ability
	Methanol	50% Methanol, 50% Unleaded Gasoline	Unleaded Gasoline	
Thiophosphoric Acid Ester	Insoluble	Insoluble	Soluble	None
Aliphatic Sulfide	Insoluble	Insoluble	Soluble	None
Polyamide	Soluble	Soluble	Soluble	None
Arylamine	Insoluble	Insoluble	Soluble	None
Succinimide	Insoluble	Insoluble	Soluble	None
Alkyl Selenide	Soluble	Soluble	Soluble	None
Dithiocarbamate	Soluble	Soluble	Soluble	None
Dicarboxylic Partial Ester	Soluble	Soluble	Soluble	None
Polyamine	Insoluble	Soluble	Soluble	None

*Inhibitors tested at concentrations of 100 ppm and 1000 ppm. 70°F, three days' exposure

Obviously there are solutions to this problem, such as plastic—lined fuel tanks to eliminate lead/tin exposure to the fuel. The retrofitting problem would be huge for the existing car population.

Another metal which is particularly susceptible to methanol corrosion is magnesium alone or as magnesium/aluminum alloys. Magnesium is increasingly being used in engine blocks and fuel pumps. It is particularly prevalent in fuel tanks for light engines such as chain saws and outboard motors. Aluminum is corroded to some extent by methanol. In the early 1940's, methanol was injected into supercharged aircraft engines for added power during take off. Corrosion problems occurred in the aluminum wing tanks so that special plastic—lined tanks had to be devised for methanol.

We have examined the susceptibility of a number of elastomers used in fuel systems towards attack by methanol/gasoline mixtures. A common problem is increased swelling of gaskets and seal material versus gasoline alone. The only test vehicle failures occurred with methyl methacrylate fuel pump housing and Viton O—ring seals.

The problems arising from water sensitivity due to air exposure and fuel system corrosion and elastomer swelling all point to the requirement for a specially designed vehicle fuel system for methanol or methanol/gasoline blends.

Summary and Conclusions

Laboratory and road tests showed methanol to be an effective octane booster. The road octane quality of three unleaded gasolines was raised two to three octane numbers.

Cold starting was found to be very difficult with straight methanol at low temperatures; a separate starting fuel system is required. However, low concentrations of methanol in gasoline do not present a problem.

Adding only 10% methanol to gasoline causes a large increase in front—end volatility. This necessitates the use of much less light gasoline blending stocks to prevent vapor lock problems in hot climates, resulting in inefficient fossil fuel utilization.

The addition of 10% methanol caused easily detectable deterioration in driveability in tests with six 1971 cars, because of methanol's "leaning" effect. Thus, 10% methanol cannot be used in gasoline without modifying the carburetors on 1971 and newer cars. High concentrations of methanol or straight methanol require extensive intake system modifications, but they offer exhaust emissions advantages because of their ability to burn at leaner mixtures than gasoline (relative to stoichiometric).

Because of its low heating value, methanol gave poorer fuel economy at all concentrations. However, methanol was found to burn more efficiently because of its combustion characteristics, particularly when operating under lean conditions. On an energy basis (miles per million Btu), the addition of methanol to gasoline caused improvements in fuel economy.

When compared to gasoline at the same relative A/F, methanol showed only one clear—cut exhaust emissions benefit — lower nitrogen oxides levels. However, methanol can be operated leaner than gasoline without causing driveability problems. At its optimum, 14% lean methanol performed better than gasoline running 5% lean (maximum for good driveability) in all emissions areas except unburned fuel in cold start tests and aldehydes.

The water sensitivity of hygroscopic mixtures of methanol and gasoline requires a separate fuel distribution system. Fuel storage in a vehicle must be protected from water absorption both in the tank and the carubretor to prevent separation into two phases, a methanol—rich lower phase and a gasoline—rich upper phase.

Corrosion and degradation problems occur where methanol/gasoline mixtures come into contact with lead, magnesium, aluminum, and some plastics. The vehicle fuel system must be constructed of methanol–resistant materials or component failure will result.

Methanol is not a useful fuel additive for existing unmodified cars. Methanol could be used effectively in special vehicles designed to handle the corrosion, water absorption, and vaporization characteristics. The cost of manufacture and distribution in a separate system that overcomes the water sensitivity problem will determine the extent of methanol's use.

References

1. G. A. Mills and B. M. Harney, "Methanol – The New Fuel from Coal," Chemtech, Jan. 1974, pp 26–31.

2. T. B. Reed and R. M. Lerner, "Methanol: A Versatile Fuel for Immediate Use," Science, Vol 182, Dec. 1973, pp 1299–1304.

3. W. D. Harris and R. R. Davison, "Methanol from Coal," Oil and Gas Journal, Dec. 1973, pp 70–72.

4. Bureau of Mines Methanol from Coal Meeting, Washington, D.C., Feb. 13, 1974.

5. "Use of Alcohol in Motor Gasoline – A Review," American Petroleum Institute, Task Force EF–12, 1971.

6. E. E. Wigg and R. S. Lunt, "Methanol as a Gasoline Extender – Fuel Economy, Emissions, and High Temperature Driveability," Paper 741008 Presented at SAE Meetings, Toronto, Oct. 1974.

7. J. C. Ingamells, "Fuel Economy and Cold Start Driveability With Some Recent–Model Cars," Paper 740522 Presented at SAE Fuels and Lubricants Meetings, Chicago, June 1974.

8. "Control of Air Pollution from New Motor Vehicles and New Motor Vehicle Engines," Federal Register, Vol 35, Nov. 1970.

Lean Combustion of Methanol-Gasoline Blends in a Single-Cylinder SI Engine*

Edward J. Canton, S. S. Lestz
and W. E. Meyer
The Pennsylvania State Univ.

IN AN ATTEMPT to increase the octane rating of lead free gasolines and possibly reduce the amount of petroleum used,* the addition of alcohols to gasoline was considered. Much previous work has been done using pure methanol (1, 2, 3, 4, 5)** as a motor fuel. A thorough review of work done on the use of alcohols as engine fuels prior to 1964 is found in Ref. 6. It was found that not much work was done using blends of methanol with gasoline (7, 8) but rather attention was focused on ethanol blends (9). No work was reported regarding emission in the ultra lean region for methanol-gasoline blends. For this reason methanol blends were chosen for the present investigation.

When using pure or blended methanol in an engine with an unmodified intake system two things happen: 1) the air fuel charge is leaned out due to the change in the stoichiometry when the charge contains methanol as compared to pure gasoline, and 2) since methanol has a latent heat of vaporization almost five times that of gasoline, its vaporization cools down the charge and effectively increases volumetric efficiency and reduces compression work (10). Both of these occurrences can be considered as decided advantages provided proper steps are taken to maintain acceptable vehicle driveability.

It is now apparent that whatever future means are used to control exhaust emissions, these means will not be permitted to compromise fuel economy. Therefore, any control method that allows stringent oxides of nitrogen, NO_x, standards to be met without an attendant increase in the production of unburned hydrocarbons, HC, and carbon monoxide, CO, is of vital importance because any such method would generally not involve a fuel economy degradation. Lean operation holds out promise of being such a method (11, 12) and for this reason the present single cylinder engine study was conducted wholly in the lean region.

* Methanol would be produced without using petroleum.

** Numbers in parenthesis refer to references at end of paper.

*Paper 750698 presented at the Fuels and Lubricants Meeting, Houston, Texas, June 1975.

ABSTRACT

Blends of up to 40% by volume methanol in a methanol-gasoline fuel blend were supplied to a single cylinder engine operating under controlled conditions. The following effects are reported as the methanol concentration increases. The lean misfire limit is extended 0.04 Ø by using a blend containing 40% methanol compared to the base fuel. It is also noted that the lean misfire limit does not vary until a blend containing greater than 20% methanol was used. Torque and thermal efficiency increase significantly. Percent by volume concentrations of carbon monoxide, carbon dioxide and oxides of nitrogen do not change, although oxides of nitrogen reported as mass per power output per hour decrease.

SAE/PT-80/19/$02.50

The thrust of this single cylinder study was directed toward a documentation of engine performance and exhaust emissions in the lean region as a function of load and fuel composition at constant speed. Previous work done on this same single cylinder engine showed that the lean misfire limit, LML, is extended by careful charge preparation (11, 13), by intensifying the intake mixture motion (14), and by using a high-energy long-duration ignition system with deep-projection, wide-gap spark plugs (11). The results of the single cylinder engine study that follow were obtained using all these previously reported techniques for extending the LML.

EXPERIMENTAL EQUIPMENT

The engine used in this study was a removable dome head (RDH) CFR single cylinder engine with a hemispherical head and overhead valves. The engine was coupled directly to a cradled electric dynamometer having a scale resolution of 0.105 ft-lb_f torque. The dynamometer was excited so as to keep the CFR engine at a constant speed of 900 rpm. Two access holes in the head of the engine provided for a spark plug and a pressure transducer.

Figure 1 shows the air-fuel system used. The air supplied is low in moisture (less than 20 grains H_2O/lb dry air) and low in hydrocarbons. Primary air heaters were used in order to completely vaporize the fuel. The fuel was injected, using an air atomizing nozzle, into the heated primary air and then flowed into a mixing tank. The mixing tank was equipped with heaters mounted on its baffles to keep the charge temperature constant at 120° F. The baffles also ensured that a highly homogeneous mixture was delivered to the engine. The air rotometers were calibrated using an accurate dry gas meter and the flow control valve arranged so that no correction was necessary at low engine pressures.

IGNITION SYSTEM SPARK PLUG AND INTAKE VALVE - The ignition system, spark plug and intake valve were the combination that provided the best LML in a study done by Ryan et al. (11). A Texaco transistorized ignition system (15) which provided a high-energy (500 mj at 900 rpm) spark of 25 CA° duration (4.63 ms) was used with a Champion spark plug that had a 0.35 in. projection into the combustion chamber and 0.055 in. gap.

Gabele (14) concluded that a vaned collar (inclined 60° from the verticle) on the intake valve appreciably extended the LML. The work of Ryan et al. (11) showed that the 60° vaned collar when used with a spark plug that protruded 0.35 in. and had a 0.055 in. gap produced a lowering trend for the HC and NO emissions and for the BSFC.

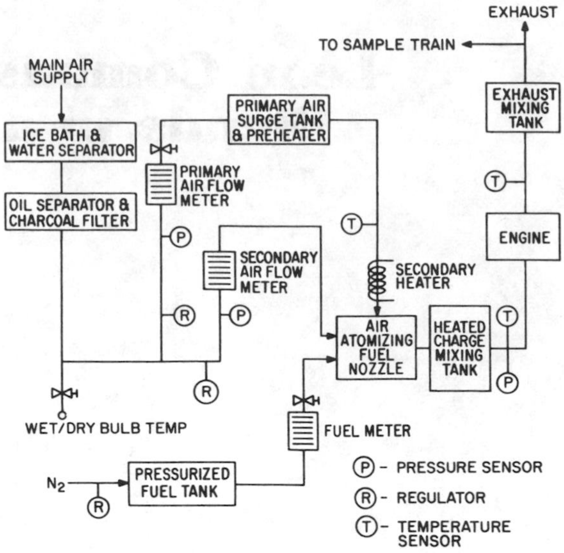

Fig. 1 - Air-fuel system

DETERMINATION OF LEAN MISFIRE LIMIT - Misfire was determined by electronically comparing the peak motor pressure to the peak pressure during engine operation. If the operational peak pressure was not slightly greater than the motored peak pressure a misfire is said to have occurred. This was done by feeding the pressure transducer signal to a charge amplifier to produce a pressure-versus-time signal and this output went to a comparator which made the comparison between peak engine operation pressure and a fixed signal corresponding to a pressure slightly above motored pressure. Details of this LML detection system are given in Ref. 11. At an engine speed of 900 rpm, 7.5 pressure peaks per second occur. The firing frequency was measured over a 10 second interval. Lear misfire limit was defined to be an engine firing frequency of 7.4 to 7.2 cycles (about 2 to 4 per cent of the time misfiring). As the LML is approached the engine firing frequency could vary from 7.2 to 7.4 cps from one counting period to the next. Therefore, on the average the LML represents a misfire frequency of 3 per cent over a 10 second counting interval.

EXHAUST GAS SAMPLING SYSTEM - The exhaust analysis system is shown in Figure 2. A continuous sample was drawn downstream of a large insulated mixing tank located in the exhaust line. The sample was first passed through an ice bath to remove most of the moisture and then split. In one leg hydrocarbons were measured with an unheated Beckman flame ionization detector, FID, which was calibrated using methane. It was found that the problem of water condensation associated with not using an ice bath for the HC line was greater than the very small decrease in HC when using an ice bath. This was determined

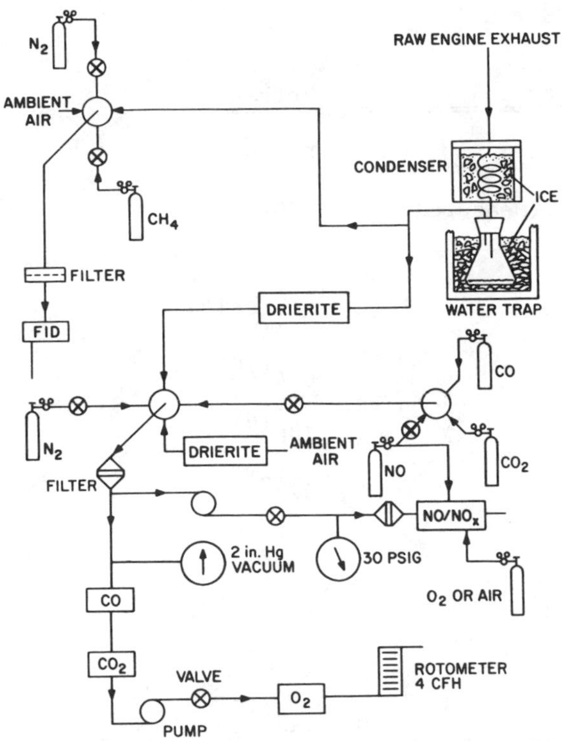

Fig. 2 - Exhaust analysis system

with the base fuel and may not be true for methanol. In the other leg the remainder of the sample was passed through a drierite column to assure that no H_2O interference occurred in the infrared analyzers. CO and CO_2 were analyzed by nondispersive infrared, NDIR, and NO_x was measured using a chemilumenescent detector. A Beckman 741 process analyzer, using an amperometric quick response cell, was utilized for O_2 measurement and calibrated using ambient air. Glass fibre filters were used on the system to protect the instruments from particulates.

FUEL MIXING - Prototype fuel supplied by EPA was used as the base fuel for this study. Properties of the prototype fuel and methanol are in Table 1. An average formula for the prototype fuel was calculated to be C_7H_{14} based on an estimated molecular weight and an analyzed weight per cent of Carbon and Hydrogen. The methanol was analyzed with a gas chromatograph and found to be extremely pure (>99.9%). The fuel was mixed by volume using volumetric flasks. Prototype fuel and methanol were added to a nitrogen purged 5 gallon can until it was about half full and then the can was sealed and agitated vigorously for one minute. It was found from test tube mixing that little agitation is necessary to make the two phase appearance disappear. This mixture was then transferred under nitrogen pressure to the fuel tank. All contact with moisture was avoided in order to reduce the chances for fuel separation. The fuel rotometer was then calibrated. For simplicity per cent volume blends of methanol in the total mixture are designated as 10M, 20M, 30M, and 40M.

EXPERIMENTAL PROCEDURE

The study was concerned with the effects of load and fuel composition on the LML and exhaust emissions. All tests were run at a compression ratio of 8 and in general with the spark timing set for maximum brake torque. For the richest WOT test points using the base fuel, the spark timing was retarded from mbt to avoid knock. Table 2 gives the engine operating parameter at which all tests were made. Before each run the analytical instruments were calibrated and the engine was allowed to run until steady state was reached. Then data were collected at various equivalence ratios, ϕ, ranging from about one to the LML for each manifold pressure using the base fuel. This procedure was then repeated for every methanol-base fuel blend. Raw data were punched on computer cards and data reduction accomplished using a digital computer. One feature of the data reduction program was the calculation of the air-fuel ratio from the exhaust gas analysis. This provided a check on the air and fuel settings for each run.

EXPERIMENTAL RESULTS AND DISCUSSION

Before giving any detailed results several interesting observations will be reported. The comparison between the measured air-fuel ratios and those calculated from exhaust gas composition were very good. The per cent difference between them generally being less than ± 5 per cent. Also for the same equivalence ratios, the exhaust gas composition was found to contain about the same amount of oxygen, as the concentration of methanol in the fuel blend was increased. The same was true for carbon dioxide.

The WOT tests demonstrated the results of octane number improvement with increasing methanol concentrations. With the base fuel, WOT operation was knock limited at $\phi \simeq 0.8$; i.e., the spark timing could not be adjusted for mbt without encountering knock. However, as the methanol concentrations were increased to 30M the engine was no longer knock limited.

As the concentration of methanol in the fuel blend was increased beyond 10 per cent, operation at light load (16.1 in. Hg manifold pressure) deteriorated. That is to say, that the LML rapidly became richer and the cycle-by-cycle variations appeared to become very severe. For the 30M and 40M blends, lean operation of the engine was virtually impossible at light load and therefore no data is reported for this condition. In fact, the general spareness of 16.1 in. Hg data is

Table 1. Fuel Properties

Fuel	Prototype	Methanol
Formula	C_7H_{14} avg.	CH_3OH
Specific Gravity*	0.728 (Measured)	.792 (10)
Molecular weight	≃ 98	32.04 (10)
Higher heating value B/lb.	18,750	9,770 (10)
A/F stoichiometric	14.8	6.4 (10)
RON	93.2	106 (6)
MON	84.7	92 (6)
Carbon % by weight	85.93**	37.5 (6)
Hydrogen % by weight	14.05**	12.5 (6)
Oxygen % by weight	- -	50.0
ASTM distillation ^oF	IP - 10% 123 50% 199 90% 325 E.P. 383	149 (10)
Reid vapor pressure	10.2	

* density of substance @ 68°F referenced to water at 39.2°F

** analysis by Galbraith Labs, Knoxville, Tennessee

due to poor engine operation at this manifold pressure. At 19.8 in. Hg manifold pressure the lean misfire limits were $\emptyset \simeq 0.63$ for the 40M blend and $\emptyset \simeq 0.49$ for the 30M blend. It appears as though the deterioration of part load performance is a function of the amount of methanol added to the base fuel.

LML VERSUS THEORETICAL AIR-FUEL RATIO - The theoretical air-fuel $A/F)_T$ ratio for the methanol-base fuel blend is determined from the following relationship

$$A/F_T = 138.2\ (1.5+9x) \Big/ (32+66x)$$

and $$x = 0.007429v \Big/ (0.02475 - 0.01732v)$$

where: v = the volume fraction of base fuel.

x = the mole fraction of base fuel.

This simple relationship was arrived from a carbon, hydrogen and oxygen balance for the complete combustion of the base fuel and methanol with air and the conversion of volume fraction to mole fraction for the liquid fuels.

Figure 3 shows the LML $\emptyset$ vs, $A/F)_T$ for each fuel blend and manifold pressures. It is interesting to note that as the amount of methanol is increased from OM up to 20M the LML does not change. Then for blends greater that 20M the LML gets leaner by about $\Delta\emptyset = .04$. Why there is no change with the addition of up to 20% methanol is not known since the 20M fuel is roughly one half methanol on a molar basis. In the work of Ebersole and Manning (4) they report the LML for pure methanol to be leaner than for iso-octane by $\Delta\emptyset$ 0.2. Also it is of interest to note that the trend toward richer values of $\emptyset$ with increasing manifold pressure is contrary to findings of Quader (12) working with propane and Most and Longwell (5) and Ebersole and Manning (4) working with iso-octane and neat methanol. The reason for this is not immediately apparent.

EFFECT OF METHANOL ON EMISSIONS - The data for all the figures that are discussed here can be found in Ref. 16. Figures 4 through 8 are presented to illustrate what are believed to be general trends. The curves

Table 2. Engine Operating Parameters

Speed, rpm		900 rpm
Spark advance, CA°		mbt**
Cooling water exit temperature		180 ± 2 F
Charge temperature		120 ± 2 F
Oil temperature		105 ± 1 F
Compression ratio, CR		8:1
Intake valve		inclined vanes. 60° from verticle
Equivalence ratio, Ø*		1.1 to 0.5
Spark plug	gap projection	0.055 in. 0.35 in.
High energy ignition system	power duration	500 mj 25 CA°
Manifold pressure		29.9 in. Hg Abs. 24.3 in. Hg Abs. 19.8 in. Hg Abs. 16.1 in. Hg Abs.

* $\phi \equiv A/F)_T \ / \ A/F)_{actual}$

** mbt; minimum spark advance for maximum brake torque.

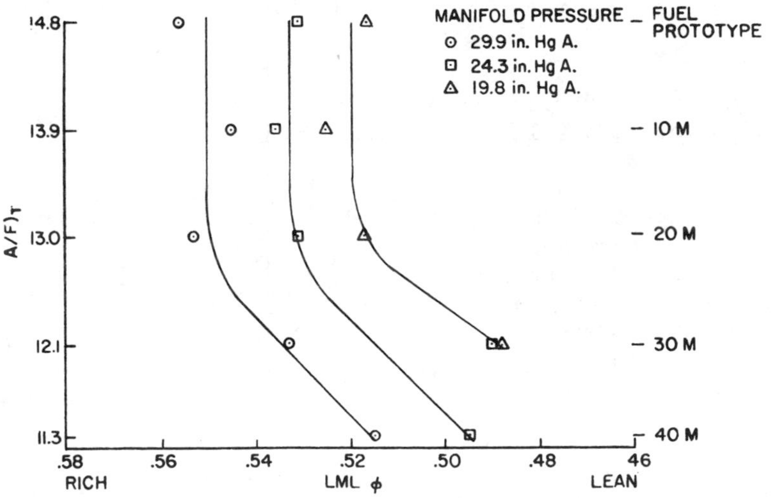

Fig. 3 - $A/F)_T$ vs lean misfire limit equivalence ratio

represent bounds for the experimental data with arrows next to an "M" always indicating the direction of increasing methanol concentration in the fuel blend.

Figure 4 shows the trend for the HC emission which is similar at other manifold pressures. As shown in Appendix A, Figure A1, the FID response to pure methanol in nitrogen is not per carbon atom as it would be with methane or propane. Since the methanol concentration in the exhaust (or the percent methanol of all hydrocarbons) is unknown, the data cannot be corrected for this effect. It was also observed that the FID response to the methanol calibration mixture was time dependent, introducing another source of error. The HC emission data are therefore only presented to show what type of trend may be found when using an FID to measure unburned HC in exhaust gas containing methanol.

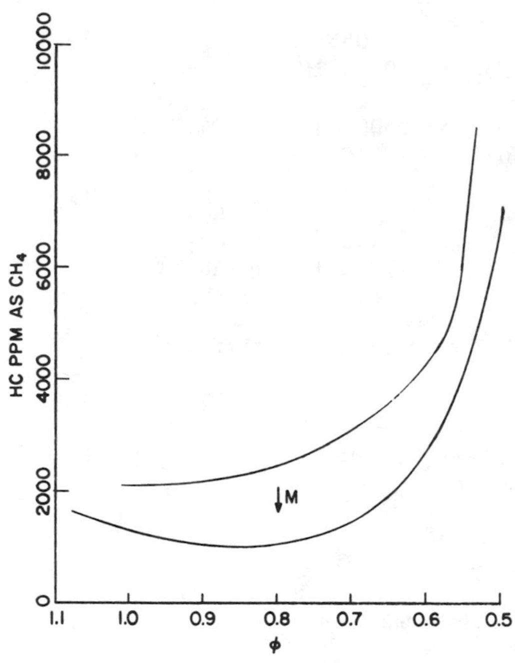

Fig. 4 - Unburned hydrocarbons vs Ø with prototype and 40M blend at 24.3 in. Hg absolute pressure

Appendix B also shows that the unburned HC measurements are open to another source of error, viz., an oxygen effect. This effect is important when operating lean because the exhaust gases will contain free oxygen. This free oxygen causes a change in the FID response (see Figures B1 and B2).

Increasing the amount of methanol in the fuel blend did not produce an appreciable effect on the NO_x emissions in the lean region. The concentration of NO_x in the exhaust gases was a strong function of equivalence ratio and manifold pressure and behaved in an expected way.

When operating leaner than Ø = 1.0 CO emissions fall rapidly. It was found that the carbon monoxide concentration is in the range of 0.05 to 0.15% which is quite low. For all the fuel blends tested CO emissions minimize for Ø = 1.0 and then remain near their minimum value as Ø decreases till the LML is reached. In the entire lean region (Ø < 1.0) CO and O_2 emissions did not appear to depend on the amount of methanol in the fuel blend.

For the same intake conditions, the energy per unit mass of charge decreases as the amount of methanol in the fuel blend increases. Therefore, the BSFC trend that appears on Figures 5 to 7 is to be expected. In the region Ø = 0.6 to 0.7 the torque curves come close to each other and the thermal efficiency is also about the same here for 0M and 40M. Therefore, the BSFC increases.

The engine was adjusted for the maximum brake torque spark advance at each condition. The mbt spark advance was not found to be a strong function of the amount of methanol in the fuel blend. It was also found that as the amount of methanol in the fuel blend increased the mbt spark advance became independent of manifold pressure and was only dependent on equivalence ratio.

Generally exhaust gas temperature became lower as the amount of methanol increased. This reduction was greatest at Ø = 1.0 and became smaller as Ø became smaller.

THE EFFECT OF METHANOL ON PERFORMANCE - Since the intake temperature was held constant and there was never any liquid fuel present in the intake charge, no increase in brake torque can be attributed to the fuel

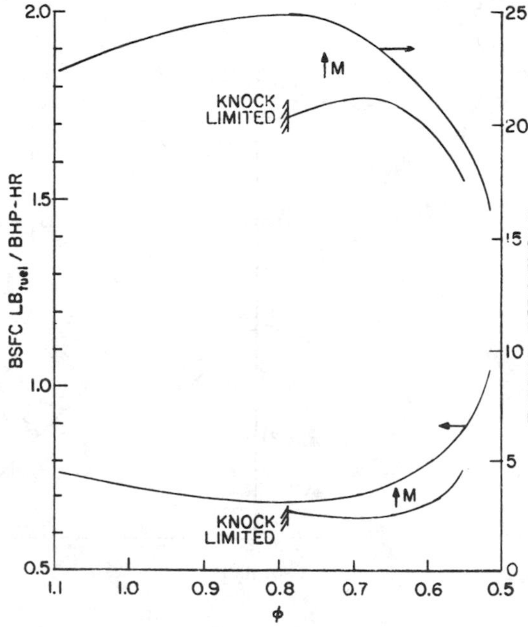

Fig. 5 - BSFC and thermal efficiency vs Ø with prototype and 40M blend at WOT

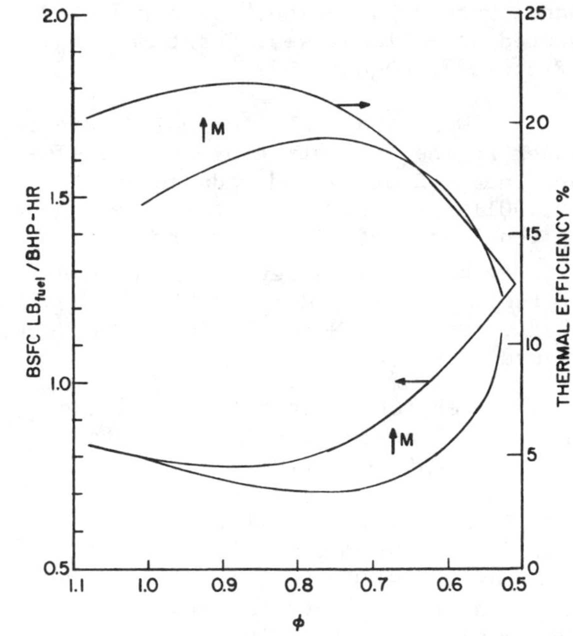

Fig. 6 - BSFC and thermal efficiency vs Ø with prototype and 40M blend at 24.3 in. Hg absolute pressure

vaporization cooling effect. Although the energy in the fuel inducted per unit volume of charge did not vary appreciably, an increase in brake torque was found to occur as the amount of methanol in the fuel blend was increased, (see Figure 8). The addition of methanol appears to enhance the combustion

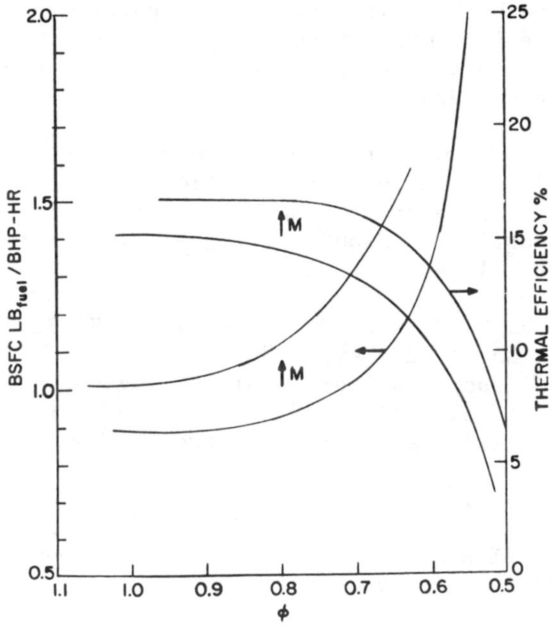

Fig. 7 - BSFC and thermal efficiency vs Ø with prototype and 40M blend at 19.8 in. Hg absolute pressure

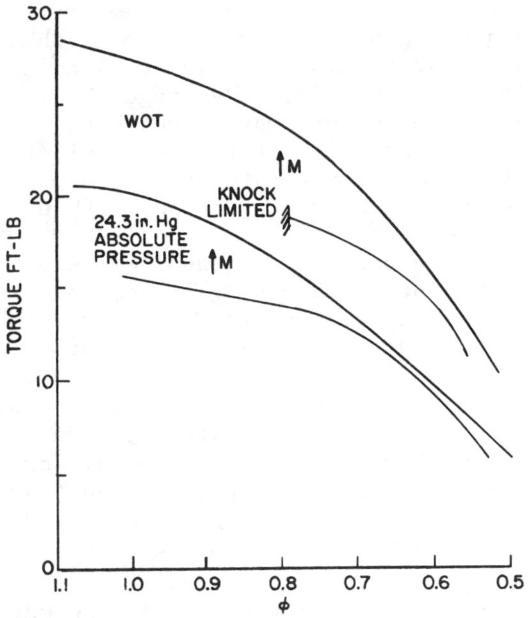

Fig. 8 - Torque vs Ø with prototype and 40M blend

process such that the thermal efficiency, Figures 5 to 7, is increased.

CONCLUSIONS

This single cylinder engine study had as its primary objective the documentation of exhaust emissions, BSFC, thermal efficiency and LML as a function of load and fuel composition. The experimental data leads to the following conclusions:

1. Methanol blends of up to 20 per cent in the base fuel yield LML's for each load that occur at similar values of Ø, and for methanol concentrations greater than 20 per cent the LML occurs at leaner values of Ø than the LML for the pure base fuel (Figure 3). However at the lightest load run (16.1 in. Hg manifold pressure) increasing the methanol concentration in the blend tended to deteriorate the LML to the point where the engine would not operate on a 40M blend.
2. Carbon dioxide and oxygen concentrations did not depend on methanol concentration or manifold pressure but were strongly dependent on equivalence ratio. Carbon monoxide was a function of manifold pressure. CO remained low, 0.05 to 0.15%, for Ø < 1.0 and did not appear to be a function of increasing methanol in the fuel blend. Nitric oxides did not appear to depend on the amount of methanol in the lean region but was a strong function of equivalence ratio. The mbt spark advance became progressively independent of the manifold pressure as the methanol concentration increased.
3. In the region 1.0 > Ø > 0.7 methanol blends have several advantages when compared

to the pure base fuel. The thermal efficiency and torque are higher for the blends than for the pure base fuel. (See Figures 5 to 8). Also there is the added advantage that, due to the octane number improvement brought about by the addition of the methanol, it is possible to always operate at mbt throughout this range of equivalence ratios without being knock limited.

4. The measurement of methanol emission with an FID presents problems. The response of the FID is not proportional to the carbon number and was observed to be time dependent. (See Appendix A).

While this single cylinder study did provide much information, it also pointed out several areas where further fundamental work is required. For example, ignition and flame speed studies for alcohol-gasoline blends would help in the understanding of the results obtained in the present study.

The effect of methanol on FID HC measurements needs to be studied and the development of an optimum instrument operating condition (if it exists) to decrease the response time and increase the methanol sensitivity would be valuable.

ACKNOWLEDGEMENTS

The authors wish to express their appreciation to the Environmental Protection Agency for financial assistance from Grants #R-802425 and T900011 administered through the Center for Air Environment Studies. Appreciation is also extended to the Mechanical Engineering Department, Center for Air Environment Studies (CAES Report No. 396-75), Texaco Research Center, and Champion Spark Plug Co. for their help in making this research possible.

REFERENCES

1. E. S. Starkman, H. K. Newhall and R. D. Sutton, "Comparative Performance of Alcohol and Hydrocarbon Fuels." Published in SP-254, "Alcohols and Hydrocarbons as Motor Fuels." New York: SAE, June 1964.

2. R. K. Pefley, M. A. Saad, M. A. Sweeney, J. D. Kilgroe and R. E. Fitch, "Study of Decomposed Methanol as a Low Emission Fuel." Final Report on Contract No. EHS-70-118, April 30, 1971, University of Santa Clara. NTIS PB-202 732.

3. H. G. Adelman, D. G. Andrews and R. S. Devoto, "Exhaust Emissions from a Methanol-Fueled Automobile." Paper 720693 presented at National West Coast Meeting, San Francisco, August 1972.

4. G. D. Ebersole and F. S. Manning, "Engine Performance and Exhaust Emissions: Methanol versus Isooctane." Paper 720692 presented at National West Coast Meeting, San Francisco, August 1972.

5. W. J. Most and J. P. Longwell, "Single-Cylinder Engine Evaluation of Methanol-Improved Energy Economy and Reduced NO_x." Paper 750119 presented at Automotive Engineering Congress, Detroit, February 1975.

6. J. A. Bolt, "A Survey of Alcohol as a Motor Fuel." Published in SP-254, "Alcohols and Hydrocarbons as Motor Fuels." New York: SAE, June 1964.

7. S. S. Hetrick, Doctoral Thesis, The Pennsylvania State University, College of Engineering, PRL-5-67 (1967).

8. J. S. Ninomiya, A. Golovoy and S. S. Labana, "Effect of Methanol on Exhaust Composition of a Fuel Containing Toluene, n-Heptane, and Isooctane." J. of Air Pollution Control Assoc., 20, 314 (1970).

9. M. W. Jackson, "Exhaust Hydrocarbon and Nitrogen Oxide Concentrations with an Ethyl Alcohol-Gasoline Fuel." Published in SP-254, "Alcohols and Hydrocarbons as Motor Fuels." New York: SAE, June 1964.

10. E. F. Obert, "Internal Combustion Engines." 3rd Edition, International Textbook Co., 1968.

11. T. W. Ryan, III, S. S. Lestz and W. E. Meyer, "Extension of the Lean Misfire Limit and Reduction of Exhaust Emissions of an SI Engine by Modification of the Ignition and Intake Systems." Paper 740105 presented at SAE Automotive Engineering Congress, Detroit, February 1974.

12. A. A. Quader, "Lean Combustion and the Misfire Limit in Spark Ignition Engines." Paper 741055 presented at SAE International Automobile and Manufacturing Meeting, Toronto, October 1974.

13. R. K. Barton, D. K. Kenemuth, S. S. Lestz and W. E. Meyer, "Cycle-by-Cycle Variations of a Spark Ignition Engine - A Statistical Analysis." Paper 700488 presented at SAE Mid-Year Meeting, Detroit, May 1970.

14. P. A. Gabele, "The Effect of Intake Valve Modification on Cycle-by-Cycle Variations in an SI Engine." M.S. Thesis, The Pennsylvania State University, March 1971.

15. Texaco Research Center, "The Texaco Transistorized Ignition System." November 1970.

16. E. J. Canton, "Lean Combustion of Methanol-Gasoline Blends in an SI Engine." M.S.

Thesis, CAES Publ. No. 394-75, The Pennsylvania State University, May 1975.

17. A. J. Andreatch and R. Feinland, "Continuous Trace Hydrocarbon Analysis by Flame Ionization." Analytical Chemistry, Vol. 32 (1960), p. 1021.

18. L. S. Ettre and H. N. Claudy, "Hydrogen Flame Ionization Detector." Chemistry in Canada, 34, September 1960.

19. W. A. Dietz, "Response Factors for Gas Chromatographic Analysis." Journal of Gas Chromatography, 5, 68 (1967).

20. Coordinating Research Council, "Oxygen Effect in Flame Ionization Response to Hydrocarbons in Automotive Engine Exhaust." March 1967.

21. C. R. Wilke, Journal of Chemistry Physics, Vol. 18 (1950) p. 517.

Appendix A

FLAME IONIZATION DETECTOR RESPONSE TO METHANOL

In view of the fact that the response of methanol in a flame ionization detector (FID) apparently varies considerably from one type to the next, it was found necessary to calibrate the Beckman Model 109 FID used for this work. It is well known that the relative response per carbon number of normal paraffins, aromatics and unsaturates does not vary appreciably from unity, usually within experimental error. When analyzing methanol, lower relative responses have been reported with variations from report to report. Adelman et al. (3) reports an FID response of unity, Ebersole and Manning (4) a response of 0.85, Andreatch and Feinland (17) a response of 0.83, Ettre and Claudy (18) a response of 0.58 and Dietz (19) a response of 0.23. Contributing factors may be, different flow rates used, fuel gas composition and adsorption on tubing surfaces.

To check the Beckman FID response a gas blend was made up in a stainless steel tank (approximately 35 litre volume) which was clean. The FID was calibrated with 1037 ppm methane in nitrogen and zeroed on nitrogen. The pure gas addition was done using an all glass high vacuum line. The procedure was to evacuate the tank to < 10 microns Hg, add the pure calibration gas to a known pressure (about 3.5 mm Hg) and then bring the total tank pressure to about 1550 mm Hg (30 psia) with the nitrogen used to zero the FID. The tank was then heated on the bottom for 24 hours to promote mixing. After the heating, the contents were allowed to reach room temperature before analysis.

Three different calibration blends were prepared to test the FID response. Before the nitrogen dilutant was added to each mixture the pure test gas was introduced. First nitrogen alone was used in order to determine the cleanliness of the tank and found to contain < 3 ppm hydrocarbons as methane. Then a blend using ultra high purity methane (Matheson) was made (2434 ppm) and the analysis was within 3% of that calculated from the partial pressure mixing. The FID response time was a few seconds. Finally the same methanol which was used in the fuel blending was degassed and added to the tank as a gas on the vacuum line. The same nitrogen addition, tank heating and cooling procedure used previously was followed before analysis. Figure A1 is the response obtained with a blend of 2178 ppm methanol in nitrogen. The relative response is 0.43 after an extremely long period of time. It is believed that at least part of this slow response is due to adsorption on the tubing surface.

Since the engine exhaust gas composition was not known only estimations can be made of the actual methanol concentration. It is estimated that the addition of methanol does not significantly reduce the unburned hydrocarbon emission. It may even increase the unburned hydrocarbons. If the use of methanol blending shows promise, further work using gas chromatographic analysis would be advisable.

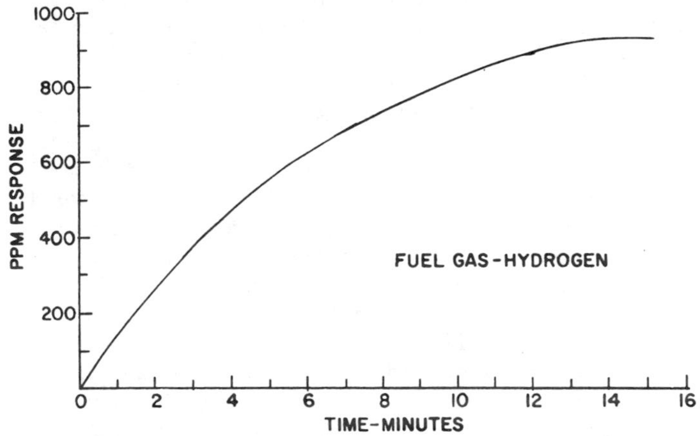

Fig. A1 - Beckman Model 109 FID response to 2178 ppm methanol in nitrogen

Appendix B

EFFECT OF OXYGEN IN A SAMPLE ON THE RESPONSE OF AN FID

Since it is well known (20) that the response of a flame ionization detector can be changed by the presence of oxygen in the flame and a flow rate variation in the sample capillary due to a viscosity change, an experiment was conducted to determine these effects. The response is also affected by the flow rates of fuel and air and by the fuel composition (hydrogen, hydrogen - nitrogen or hydrogen - helium). These tests were conducted using the Beckman Model 109 FID fueled with hydrogen. The gas flow rates are unknown, but the regulator pressure for each gas was the same during these tests as during the engine exhaust sampling.

The hydrocarbon used was methane. It must be noted that if other hydrocarbons are used, the oxygen effect can be different.

A dynamic gas blending system was constructed to check the oxygen effect. Here, three gases, 10.2% methane in nitrogen, nitrogen, and air, were blended to give the same ppm of methane as the percent oxygen was varied. Each gas was measured by a rotometer calibrated with the respective gas at the pressure and temperature for the test conditions using a wet test meter or soap bubble flow meter. The nitrogen and air flows were balanced to give the desired oxygen concentration while keeping the total flow rate constant, thereby leaving the methane concentration constant at each concentration. These gases were then fed into a cross connector to mix and then through a 0.25 inch o.d. tube to the FID.

The FID was zeroed on nitrogen and spanned on 1037 ppm methane in nitrogen. The FID response was three ppm when the blending air (compressed air) was sampled, which indicates negligible effect on the zero point.

The test results are presented in Figures B1 and B2. As the data shows there is a decrease in the FID response which percentage wise does not depend on the baseline methane concentration. About 3% of this decrease, with a sample containing 20% oxygen, can be attributed to the reduction in sample flow rate due to the increase in gas viscosity according to a Coordinating Research Council Report (20). If the viscosity effect is assumed to be proportional to the oxygen concentration, the net effect of the oxygen in the flame is a 0 to 4% decrease in response.

When sampling raw engine exhaust the other major constituent is CO_2. Table B1 shows the concentration of the major components in engine exhaust for different equivalence ratios. Table B2 shows the viscosity change, with respect to nitrogen, for various engine $\emptyset$ and a gas blend of 80% N_2 and 20% O_2. Units are not shown for viscosity since the only concern is with changes. The mixture viscosity was calculated by the method of Wilke (21). As $\emptyset$ goes richer the viscosity decrease shown in Table B2 would result in a response increase of the FID.

From these data it is estimated that the FID response to methane can vary from a 2% increase at $\emptyset = 1.0$ to a 3% decrease at $\emptyset = 0.5$.

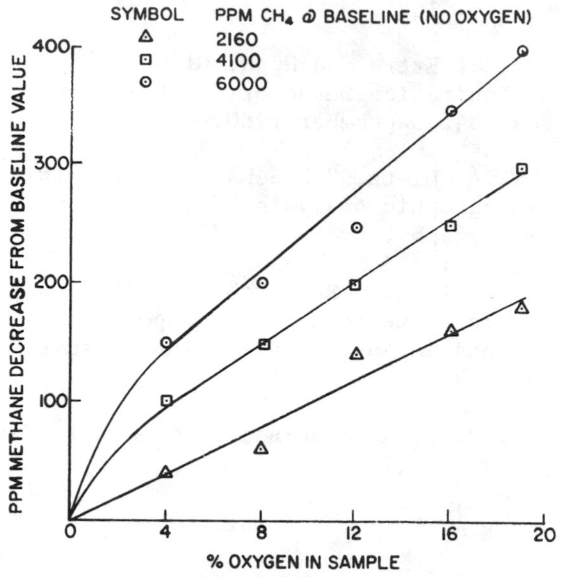

Fig. B1 - Effect of oxygen in a methane sample on the ppm response of a Beckman FID

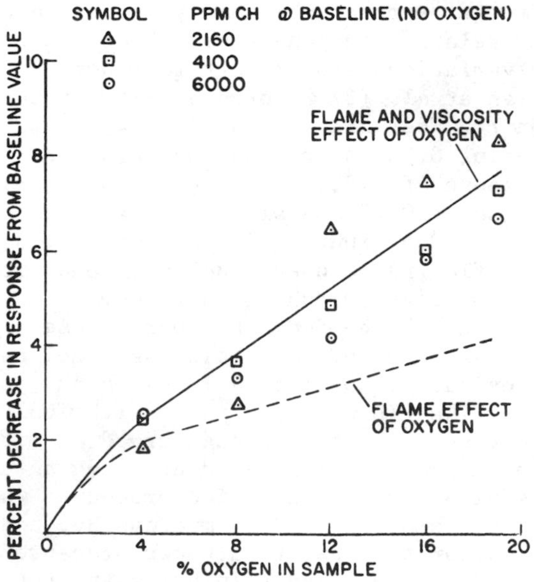

Fig. B2 - Effect of oxygen in a methane sample on the percent decrease in response of a Beckman FID

Table B1

Exhaust Emission Percent Values

Gas	Viscosity	Percent Volume Concentration at Ø = 1.0	Ø = 0.75	Ø = 0.5
N_2	4.32	85	83	80
O_2	4.99	1	6	13
CO_2	3.62	14	11	7

Table B2

Viscosity Variation for Various Engine Equivalence Ratios

	Ø = 1.0	Ø = 0.75	Ø = 0.5	Calibration Blend 80% N_2 - 20% O_2
Viscosity	4.202	4.259	4.34	4.454
Viscosity % change from N_2	-2.7	-1.4	+0.5	+3.1

A Single-Cylinder Engine Study of Methanol Fuel-Emphasis on Organic Emissions*

David L. Hilden and Fred B. Parks
Research Labs.
General Motors Corp.

ALTERNATIVE AUTOMOTIVE FUELS have been sought to reduce exhaust pollutants. The escalating price of foreign crude oil and the United States' desire for energy independence have recently added two more reasons for seeking alternative fuels. In the United States, the most abundant potential source of automotive fuels is coal. Recent studies (1-3)* show that the most feasible liquid fuels which can be produced from coal are synthetic gasoline and distillates, and methanol. Vehicle manufacturers have considerable experience with gasoline and distillates derived from petroleum. This experience should be largely applicable to their synthetic counterparts (4). Methanol's potential as an automotive fuel, however, is less well understood. There may be supply and economic problems if methanol is used as an automotive fuel. However, this paper only considers certain engine-related areas.

Fueling entirely with methanol looks promising since, researchers generally agree that engine efficiency and power are comparable to, or even improved over that of gasoline (5-8). Furthermore, NO_x exhaust emissions can be significantly reduced both at a given equivalence

* Numbers in parentheses designate References at end of paper.

*Paper 760378 presented at the Automotive Engineering Congress and Exposition, Detroit, Michigan, February 1976.

ABSTRACT

Exhaust emission and performance characteristics of a single-cylinder engine fueled with methanol are compared to those obtained either with gasoline or a methanol-water blend. Our measurements of engine efficiency and power, and CO and NO_x emissions agree with trends established in the literature. Consequently, the emphasis is placed on organic emissions (unburned fuel including hydrocarbons, and aldehydes), an area in which there is no consensus in the literature.

In all cases with methanol fueling, the unburned fuel (UBF) emissions were virtually all methanol as opposed to hydrocarbon compounds. Without special measures to overcome methanol's large heat of vaporization, UBF emissions were four times greater with methanol than those with gasoline. Similarly, aldehyde emissions were an order of magnitude greater with methanol. These high levels of organic emissions with methanol were related to inadequate fuel-air mixture preparation, which was caused by methanol's large heat of vaporization. Modifying the single-cylinder engine intake system to improve vaporization reduced UBF emissions 80 to 90 percent with methanol and 30 to 50 percent with gasoline. Aldehyde emissions were also significantly reduced by improving mixture preparation, but remained three to four times greater for methanol than for gasoline.

Blending 10 percent water with methanol resulted in: 1) reduced engine efficiency and power, 2) increased UBF emissions, 3) no measurable effect on aldehyde and CO emissions, and 4) reduced NO_x emissions. Our tests indicate that the advantages of blending water with methanol are outweighed by the disadvantages.

SAE/PT-80/19/$02.50

ratio (Ø),** and by exploiting the lean operating ability of methanol (7-9). However, some important problems remain with methanol fueling.

Due to the differences in both physical and chemical properties between methanol and gasoline, the usual technique*** for measuring organic exhaust emissions with gasolines is not appropriate for methanol. Some investigators have acknowledged this problem and qualified their emission measurements (6). Other workers have recognized the pitfalls and simply speculated on the true emissions picture (8). A third group has used gas chromatography (GC) to determine even the minor organic exhaust components (10,11). However, GC has not been applied over a wide range of engine variables and test conditions. Finally, aldehyde exhaust emissions require further study since they are probable products of incomplete methanol combustion (12) and known to be atmospheric pollutants (13).

This study contributes to the knowledge of methanol fueling by concentrating on the area of most uncertainty, i.e., organic exhaust emissions (unburned fuel, hydrocarbons, and aldehydes). Supplementary data on engine efficiency and power, and NO_x and CO emissions appear in Appendices A and B, respectively. All results are compared to those obtained with a reference gasoline under the same test conditions. Also, a fuel consisting of 90 percent methanol-10 percent water (M10W) was tested because of reported NO_x (8) and aldehyde (7) exhaust emission advantages.

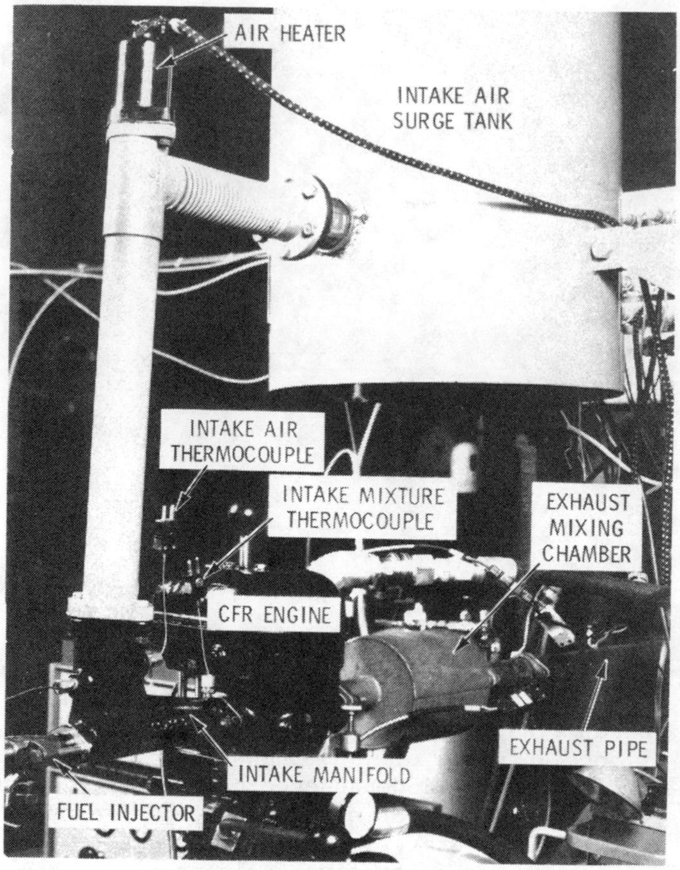

Fig.1 -Single-cylinder engine with standard mixture preparation system

EXPERIMENTAL APPARATUS AND PROCEDURE

TEST ENGINE - Tests were conducted with a single-cylinder, ASTM-CFR split-head variable-compression ratio engine. The bore was 82.6 mm and the stroke was 114.3 mm, giving a displacement of 0.61 ℓ. Engine speed was held constant by a 29.8 kW electric dynamometer.

TEST FUELS - The engine was fueled either with a reference gasoline (clear Indolene), commercial grade methanol, or a blend of 90 percent methanol-10 percent water (M10W).

FUEL-AIR MIXTURE PREPARATION SYSTEMS - During the course of this study, it was recognized that fuel-air preparation can strongly influence organic emissions. Consequently, different intake manifold systems were used to provide two degrees of fuel-air preparation. The intake manifold which is standard equipment for the CFR engine was used to provide "standard" mixture preparation (Figure 1). With that system, the fuel injector was mounted horizontally in the elbow of the manifold about 0.25 m upstream of the intake valve. The sole source of controlled heat with this system was a 1 000 watt electric heater located between the intake air surge tank and the fuel injector. This system was "standard" only in that it is a common arrangement using "off-the-shelf" equipment. It should not be construed as the optimum intake system, even for gasoline. Note in Figure 1 that two thermocouples were located in the manifold. One, located just upstream of the fuel injector, measured only air temperatures. The other thermocouple, located downstream of the fuel injector at the exit plane of the intake manifold, measured fuel-air mixture temperatures.

After completion of tests with the standard manifold, the intake system was modified to improve vaporization of the methanol - a difficult task since, on an equal fuel energy basis, approximately nine times more heat is required to vaporize methanol than gasoline. A side-by-side schematic comparison of the standard and improved mixture preparation systems is shown in Figure 2. The first of two modifications was to place inside the intake manifold a heat exchanger through which low-temperature steam could flow at controlled rates. This change provided the means for gross adjustment of

** Equivalence Ratio (Ø) = $\dfrac{\text{actual fuel-air ratio}}{\text{stoichiometric fuel-air ratio}}$

*** A system consisting of water traps ahead of a flame ionization detector which is calibrated with a paraffinic hydrocarbon.

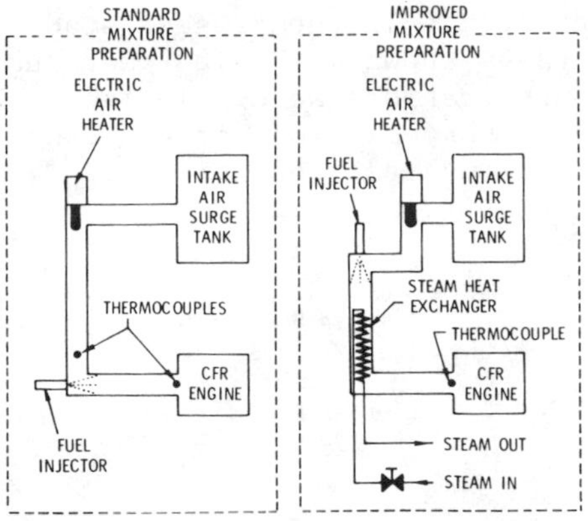

Fig.2 -Standard and improved mixture preparation systems

large quantities of heat to the fuel-air mixture. The helical heat exchanger was constructed of 1/4 inch copper tubing. Because of interference with the heat exchanger, the thermocouple formerly used to measure intake air temperature was removed. However, the thermocouple at the exit plane of the manifold was retained. The second modification was to relocate the fuel injector so that fuel was discharged down the axis of the heat exchanger. This relocation more than doubled the time available for fuel vaporization prior to induction into the combustion chamber. The electric air heater was retained as a fine adjustment of heat flow to the air stream.

TEST CONDITIONS - The engine was operated at constant speed, constant airflow, and MBT spark timing. Specific operating conditions are shown in Table 1. For tests with the modified manifold, the temperature of the fuel-air mixture at the exit plane of the intake manifold was controlled at 60°C. The choice of 60°C was based on preliminary tests at various Ø's with methanol fuel. Independently of Ø, unburned fuel emissions increased rapidly as the temperature was allowed to drop below 60°C. Conversely, these emissions were reduced only slightly at higher temperatures. Although possibly characteristic only of our apparatus and test conditions, a mixture temperature of 60°C was optimum with regard to unburned fuel emissions since additional heating of the mixture provided only slight emission benefits.

All fuels were tested over a range of Ø's from stoichiometric to the lean limit. The lean limit was defined as the leanest mixture for which no misfire occurred. Misfires were detected both audibly and by sudden increases in unburned fuel emissions.

MEASUREMENT OF EXHAUST EMISSIONS

GENERAL - Concentrations of O_2, CO_2, CO, and NO_x were measured in the dried exhaust gas as shown in Figure 3. Equivalence ratio was determined by exhaust gas (CO, CO_2, O_2) analysis. For the purpose of reporting emissions, concentrations of CO and NO_x were converted to mass specific units.

Organic emissions from gasoline-fueled engines are usually described as hydrocarbons. That description would be incorrect for methanol since a large fraction of the organic emissions is expected to be oxygenated. Due to this aspect of methanol fueling, organic emissions require different nomenclature, special measurement techniques, and carefully defined reporting procedures. In this report, the organic emissions have been classified as either unburned fuel or aldehydes. Unburned fuel (UBF) emissions include any portion of the fuel which escapes reaction, plus hydrocarbon compounds either generated in the combustion process or resulting from oil ingestion. (Note that for gasoline fueling UBF and hydrocarbons are virtually synonymous.)

UNBURNED FUEL EMISSIONS - Because most UBF

Table 1 - Test Conditions

Test Condition	Standard Mixture Preparation	Improved Mixture Preparation
Airflow, kg/min	0.206	0.206
Engine Speed, r/min	1200	1200
Spark Timing	MBT*	MBT
Intake Air Temperature, °C	52	not controlled
Intake Mixture Temperature, °C	not controlled**	60
Compression Ratio	8	8
Intake Air Humidity, g/kg	1.9 - 2.4	1.9 - 2.4
Coolant Temperature, °C	82	82
Oil Temperature, °C	82	82
Exhaust Pressure, kPa	1.7	1.7

* Minimum advance for best engine torque.

** This temperature was measured to be 9 ± 5°C for methanol fueling, whereas for gasoline the corresponding value was 28 ± 5°C.

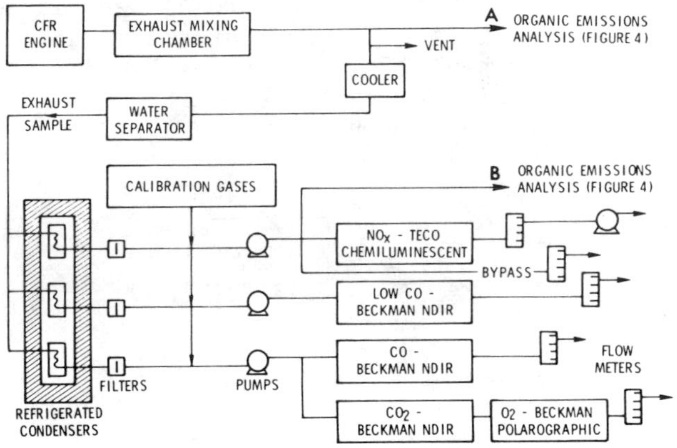

Fig.3 -Apparatus for analysis of O_2, CO_2, CO, and NO_x in the exhaust

emissions with methanol fuel are expected to be water soluble, they must be measured prior to any water condensation in the exhaust sample. Thus, a system like the heated flame ionization detector (HFID) shown in Figure 4 becomes essential. (Emissions from gasoline combustion were also measured with the HFID to ensure a consistent basis of comparison.) All exhaust samples, including those for the HFID, passed through the exhaust mixing chamber shown in Figure 3. The volume of this chamber was approximately three times larger than the engine displacement and served two functions. First, it allowed the organic compounds in the exhaust to complete their after-reactions. That the after-reactions were indeed complete was confirmed by sampling at various locations downstream of the chamber and finding the UBF emissions practically unchanged. The second function of the exhaust mixing chamber was to guarantee adequate mixing of the exhaust species, thus ensuring a representative exhaust sample. Beyond the exhaust mixing chamber, the sample line to the HFID (Figure 4) was heated to 140 ± 5°C to prevent any water condensation and the potential loss of water-soluble exhaust components prior to analysis.

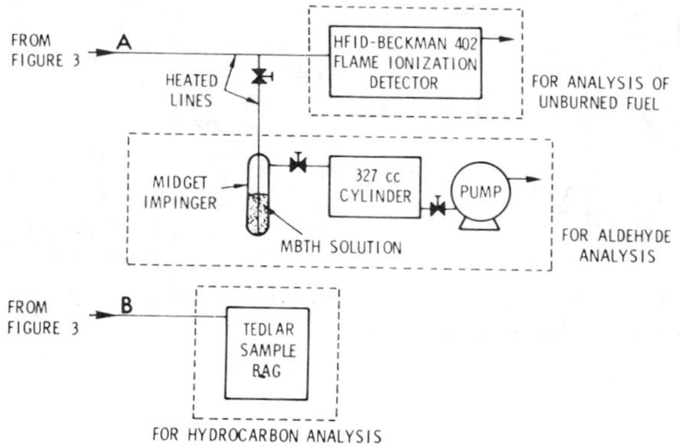

Fig.4 -Apparatus for analysis of organic emissions in the exhaust

The HFID was fueled with a 40 percent H_2/ 60 percent N_2 mixture, a common fuel for flame ionization detectors. Nitrogen was used to zero the analyzer. Calibration gases for the HFID must be judiciously selected because the HFID does not respond alike to the carbon atoms in all organic molecules. On the one hand, formaldehyde, which is expected to be the most abundant exhaust aldehyde, is not detected by the HFID (14) and must be determined independently. On the other hand, the HFID detects the carbon atom in methanol, but with less sensitivity than for paraffinic carbon atoms (see Appendix C). Because of the variation in response exhibited by the HFID, a true measure of UBF emissions requires a calibration gas representative of the exhaust species which are being detected. In the present study, the calibration gases shown in Table 2 were used. These concentrations were selected to be similar to those of UBF emissions in the exhaust sample for the various test conditions.

Table 2 - Calibration Gases for the Heated Flame Ionization Detector

	Standard Mixture Preparation	Improved Mixture Preparation
	(Nominal Concentration, ppm in Nitrogen*)	
Propane	897	203
Methanol	2060	1040

* All calibration gases were prepared and analyzed by Scott Environmental Technology, Inc. The concentration of each gas was certified within ±2 percent. Additionally, the concentration of the 2060 ppm methanol gas was confirmed at the General Motors Research Laboratories using gas chromatography.

UBF concentrations were converted to mass specific units. For gasoline fueling, the standard practice of reporting emissions as mass specific hydrocarbons (propane equivalent) was followed. With methanol fueling, emissions were reported as mass specific unburned methanol. Use of the methanol-in-nitrogen calibration gas assumes that methanol is practically the only component in the exhaust of a methanol-fueled engine detected by the HFID. The validity of that assumption was ascertained by measuring hydrocarbons in the dried exhaust sample using the apparatus depicted in Figures 3 and 4.

HYDROCARBON EMISSIONS - The exhaust was collected in a Tedlar sampling bag and analyzed for hydrocarbons with a Perkin-Elmer Model 3920 gas chromatograph. The exhaust hydrocarbons were separated into individual species on a column containing Chromosorb 102 (80/100 mesh range). Since the gas chromatograph was remote from the engine, a delay of at least one-half hour occurred before the exhaust sample was analyzed. Consequently, composition changes occurring within that time period would go undetected. Analyses at various periods longer than one-half hour showed no significant change in hydrocarbon composition.

ALDEHYDE EMISSIONS - As shown in Figure 4, aldehyde emissions were measured by passing the exhaust sample through a heated (130°C) Teflon line to a midget impinger containing a solution of 3-methyl-2-benzothiazolone hydrazone (MBTH). The flowrate through the impinger was approximately 30 cm^3/minute. At this rate, the removal of aldehydes was essentially complete. The water-insoluble exhaust components passed through the impinger into an initially evacuated 327 cm^3 cylinder, used to control the volume of the sample analyzed.

The colorimetric MBTH method (15) for analysis of aliphatic aldehydes was used. The analytical procedure was tested for interference from methanol which, like many aldehydes, is very water soluble. For methanol concentrations comparable to those measured in the engine exhaust, no interference was observed.

REPEATABILITY OF ORGANIC EMISSION MEASUREMENTS - In addition to organic exhaust emissions, NO_x and CO emissions, and engine efficiency and power were measured at all test conditions. Because these latter emission and engine results were similar to published information (6,8,9), they have been placed in Appendices A and B, respectively. Thus, we are able to focus this paper on organic emissions. Because of this emphasis, the repeatability of the organic emission data deserves special consideration.

During the five-month test period, various organic emission measurements were conducted to assess repeatability. Typical error bands for UBF emission measurements are shown in Figure 5. Averaging all repeated tests for UBF, the standard deviation was less than 4 percent of the mean. Similar analyses of aldehyde and hydrocarbon emissions showed 5 and 7 percent repeatability, respectively. These organic emissions are generally compared at various Ø's which were determined with an accuracy of ±2 percent.

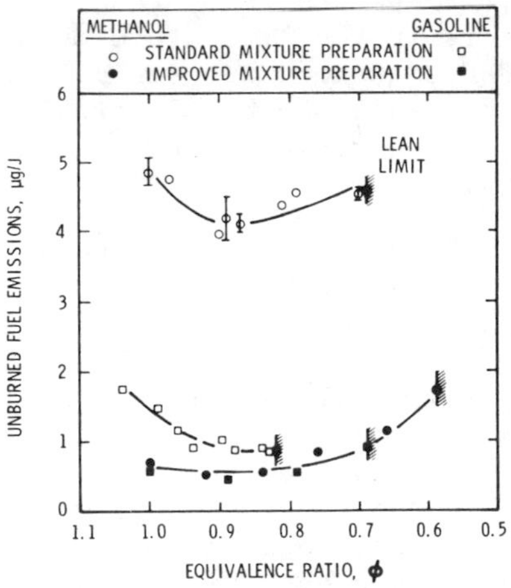

Fig.5 -Mixture preparation effects on UBF exhaust emissions with methanol and gasoline fuels

ORGANIC EMISSIONS - RESULTS AND DISCUSSION

HYDROCARBON EMISSIONS - As previously stated, UBF emissions with gasoline fueling were assumed to be all hydrocarbons which could be compositely represented by a propane calibration gas. With methanol fueling, it was necessary to ascertain the relative amounts of unburned methanol and hydrocarbon compounds in the exhaust. If the hydrocarbon fraction was large, use of methanol as the calibration gas for the HFID would yield erroneous results.

Hydrocarbon emissions with methanol fueling were analyzed with a gas chromatograph for both the standard and improved mixture preparation systems at the four equivalence ratios shown in Table 3. These Ø's were selected to represent the range for which UBF emissions were evaluated. Hydrocarbons identified in the chromatograms included: methane, acetylene, ethylene, allene, propylene, propane, and a C_4 olefin. In Table 3, the sum of these hydrocarbons has been divided by the total UBF emissions at the various test conditions. (UBF emissions will

Table 3 - Relative Hydrocarbon Content of UBF Emissions with Methanol Fuel

Equivalence Ratio (Ø)	Hydrocarbon Emissions / Unburned Fuel Emissions (by mass): Standard Mixture Preparation	Improved Mixture Preparation
1.05	0.006	0.02
0.96	0.0045	0.016
0.87	0.0017	0.017
0.79	0.0018	0.01

be discussed in later paragraphs.) Hydrocarbon emissions generally were a greater fraction of the UBF emissions as Ø was increased and as the fuel-air mixture preparation was improved. For the case where the fraction of hydrocarbon in UBF was greatest at 0.02 (Ø = 1.05, Improved Mixture Preparation), the error in UBF emissions due to the hydrocarbons would be less than one-half percent. This slight error results from the difference in the HFID response between methanol and propane (see Appendix C). Hence, with methanol fueling, total UBF emissions can be accurately measured with a HFID which is calibrated with a methanol-in-nitrogen mixture.

UNBURNED FUEL EMISSIONS - For both mixture preparation systems, UBF emissions with methanol and gasoline fuels are shown in Figure 5. The profiles of UBF versus Ø exhibit conventional features (12) which are independent of fuel and mixture preparation. UBF emissions declined as Ø was reduced from stoichiometric. After reaching minimum values between Ø = 0.9 and Ø = 0.8, UBF emissions generally increased with further reductions in Ø. This last trend was not observed with standard mixture preparation and gasoline because lean operating ability was restricted.

Looking specifically at the case of standard mixture preparation in Figure 5, UBF emissions with methanol were about four times greater than those with gasoline. This large difference in the mass of UBF emissions has not been previously noted in the literature.

Our results most closely agree with those of the single-cylinder engine experiments of Harrington and Pilot (6). They found UBF emissions from methanol combustion, on a "ppm C" basis, to be either equal to or lower than those with gasoline for stoichiometric and lean mixtures. Note that when "ppm C" emissions are equal with methanol and gasoline fuels, the actual mass of UBF emissions with methanol is 2.3 times those with gasoline.* Furthermore, Harrington and Pilot acknowledged that their UBF measurements with methanol fuel represent lower limits since no compensation was made for the difference in HFID response between the exhaust species with methanol fueling and those with gasoline fueling. The data in Appendix C suggest that this causes no more than a 25 percent error. Therefore, had Harrington and Pilot measured and presented UBF emissions in the same manner as that used in this paper, they would have concluded that UBF exhaust emissions for methanol might be as much as three times greater than those for gasoline.

Higher UBF emissions with methanol have also been found in multicylinder engine and vehicle tests. Ingamells and Lindquist (16) modified a compact car to operate with stoichiometric or leaner mixtures using either gasoline or methanol as fuel. For the 1972 Federal Test Procedure (FTP), UBF emissions on a mass basis with methanol were about twice as high as those with gasoline. Adelman, et al., (10) found similar results with a specially modified compact car fueled with methanol and driven on a cold-start cycle. However, we would be remiss if we left the impression that all multicylinder engine and vehicle tests (or even all other single-cylinder engine tests) found greater UBF emissions with methanol than with gasoline.

Fleming and Chamberlain (5) found UBF emissions (mass basis) from a V-8 engine operated at steady state to be 23 percent lower with methanol than with gasoline. Even lower UBF emissions were reported by Bernhardt (7), who modified a Volkswagen to accommodate the gross differences in both combustion stoichiometry and heat of vaporization between methanol and gasoline. He reported UBF emissions (volume basis) with methanol to be 0.1 to 0.25 times those with gasoline.

Such reports prompted us to consider which conditions of the standard mixture preparation tests could be responsible for the high levels of UBF emissions observed with methanol fuel. Two tests which involved varying inlet air and coolant temperatures (summarized in Appendix D) led us to believe that improved mixture preparation, to compensate for methanol's large heat of vaporization, would reduce UBF emissions. The results of such an improvement in mixture preparation with both gasoline and methanol are also shown in Figure 5.

With improvements in mixture preparation for the gasoline-fueled CFR engine, reductions in UBF exhaust emissions varied from 50 percent at $\phi \simeq 1.0$ to 30 percent at $\phi = 0.82$ (the gasoline lean limit with standard mixture preparation). In terms of controlling NO_x emissions (Appendix B), a significant by-product of improved mixture preparation was the extension of the gasoline lean limit to $\phi = 0.69$. However, as shown in Figure 5, exploiting that lean operating ability would significantly increase the UBF emissions from their minimum value. Methanol's large heat of vaporization was believed responsible for mixture preparation problems and attendant high UBF emissions. Consequently, improved mixture preparation was expected to produce more dramatic UBF reductions with methanol than with gasoline. That expectation was realized. Improved mixture preparation reduced UBF emissions by 80 to 90 percent, thus eliminating any difference in UBF emissions between the fuels. Finally, note that the lean operating ability with methanol fuel was likewise extended with improved mixture preparation. However, as with gasoline, UBF emissions increased for mixtures leaner than $\phi = 0.85$.

Thus, it appears that a major portion of the conflicts in the literature over differences in UBF emissions between methanol and gasoline may be explained by differences in fuel-air mixture preparation among the various experiments.

ALDEHYDE EMISSIONS - A number of studies with methanol have considered the problem of aldehyde exhaust emissions, but the results are not consistent. For example, Ingamells and Lindquist (16) modified a 1971 fuel-injected compact car to operate on methanol. They found the aldehyde exhaust emissions "were about the same" for methanol as for gasoline. Bernhardt (7), on the other hand, observed the aldehyde exhaust concentration to be at least two times greater for methanol than for gasoline with a water-cooled Volkswagen engine which utilized a modified carburetor and heated intake manifold. The experiments of Ebersole and Manning (9) (using a CFR engine with a pressurized fuel system and heated manifold) and of Adelman, et al., (10) (using a 1970 AMC Gremlin with modified carburetor and manifold) also indicate that aldehyde emissions are more of a problem with methanol than with gasoline. What isn't clear, however, is the magnitude of the problem.

Aldehyde exhaust emissions were determined during the course of the present study in an attempt to explain the variance in reported results. Data were taken with both of the fuel-air mixture preparation systems in the expecta-

---

*

	Methanol	Gasoline (typical)
Chemical formula	CH_3OH	$CH_{1.85}$
$\frac{\text{Molecular weight}}{\text{Carbon atom}} \left(\frac{MW}{C}\right)$	32	13.85

$$\left(\frac{MW}{C}\right) \text{methanol} \Big/ \left(\frac{MW}{C}\right) \text{gasoline} = 32/13.85 = 2.3$$

tion that aldehyde emissions would decrease with improved vaporization, as did UBF emissions. The premise that UBF and aldehyde emissions are closely linked is based on the assumption that aldehydes are produced by partial oxidation of UBF (12).

Mass specific aldehyde emissions (on a formaldehyde basis) are plotted versus ϕ in Figure 6. Results span a large range, so the ordinate in Figure 6 has a logarithmic scale to facilitate comparison. Consider first the results derived with the standard intake system.

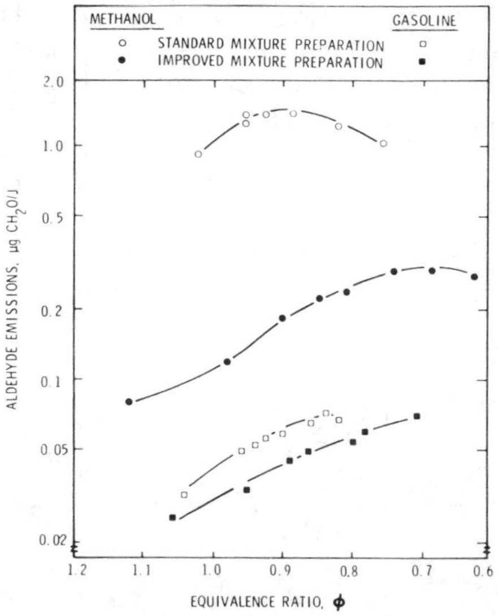

Fig.6 -Mixture preparation effects on aldehyde exhaust emissions with methanol and gasoline fuels

In the case of gasoline fueling, aldehyde emissions increased as the engine ran leaner. This increase was in contrast to the decrease in total UBF emissions (Figure 5) over the same operating range. Wodkowski and Weaver (17) and Bernhardt (7) have observed a similar trend. According to Wodkowski and Weaver, increased aldehyde emissions follow from increased spark advance as well as decreased equivalence ratio. In the present work, increased spark advance accompanies decreased equivalence ratio since MBT spark timing was used throughout.

Aldehyde emissions produced by methanol fueling with the standard manifold (top line of Figure 6) increased as the engine ran leaner than stoichiometric, and reached a maximum at $\phi = 0.88$. Unlike the results for gasoline fueling, emissions decreased for ϕ less than 0.88. By operating close to the lean limit, the mass specific aldehyde emissions were reduced approximately 25 percent from the maximum. This reduction is surprising since other methanol studies indicate that aldehydes do not decrease as the lean operating limit is approached (7, 9). We have no explanation for the contradiction, although mixture preparation appears to be involved as will be shown later.

With the standard intake system, aldehyde emissions were an order of magnitude greater with methanol than with gasoline. This is a much larger fuel effect than heretofore reported. A large part of the difference disappeared when the standard intake system was replaced by the modified one.

With gasoline, modifying the standard manifold reduced aldehyde emissions approximately 30 percent (comparing the bottom two lines of Figure 6). This is roughly the same reduction as that observed for UBF. At most ϕ's with methanol fueling, the aldehyde emissions decreased 75 to 90 percent with improved mixture preparation. Again, the reduction in aldehyde emissions closely followed the 80 to 90 percent decrease in UBF. The premise presented earlier, that reductions in aldehyde emissions will parallel reductions in UBF, appears to be correct at a given ϕ.

The shape of the aldehyde emissions curve for methanol fueling with the modified intake system is very similar to that of either of the aldehyde emission curves for gasoline. Apparently, the maximum in these emissions, observed when methanol was used with the standard intake system, stemmed from incomplete mixture preparation. However, the actual mechanism by which mixture preparation can affect the shape of these curves is not considered here.

Although the modified intake system led to virtually identical mass specific UBF emissions for methanol and gasoline, aldehyde emissions remained roughly four times higher for methanol. Ebersole and Manning (9), who tested a single-cylinder engine at 1000 rpm with both methanol and isooctane, found virtually the same aldehyde emissions for isooctane as we found for gasoline with the modified manifold. However, their results for methanol fueling were 25 to 45 percent below the aldehyde emissions presented here. The source of this discrepancy is not known.

It seems certain, based on this and earlier studies, that aldehyde exhaust emissions are greater for methanol fueling than for gasoline. Unfortunately, the difference in the magnitude of these emissions is still not entirely clear. We have, however, shown that variations in the quality of fuel-air mixture preparation play a very important part in determining the level of aldehyde emissions, especially for methanol fueling.

The increased aldehyde emissions associated with methanol fueling appear to be a significant disadvantage due both to their photochemical reactivity and toxicity. One suggested way around the problem is to add water to the methanol. The implications of this approach are considered in the next section.

WHAT ARE THE CONSEQUENCES OF FUELING WITH A BLEND OF METHANOL AND WATER?

Recent studies indicate possible emission reductions when water (up to 30 volume percent) is added to methanol. Most and Longwell (8), for example, observed reduced NO_x emissions from a CFR engine fueled with a methanol-water blend rather than with methanol alone. Unfortunately, neither the spark timing was optimized for their tests, nor the UBF emissions measured. Bernhardt (7) has reported decreased exhaust aldehyde emissions for methanol-water as compared to methanol. However, he tested at only one equivalence ratio.

To gain a more thorough understanding of the effect of water addition, a fuel consisting of 90 (volume) percent methanol and 10 percent water (M10W) was tested. The experiment was carried out with the standard fuel-air mixture preparation system under the operating conditions given in Table 1. Efficiency and power as well as CO, NO_x, UBF, and aldehyde exhaust emissions were determined. Also, some additional data (on aldehyde emissions) were taken with the modified mixture preparation system. The results are described below.

EFFICIENCY AND POWER - The indicated thermal efficiency at a specific equivalence ratio (Figure 7) was 2 to 3 percent less for M10W than

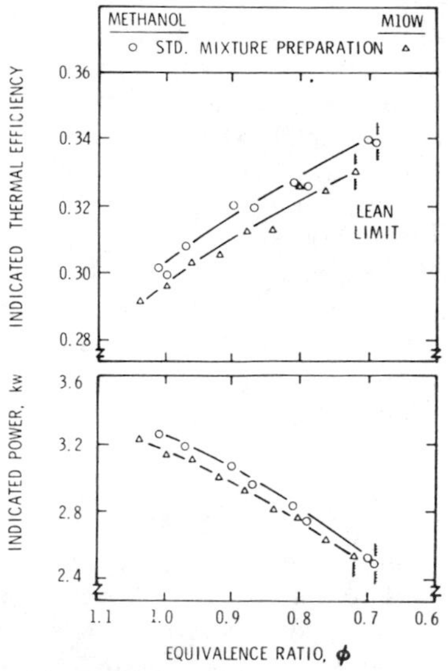

Fig.7 -Engine efficiency and power with methanol and 90% methanol-10% water fuels

for methanol. Most and Longwell (8) have also reported decreased engine efficiency for methanol-water blends. However, as mentioned above, their data were taken with constant spark timing and, therefore, are not directly comparable to the results presented here. With either methanol or M10W, maximum efficiency was achieved at the lean limit. Methanol operated out to a lean limit of ϕ = 0.69 yielding a thermal efficiency of 0.34. For M10W the corresponding values were 0.72 and 0.33, respectively.

Indicated power output for the methanol-water blend (Figure 7) decreased 2 to 3 percent (at a given ϕ) relative to methanol, as it must to parallel the drop in thermal efficiency. Bernhardt (7) has also reported decreased power output with addition of water to methanol. A similar effect may be seen when water is added to gasoline (18). Possible reasons for this effect have been discussed by Obert (19).

EMISSIONS - CO exhaust emissions at a given equivalence ratio were virtually equal for methanol and M10W. Since the results with methanol are discussed in Appendix B, they will not be mentioned here. However, NO_x exhaust emissions were significantly different for the two fuels (see Figure 8). NO_x emissions went down, as expected, when combustion temperatures were lowered by adding water. Ten percent water decreased the maximum NO_x emissions value by 34 percent. The engine could operate significantly leaner (0.03 equivalence ratio) with

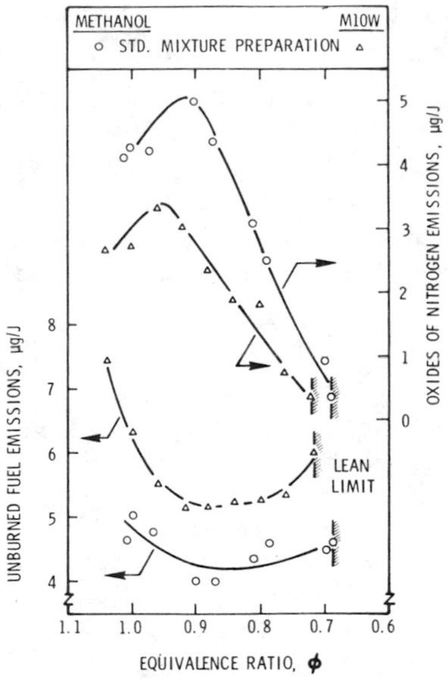

Fig.8 -NO_x and UBF exhaust emissions with methanol and 90% methanol-10% water fuels

methanol than with M10W. As a consequence the lean limit NO_x exhaust emissions were virtually equal for the two fuels. Most and Longwell (8) found a similar result when they compared NO_x emissions from a single-cylinder engine (7.82 compression ratio) fueled with methanol to those of methanol containing 20 percent water.

Figure 8 also includes results for UBF emissions. The two curves are similar in shape but UBF emissions were greater by approximately 20 percent (at a given ϕ) with the blend. This increase may be explained by the added diluent which increases quench layer thickness as it decreases combustion temperatures. Bernhardt (7), however, observed virtually no change in UBF emissions (concentration basis) when up to 10 percent water was added to methanol.

A large part of this difference in results may be explained by the decreased power output (approximately 10 percent) which Bernhardt observed with water addition. The mass specific emissions reported in this paper reflect that drop, while the concentration units used by Bernhardt do not.

Bernhardt also reported a reduction of approximately 40 percent in the concentration of aldehyde emissions when methanol fuel was diluted by 10 percent water. Figure 9 shows aldehyde emissions for the present study, including data taken with both standard and modified intake manifolds. There was no significant difference in mass specific aldehyde emissions between methanol and M10W, regardless of the fuel-air mixture preparation system which was employed. Again, part of the difference in results between the present work and that of Bernhardt can be explained by the difference between mass specific and concentration units mentioned in the previous paragraph. However, the reasons for the major part of the difference are not known.

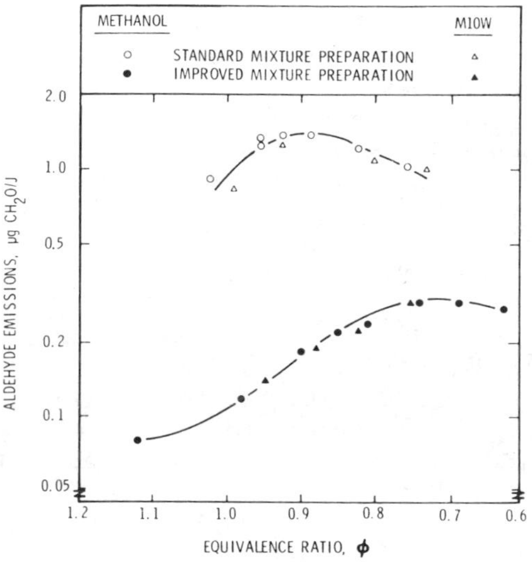

Fig.9 -Aldehyde exhaust emissions with methanol and 90% methanol-10% water fuels

From the standpoint of emission control, the one advantage of fueling with the methanol-water blend, rather than with methanol alone, was reduced NO_x emissions (at a given ϕ). However, this advantage disappeared when the engine was operated close to its lean limit, since methanol will burn leaner than M10W. For these lean operating conditions, NO_x, CO, and aldehyde exhaust emissions and power output were practically unchanged from one fuel to the other. However, fueling with methanol rather than M10W had the advantage of lower UBF emissions and higher efficiency.

CONCLUDING REMARKS

The principal points addressed in this paper are: 1) the types and levels of organic emissions to be expected from methanol-fueled engines, 2) the effects of mixture preparation on organic emissions, and 3) the possible benefits of blending water with methanol.

Without special consideraton of methanol's large heat of vaporization, both UBF and aldehyde mass emissions could be many times greater than those with gasoline. It is, however, unlikely that automobile engines would, or even could, be designed for methanol fueling without considering mixture preparation problems. As a consequence, the organic emissions picture of specially modified vehicles fueled with methanol would likely not be as bleak as indicated by our single-cylinder engine results with standard mixture preparation. On the other hand, it is unlikely that the fuel-air mixture could be better prepared on a vehicle than in our single-cylinder engine with improved mixture preparation. In fact, the test conditions were empirically selected as those for which additional manifold heat would not significantly reduce UBF emissions. As a result, fueling automobiles with methanol instead of gasoline would be expected to produce equal or greater levels of UBF emissions and a manyfold increase in aldehyde emissions.

It isn't clear that the potentially higher levels of organic emissions constitute pollution problems. The atmospheric reactivity of lower alcohols is believed low (13). Aldehydes, on the other hand, are reactive (13) and must be considered a liability of methanol fueling.

The addition of 10 percent water to methanol caused reductions in NO_x emissions but not aldehydes. The NO_x advantage must be weighed against penalties such as increased UBF emissions and lower engine efficiency. For our choice of test conditions, the advantages of blending water with methanol appear outweighed.

Technical factors such as cold starting, driveability, and fuel tank capacity to maintain constant range may influence the extent to which methanol will be used in spark ignition engines. In general, the present study has not identified any exhaust emission factors which would preclude its use. Engine efficiency and exhaust pollutants should not be the limiting factors in the use of methanol. Rather, the extent to which methanol will actually be used in automotive engines will probably be decided not in the combustion chamber, but in political and economic chambers.

ACKNOWLEDGMENTS

We acknowledge the assistance of Mr. M. J.

Bartch, Fuels and Lubricants Department, General Motors Research Laboratories, in maintaining the apparatus and conducting the experiment. Gas chromatography analyses by Mr. Robert Halsall, also of the Fuels and Lubricants Department, are appreciated.

REFERENCES

1. J. Pangborn and J. Gillis, "Alternative Fuels for Automotive Transportation - A Feasibility Study," Vols. 1-3, prepared for U. S. Environmental Protection Agency, Publication No. EPA-640/3-74-012-a, b, and c, July 1974.

2. F. H. Kant, R. P. Cahn, A. R. Cunningham, M. H. Farmer, W. Herbst, and E. H. Manny, "Feasibility Study of Alternative Fuels for Automotive Transportation," Vols. 1-3, prepared for U. S. Environmental Protection Agency, Publication No. EPA-460/3-74-009-a, b, and c, June 1974.

3. R. T. Johnson, "Energy and Synthetic Fuels for Transportation," presented to the Central States Combustion Institute Symposium, Madison, Wisconsin, 1974.

4. R. W. Hurn, "Characteristics of Conventional Fuels from Nonpetroleum Sources - An Experimental Study," presented at General Motors Research Laboratories Symposium - Future Automotive Fuels - Prospects, Performance, and Perspective, Warren, Michigan, October 1975.

5. R. D. Fleming and T. W. Chamberlain, "Methanol as Automotive Fuel, Part I - Straight Methanol," SAE Paper 750121, February 1975.

6. J. A. Harrington and R. M. Pilot, "Combustion and Emission Characteristics of Methanol," SAE Paper 750420, February 1975.

7. W. E. Bernhardt, "Engine Performance and Exhaust Emission Characteristics from a Methanol-Fueled Automobile," presented at General Motors Research Laboratories Symposium - Future Automotive Fuels - Prospects, Performance, and Perspective, Warren, Michigan, October 1975.

8. W. J. Most and J. P. Longwell, "Single-Cylinder Engine Evaluation of Methanol - Improved Energy Economy and Reduced NO_x," SAE Paper 750119, February 1975.

9. G. D. Ebersole and F. S. Manning, "Engine Performance and Exhaust Emissions: Methanol versus Iso-octane," SAE Paper 720692, August 1972.

10. H. G. Adelman, D. G. Andrews, and R. S. Devoto, "Exhaust Emissions from a Methanol-Fueled Automobile," SAE Paper 720693, August 1972.

11. R. K. Pefley, M. A. Saad, M. A. Sweeney, and J. D. Kilgroe, "Performance and Emission Characteristics Using Blends of Methanol and Dissociated Methanol as an Automotive Fuel," Paper 719008, Intersociety Energy Conversion Engineering Conference, 1971.

12. D. J. Patterson and N. A. Henein, Emissions from Combustion Engines and Their Control, Ann Arbor Science Publishers, Inc., Ann Arbor, Michigan, 1972.

13. A. P. Altshuller and I. R. Cohen, "Structural Effects on the Rate of Nitrogen Oxide Formation in the Photo-Oxidation of Organic Compound-Nitric Oxide Mixtures in Air," Journal of Air and Water Pollution, Vol. 7, 1963.

14. Private Communication from Tony Christopher, Beckman Instruments, Fullerton, California, April 1975.

15. E. Sawicki, T. R. Hauser, T. W. Stanley, and W. Elbert, Analytical Chemistry, Vol. 33, 93, (1961).

16. J. C. Ingamells and R. H. Lindquist, "Methanol as a Motor Fuel or a Gasoline Blending Component," SAE Paper 750123, February 1975.

17. C. S. Wodkowski and E. E. Weaver, "The Effects of Engine Parameters, Fuel Composition, and Control Devices on Exhaust Aldehyde Emissions," presented at the West Coast Air Pollution Control Association Meeting, October 1970.

18. S. S. Lestz and W. E. Meyer, "The Effect of Direct-Cylinder Water Injection on Nitric Oxide Emission from an SI Engine," Proc. 14th FISITA Congr., London, England, June 1972.

19. E. F. Obert, Internal Combustion Engines and Air Pollution, Intext Educational Publishers, New York, New York, 1973.

20. "Basic Considerations in the Combustion of Hydrocarbon Fuels with Air," NACA Report 1300, 1959.

21. C. F. Taylor and E. S. Taylor, The Internal Combustion Engine, Second Edition, International Textbook Company, Scranton, Pennsylvania, 1970.

22. Private Communication with W. A. Daniel, General Motors Research Laboratories, December 1975.

Appendix A

ENGINE EFFICIENCY AND POWER WITH METHANOL AND GASOLINE FUELS

The efficiency and power traits of the CFR engine fueled either with methanol or gasoline are presented in Figures A-1 and A-2, respectively. The effect of improved mixture preparation was the same for both fuels; therefore, observed differences between fuels apply independently of the mixture preparation system. In general, our results are in agreement with other comparisons of the two fuels (5,6). Efficiency and power were greater with methanol than with gasoline.

EFFICIENCY - The engine efficiency data (Figure A-1) fall into two adjacent bands. The upper band, representing methanol, shows a 2-3 percent improvement relative to gasoline at the same Ø. Such improvements can be predicted for

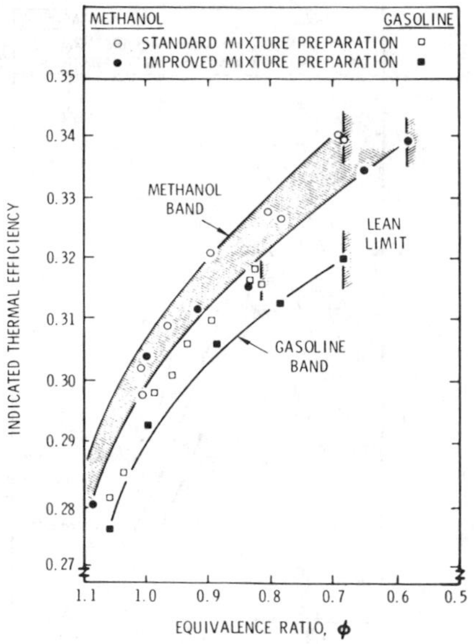

Fig.A-1 -Engine efficiency with methanol and gasoline fuels-standard and improved mixture preparation systems

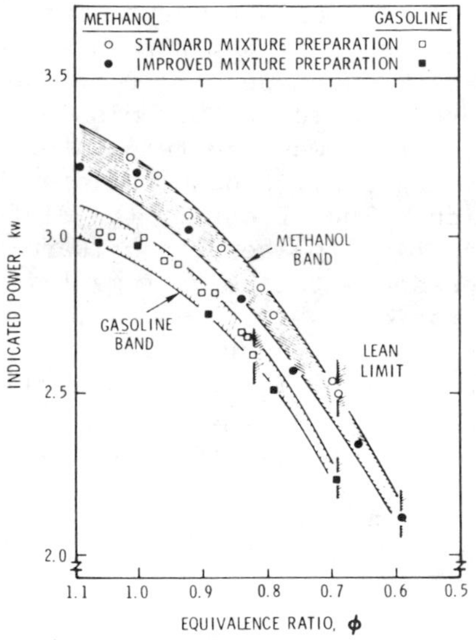

Fig.A-2 -Engine power with methanol and gasoline fuels-standard and improved mixture preparation systems

methanol because of reduced heat transfer from the combustion gases to the chamber boundaries [lower combustion temperatures and higher flame speeds (20)] and because of more moles of combustion products per mole of air. In addition, methanol has another efficiency advantage over gasoline; the extended lean operating ability of methanol resulted in a maximum efficiency of 0.34, a 6-7 percent improvement over that with gasoline.

As shown in Figure A-1, improved mixture preparation both extended the lean limit and slightly reduced engine efficiency at a given Ø. A possible explanation for the reduced efficiency was increased heat transfer associated with higher combustion temperatures (21). The system for improved mixture preparation increased mean gas temperatures by vaporizing the fuel outside the combustion chamber. The NO_x data (Appendix B) also support the premise of higher combustion temperatures with the modified intake system.

POWER - Indicated power, the sum of friction and brake powers, is plotted for both fuels in Figure A-2. The difference in power output between the fuels is partially explained by the difference in energy input between methanol and gasoline.* Greater energy input with methanol accounts for approximately two-thirds of the 8 percent power increase, while the remaining power increase with methanol is attributed to improved engine efficiency. Although the power increase with methanol is real, the division of the 8 percent into energy inputs and efficiency gains is somewhat arbitrary. It depends on the heating values ascribed to methanol and Indolene, which vary depending on the reference cited.

Appendix B

CO AND NO_x EXHAUST EMISSIONS WITH METHANOL AND GASOLINE FUELS

CO EMISSIONS - Mass specific CO emissions from the CFR engine equipped either with the standard or the improved mixture preparation system, and fueled with either methanol or gasoline are shown in Figure B-1. Note that the ordinate is logarithmic to accommodate the broad range of CO emissions observed under different test conditions. At first glance, the data in Figure B-1 appear to fall in one broad band that is generally characteristic of CO emissions as a function of Ø. However, scrutiny of the data shows that, under lean conditions, the fuels and the mixture preparation systems did affect the CO emissions.

With standard mixture preparation, CO emissions appear slightly higher with methanol than with gasoline. This was not unexpected since, with lean mixtures, CO emissions are probably the result of partial oxidation of unburned fuel

* Fuel Property	Methanol	Indolene
stoichiometric fuel-air ratio	0.155	0.069
lower heating value (kJ/g)	19.91(6)	42.42(6)
kJ of fuel energy per gram of stoichiometric air	3.086	2.928

Relative Energy Input (Methanol/Indolene)

$$= \frac{3.086}{2.928} \times 100 = 1.054$$

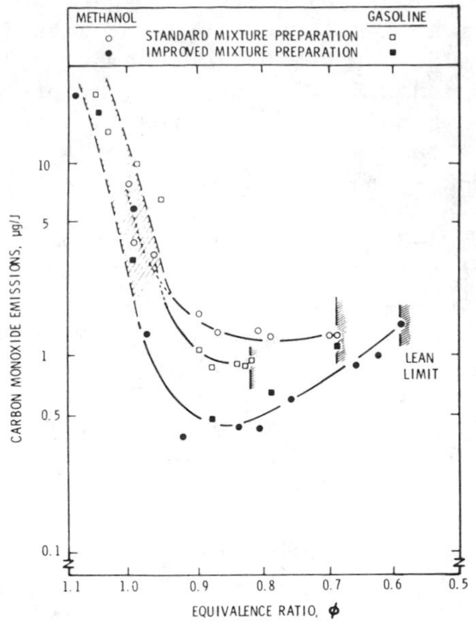

Fig.B-1 -Mixture preparation effects on CO exhaust emissions with methanol and gasoline fuels

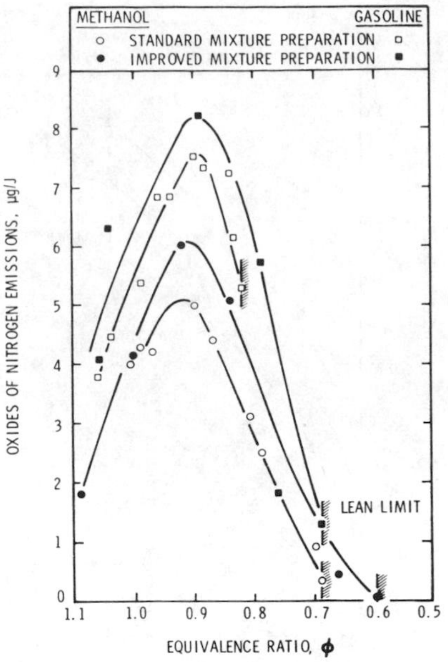

Fig.B-2 -Mixture preparation effects on NO_x exhaust emissions with methanol and gasoline fuels

from the quench layer (22), and the UBF emissions with methanol were greater than those with gasoline (see the body of this paper).

Improved mixture preparation reduced CO emissions with both fuels. Relative to emissions levels with standard mixture preparation, the minimum CO emissions with methanol were reduced 70 percent; with gasoline the reduction was 50 percent. As a result of the reductions, CO emissions with both fuels were comparable during lean engine operation. The observation of reduced CO emissions with improved mixture preparation is consistent with similar effects of mixture preparation on UBF emissions.

NO_x EMISSIONS - NO_x emissions for the various test conditions, plotted in Figure B-2, show well established trends (12).

Considering first the case of standard mixture preparation, methanol fueling considerably reduced NO_x emissions relative to gasoline. The peak NO_x emissions with methanol were about 33 percent below the peak NO_x value with gasoline. This reduction is believed to result primarily from the lower flame temperature of methanol as compared to that of gasoline (20). Not only does methanol vapor have a lower adiabatic flame temperature, but the large heat of vaporization of methanol also cools the mixture and, hence, tends to further lower the combustion gas temperature. The advantage of lean operation is also shown in Figure B-2. At the lean limit for methanol ($\phi = 0.69$), NO_x emissions were 90 percent lower than those for gasoline at its lean limit ($\phi = 0.82$).

Improved mixture preparation increased the maximum NO_x emissions with gasoline and methanol, 15 and 20 percent, respectively. This trend was expected since the higher temperature fuel-air mixture increased flame temperatures. Additionally, it was not surprising that the NO_x increase was a higher percentage with methanol than with gasoline. Part of the reason methanol produced lower NO_x emissions with the standard manifold was the temperature depression resulting from methanol's large heat of vaporization. The effect of the difference in heat of vaporization between methanol and gasoline was erased with the system of improved mixture preparation.

Improved fuel vaporization also extended the lean combustion limits. The lean operating limit of gasoline was extended from 0.82 to 0.69. Concomitant with that extension, the lean limit (minimum) NO_x emissions were reduced 75 percent. The methanol lean limit was extended from 0.69 to 0.59 and NO_x was reduced 83 percent.

Appendix C

FLAME IONIZATION DETECTOR RESPONSE TO METHANOL AND PROPANE

"Both <u>speed and magnitude</u> of analyzer response are affected by the type of hydrocarbon in the sample. Magnitude of the analyzer response to an atom of carbon depends on the chemical environment of this atom in the molecule." This quote from the Operator's Manual for the Beckman Model 402 Hydrocarbon Analyzer indicates one problem associated with unburned fuel measurement, i.e., calibration gas selection. Two basic options exist for measuring unburned fuel emissions (UBF) in the exhaust of a methanol-fueled engine. A calibration gas composed of methanol in an inert diluent can be used. Alternatively, the analyzer can be calibrated with a conventional hydrocarbon gas standard, such as propane, and corrected for the relative response* to methanol. The former approach is recommended since indiscriminate application of the latter introduces significant error.

The relative response of two flame ionization detectors to methanol was determined. A Beckman Model 402 (HFID) included provisions for heating the exhaust sample stream to prevent any condensation. A Beckman Model 400 (FID) did not include heated sample lines and is often used to analyze dried exhaust samples. Fuel (40% H_2/60% N_2), air, and exhaust sample were flowed to the analyzers at rates suggested in their respective Operator's Manuals. To avoid any nonlinearity associated with switching scales, no changes were made in sensitivity over the course of the measurements.

Four calibration gases, two methanol-in-nitrogen and two propane-in-nitrogen (Table C-1), were separately flowed to the two FID's through their respective sample ports. The reported concentration of each gas was certified to be within ±2 percent. For all tests, the FID's were zeroed on nitrogen and then calibrated on the gas containing 898 ppm propane. The responses of the analyzers were then recorded for the remaining three calibration gases. The values in Table C-1 represent the the average of two separate tests conducted the same day. Results were repeatable within one percent.

The HFID demonstrated excellent linearity of response to propane calibration gases. The relative response to the 208 ppm propane gas was within the 2 percent error band of nominal concentration. The average relative response of the HFID to methanol was 0.765. Additionally, there was a hint of nonlinearity with the methanol calibration gases since a 7 percent difference in the relative response between the two concentration levels was observed. Tests conducted with different fuel, air, and sample flowrates showed that the relative response of the detector could range as high as 0.88. The Model 400 showed excellent linearity to the methanol calibration gas as well as to propane. However, it's average relative response to methanol was 0.885, 16 percent higher than the value obtained with the HFID. These values for the relative response to methanol are reasonably consistent with values reported elsewhere (5,7,9).

Because analyzer response to methanol varies significantly, indiscriminate application of relative response correction factors from the literature can lead to errors.

Table C-1 - Flame Ionization Detector Response to Methanol and Propane

		Propane (Nominal Concentration of Calibration Gases, ppm in Nitrogen)		Methanol		
		898	208	2060		1040
Beckman Model 402 HFID	Response	*	203	1620		774
	Relative Response	1.00	0.98	0.79	Avg 0.765	0.74
Beckman Model 400 FID	Response	*	209	1854		900
	Relative Response	1.00	1.00	0.90	Avg 0.885	0.87

* Instrument calibration on this gas

* Relative response is defined as the ratio: FID response to the carbon atom in the molecule of interest over the response to the carbon atom of an aliphatic compound.

Appendix D

SUPPLEMENTARY VAPORIZATION TESTS

At the conclusion of tests with standard mixture preparation under the operating conditions of Table 1, it was hypothesized that inadequate vaporization of the methanol fuel was leading to high levels of UBF emissions. That hypothesis was based primarily on the fact that on an equal fuel energy basis, nine times more heat is required to vaporize methanol than gasoline. Two supplementary tests were conducted with the standard manifold design which support that hypothesis.

One way chosen to improve vaporization was increased heat flow to the liquid fuel droplets in the intake manifold. To accomplish this, increased heat was supplied to the combustion air prior to mixing with the fuel. The inlet air temperature was varied over the approximate range of 25 to 130°C with $\emptyset$ = 0.87 for both methanol and gasoline. All other test conditions of Table 1 were held constant. UBF emission results are shown in Figure D-1.

Considering first the results with gasoline fueling, the UBF emissions were essentially unaffected by changing the intake air temperature. With methanol fueling, the results are quite different. As the intake air temperature was

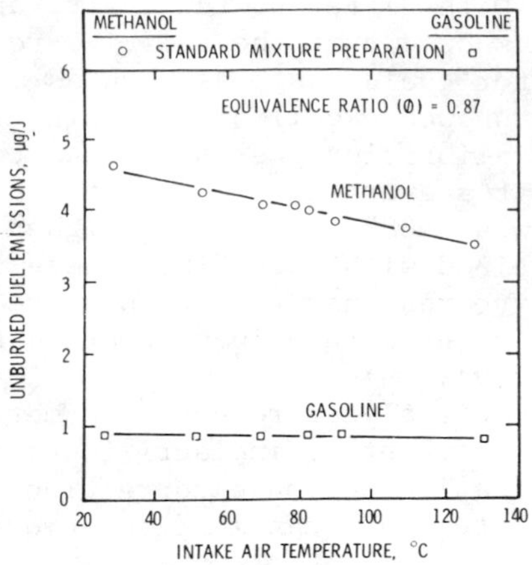

Fig.D-1 -Intake air temperature effects on UBF exhaust emissions with methanol and gasoline fuels

increased from 28 to 128°C, UBF emissions were reduced 22 percent. Furthermore, the reduction was almost linear, indicating further decreases in UBF would be possible with higher air temperatures.

A second supplementary test was conducted to examine the effect of coolant temperatures on UBF emissions. Higher coolant temperatures increase combustion chamber wall temperatures, thus providing additional heat transfer to the intake charge. For this test, the temperature of the coolant (water with rust inhibitor) was varied over the approximate range of 65 to 115°C, while holding Ø constant at 0.87. UBF emission results are shown in Figure D-2. With

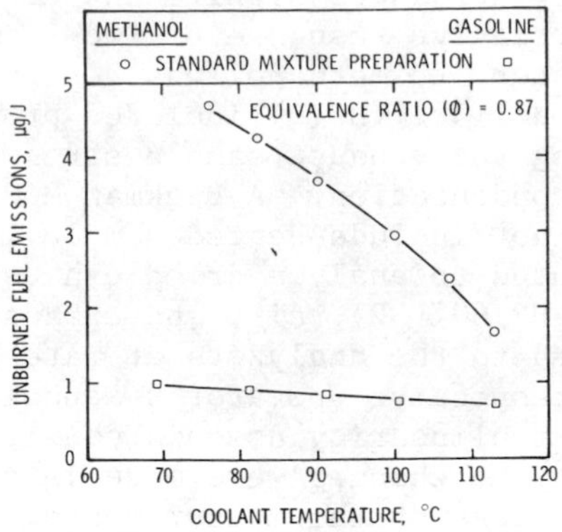

Fig.D-2 -Engine coolant temperature effects on UBF exhaust emissions with methanol and gasoline fuels

gasoline, UBF emissions decreased 25 percent as the coolant temperature was increased from 68 to 113°C. Over a slightly smaller range (76 to 113°C), UBF emissions with methanol fuel decreased 65 percent.

Apparently, the standard mixture preparation system with moderate inlet air and coolant temperatures does not provide sufficient heat to fully vaporize the methanol fuel. Both by increasing inlet air and coolant temperatures, significant reductions in UBF emissions were obtained with methanol fuel.

Development of a Pure Methanol Fuel Car*

Holger Menrad, Wenpo Lee
and Winfried Bernhardt
Volkswagenwerk AG (Germany)

ABSTRACT

Methanol as a fuel for spark ignition engines offers a lot of advantages in comparison to gasoline. Results of a prototype passenger car fueled with pure methanol show promising aspects of lower energy consumption, higher energy output and more favorable emission figures.
Modifications on the engine are limited, the unfavorable cold start and warm up behavior of pure methanol can be eliminated by the use of suitable additives.

CORRESPONDING RESULTS of energy research all over the world point out alcohols, mainly methanol (1)*). Basic work in the application of methanol as a fuel was done in several institutions (1-7). The necessary modifications to the engine and to the vehicle may be derived from the chemical and physical data of methanol (1,8-14). The most influential technical facts in the application of methanol as an automotive fuel are:

- low heating value,
- high latent heat of vaporization,
- low vapor pressure

in comparison to normal gasoline.

In order to get more information on the automotive application of methanol, a project was started to develop a methanol fueled prototype car. This prototype is expected to be as similar as possible to a current model car. It should be the basis for further tests and for comparison with a current production gasoline engine.

This project was carried out with the targets of the development tendency of modern automotive techniques in mind:

- low energy consumption,
- low emission rates,
- high power output,
- satisfactory cold start behaviour
- good driveability,
- low production costs,
- low maintenance requirements.

The work was carried out in several steps which are discussed in this paper:

1. Fundamental research on a single-cylinder engine,
2. Basic work on four-cylinder engines,
3. Adjustment on four-cylinder engines,
4. Cold start and driving tests,
5. Emission tests,
6. Durability tests.

Supplementary laboratory work was done on fuels, lubricants, and materials.

FUNDAMENTAL RESEARCH ON A SINGLE-CYLINDER ENGINE.

Based on the high octane rating of methanol the most favorable compression

*) Numbers in parantheses designate references at end of paper

*Paper 770790 presented at the Passenger Car Meeting, Detroit, Michigan, September 1977.

SAE/PT-80/19/$02.50

ratio was determined according to the characteristics

- efficiency,
- emission values,
- stress on engine parts.

The decision for the upper compression ratio limit was taken from single-cylinder engine test results. This engine is similar to the production engine in the main parts. Cylinder, piston, cylinder head, valve timing etc. are the same. Compression ratio can be varied while other test conditions can be kept constant. Therefore, the characteristics of the full scale engine can be estimated from these single-cylinder experiments.

Fig. 1 shows power and brake thermal efficiency (5) as functions of compression ratio for a stoichiometric (corresponding to full load conditions) and a lean (part load) mixture.*) (4). For both air fuel ratios best efficiency can be expected with a compression ratio between 12 and 13. As shown in the upper part of the figure, power output for the stoichiometric mixture is still increasing above these values, but only insignificantly.

In additional work on the single-cylinder engine other measures were investigated for their usability, such as EGR systems with hot exhaust gas in order to increase mixture temperature.

BASIC WORK ON FOUR-CYLINDER ENGINES.

On the test bench for four-cylinder engines basic experiments from the single cylinder engine were continued in order to get quantitative results of the potential of the applied measures. In this work the hot EGR system did not prove to be successful, because with lean mixtures and part load conditions misfiring could not be prevented.

COMPARISON WITH GASOLINE ENGINE - Comparing results from the same engine with equal compression ratio running with gasoline and pure methanol showed better efficiency, lower emission rates, and higher power output for the methanol fueled engine.

Fig. 2 indicates lower specific energy consumption SEC for methanol at part load and diffrent A/F ratios. For Methanol the course of HC emissions indicates the lean limit is more extended than for gasoline. The quantitative comparison of HC (UBF = unburnt fuel) emissions from methanol and gasoline engines is explained later.

---

* A/F ratio is given as air fuel equivalence ratio AFER, i.e. actual air flow through the engine, devided by the stoichiometric necessary air flow. AFER greater than 1 means lean mixture.

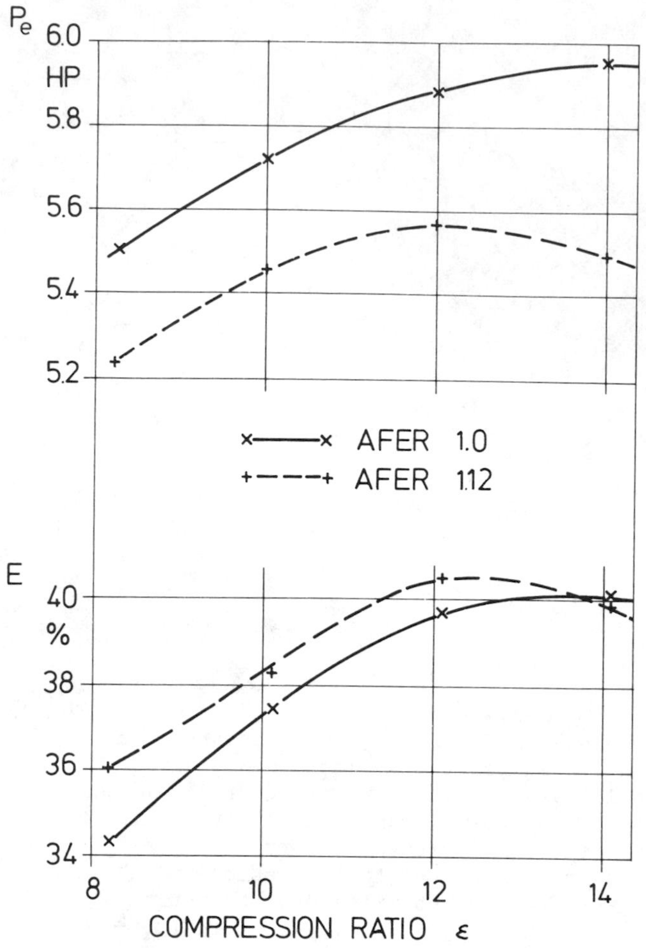

Fig. 1 - Performance Pe and brake thermal efficiency E for different air fuel equivalence ratio AFER versus compression ratio full load 2000 rpm, single-cyl eng spark timing for optimal power output

INFLUENCE OF COMPRESSION RATIO - The continuation of fundamental research studies had mainly the objective of evaluating the influence of higher compression ratio on methanol combustion quantitatively. Consequently, research was carried out on engines of the same type having compression ratios of 8.2, 9.6 and 13. As mentioned previously, the upper compression ratio limit resulted from the single cylinder engine tests. All tests discussed here were carried out with the same type of carburetor. In this connection the influence of

- air fuel mixture
- ignition timing
- mixture temperature
- compression ratio

was investigated at certain part and full throttle conditions. Some significant results are described below.

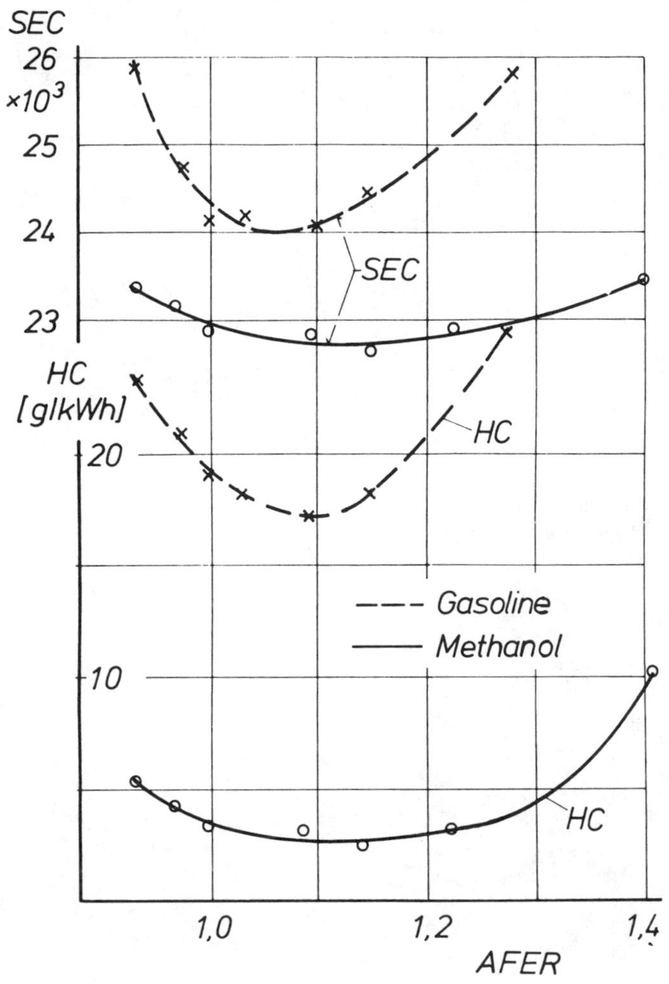

Fig. 2 - Specific energy consumption SEC and HC (UBF) emissions versus AFER, equivalence ratio Ø, rsp. Road load 40km/h (25 mph), n = 1410 rpm, Md = 18 Nm, spark timing $\alpha_z = 27^{o}$ BTDC

Part Load - Fuel Consumption -
At part throttle the fuel consumption seems to be one of the most important factors.

Fig. 3 shows the specific energy consumption SEC as a function of the A/F ratio for two part throttle points, 60 km/h (37 mph) and 100 km/h (62 mph). In each case min. fuel consumption spark timing was chosen. The figure shows quite clearly that significant savings in fuel consumption through a high compression ratio were determined mainly at lean A/F ratios. This figure also indicates that the minimum fuel consumption is displaced toward the lean side somewhat as the compression ratio increases.

The possible savings in fuel consumption for the previously mentioned part load points as a function of the compression ratio are

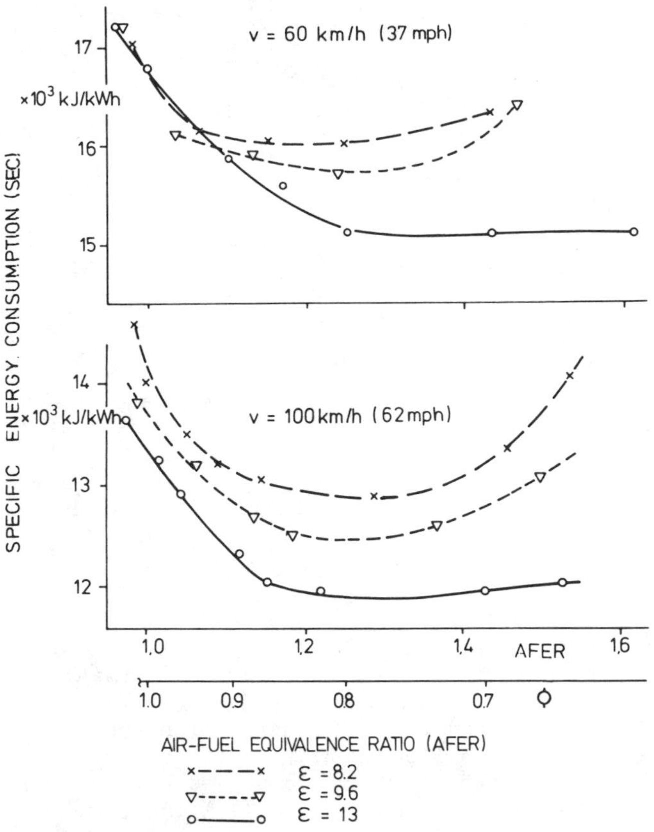

Fig. 3 - Specific energy consumption SEC versus AFER at different compression ratios. Road load 60km/h (37 mph), n = 2120 rpm, Md = 23.8Nm, 100 km/h (62 mph), n = 3530 rpm, Md = 42.6 Nm, spark timing for minimum fuel consumption

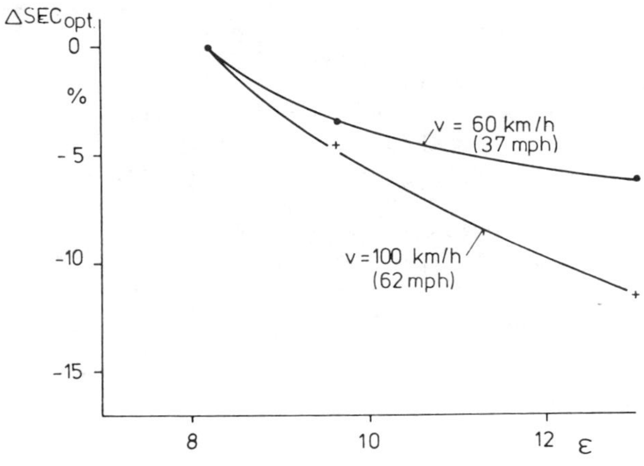

Fig. 4 - Possible savings in fuel consumption Δ SEC by increased compression ratio ε , road load 60 km/h (37 mph) and 100 km/h (62 mph), as Fig. 3

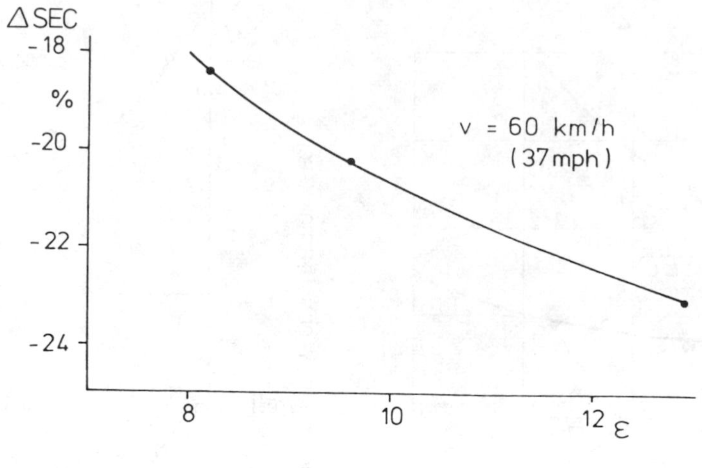

Fig. 5 - Percentual decrease in fuel consumption Δ Sec in relation to a production gasoline engine, road load 60 km/h (37 mph)

presented in Figure 4. Comparing the most favorable fuel consumption of an engine with a normal compression ratio of ξ = 8.2, at 60 km/h (37 mph) the fuel consumption can be reduced up to 5 % by increasing the compression ratio up to 13.0, while at 100 km/h (62 mph) savings of more than 11 % can be reached. However, in this case the influence of an unfavorable air fuel mixture distribution at low loads has to be considered.

In summary, Figure 5 shows the improvement of road load fuel economy at 60 km/h (37 mph) with increasing compression ratio for methanol fueled engines in comparison to the current production gasoline engine. Depending on overall optimized values the gain in efficiency ranges from 18 to 23 % according to compression ratio.

Full Load Performance - In the full throttle range the inrease in compression ratio led, in addition to fuel consumption savings, to a considerable plus in performance output, which is illustrated in Figure 6 for 2000 rpm as a function of spark timing.

For comparison here, too, the values for gasoline operation are also presented. At this load point, the energy savings by using methanol in place of gasoline can be doubled by increasing the compression ratio from 8.2 to 13. More definite is the increase in power output resulting from the higher compression ratio.

Emissions - CO emissions in the interesting ranges of A/F ratio and ignition timing did not show significant difference to a gasoline fueled engine. Furthermore, an influence by changing the compression ratio could not be observed. A distinct decrease in CO emissions, however, can be expected by

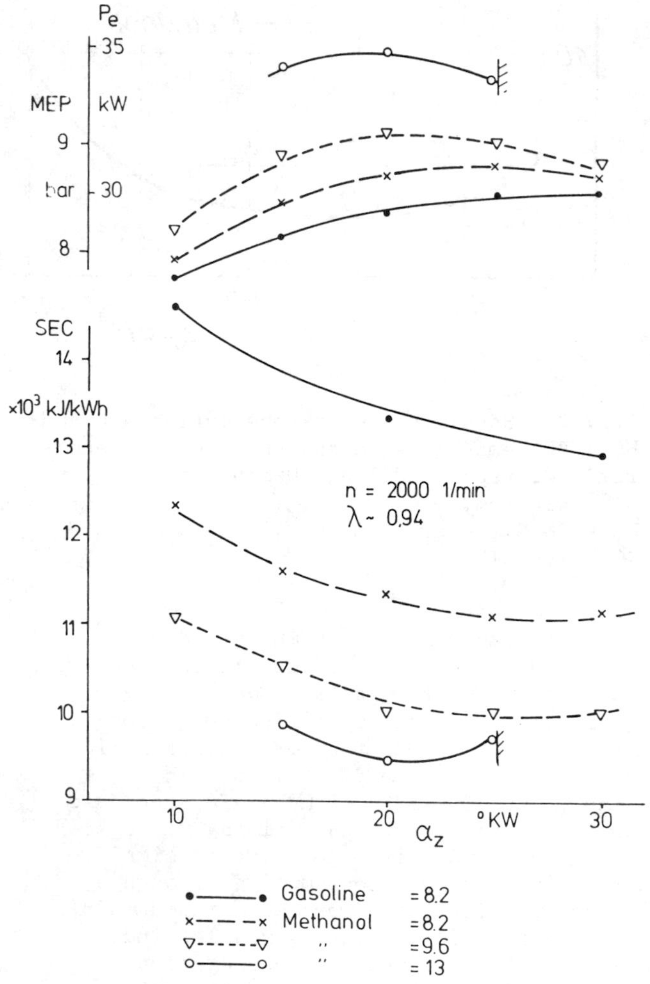

Fig. 6 - Mean effective pressure MEP and specific energy consumption SEC versus spark timing α_z full throttle, 2000 rpm, AFER ~ 0,94

the extension of the lean running limit for methanol.

HC emissions should be understood as unburnt fuel components (UBF) (3). A comparison with gasoline engines with regard to HC emissions is difficult, since the unburnt residual components of both fuels belong to different groups of species. A considerable share of methanol derived HC's is water soluble. Therefore heated FID (HFID) results show more reliable values. Hence the exhaust gas sample system was equipped with a heating system together with HFID.

In laboratory tests it was found that the normal propane calibrated HFID had a response factor of 0.80 to 0.85 to methanol. Generally HC (UBF) emissions of methanol fueled engines are lower than these of gasoline engines, fig. 2.

Emissions of aldehydes are not indicated by FID therefore wet chemistry techniques (MBTH) have to be used. Emissions of aldehydes - mainly consisting of formaldehyde - showed higher values than gasoline engines. However several reduction methods can be applied (13,14).

Emissions of NOx with CLD analysis showed only a small, negligible response to methanol vapor (3). Specific NOx emissions showed the usual dependency on A/F ratio on constant spark timing and vice versa as plotted in Fig. 7a for part load 60 km/h (37 mph).

In comparison to the gasoline engine the methanol engine has definitely more favorable NOx values for the same compression ratio, while the high compression methanol engine showed very similar values to the normal gasoline engine. However all these values are figured for constant ignition timing. Due to the high compression ratio less ignition timing advance is favorable which results in much lower NOx values, without any mentionable reduction in fuel consumption. A further advantage is to be expected with methanol from running with leaner A/F ratios. This is illustrated in Fig. 7b showing NOx emissions and fuel consumption depending on AFER and spark timing.

FINAL ADJUSTMENT OF COMPLETE ENGINE

While in previous fundamental research work the ignition timing and A/F ratio were adjusted to optimal values for the various load points with little consideration for the feasibility, now engine test bench results from finally adjusted engines with

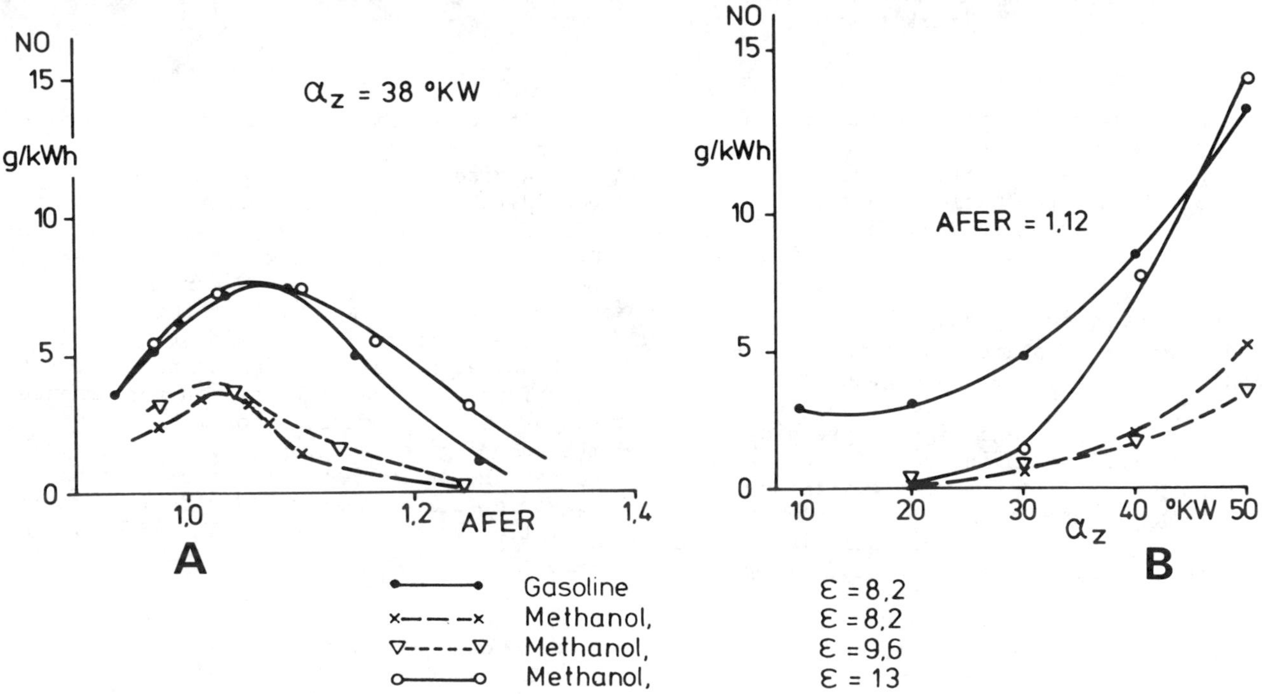

Fig. 7A - NO emissions at road load 60 km/h (37 mph): A) - versus AFER, B) - versus spark timing

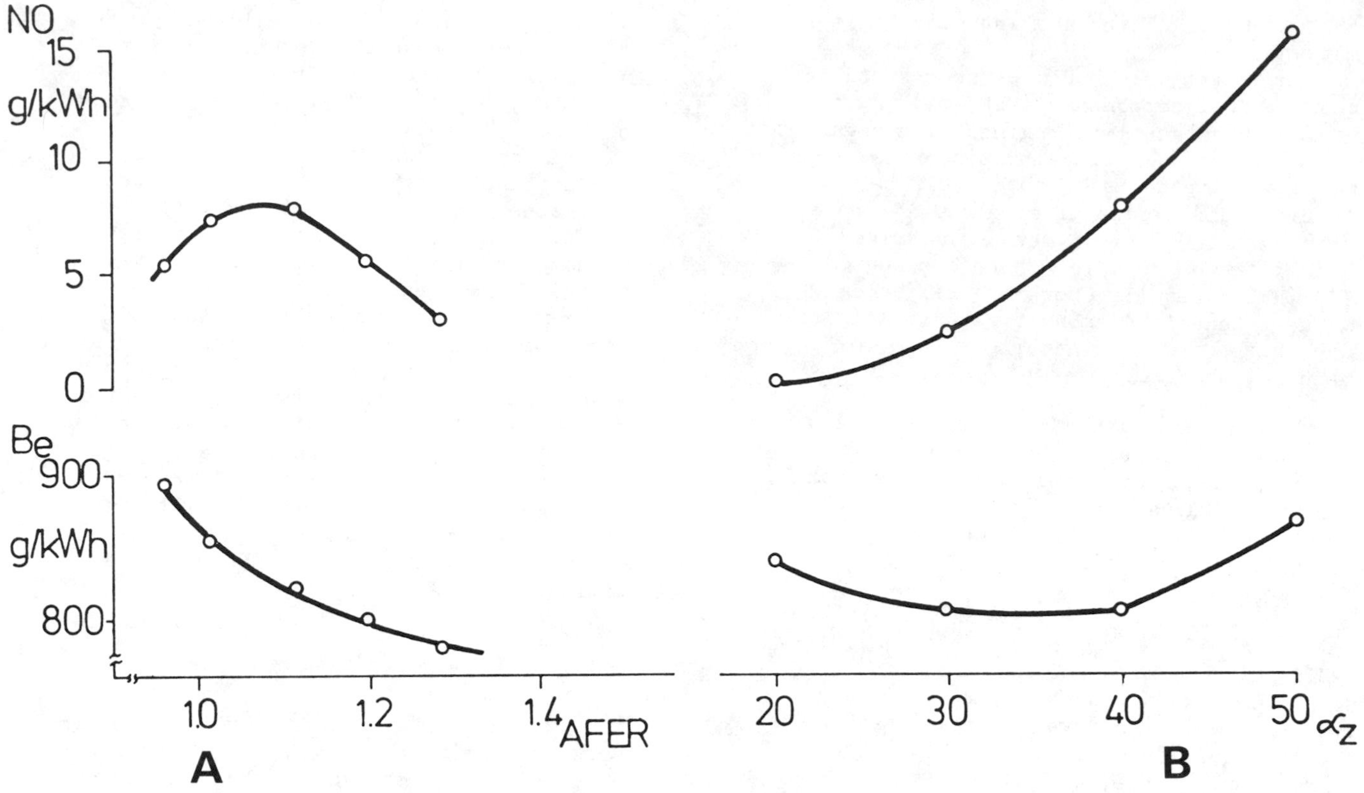

Fig. 7B - NO emissions and fuel consumption of a high compression methanol engine at road load 60 km/h (37 mph): A) - versus AFER, B) - versus spark timing

fixed carburetor and ignition timing settings will be discussed. The results of this methanol prototype engine closely related to the production type enable direct comparisons to the current production gasoline engine.

FUEL CONSUMPTION AND POWER OUTPUT - The road load fuel consumption of the prototype engine is illustrated in Fig. 8. The figure showes the volumetric fuel consumption of methanol for normal and high compression ratios, with gasoline engine consumption data for comparison. In accordance with the lower heating value of methanol the volumetric consumption is higher. However, there is a gain due to the higher degree of efficiency from methanol combustion as illustrated in Fig. 9. The improvement of specific energy consumption SEC for methanol engines in comparison to gasoline engines is more clearly illustrated in Fig. 10. It shows the percentage of improvement in relation to gasoline. Thus, a mean value of about 8 % can be calculated for road load operation at normal compression ratio. These savings could be brought up to almost 15 % on an engine with a higher compression ratio.

A rise in compression ratio leads to a considerable increase in power output in addition to the definite energy consumption decrease. Fig. 11 shows the performance output and the specific energy consumption versus engine speed at full throttle for gasoline at normal compression ratio and methanol at normal and high compression ratios. Conspicuous is the high gain in performance output of more than 20 % for methanol with the same engine equipment, i. e. same type of carburetor and exhaust system. The calorimetric fuel consumption for full throttle operation reached values which are very close to diesel engine data.

REASONS FOR HIGHER EFFICIENCY AND INCREASED POWER OUTPUT - One of the reasons for significant benefits in performance and efficiency can be recognized from temperatures of air/fuel mixture and exhaust gas. Fig. 12 shows the mixture temperature of the same type of engine for gasoline and methanol.

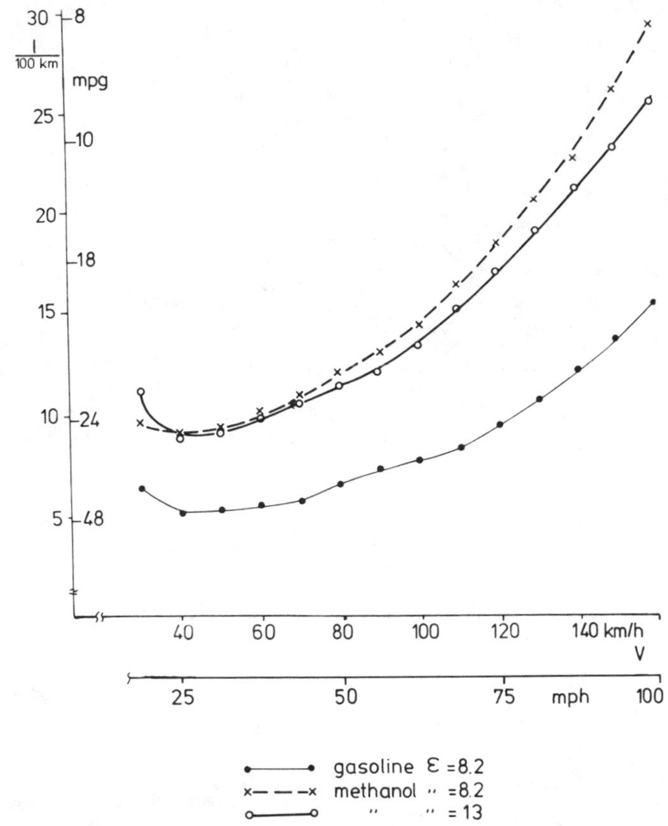

Fig. 8 - Road load quantity consumption

Due to the higher heat of evaporation together with the larger mass consumption of methanol , the temperature is about 20° C lower and, therefore, results in an increase of volumetric efficiency, Fig. 13. Furthermore exhaust gas temperature is lower in spite of higher engine performance, Fig. 14. This indicates a more favorable thermodynamic process cycle.

In accordance to the well known effect for spark ignited engines, Figure 15 illustrates the decrease of exhaust gas temperature with increasing power output from higher compression ratio.

It can be seen in Figures 16 and 17 that at part load conditions, mixture temperature and exhaust gas temperature are also considerably lower than those for gasoline. The lower intake temperature points out driveability problems when the standard intake heating system is used unmodified. On the equal fuel energy basis, about seven times more heat is required to vaporize methanol than gasoline. Therefore, an effective improvement at this point is difficult since the large need for heating energy can hardly be satisfied.

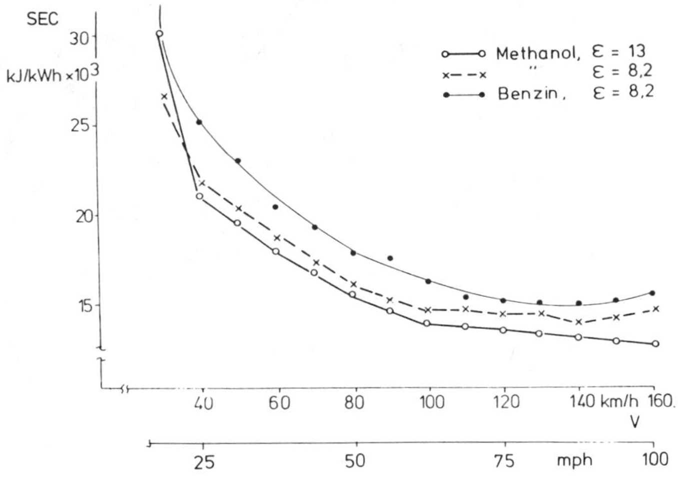

Fig. 9 - Road load specific energy consumption SEC

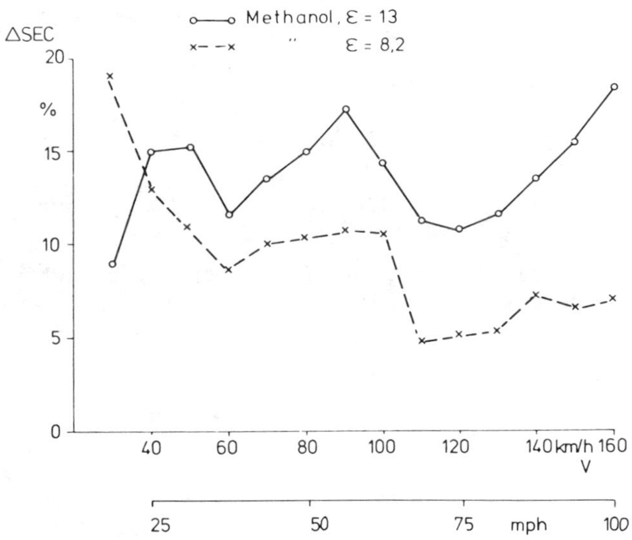

Fig. 10 - Improvement of energy consumption, methanol versus gasoline, road load operation

FUEL DISTRIBUTION - A referred, mixture temperature of the methanol engine is considerably lower. Thus, this results for carburetor engines in poorer cylinder to cylinder A/F ratio distribution. Fig. 18 shows these values for different part load and full load conditions in comparison to a gasoline engine. The maldistribution for methanol limits the possible lean A/F values at part load. At full load conditions no difference in engine running behavior could be observed.

Better fuel distribution from the original state could be realized by an improved mixture heating system in connection with

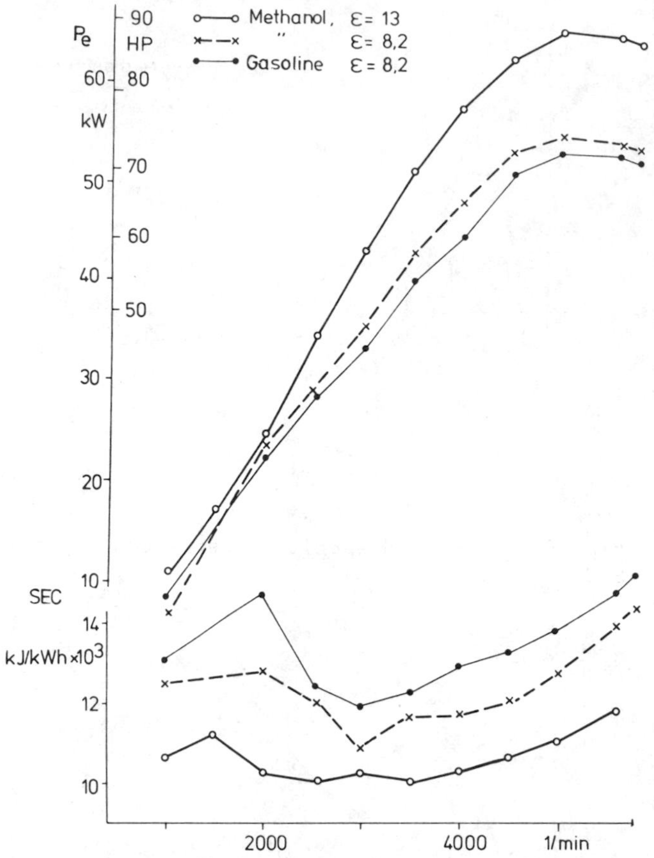

Fig. 11 - Performance Pe and specific energy consumption SEC, full throttle

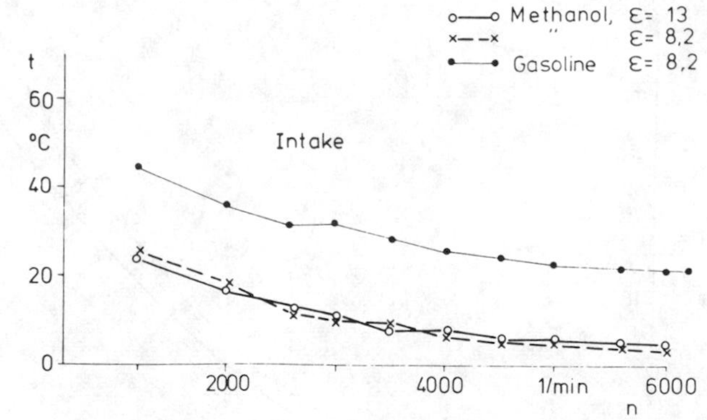

Fig. 12 - Mixture temperature in intake system, full throttle operation

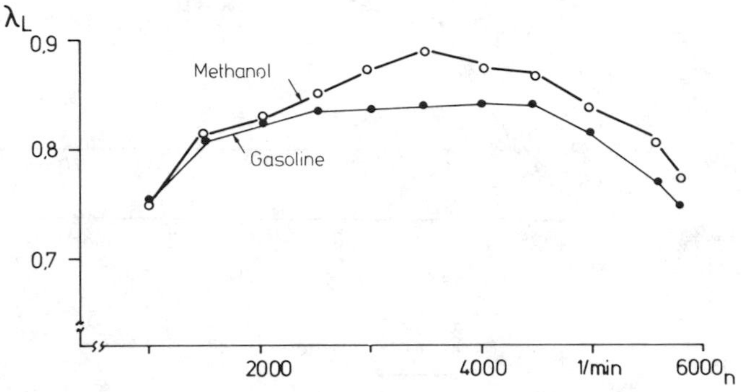

Fig. 13 - Volumetric efficiency λ_L of the same engine fueled with methanol or gasoline, full load conditions

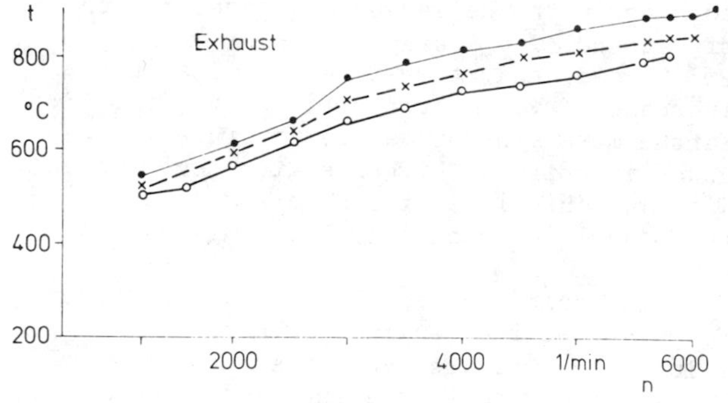

Fig. 14 - Exhaust gas temperature, full throttle operation, same parameters as Fig. 12

liners on the bottom of the intake manifold in order to guide the liquid fuel equally to the cylinders.

EMISSIONS - The exhaust emissions from engines with final adjustment of carburation and spark timing confirm the results of the fundamental evaluation mentioned. As CO emissions mainly depend on A/F ratio, they decrease with a leaner mixture and are not influenced by a change of compression ratio. As expected the increase of compression ratio shows, however, a definitely increasing trend to HC (UBF) emissions, as well as at part load and full load conditions. Fig. 19a shows this tendency for different loads. However, optimizing work on combustion chamber with regard to a reduction of the HC emissions could not be finalized.

In the final tuning of the engine no significant influence of compression ratio on NO_x emissions was to be observed (Fig. 19b), because - in contrast to reference (15) - an optimal tuning of A/F ratio and ignition timing could be realized in accordance with Figure 7.

COMPARISON WITH OTHER ENGINES - To summarize the results from the tuning work, Fig. 20 indicates the A/F ratio at part load conditions for the high compression methanol engine in comparison with the production gasoline engine. At low load the methanol engine is evidently running on leaner mixtures. Power output at full load condition requires nearly the same rich A/F ratio for both fuels. Limited possibilities in adjustment of

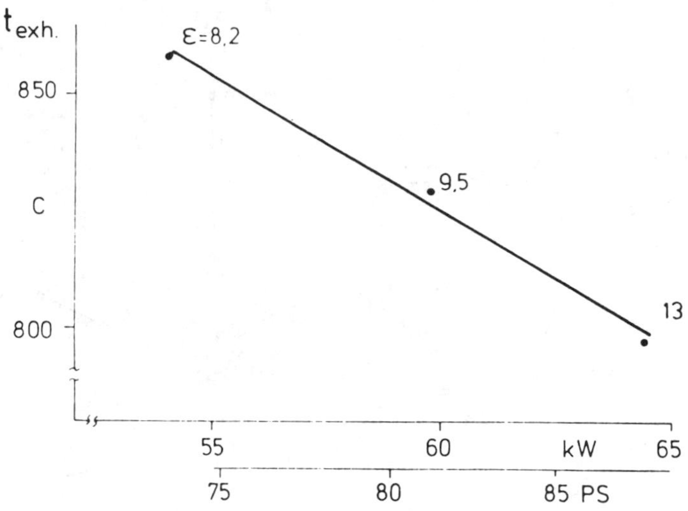

Fig. 15 - Decrease of exhaust gas temperature of methanol engines with increasing performance from higher compression ratio ε

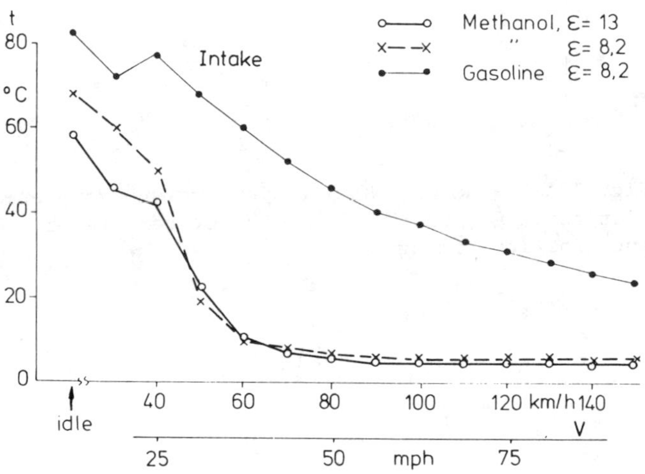

Fig. 16 - Mixture temperature in intake system, road load conditions

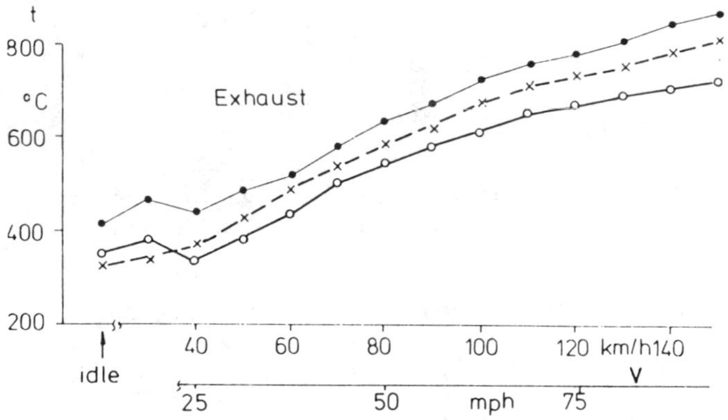

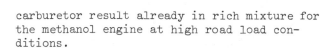

Fig. 17 - Exhaust gas temperature, road load conditions, same parameters as in Fig. 16

carburetor result already in rich mixture for the methanol engine at high road load conditions.

In order to give more information on road load fuel economy (energy basis), in Fig. 21 the specific energy consumption of a production gasoline engine, of the high compression methanol prototype engine, and of a diesel engine of very similar engine type are compared in the same vehicle. As it can be seen, the methanol engine values are close to those of the diesel engine, whereby at high load ranges the values of the diesel engine are reached.

COLD START AND DRIVING BEHAVIOR.

In order to develop a methanol fueled vehicle, adjustment work focused on good driveability and acceptable cold start behavior.

Cold start - One of the severest problems inhibiting future application of methanol as an automotive fuel is its poor cold start performance. Without any additional measures the cold start limit under normal conditions is about + 10° C.

Several measures were tested to improve this inadequate performance. In the applica-

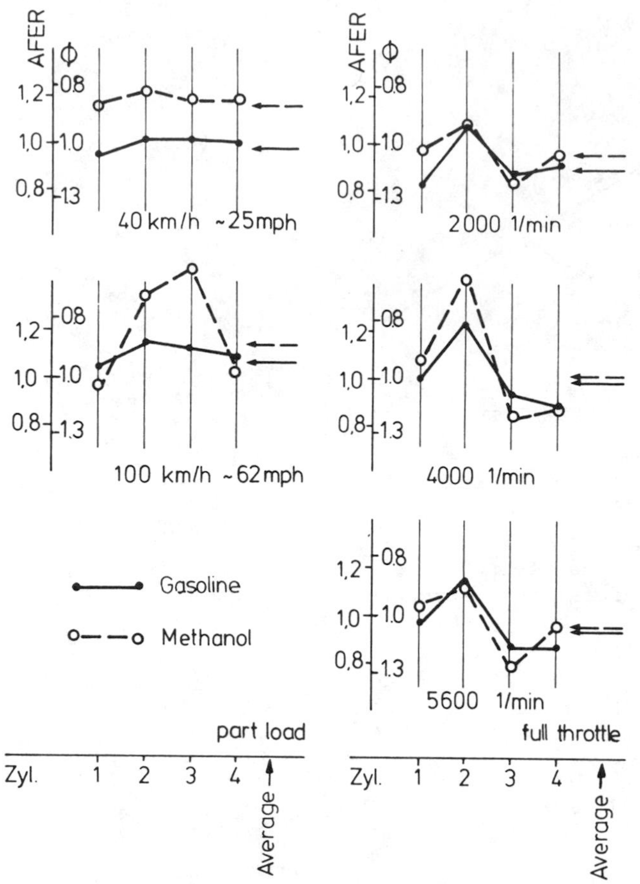

Fig. 18 - Cylinder to cylinder variations of A/F ratio (AFER) for part and full load operation

tion of electric heating the limited amount of energy available onboard the vehicle must be used as effectively as possible.
Thus it seems recommendable to heat only the fuel, not the fuel/air mixture. The starting limit was dropped considerably by heating and evaporating a small quantity of fuel electrically before starting the engine. A disadvantage could be the waiting time of 30 to 45 seconds before starting (heating performance 200 Watts). Utilizing the electric heating device the engine was started in tests down to about - 8° C. But the engine's idle behavior immediately after engine cold start was not satisfactory. The heating performance, therefore, has to be improved and other approaches to promote fuel evaporation during cold start and warm-up phase have to be tested (16).

Tests with totally atomized fuel did not improve the starting behavior to any great degree. Another approach is the application of a separate starting fuel which lead in every case to an usuable, but rather expensive solution because an additional supply metering device was necessary. Warm-

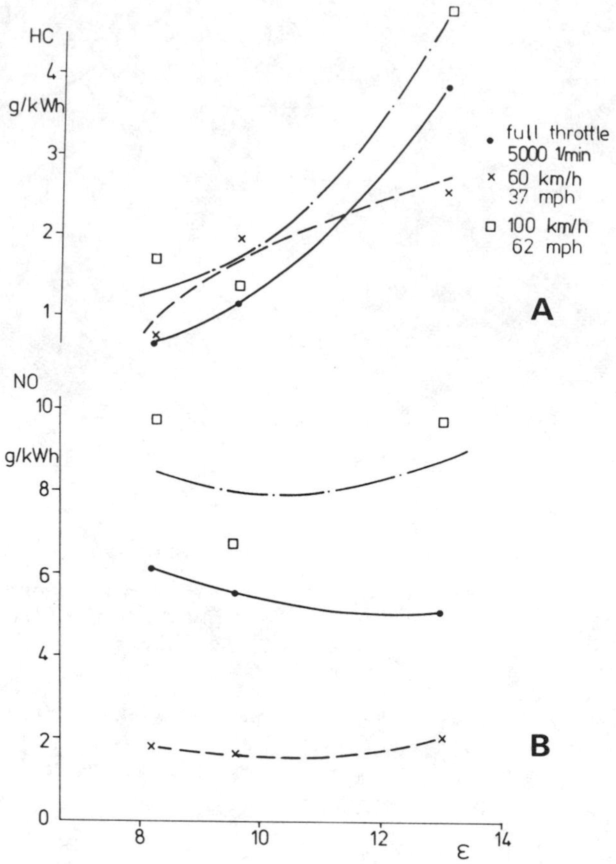

Fig. 19A&B - HC and NO emissions versus compression ratio ε, final adjustment of carburetor and ignition timing

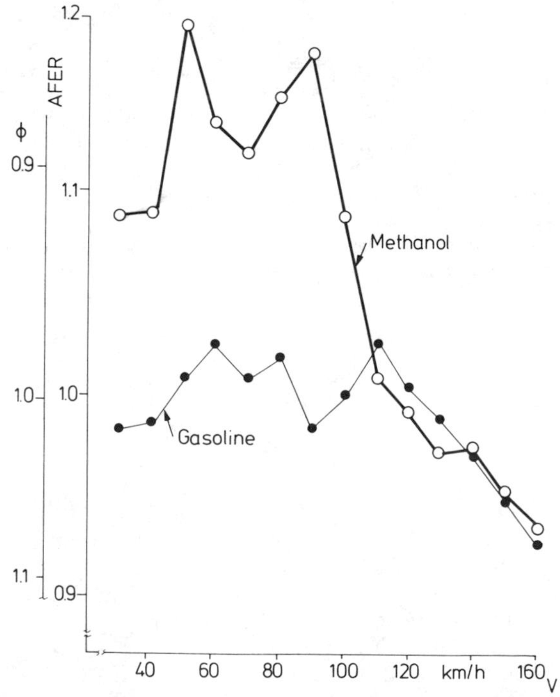

Fig. 20 - A/F ratio (AFER) of prototype methanol engine compared with a production gasoline engine, road load

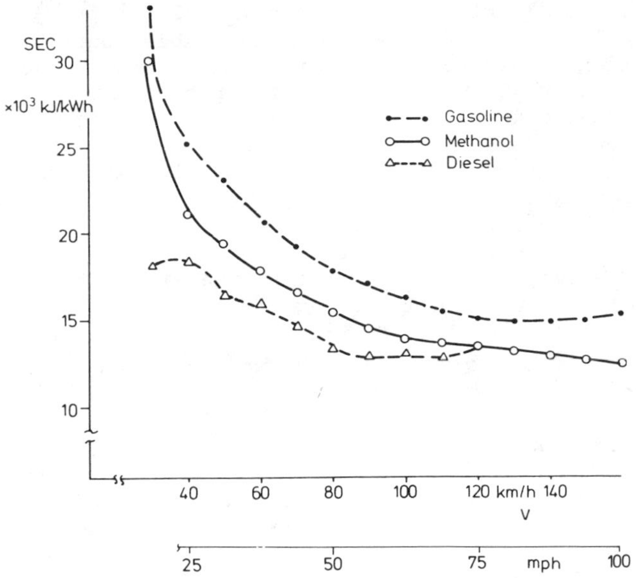

Fig. 21 - Specific energy consumption of a gasoline, methanol and diesel powered car of the same type at road load

up driveability problems can be solved by adding additives to straight methanol. This has proven to be the most favorable measure. Fig. 22 gives an example for one possible additive (i-pentane). It can be seen that the necessary quantity for cold start depends on the ambient temperature (17). Below + 10° C the addition of a suitable additive is necessary. Adding 8 % of this volatile component ensures cold starts down to - 20° C. Using additives of this volatile behavior perfect starts were realized down to temperatures around - 25° C both in cold climate chambers and also in road tests in Sweden in February in 1976 and 1977.

Table 1 gives a survey on successfully tested additives.

However, the comparatively high amount of liquid fuel in the air fuel mixture requires carefully tuned automatic choke devices in order to prevent failures from wet spark plugs during the start operation and in order to avoid to much oil dilution in the crankcase.

DRIVEABILITY DURING WARM-UP PHASE - Many of the measures mentioned above to improve the cold start behavior itself could also be applied satisfactorily to improve the warm-up driveability. Here, too, the adding of volatile additives to the methanol proved to be the best measure. It was evident that a quantity of 8 - 10 % vol would

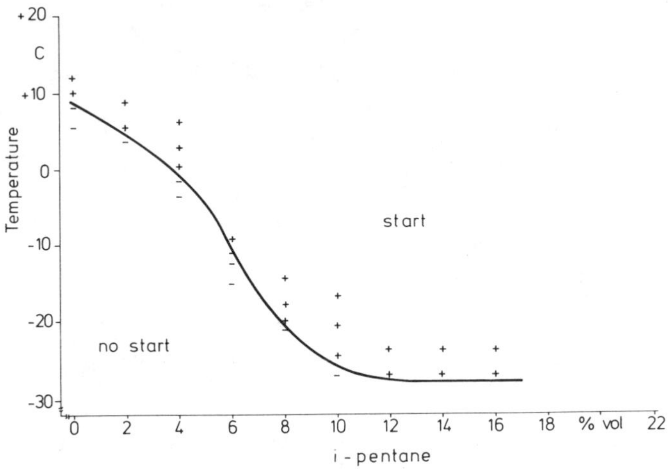

Fig. 22 - Necessary quantity of additive for cold start i- pentane

Table 1: Tested start additives for cold start at - 15° C. (Cranking time < 3 sec.)

additives	boiling point °C	necessary quantity for cold start %	RVP of the mixture bar
1. i-pentane	+ 28	6	0,72
2. 40/60 gasoline	+ 40 ÷ 60	7,5	0,72
3. dimethyl ether	- 28	5,6	0,90

be sufficient also. Fig. 23 shows the distillation curve of a succesfully tested fuel.

WARMED-UP DRIVEABILITY - The carburetor and ignition timing settings determined on the test bench were kept for road testing to a great extent. In doing so the standard heating of the air/fuel mixture by the engine's coolant was applied. In this case warmed-up driveability of the test car fueled by pure methanol (without additives) was equivalent to that obtained with gasoline.

SPECIAL TESTS - The feasibility of the methanol based fuel recommended here was also tested in summer tests with regard to hot starting behavior and high temperature drive-ability (vapor lock). No major difficulties were encountered even with fuels with high RVP values up to 1.15 bar (16.7 psi). The high vaporization heat of the fuel helps greatly in reducing the problem of vapor

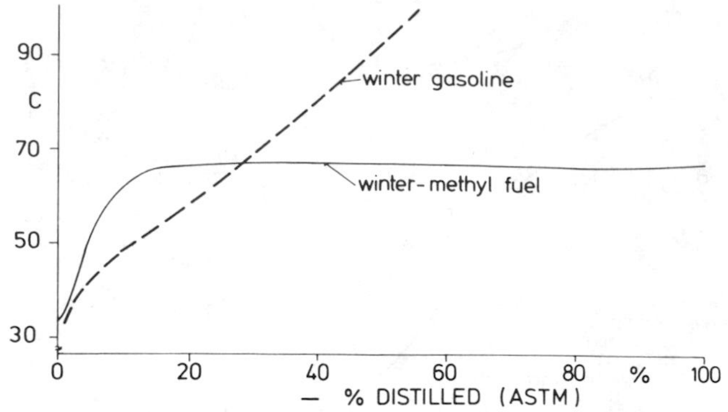

Fig. 23 - Destillation curve of methanol with proper cold start additives

lock in critical driving modes as well as the high mass flow of fuel.

A number of tests are still being performed to determine the compatibility of materials with methanol and methyl fuels. This concerns plastics, as well as various metals, for example die-cast zinc.

DESCRIPTION OF THE PROTOTYPE VEHICLE

The methanol prototype vehicle is a small passenger car (VW Rabbit) with a 1.6 l four-cylinder engine with a compression ratio of 12.5 and a single barrel carburator.

Besides careful tuning, the utilization of different modifications in carburator and ignition system and the replacement of some plastic parts in the fuel line system no other major modifications were necessary. Several of these prototype cars are now in durability tests under various conditions. Up to now 100,000 miles have been covered with no serious damage.

SUMMARY AND CONCLUSIONS

The development of a methanol fueled prototype car points out that a passenger car with a spark ignition engine will be a low-emission, high energy-economy and high performance power system of the same usability as modern gasoline vehicles, including cold start and warm up behaviour. Where as the volumetric fuel consumption is higher due to the lower energy content of methanol, the energy-based fuel economy will be considerably better. Low additional production costs are expected, since the vehicle will be very similar to present day production gasoline cars.

ACKNOWLEDGEMENTS

The authors wish to thank Mr. W. Geffers and Dr. A. König for their helpful collaboration. The BASF, Ludwigshafen, and VEBA-Chemie, Gelsenkirchen, have given a valuable contribution in laboratory fuel work. The Robert Bosch GmbH Stuttgart, supported the project in the field of ignition systems, the Deutsche Vergaser GmbH, Neuß in the field of carburation and the Deutsche Shell, Hamburg, gave assistance in solving knocking problems.

The project was sponsored by the Ministery of Research and Technolgy (BMFT) of the Federal Republic of Germany under grant 522-7291-TV 7525. This support is gratefully acknowledged.

REFERENCES

1. "Neuen Kraftstoffen auf der Spur" , Herausgeber: Bundesminister für "Forschung und Technologie," Verlag TÜV Rheinland GmbH, Köln, 1974.

2. R. M. Tillman, O. L. Spilman and J. M. Beach, "Potential for Methanol as an Automotive Fuel", SAE 750 118 (1975).

3. R. K. Pefley et. al., "Characterization and Research Investigation of Methanol and Methyl Fuels in Automobile Engines"
Report ME 76-2, The University of Santa Clare, Calif, Oct. 1976.

4. A. Koenig ,W. Lee, W. Bernhardt, "Technical and Econimical Aspects of Methanol as an Automotive Fuel" SAE 76 05 45 (1976)

5. W. Lee, W. Geffers, "Engine Performance and Exhaust Emission Characteristics of Spark-Ingnition Engines Burning Methanol and Methanol-Gasoline Mixtures". AI ch E Meeting Boston, Sept. 1975.

6. W. E. Bernhardt, A. Koenig, W. Lee and H. Menrad, "Recent Progress in Automotive Alcohol Fuel Application". Fourth Int. Symp. on Automotive Propulsion systems, April 17.-22., 1977, Washington for D-C.

7. M. Schaffrath, "Die motorischen Eigenschaften von Methanol", Erdoel und Kohle, Bd. 29, Heft 2, Feb. 1976, p. 64-69.

8. Wilke, W. "Methanol als Kraftstoff", Öl und Kohle, 13 (1937), 42, S. 1030/1038.

9. Wolf, W. "Alkohole und ihre motorische Verbrennung", COMPENDIUM 74175, Erdöl und Kohle, Erdgas, Petrochemie, S. 666 - 686. Industrieverlag von Hernhausen KG, D - 7022 Leinfelden.

10. K. Oblaender, J. Althoff, A. Magel "Methanol als Alternativkraftstoff für Ottomotore". ATZ 78 (1976) p. 113-114.

11. R. M. Tillmann, O. L. Spilmann, J. M. Beach, "Potential for Methanol as an Automotive Fuel". SAE 750 118.

12. R. D. Pleming, T.W. Chamberlain, "Methanol as Automotive Fuel", SAE 750 121.

13. R. W. Hurn, T. W. Chamberlain, "Fuels and Emissions", SAE 140 694.

14. H. G. Adelman, R. K. Pefley, "Methanol as an Automotive Fuel". AI ch E Meeting Boston, Sept. 7-10, 1975.

15. W. Bernahrdt, A. Koenig, "Engine Performance and Exhaust Emission Characteristics of Methanol Fueled Spark-ignition Engines". European Automotive Symposium, Paris, Nov. 13-15, 1975.

16. W. Bernhardt, "Combustion Technologiy for the Improvement of Engine Efficiency and Emission characteristics". 16. Symp. on Combustion, Mass. Inst. of Technology, Cambridge, Mass. August 15-21, 1967.

17. W. Bernhardt, A. Koenig, W. Lee, H. Menrad, "Recent Progress in Automotive Alcohol Fuel Application". Fourth Int. Symposion on Automotive Propulsion Systems, Washington April 17-22, 1977.

18. K. Starke, Oppenländer, Kraft, "Zusatzstoffe für Methanol zur Verbesserung der Kaltstarteigenschaften methanolbetriebener Ottomotoren". International Report BASF, Feb. 77.

Effect of Compression Ratio on Exhaust Emissions and Performance of a Methanol-Fueled Single-Cylinder Engine *

Norman D. Brinkman
General Motors Research Labs.

METHANOL, because it can be made from coal (1,2)* or other nonpetroleum sources (3), is a potential alternative fuel for automobiles. In evaluating methanol's potential as a fuel, various studies have compared methanol to gasoline in single-cylinder engines (4-7), multi-cylinder engines (8,9), and vehicles (10-12). These studies have shown that not only is methanol a feasible automotive fuel, but that engine efficiency increases and NO_x emissions decrease relative to gasoline.

Another advantage of methanol is its high octane number (4), which suggests its use with high compression ratio engines. Increased compression ratio (CR), while increasing efficiency, is also expected to increase exhaust emissions (13), particularly NO_x. In previous studies of methanol fueling, inconsistent effects on NO_x (7,9), negligible effect on NO_x (4), and even decreases in NO_x (6,14) have been found with increased CR. Furthermore, one group of authors has recently published results (11) which conflict with their earlier findings (6,14).

To resolve the reported anomalies, the effect of CR on NO_x emissions was investigated

* Numbers in parentheses designate References at end of paper.

*Paper 770791 presented at the Passenger Car Meeting, Detroit, Michigan, September 1977.

ABSTRACT

One of the reasons methanol is considered an attractive alternative fuel for automobiles is its high octane quality, which may allow the use of high compression ratio (CR) engines. To evaluate compromises between engine efficiency and exhaust emissions, a methanol-fueled single-cylinder engine was run at CR's from 8 to 18. At each CR, engine speed and airflow were constant at 1200 rpm and about half throttle, respectively; equivalence ratio (Ø) was varied from 0.7 to 1.1; and spark timing was varied from best power (MBT) to 10° retarded. Knock was observed only at CR = 18 with MBT spark timing.

Increasing CR from 8 to 18 while maintaining MBT spark timing increased efficiency about 16 percent, but also increased NO_x and unburned fuel (UBF) emissions. Some previous studies have reported decreased NO_x emissions with increased CR, possibly because MBT spark timing was not maintained. Results of this study indicate that constant NO_x emissions can be maintained by retarding spark timing while increasing CR to improve efficiency. Retarding spark timing, however, only marginally reduced UBF emissions.

Vehicle tests are necessary to define the optimum CR for methanol fueling because exhaust emission trends and knocking tendency may be different than those observed with this single-cylinder engine.

SAE/PT-80/19/$02.50

by operating a single-cylinder engine at CR's from 8 to 18 at both MBT and retarded spark timing. A second objective, one which had not been explored in previous high CR studies of methanol, was to determine how exhaust emission constraints affect the potential efficiency improvement with increased CR. Similar studies with gasoline fueling (15,16) indicate that, with exhaust emission control other than use of a catalytic converter, increasing CR did not significantly increase efficiency.

Although aldehyde emissions and lean operating limits may also be affected by increasing CR, neither was evaluated in this study.

EXPERIMENTAL

FUEL AND ENGINE - Neat methanol fuel was used in a Waukesha removable dome head (RDH), single-cylinder engine. This engine's combustion chamber is hemispherical, the piston is domed, and the spark plug is centrally-located. Because CR could be increased up to 18, the RDH engine was used instead of the more-common Cooperative Fuel Research (CFR) engine, which has a maximum CR of 12. The bore and stroke of the RDH engine were 96.8 and 92.1 mm, respectively, giving a displacement of 0.68 L.

As in a previous single-cylinder engine study of methanol (5), the intake system was modified to improve methanol vaporization. With the modified system, shown schematically in Figure 1, the air-fuel mixture is heated to 60°C. Without increased heating of the methanol-air mixture, vaporization is poor and UBF emissions are high (5).

OPERATING CONDITIONS - During this study, engine speed and airflow were constant. Values for these and the other constant operating conditions are listed in Table 1. These test conditions, similar to those used in previous single-cylinder engine studies of alternative fuels (5,17), roughly correspond to the 24-48 km/h, part-throttle acceleration on the Federal emission test cycle.

Values for the variables in this study - compression ratio, equivalence ratio (Ø),* and spark timing - are listed in Table 2. CR varied from 8 (that of most present vehicle engines) to 18 (highest attainable with the RDH engine). Equivalence ratio ranged from 0.7 to 1.1, and spark timing varied from MBT (minimum advance for best torque) to 10° retarded from MBT. MBT spark timing data are shown in Figure 2.

Each of the 5 Ø's and 6 CR's was tested using the first three spark timing values listed in Table 2 (MBT, 2° retarded, and 5° retarded). In addition, at Ø = 0.9 (the Ø which gave highest NO_x emissions), spark timing was also retarded 10° from MBT.

Trace knock occurred when the engine operated at CR = 18 with MBT spark timing. Knock did not occur at any other test condition. In previous studies with methanol in CFR engines (4,7), knock was encountered at CR's of 10-12. Two factors could account for the fact that higher knock-free CR's were achieved in this study. First, the speed and load which were used may have been less con-

---
* Ø = (stoichiometric air-fuel ratio)/(actual air-fuel ratio)

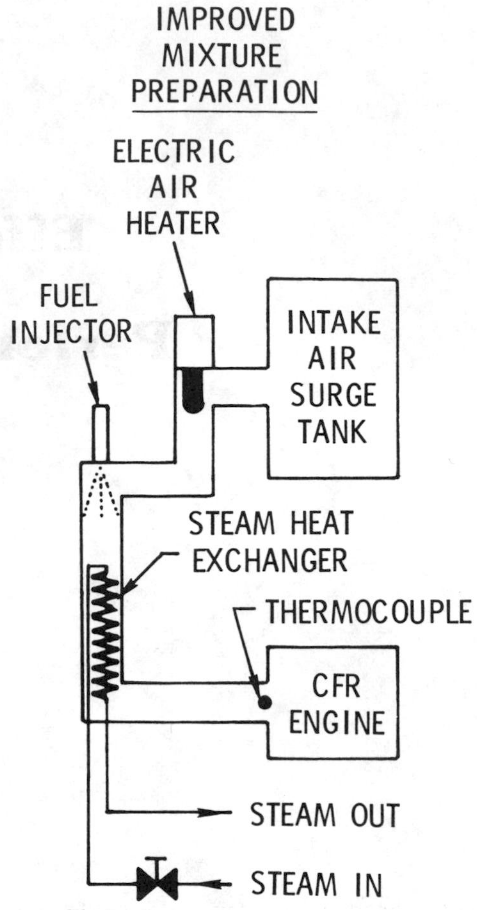

Fig. 1 - Intake mixture preparation system

Table 1 - Test Conditions

Airflow, kg/min	0.206*
Engine Speed, rpm	1200
Intake Mixture Temperature, °C	60
Coolant Temperature, °C	82
Oil Temperature, °C	82
Exhaust Pressure, kPa	3.4

*Intake vacuum was 25-35 kPa, which corresponds to about one-half throttle opening.

Table 2 - Values for Test Variables

Compression Ratio	Equivalence Ratio (Ø)	Spark Timing
8	0.7 (lean)	MBT
10	0.8	2° Retarded
12	0.9	5° Retarded
14	1.0	10° Retarded*
16	1.1	
18		

*Used only when Ø = 0.9

ducive to knock than those used previously. Secondly, the RDH engine, because of its hemispherical combustion chamber and central plug location, may not be as susceptible to knock as is the CFR engine. According to Caris, et.al. (18), concentrating the clearance volume around the point of ignition tends to reduce an engine's octane requirement.

EXHAUST EMISSIONS - Exhaust concentrations of O_2, CO_2, CO, and NO_x (reported as NO_2) were measured using the apparatus described previously (5). Unburned fuel (UBF) emissions, which include both unburned methanol and any hydrocarbons formed during combustion, were measured with a heated flame ionization detector, and are reported as unburned methanol. The exhaust concentrations measured for each pollutant were converted to mass emissions per unit energy output of the engine, and are presented as such. Aldehyde emissions, although significant with methanol fueling (5,11), were not measured.

Using measured exhaust gas concentrations, Ø's were computed by carbon and oxygen balance. Deviation of Ø from the nominal values never exceeded 2 percent.

EXPERIMENTAL DESIGN AND REPEATABILITY - Shown in Table 3 is the experimental design, adapted from a randomized complete block design (19). All CR's were tested with one Ø (chosen randomly) before another Ø was selected. By conducting the experiment in this manner, all CR's for a single Ø could be tested in one day, thus minimizing the influence of test variability on the observed CR effects.

In addition, to simplify test operation, spark timing was varied while holding both Ø and CR constant. Table 3 shows the order in which spark timings were tested.

Analyses of variance were performed using the observed data and, thus, residual mean squares were obtained for each of the measured parameters. Standard error estimates were then computed (19), using these residuals, and are listed in Table 4.

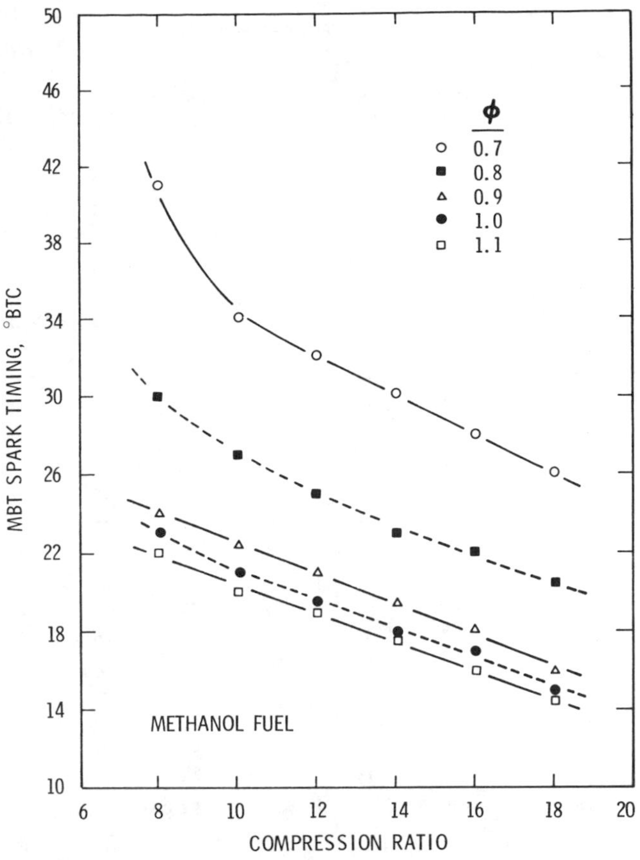

Fig. 2 - Effects of equivalence ratio and compression ratio on MBT spark timing

These errors were sufficiently small and did not, in most cases, mask CR, Ø, and spark timing effects.

RESULTS AND DISCUSSION

To determine the effect of compression ratio on NO_x emissions with methanol fueling, and to determine the tradeoffs between engine efficiency and exhaust emissions control, a single-cylinder engine was operated at a variety of CR's, Ø's, and spark timings. Spark timing and equivalence ratio were varied because both are very powerful tools for control of exhaust emissions. Observed effects on power and efficiency, NO_x emissions, and UBF emissions are discussed here. CO emissions, which were not greatly affected by either CR or spark timing, are given in Appendix A.

POWER AND EFFICIENCY - MBT Spark Timing - As expected, power and efficiency, shown in Figures 3 and 4, respectively, increased as CR increased at constant Ø. With CR = 18 instead of 8, power and efficiency improved about

Table 3 - Experimental Design

Compression Ratio	Spark Timing*	Run Number Ø=0.7	Ø=0.8	Ø=0.9	Ø=1.0	Ø=1.1
8	0	67	46	81	10	25
8	2	69	48	83	12	27
8	5	68	47	82	11	26
8	10			84		
10	0	55	40	93	1	22
10	2	57	42	95	3	24
10	5	56	41	94	2	23
10	10			96		
12	0	64	49	77	7	34
12	2	66	51	79	9	36
12	5	65	50	78	8	35
12	10			80		
14	0	58	37	85	16	28
14	2	60	39	87	18	30
14	5	59	38	86	17	29
14	10			88		
16	0	70	43	73	4	19
16	2	72	45	75	6	21
16	5	71	44	74	5	20
16	10			76		
18	0	61	52	89	13	31
18	2	63	54	91	15	33
18	5	62	53	90	14	32
18	10			92		

* Degrees retarded from MBT

16 percent.* Caris (20) found similar improvement using a high CR, gasoline-fueled engine.

Figures 3 and 4 also show that power and efficiency responded in the expected manner to changes in Ø. Power was highest at Ø = 1.1 and decreased as the mixture leaned. Conversely, efficiency was highest at Ø = 0.7 and decreased as the mixture richened. Operation at Ø = 0.7 instead of Ø = 1.1 decreased power about 25 percent, but increased efficiency about 15 percent.

Retarded Spark Timing - The effect of retarding spark timing on power and efficiency at Ø = 0.9 and CR = 8 is shown in Figure 5. These data are typical of those for other Ø's and CR's. With spark timing retarded 2, 5, and 10° from MBT, power and efficiency decreased 0.2, 1.5, and 4 percent, respectively. This spark timing effect was expected and

* Efficiency is expressed in this report as a fraction, not as a percent. Percent changes in efficiency were computed as follows:

$$\frac{\text{Eff.}_2 - \text{Eff.}_1}{\text{Eff.}_1} \ (100\%)$$

Table 4 - Standard Error Estimates

Measured Parameter	Standard Error Estimate (percent of mean)
Power	0.3
Efficiency	0.4
CO emissions	8
NO_x emissions	10
UBF emissions	7

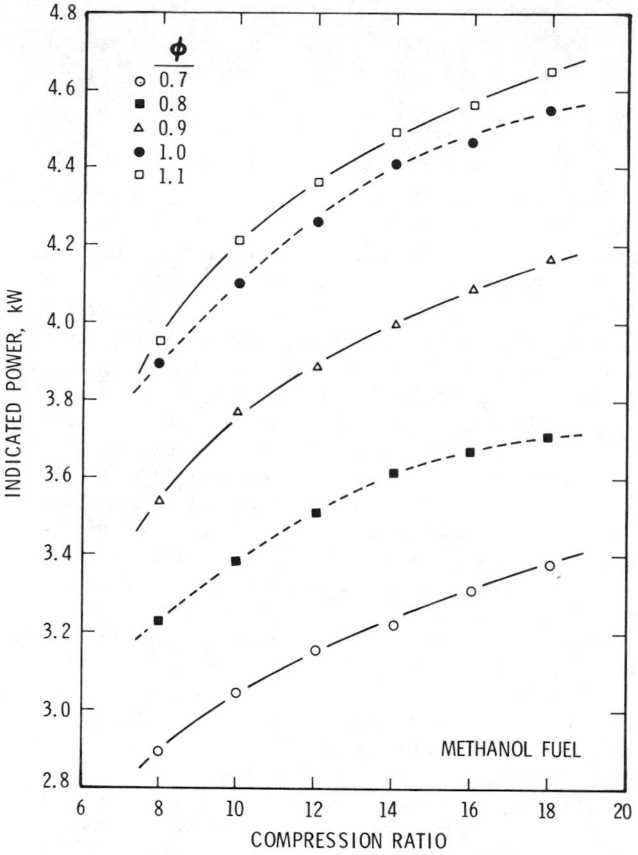

Fig. 3 - Effects of equivalence ratio and compression ratio on power at MBT spark timing

compares favorably with previously-published data (21).

NO_x EMISSIONS - MBT Spark Timing - Figure 6 shows the effect of CR on NO_x emissions at MBT spark timing. For each Ø, NO_x emissions at CR's greater than 8 were higher than those at CR = 8. Furthermore, regardless of the CR used for reference, NO_x emissions never significantly decreased as CR increased. Although these results were expected (13), they conflict with some published data for methanol-fueled engines (6,14).

The relative magnitude of the CR effect on NO_x was different for the different Ø's. As CR increased from 8 to 18, NO_x emissions increased 200 percent at Ø = 0.7, but only

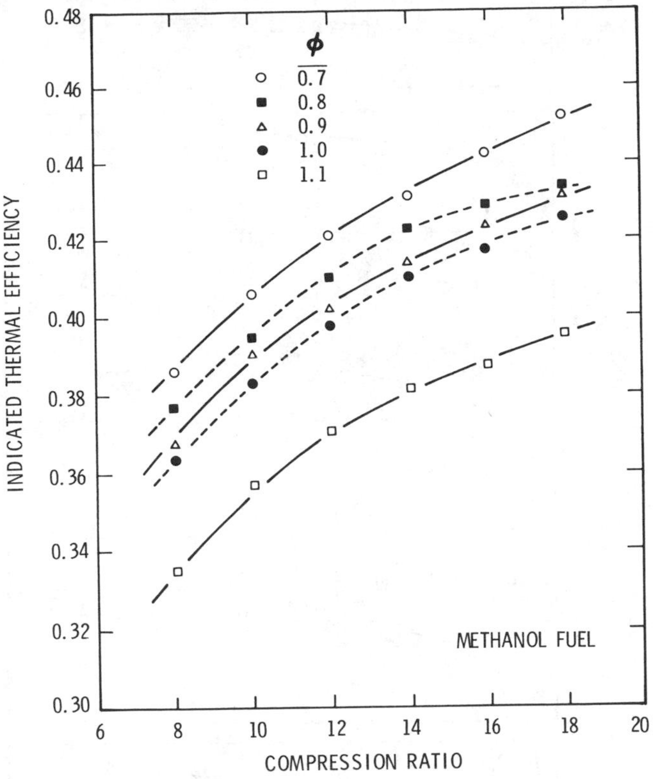

Fig. 4 - Effects of compression ratio and equivalence ratio on efficiency at MBT spark timing

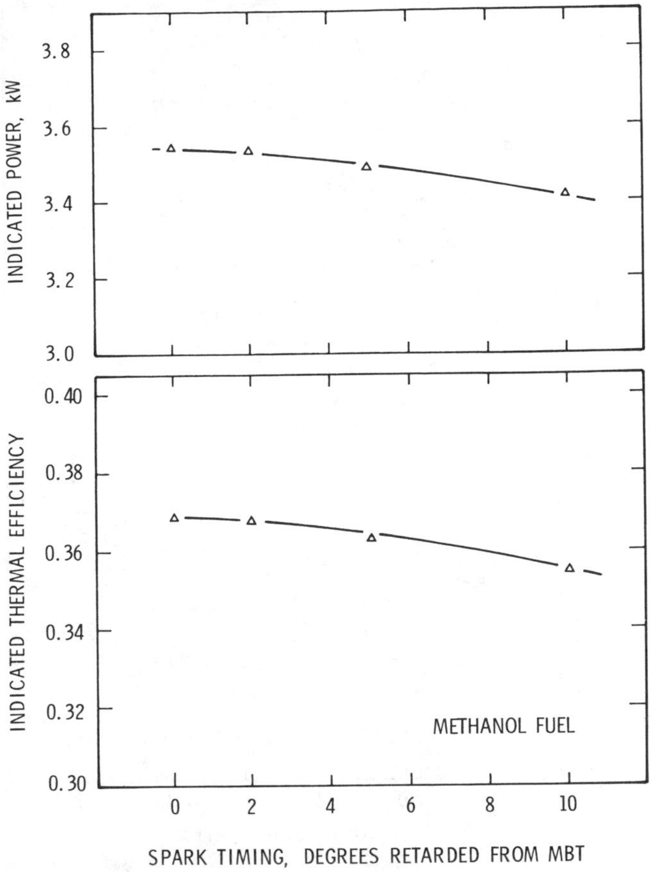

Fig. 5 - Effect of retarding spark timing on power and efficiency at Ø = 0.9 and CR = 8

15 percent at Ø = 1.1. Increasing CR from 8 to 18 at Ø's of 0.8, 0.9, and 1.0 resulted in NO_x increases of 90, 69, and 15 percent, respectively. Thus as Ø increased, the effect of CR on NO_x decreased.

At high CR, higher combustion temperature, resulting from decreased residual dilution and increased burn rate, increases the formation of NO_x. Decomposition reactions, which decrease NO_x concentrations, are also accelerated by increased combustion temperature (22). Thus, the net effect of CR on NO_x emissions is dependent on the relative influence of the NO_x formation and decomposition reactions.

At air-fuel mixtures as lean as Ø = 0.7, essentially no decomposition occurs (22). Therefore, increasing CR at Ø = 0.7 causes an increase in NO_x emissions consistent with the increased rate of the formation reaction. As Ø increases, decomposition reactions become more important (22,23). Higher combustion temperatures, caused by increased CR, accelerate these decomposition reactions and tend to offset the corresponding increase in NO_x formation. As a result, the effect of CR on NO_x emissions decreases as the air-fuel mixture is richened.

The data in Figure 6 also conform to the well-established trend of NO_x emissions versus Ø. NO_x emissions were generally highest at Ø = 0.9, and decreased as the mixture was either richened or leaned. When Ø decreased from 0.9 to 0.7, NO_x emissions decreased 65-85 percent, depending on CR.

Retarded Spark Timing - Figure 7 shows how NO_x emissions were affected by retarding the spark by 2, 5, and 10° relative to MBT. At each combination of Ø and CR, NO_x emissions decreased as spark timing was retarded. A decrease in both combustion temperature and the time available for NO_x formation causes NO_x emissions to decrease as spark timing is retarded.

The magnitude of the spark retard effect depended, as shown in Figure 7, on both CR and Ø. With spark timing retarded 5°, for example, the NO_x decrease ranged from about 15 percent at Ø = 1.0 and CR = 8, to 55 percent at Ø = 0.7 and CR = 18. Without more extensive analysis, such as combustion modeling, these varying effects of retarded spark timing on NO_x emissions cannot be adequately explained. Apparently, combustion temperature and the time available for combustion are affected by spark retard differently at the

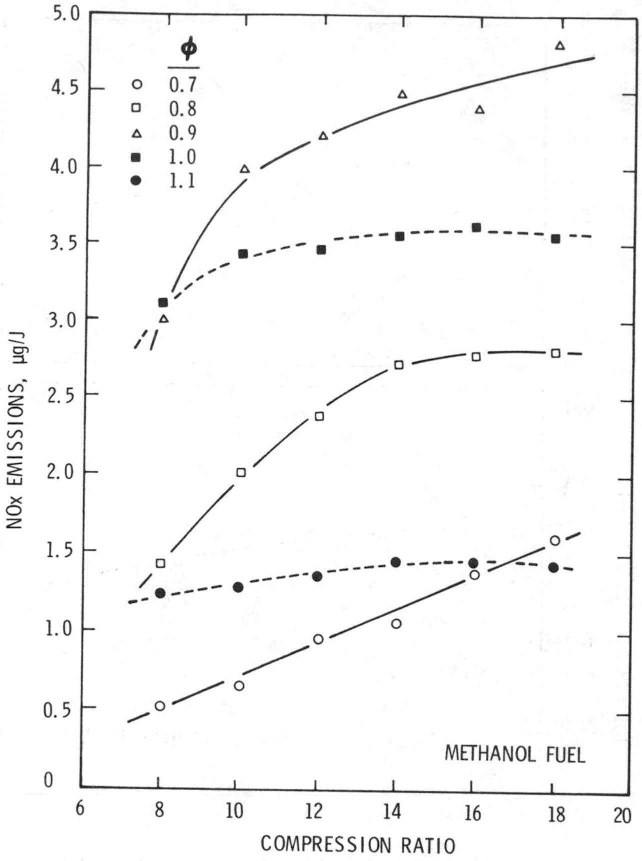

Fig. 6 - Effects of compression ratio and equivalence ratio on NO_x emissions at MBT spark timing

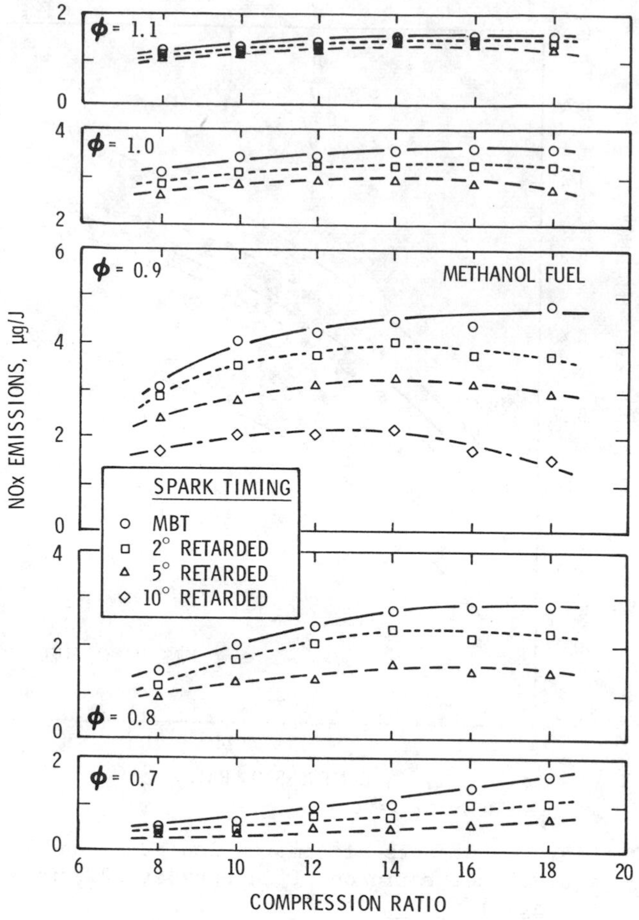

Fig. 7 - Effects of retarding spark timing on NO_x emissions

different Ø's and CR's. Factors important in explaining the CR effect on NO_x at different Ø's, such as burn rate and the relative importance of decomposition, probably also influence the spark retard effect on NO_x.

The effect of CR on NO_x emissions can, as indicated by the data in Figure 7, be strongly influenced by the selection of spark timing. At Ø = 0.9 and CR = 18 (MBT spark timing), for example, retarding spark timing 5° and decreasing CR to 8 each reduced NO_x emissions similarly. In addition, operation at a fixed spark retard can affect the NO_x versus CR curve. With 2, 5, or 10° retard at Ø's of 0.8, 0.9, or 1.0, NO_x emissions reached a maximum at CR's below 18. Some of the inconsistent and surprising effects of CR on NO_x which have been reported in the literature may have occurred because MBT spark timing was not maintained.

Tradeoffs Between NO_x Emissions and Efficiency - As discussed previously, retarding spark timing can control the NO_x increase with increasing CR. Earlier, though, it was shown that efficiency also decreases as spark timing is retarded from MBT. The purpose of this section is to answer the following question: will efficiency increase if, while increasing CR, spark timing is retarded to control NO_x emissions?

To answer this question, NO_x emissions have been plotted versus efficiency. Figure 8 is such a plot for Ø = 0.9. Note that lines of constant spark timing (relative to MBT) are solid, and lines of constant CR are dashed. When CR is increased from 8 to 18 while maintaining MBT spark timing, NO_x emissions and efficiency increase, as described previously. Now consider, instead, increasing compression ratio while retarding spark timing to maintain NO_x emissions at the level observed using CR = 8 with MBT timing (3 μg/J). This is indicated by the dotted line in Figure 8. By simultaneously adjusting CR and spark timing in this manner, efficiency can be increased without increasing NO_x emissions.

Figure 9 is a comparison of the efficiency improvement with increased CR at Ø = 0.9 using two different spark timings: MBT, and that required to maintain NO_x at the level for CR = 8 with MBT spark timing. If NO_x emissions are controlled while increasing CR, less

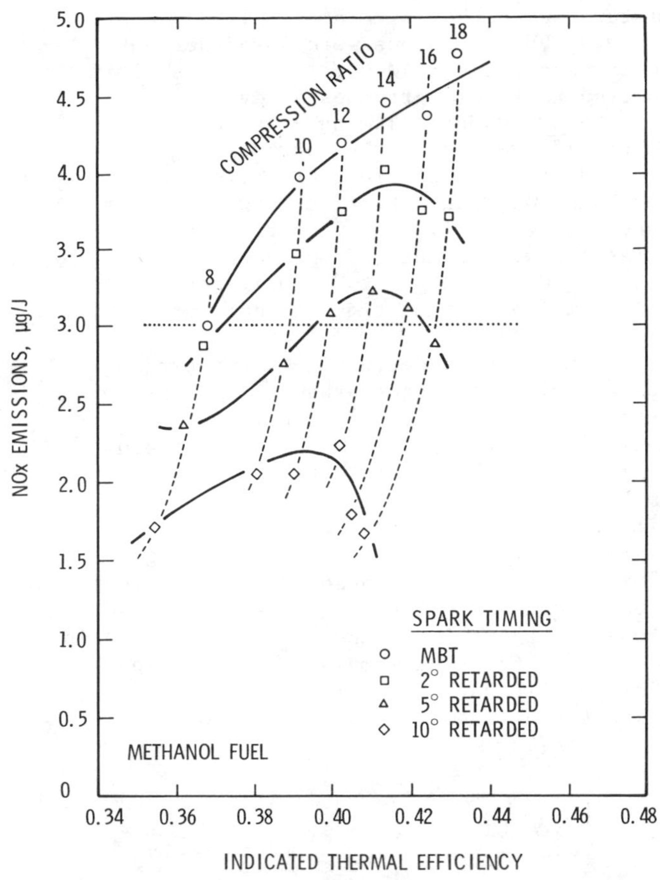

Fig. 8 - Tradeoffs between NO_x emissions and thermal efficiency at $\emptyset = 0.9$

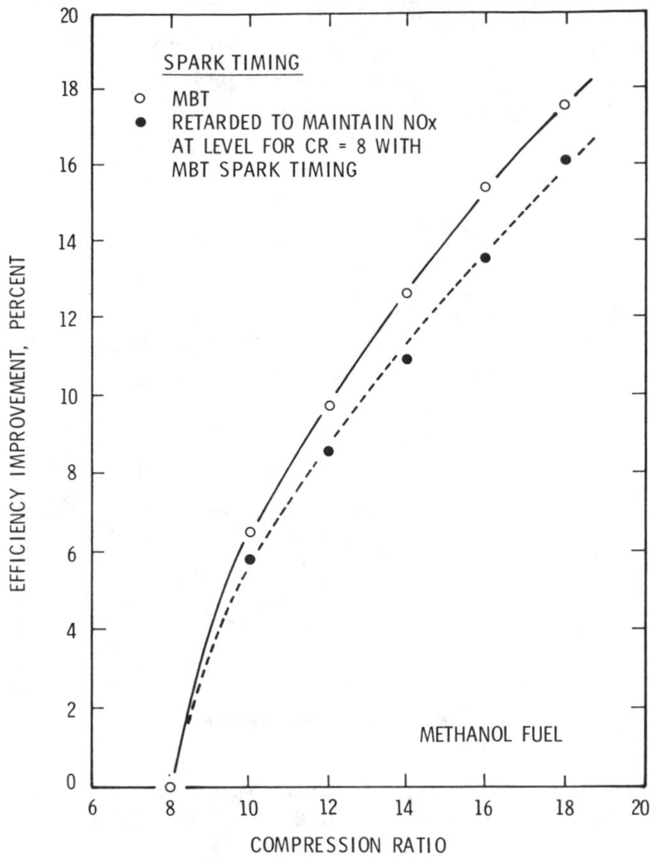

Fig. 9 - Efficiency improvement with increasing compression ratio at $\emptyset = 0.9$

improvement in efficiency is realized than that obtained with MBT spark timing. The difference, however, is minor compared to the total improvement in efficiency with increased CR. For example, efficiency was about 18 percent higher at CR = 18 instead of 8 using MBT spark timing and about 16 percent higher using the spark timing for constant NO_x.

According to Figure 8, if increased CR is accompained by additional retarding of spark timing, NO_x emissions can be reduced while increasing efficiency. At CR = 18 with spark timing retarded 10° from MBT, for example, NO_x emissions were about 40 percent lower, but efficiency was about 10 percent higher than that at CR = 8 with MBT spark timing.

Crossplots of NO_x emissions and efficiency for $\emptyset$'s other than 0.9 are shown in Figure 10. These results are similar to those for $\emptyset = 0.9$: efficiency increases with increasing CR even if NO_x emissions are controlled by retarding spark timing. NO_x emissions could alternatively be controlled by exhaust gas recirculation. Apparently, NO_x emission constraints will not be the limiting factor on increasing CR of a methanol-fueled engine. A recent analytical study (24) had similar findings.

UBF EMISSIONS - MBT Spark Timing - UBF, emissions, shown in Figure 11, increased with increasing CR. The largest relative increase occurred at $\emptyset = 0.9$, where UBF emissions were 150 percent greater at CR = 18 than at CR = 8. At $\emptyset = 1.1$, on the other hand, UBF emissions at CR = 18 were only about 60 percent greater than those at CR = 8.

Due to increases in combustion temperature and pressure, quench layer thickness probably decreases as CR increases (13). The density of UBF in the quench layer, however, increases as CR increases. Thus, the net contribution of quench to increased UBF with increasing CR cannot be determined. The major reason for the observed increase in UBF emissions is probably decreased oxidation in the exhaust mainfold at higher CR. Because efficiency improves as CR increases, exhaust temperature decreases, so fewer of the UBF emissions are oxidized in the exhaust manifold.

Figure 11 also shows that UBF emissions were lowest at $\emptyset = 0.9$, and increased as the

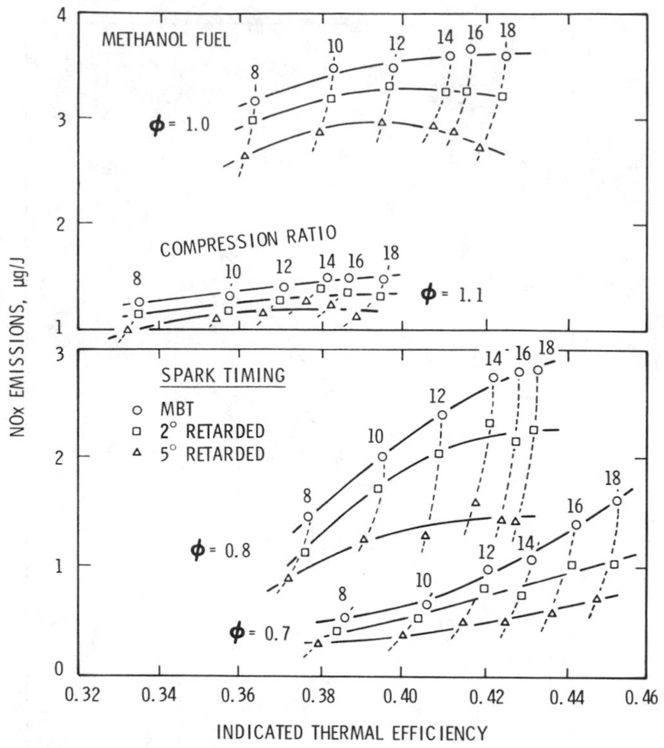

Fig. 10 - Tradeoffs between NO_x emissions and efficiency at Ø = 0.7, 0.8, 1.0, and 1.1

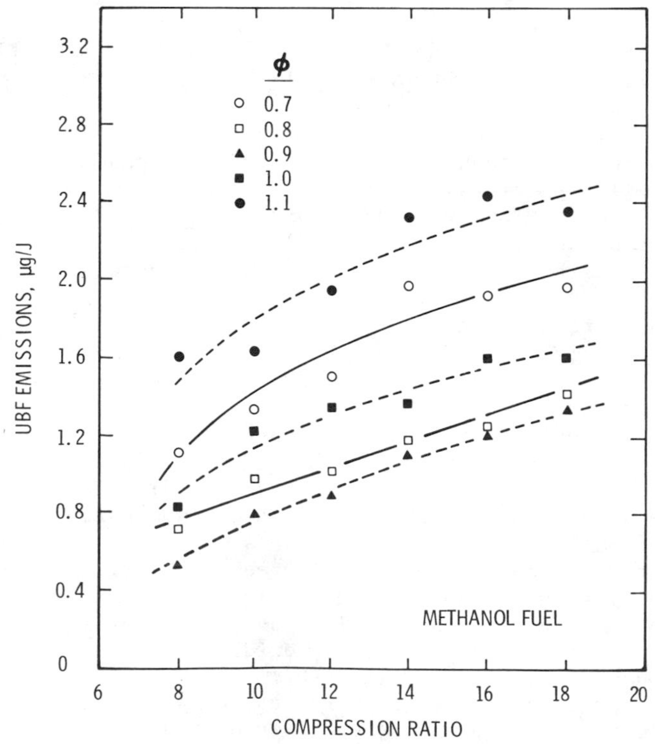

Fig. 11 - Effects of compression ratio and equivalence ratio and UBF emissions at MBT spark timing

mixture was either richened or leaned. At Ø = 0.9, UBF emissions were 40-65 percent lower than those at Ø = 1.1. This is similar to results from a previous study (5).

Retarded Spark Timing - As shown in Figure 12, retarding the spark timing either reduced or did not affect UBF emissions, depending on Ø and CR. The largest effect occurred at Ø = 1.0 and CR = 8, where UBF emissions were reduced 25 percent with 5° retard. Conversely, at Ø = 0.7, retarding spark timing did not significantly affect UBF emissions.

Similarly to decreasing CR, retarding spark timing increases exhaust temperature which speeds oxidation of UBF in the exhaust manifold. A typical relationship between exhaust temperature and UBF emissions is shown in Figure 13, using data obtained at Ø = 0.9. First, consider the data at CR = 18 with various spark timings. As spark timing was retarded from MBT, exhaust temperature increased and UBF emissions decreased. Also shown in Figure 13 is the effect of decreasing CR from 18 to 8 while maintaining MBT spark timing. Although the relationship between exhaust temperature and UBF emissions was

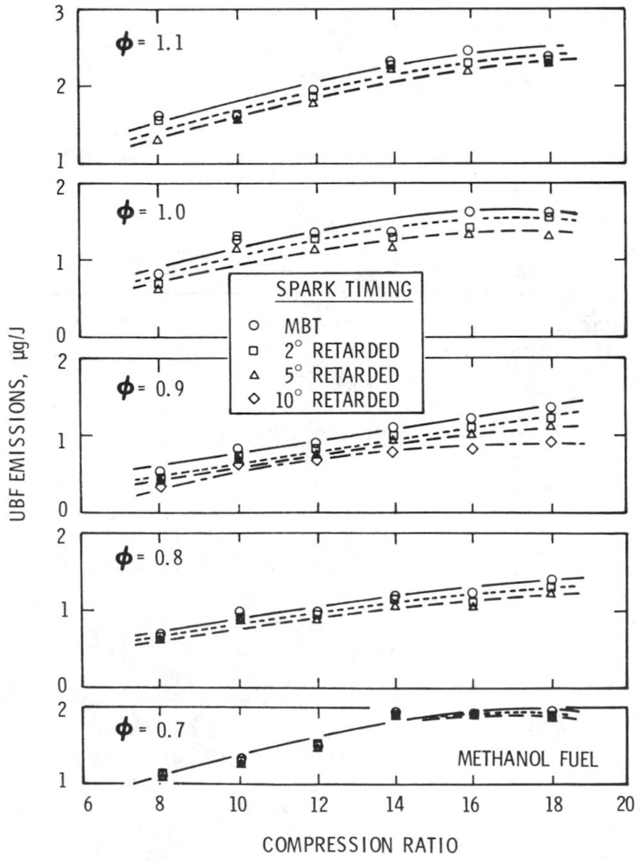

Fig. 12 - Effects of retarding spark timing on UBF emissions

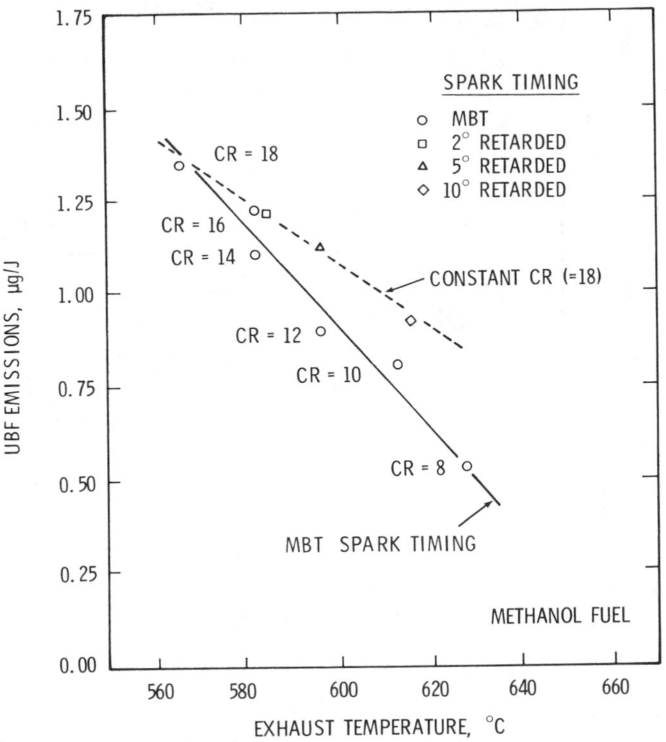

Fig. 13 - Influence of exhaust temperature on UBF emissions at Ø = 0.9

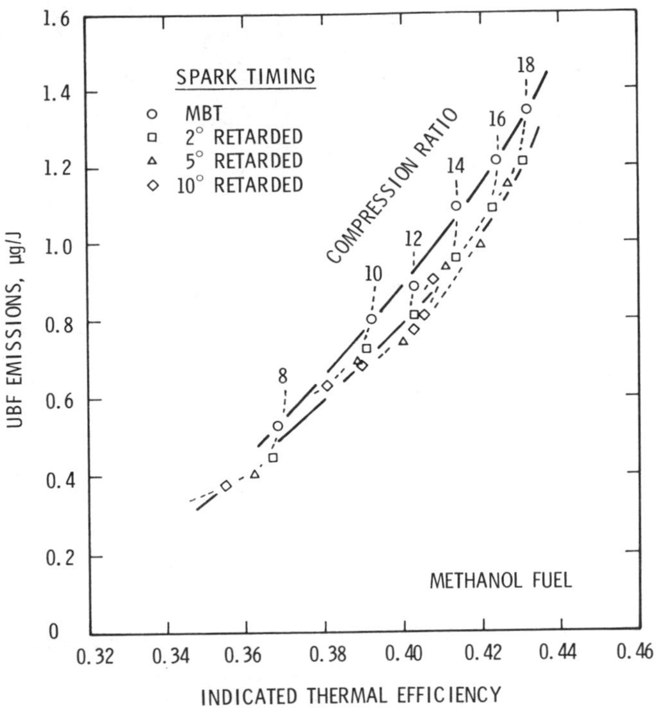

Fig. 14 - Tradeoffs between UBF emissions and thermal efficiency at Ø = 0.9

qualitatively the same as that for retarding spark timing, the quantative relationship (slope of the line) was different. This indicates that factors in addition to exhaust temperature are involved in the effects of both spark timing and CR on UBF emissions.

Changes in spark timing may also affect the amount of UBF emissions resulting from wall quenching. Because combustion temperature and pressure decrease when spark timing is retarded, quench layer thickness increases (13). In addition, with retarded spark timing, surface area of the combustion chamber increases because combustion occurs later in the expansion stroke. Each of these two factors increases quench volume. Density of the quench layer, however, decreases because pressure decreases when spark timing is retarded. Thus, the net effect of retarding spark timing on the mass (product of density and volume) of UBF emissions from the quench layer cannot be determined without additional information.

At Ø = 0.7, spark timing had essentally no effect on UBF emissions (Figure 12). Quader (25) has found that, especially at lean Ø's, retarding spark timing can increase the occurrence of partial burn cycles and, thus, also UBF emissions. This would counteract the exhaust temperature effect and may explain the observed results at Ø = 0.7.

Tradeoffs Between UBF Emissions and Efficiency - UBF emissions are plotted versus efficiency in Figures 14 and 15. Results for Ø = 0.9 (Figure 14) will be discussed first.

Efficiency and UBF emissions both increased when CR increased at MBT spark timing. However, unlike the NO_x case, spark retard only marginally reduced UBF emissions. With spark timing retarded 10° from MBT at CR = 10, for example, UBF emissions exceeded the level for CR = 8 with MBT spark timing.

Furthermore, extrapolation of the curves in Figure 14 indicates that retarding spark timing in excess of 10° would not be desirable. If spark timing were retarded sufficiently to reduce UBF emissions to the level for CR = 8 and MBT spark timing, efficiency would probably be no better than that at CR = 8. The efficiency loss due to retarding of the spark timing to control UBF emissions could apparently offset the efficiency gain obtained with increased CR.

Similar plots for Ø's other than 0.9 are shown in Figure 15. At Ø = 1.0, retarding the spark was somewhat more effective than at Ø = 0.9, but UBF emission levels at CR's greater than 8 were still in excess of those at CR = 8. For Ø's of 0.7, 0.8, or 1.1, the effectiveness of retarding the spark for UBF emission control was not better than that at Ø = 0.9. Efficiency cannot be improved if CR

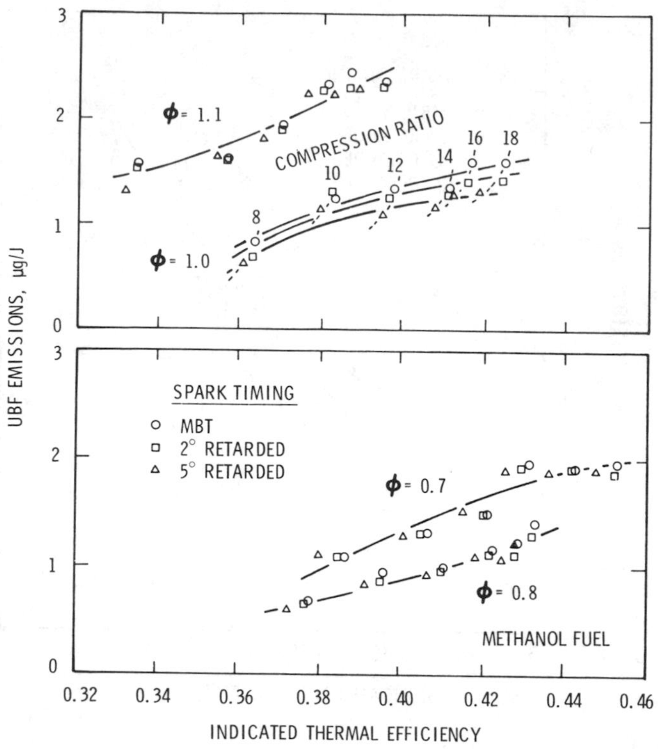

Fig. 15 - Tradeoffs between UBF emissions and efficiency at Ø = 0.7, 0.8, 1.0, and 1.1

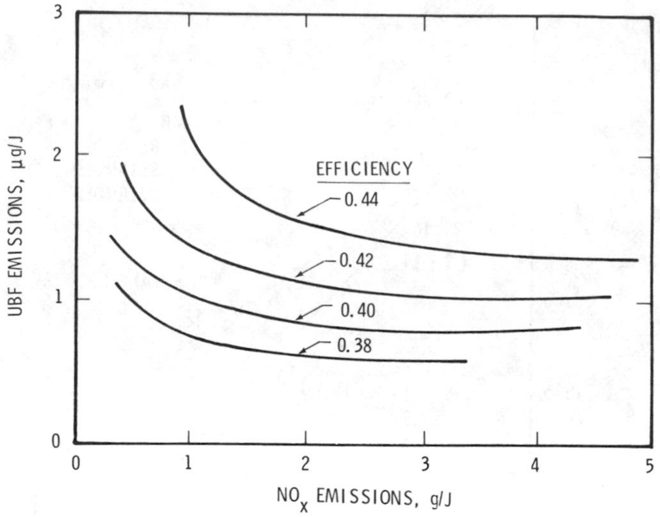

Fig. 16 - Cross-plot of UBF and NO_x emissions at various efficiency levels

is increased while retarding spark timing to maintain constant UBF emissions. However, as discussed previously (12), unburned methanol emissions are less photochemically reactive than unburned gasoline emissions, so the seriousness of higher UBF with methanol is not presently known.

TRADEOFFS AMONG NO_x EMISSIONS, UBF EMISSIONS, AND EFFICIENCY - Tradeoffs between NO_x emissions and efficiency and tradeoffs between UBF emissions and efficiency have each been evaluated in previous sections. Here, the concept is expanded by evaluating tradeoffs among all three: NO_x emissions, UBF emissions, and efficiency. Figure 16 is a cross-plot of UBF and NO_x emissions at various efficiency levels. In order to simplify the plot, none of the individual data points are shown, and results at Ø = 1.0 and 1.1 are not included. Due to low efficiency and high CO emissions, Ø's of 1.0 or greater are not of much interest.

Two points can be made using Figure 16. First, as efficiency increases, exhaust emissions also increase. At a constant NO_x level of 3 μg/J, for example, a 16 percent increase in efficiency results in a 130 percent increase in UBF emissions. At 1 μg/J NO_x, a 16 percent increase in efficiency results in a 200 percent increase in UBF emissions. The second point is that at any given level of efficiency, an inverse relationship between UBF and NO_x emissions exists. If the engine is calibrated for low NO_x emissions, high UBF emissions are observed. Similarly, when UBF emissions are low, NO_x emissions are high.

These tradeoffs are similar to those found previously for gasoline fueling (16,26). Thus, with methanol as with gasoline fueling, the maximum efficiency depends on the level at which exhaust emissions are controlled. In order to determine the CR, Ø, and spark timing which produce maximum efficiency while meeting exhaust emission standards, work must be done with a vehicle.

CAUTION ON APPLYING RESULTS TO VEHICLES - A recent methanol study (12) indicated that vehicle results may not always agree with those obtained in single-cylinder engines, primarily because single-cylinder engine tests are normally conducted at steady conditions, whereas vehicles are operated under transient conditions. It is not possible to assume, therefore, that the results from this study will absolutely predict what would occur in a vehicle.

Knocking tendency, for example, may be different in a vehicle than that observed in this study. In addition to variables such as CR, Ø, and spark timing, knock is influenced by both engine speed and load. The single-cylinder engine used in this study, while operating at approximately half throttle and 1200 rpm, was free of knock at CR's of 16 and less. A vehicle, however, may knock with methanol at CR's below 16, because it must operate at loads up to WOT and speeds up to 4000 rpm. Also, combustion chamber design and deposit level will influence the occurrence of knock.

The effects of engine variables on exhaust emissions may also be different in a vehicle than those observed in this study. A previous study (12) showed, for example, that NO_x versus Ø relationships can be quite different in a vehicle than in a single-cylinder engine. Vehicle tests are needed, therefore, to supplement this single-cylinder engine study.

SUMMARY

Using methanol fuel, a single-cylinder RDH engine was operated at CR's up to 18 to investigate tradeoffs between exhaust emissions and efficiency. Although trace knock occurred with MBT spark timing at CR = 18, the engine did not knock at CR's of 16 or less during the 1200 rpm half throttle test condition used for the study.

At MBT spark timing, efficiency and power increased about 16 percent as CR increased from 8 to 18. Retarding spark timing 5° from MBT decreased efficiency and power about 2 percent.

NO_x emissions never decreased as CR increased while MBT spark timing was maintained. Instead, NO_x emissions at CR = 18 were, depending on Ø, 15-200 percent greater than those at CR = 8. This conflicts with results of some previous studies, possibly because MBT spark timing was not maintained in those studies. When NO_x emissions were reduced at the higher CR's by retarding spark timing, efficiency was still greater than that at CR = 8 with MBT spark timing. At Ø = 0.9, for example, with CR = 18 and spark timing retarded 10° from MBT instead of CR = 8 and MBT spark timing, NO_x emissions decreased 40 percent while efficiency increased 10 percent.

Unburned fuel (UBF) emissions increased 60-150 percent as CR increased from 8 to 18. Retarding the spark timing, though, only marginally reduced UBF emissions.

Tradeoffs among NO_x emissions, UBF emissions, and efficiency with methanol fueling were similar to those previously reported for gasoline-fueled engines. However, vehicle tests are necessary to determine the CR, Ø, and spark timing which provide maximum efficiency while satisfying exhaust emission constraints. Apparently, compression ratio will be limited by constraints of UBF emissions or knock, but not NO_x emissions.

ACKNOWLEDGMENT

It was my pleasure, during this experiment, to be assisted by M. J. Bartch, who instructed me in the art of taking data on a single-cylinder engine and kept all equipment operating properly. I am also indebted to B. D. Peters and A. A. Quader for their assistance in interpreting the data.

REFERENCES

1. J. Pangborn and J. Gillis, "Alternative Fuels for Automotive Transportation - A Feasibility Study," Vols. 1-3, prepared for U.S. Environmental Protection Agency, Publication No. EPA-640/3-74-012-a, b, and c, July 1974.

2. F. H. Kant, R. P. Cahn, A. R. Cunningham, M. H. Farmer, W. Herbst, and E. H. Manny, "Feasibility Study of Alternative Fuels for Automotive Transportation," Vols. 1-3, prepared for U.S. Environmental Protection Agency, Publication No. EPA-460/3-74-009-a, b, and c, June 1974.

3. E. Faltermayer, "The Clean Synthetic Fuel That's Already Here," Fortune, September 1975, pp. 147-154.

4. W. J. Most and J. P. Longwell, "Single-Cylinder Engine Evaluation of Methanol - Improved Energy Economy and Reduced NO_x," Paper 750119, presented at SAE Automotive Engineering Congress and Exposition, Detroit, Michigan, February 1975.

5. D. L. Hilden and F. B. Parks, "A Single-Cylinder Engine Study of Methanol Fuel - Emphasis on Organic Emissions," Paper 760378, presented at SAE Automotive Engineering Congress and Exposition, Detroit, Michigan, February 1976.

6. A Koenig, W. Lee, and W. Bernhardt, "Technical and Economic Aspects of Methanol as an Automotive Fuel," Paper 760545, presented at SAE Fuels and Lubricants Powerplant Meeting, St. Louis, Missouri, June 1976.

7. J. A. LoRusso and R. J. Tabaczynski, "Combustion and Emissions Characteristics of Methanol, Methanol-Water, and Gasoline-Methanol Blends in a Spark Ignition Engine," Paper 769019, presented at the Eleventh IECEC Conference, Lake Tahoe, Nevada, September 1976.

8. R. D. Fleming and T. W. Chamberlain, "Methanol as Automotive Fuel, Part I - Straight Methanol," Paper 750121, presented at SAE Automotive Engineering Congress and Exposition, Detroit, Michigan, February 1975.

9. J. R. Allsup, "Experimental Results Using Methanol and Methanol/Gasoline Blends as Automotive Engine Fuel," Report No. BERC/RI-76/15, Energy Research and Development Administration, Bartlesville, Oklahoma, January 1977.

10. J. C. Ingamells and R. H. Lindquist, "Methanol as a Motor Fuel or a Gasoline Blending Component," Paper 750123, presented at SAE Automotive Engineering Congress and Exposition, Detroit, Michigan, February 1975.

11. W. E. Bernhardt, A. Koenig, W. Lee, and H. Menrad, "Recent Progress in Automotive Alcohol Fuel Application," Presented at

NATO/CCMS Fourth International Symposium on Automotive Propulsion Systems, Arlington, Virginia, April 1977.

12. N. D. Brinkman, "Vehicle Evaluation of Neat Methanol - Compromises Among Exhaust Emissions, Fuel Economy, and Driveability," Presented at NATO/CCMS Fourth International Symposium on Automotive Propulsion Systems, Arlington, Virginia, April 1977.

13. D. J. Patterson and N. A. Henein, "Emissions from Combustion Engines and Their Control," Ann Arbor Science Publishers, Inc., Ann Arbor, Michigan, 1972.

14. W. E. Bernhardt, "Combustion Technology for the Improvement of Engine Efficiency and Emission Characteristics," Presented at the Sixteenth Symposium on Combustion, Massachusetts Institute of Technology, Cambridge, Massachusetts, August 1976.

15. J. J. Gumbleton, G. W. Niepoth, J. H. Currie, "Effect of Energy and Emission Constraints on Compression Ratio," Paper 760826, presented at SAE Automobile Engineering Meeting, Dearborn, Michigan, October 1976.

16. R. E. Baker, E. E. Daby, and J. W. Pratt, "Selecting Compression Ratio for Optimum Fuel Economy with Emission Constraints," Paper 770191 (part of SP-414) presented at SAE Automotive Engineering Congress and Exposition, Detroit, Michigan, February 1977.

17. F. B. Parks, "A Single-Cylinder Engine Study of Hydrogen-Rich Fuels," Paper 760099, presented at SAE Automotive Engineering Congress and Exposition, Detroit, Michigan, February 1976.

18. D. F. Caris, B. J. Mitchell, A. D. McDuffie, and F. A. Wyczalek, "Mechanical Octane for Higher Efficiency," SAE Transactions, Vol. 64, 1956, pp. 76-96.

19. V. L. Anderson and R. A. McLean, "Design of Experiments: A Realistic Approach," Marcel Dekker, Inc., New York, New York, 1974.

20. D. F. Caris and E. E. Nelson, "A New Look at High Compression Engines," Paper #61A presented at SAE Summer Meeting, Atlantic City, New Jersey, June 1958.

21. W. L. Aldrich, "Matching Compression Ratio and Spark Advance to Engine Octane Requirements," General Motors Engineering Journal, October-November 1956, p. 29.

22. P. Blumberg and J. T. Kummer, "Prediction of NO Formation in Spark-Ignition Engines," Combustion Science and Technology, Vol. 4, 1971, pp. 73-95.

23. W. R. Aiman, "A Critical Test for Models of the Nitric Oxide Formation Process in Spark-Ignition Engines," Presented at the Fourteenth Symposium on Combustion, Pennsylvania State University, August 1972.

24. L. H. Browning and R. K. Pefley, "Computer Predicted Compression Ratio Effects

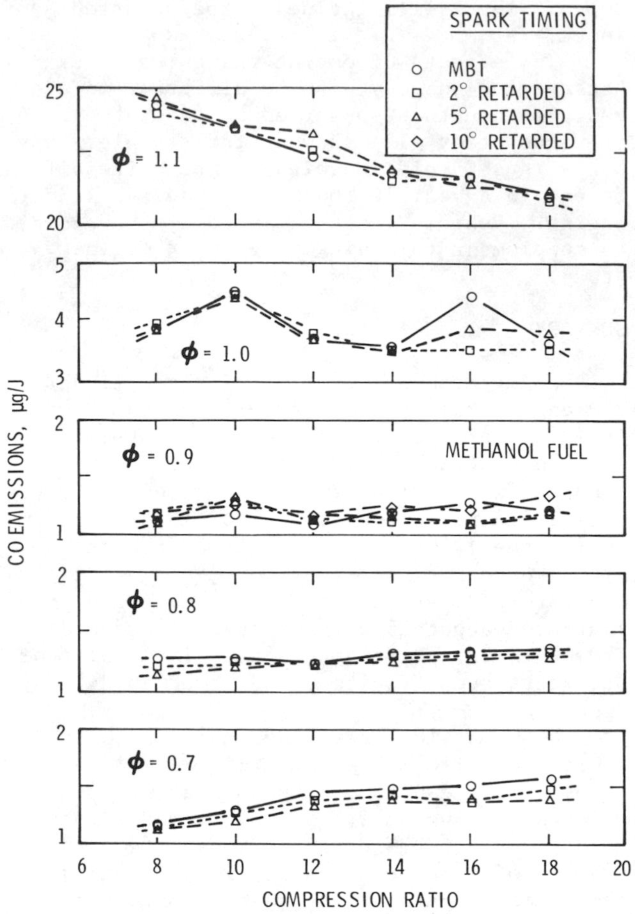

Fig. A-1 - Effects of equivalence ratio, compression ratio, and spark retard on CO emissions

on NO_x Emissions from a Methanol Fueled SI Engine," Presented at the Twelfth IECEC Conference, Washington, D.C., August-September 1977.

25. A. A. Quader, "What Limits Lean Operation in Spark Ignition Engines - Flame Initiation or Propagation?" Paper 760760, presented at SAE Automotive Engineering Meeting, Dearborn, Michigan, October 1976.

26. J. J. Gumbleton, R. A. Bolton, and H. W. Lang, "Optimizing Engine Parameters with Exhaust Gas Recirculation," Paper 740104, presented at SAE Automotive Engineering Congress, Detroit, Michigan, February-March 1974.

Appendix A

CO EMISSIONS

CO emissions results are plotted in Figure A-1. As CR increased from 8 to 18, CO emissions decreased about 15 percent at $\phi = 1.1$, were not significantly affected at $\phi = 1.0$ or at $\phi = 0.9$, increased about 8 percent at $\emptyset$ = 0.8, and increased about 35 percent at $\emptyset$ = 0.7. Mass specific CO emissions at $\emptyset$ = 1.1 decreased because power increased as CR increased; CO concentration at $\emptyset$ = 1.1 was not greatly affected by CR. At $\emptyset$ = 1.0 and 0.9, the effects of CR are masked by the large CO emission variability resulting from small variation in $\emptyset$. At $\emptyset$ = 0.8 and 0.7, excess oxygen is available, so CO, like UBF, is oxidized in the exhaust manifold. When CR increased, causing exhaust temperature to decrease, less CO was oxidized and CO emissions increased.

The expected effects of $\emptyset$ on CO emissions were observed. CO decreased as $\emptyset$ decreased from 1.1 to 0.9, but remained essentially constant as $\emptyset$ decreased from 0.9 to 0.7.

Retarding the spark timing had, in general, little effect on CO emissions.

Methanol as a Fuel: A Review with Bibliography *

David L. Hagen
Dept. of Mech. Engrg.
Univ. of Minnesota

ABSTRACT

A survey of recent studies and research on methanol is given within the historical context. Fuel related properties are reviewed and compared with isooctane. Combustion emissions and their variation with temperature and fuel preparation are similarly compared. Uses of methanol as a combustion fuel and recent tests in boilers, turbines, conventional and stratified charge Otto engines, and Diesel engines are discussed emphasizing comparative efficiencies. Current developments on the uses of methanol directly and indirectly in fuel cells and as a feedstock for single cell protein are examined.

The relevant biological, physical and chemical hazards of using methanol as a fuel are discussed together with safety precautions and treatment. A comprehensive bibliography is given covering the above topics.

METHANOL IS A clean synthetic fuel and chemical feedstock that can be made from a wide variety of renewable as well as conventional material and energy sources and applied to an equally wide variety of uses. A number of factors have converged in the last few years to propel methanol to the forefront of the discussion on synthetic fuels. The OPEC cartel's embargo during 1973 brought into vivid focus the energy "crisis" of the world's rapidly dwindling petroleum supplies and the need to develop alternate sources within the next two decades. Environmental considerations and the difficulties of reducing automotive and power plant emissions have emphasized the comparative cleanliness of methanol. Economic considerations show methanol competing favorably with gasoline and other synthetic fuels in predicted costs and efficiencies of synthesis and combustion.

The capacity to produce synthetic methanol in the US alone has arisen from about 10,000 tons in 1927 to 4 million tons in 1977. Capacity has been increasing at 10% per year over the last decade. With the prospects of producing synthetic liquids from coal and renewable resources on a large scale, methanol has been proposed to replace gasoline and natural gas and it could thus become the most widely used clean fuel.

The following discussion will review the historical interest in the relevant physical and combustion properties, applications and hazards of using methanol as a fuel.

HISTORICAL PERSPECTIVE

The early uses of methanol as a fuel are lost in antiquity, but probably arose in conjunction with the production of charcoal by condensing the vapors driven off. By the middle of the nineteenth century methanol was well established in France as a clean fuel for cooking and heating. It was used a a fuel for lighting until replaced by the more luminescent kerosene around 1880. (220)* Alcohols were used in vehicles in the early part of this century, until low cost gasoline forced them off the market.

* Numbers in parenthesis designate References within the Bibliography at the end of paper.

*Paper 770792 presented at the Passenger Car Meeting, Detroit, Michigan, September 1977.

SAE/PT-80/19/$02.50

During the second world war, petroleum was in short supply especially in Europe, so many vehicles were run on the fumes from wood burners. These vapors contained methanol as well as carbon monoxide and hydrogen. Synthetic methanol was also used in Germany consuming 70,000 tons in 1937 for instance. (195, 226) Racing cars have continued using methanol as a fuel of choice because of the increased power obtainable from the same engine over gasoline. (114, 216)

Shortage of petroleum fuels have frequently concentrated attention on alternate synthetic fuels, especially the alcohols. In the early 1920's for instance there was considerable concern over the imminent shortage of petroleum, and intense research and discussion of alcohols until large petroleum resources were discovered. (114) Numerous other occasions have arisen where shortages of petroleum have existed from political, economic or geographic reasons, and alcohols have been used for fuel. Blends of ethanol in gasoline are in commercial use in Brazil, South Africa, and Cuba. During this last decade we have seen a rapidly increasing evidence for and an awareness of the rapid depletion of petroleum fuels worldwide during this generation, and consequently an increasingly intensive search for alternate synthetic fuels, of which methanol is one of the most promising contenders.

In 1964 the Society of Automotive Engineers published a survey of the use of alcohols as automotive fuels. (241) This was followed in 1971 by an American Petroleum Institute review of the use of alcohols in gasoline, concentrating primarily on ethanol. (14) In the following years, the Environmental Protection Agency (EPA) sponsored studies of alternative fuels by Esso (now EXXON Corp.), (167) the Institute of Gas Technology, (202) and the Aerospace Corp. (175) as part of the effort to decrease pollution. The OPEC cartel oil embargo of 1973 followed by a timely article by Reed and Lerner (220) touched off a flurry of interest in methanol. The following summer the Engineering Foundation held a conference exclusively on methanol as an alternate fuel. (91) As part of a large AEC study, Lawrence Livermore Labs made an extensive survey of the practicalities and possibilities of using methanol for fuel, and proposed various strategies and testing programs. (16)

Mounting piles of urban refuse and little room to bury it in, along with environmental, energy and economic incentives inspired the City of Seattle to conduct an in depth study into the practicalities of converting its municipal wastes to fuel (methanol) or fertilizer (ammonia). (183) As part of this study, Cassady of Mathematical Sciences made a detailed survey of the use of methanol as an automotive fuel and of the current methanol literature. (54, 55)

Europe is far more dependent on imported petroleum than North America. In 1974 this acute awareness brought about by the embargo spurred the German Federal Ministry for Research and Technology to sponsor a comprehensive study of the production, distribution, uses, economics, and dangers of methanol and hydrogen as a fuel. (100) Volkswagen's Research and Development division has continued on from this by testing methanol and methanol blends in conventional and advanced engines as well as in their fleets of vehicles.

During 1975 there was an increasing number of studies on methanol and its use in gasoline, with a greater awareness of the difficulties and problems as well as the increasing advantages. An increasing emphasis was placed on alternative fuels and methanol at meetings held by the American Chemical Society as well as the Society of Automotive Engineers. (e.g. 193) Lawrence Livermore Labs continued their studies of methanol in engines. (68, 69) On the basis of the feasibility studies sponsored by the EPA, the US Energy Research and Development Administration (ERDA) began sponsoring efforts at the University of Santa Clara (cf R. K. Pefley), the University of Miami (cf R. R. Adt) and the ERDA Bartlesville Energy Research Center (cf R. Hurn and J. R. Allsup) to characterize in detail the performance of methanol and gasoline-methanol blends in IC engines. Johnson and Riley also completed an evaluation of methanol gasoline blends for the US Department of Transportation, (161) and the Mitre Corporation completed a major survey of alcohol fuel technology. (291, 292)

The Swedish Methanol Development Co., in 1976, sponsored the first international conference on methanol coupled with a symposium at the Royal Swedish Academy of Sciences. (233, 290) The American Petroleum Institute made another survey of methanol as a fuel. (15) The State of California is agressively pursuing the development of alternative fuels and legislation has already been proposed to mandate the use of methanol in gasoline. (116)

ERDA is now sponsoring a number of further projects on methanol and methanol blends: the US Army Fuels and Lubricants Research Laboratory (cf S. J. Lestz) is investigating the effects on lubricants; the Union Oil Company of California (cf E. L. Wiseman) is studying the known problems of these fuels in depth and studying the effects of varying fuel formulations; at the University of Michigan (cf D. E. Cole) engine and fuel system problems and potential modifications are being studied; and the Stanford Research Institute (cf J. A. Russell) is studying the economics and viability of trade offs in production, fuel composition, and engine performance. ERDA has also begun a quarterly "Alternative Fuels Utilization Report" edited by the Mueller Associates Inc., of Baltimore, Maryland, to keep researchers and the public up to date on developments in the field.(331)

The planning committee set up at the International Conference on Methanol in Stockholm has announced an "International Symposium on Alcohol Fuel Technology-Methanol and Ethanol" to be held November 21-23, 1977 at Wolfsburg, Federal Republic of Germany, which will be hosted by Volkswagonwerk (Chairmen of the Conference are Bernhardt, Lee, and Beckmann). Gorham International, Inc. (Gorham, Maine) is planning a conference "The Future of Methanol as a Motor Fuel" the winter of 1977 (Co-ordinator: A. G. Keene). The number of papers on methanol presented at automotive and fuel related conferences is rapidly growing, with prospects for still more.(322-350)

The continuing depletion of petroleum fuels, price increases and international political tensions as well as the potential of a renewable fuel supply promise growing development of synthetic fuels. Some have suggested that initially they may be made from remote sources of natural gas, from municipal wastes or coal, and eventually from renewable sources.(80,183,227) Electric utilities may well be the first to use methanol as a fuel on a large scale in their peak load turbines. Because of distribution problems initial automotive use will probably be in fleets or regions where travel is restricted, or in blends with gasoline. Eventually use as a pure fuel will become practical as supply, distribution, combustion and economic problems are overcome.

FUEL RELATED PROPERTIES

Methanol (CH_3OH) may be considered as two molecules of hydrogen chemically liquified by a molecule of carbon monoxide. It is a small molecule with combustion properties similar to these gases and distinct from the hydrocarbon molecules such as isooctane (C_8H_{18}). The oxygen constitutes 50% of its weight and forms an hydroxyl group in the molecule making it strongly polar as compared with the nonpolar hydrocarbon fuels. These three factors cause the major differences between the properties of methanol as a fuel compared with conventional petroleum hydrocarbon fuels.

STOICHIOMETRY OF OXIDATION - The essential use of a fuel is to store and transport energy. The energy stored in chemical fuels is usually released through controlled oxidation. This can be either through the usual combustion in air (in boilers or internal combustion engines) via electrochemical oxidation in fuel cells, or metabolically as food.

The stoichiometric equations for combustion (or oxidation) are:

Methanol $CH_3OH + 1.5\ O_2 = CO_2 + 2H_2O$

Isooctane $C_8H_{18} + 12.5\ O_2 = 8CO_2 + 9H_2O$

Here isooctane (2,2,4 trimethylpentane) is used as the standard hydrocarbon to compare fuel properties. The relative mole ratios of the reactants and products, and the respective mass ratios of the two reactions are given in Table 1 for oxidation in oxygen and air.

HEATS OF COMBUSTION (OXIDATION) - Discussion of the comparative efficiencies and power available using methanol versus petroleum fuels has often been clouded by comparisons involving unequal conditions and different or unstated assumptions. The heat of combustion usually used in comparing fuels is the "lower" value of combustion to water vapor rather than the "higher" value to liquid water used in standard thermodynamic tables. (see Table 2.)

In accurate work, these values will have to be changed to accommodate different initial temperatures. The heat of combustion of liquid methanol for instance will increase or decrease 0.0812 kJ/mol per K increase or decrease in the initial conditions, while gaseous methanol will change 0.0439 kJ/mol K because of the heat capacity effect. Note that since methanol contains 50% oxygen by weight, it has only 45 to

Table 1 - Stoichiometric Oxidation Quantities

Fuel	Methanol				Isooctane			
Oxidant	O_2		Air*		O_2		Air	
	moles	kg	moles	kg	moles	kg	moles	kg
Fuel	1.0	1.000	1.0	1.0	1.0	1.0	1.0	1.0
Oxidant	1.5	1.498	7.295	6.549	12.5	3.501	60.789	15.306
Reactants	2.5	2.498	8.295	7.549	13.5	4.501	61.789	16.306
Products	3.0	2.498	8.795	7.549	17.0	4.501	65.289	16.306
% Increase	20.0		6.0		25.9		5.7	

* Assuming 50% Humid Air: 76.658% N_2, 20.563% O_2, 1.826% H_2O, 0.917% A, 0.032% CO_2.

Mol Wt. CH_3OH = 32.042, C_8H_{18} = 114.233, Equivalent Mol Wt. Air = 28.763

Table 2 - Heats of Combustion $\Delta H_c°$ at 25°C**

	LIQUID		GAS	
	kJ mol^{-1}	MJ kg^{-1}	kJ mol^{-1}	MJ kg^{-1}
Methanol CH_3OH, M. W. = 32.042				
Higher: to H_2O (l)	-726.13	-22.662	-764.08	-23.846
Lower: to H_2O (g)	-638-11	-19.925	-676.05	-21.099
Isooctane C_8H_{18}, M. W. = 114.233				
Higher: to H_2O (l)	-5465.5	-47.849	-5503.2	-48.179
Lower: to H_2O (g)	-5069.3	-44.381	-5107.0	-44.711

Sources: Wilhoit and Zwolinski 1973 (277) and Rossini et al. 1953 (289)
**A conversion factor of 1 cal (thermochemical) = 4.1840 J was assumed. I cal (International Table) = 4.18680 J, 1 Btu (Int'l Table) = 1055.056 J, 1 Btu (Int'l Table)/lb = 2,326.00 J/kg. (294)

Table 3 - Energy Densities of the Fuel Charge at 25°C

Gaseous Oxidant +	Liquid Fuel Volume	Liquid Fuel Energy Density	Gaseous Fuel Volume	Gaseous Fuel Energy Density
	m^3 kg^{-1}	MJ m^{-3}	m^3 kg^{-1}	MJ m^{-3}
Methanol + O_2	36.743	-17.367	61.160	-11.054
Isooctane + O_2	305.96	-16.568	330.26	-15.464
Methanol + Air	178.50	-3.5747	202.93	-3.3314
Isooctane + Air	1487.73	-3.4074	1511.6	-3.3785

$\Delta H°_c$ to H_2O (g) Methanol -638.11 MJ kg^{-1} mol^{-1} -676.05 MJ kg^{-1} mol^{-1}
Isooctane -5069.3 MJ kg^{-1} mol^{-1} -5107.0 MJ kg^{-1} mol^{-1}

Air @ 50% Humidity. Assumed ideal at 25°C 1 Vol O_2 = 4.8631 Vol Air
Ideal Volume = 24.464 l mol^{-1} at 25°C, 1 atm, ρ(25°C): CH_3OH = 786.6 kg m^{-3},
C_8H_{18} = 687 kg m^{-3}; Vol CH_3OH(l) = .0473 m^3 kg^{-1} mol^{-1}, Vol C_8H_{18}(l) = .166 $m^3 kg^{-1} mol^{-1}$

48% of the specific (per kg) heat of combustion of isooctane.

ENERGY DENSITIES OF FUEL - OXIDENT CHARGE - To estimate the energy densities of the fuel charge introduced into an internal combustion (IC) engine, the lower heats of combustion were used along with ideal volumes of stoichiometric quantities of oxygen and 50% relative humidity air. (See Table 3) The liquid fuel represents fuel injection with no evaporation while the gaseous fuel represents carburation with complete evaporation, all mixtures taken at 25°C.

The fuel charge using methanol shows a 4.8% greater energy density than with isooctane using fuel injection in oxygen but 28.5% less if this is vaporized and heated to 25°C. When the mixtures are diluted with the nitrogen in air, these figures are changed to 4.9% greater and 1.4% less respectively. Introduction of equal volumes of fuel air charge would give correspondingly greater or smaller energy rates into the engine and thus maximum power or volumetric efficiency obtainable by the engine.

HEAT OF VAPORIZATION $\Delta H_v°$ - The hydroxyl (-OH) group in methanol makes it highly polar (similar to water H-OH) with strong hydrogen bonding between molecules. This gives methanol a specific latent heat of vaporization of 1.167 kJ/kg at 25°C (277) which is 3.8 times as large as that of isooctane at 307.7 kJ/kg at 25°C. (289) This high heat of vaporization causes a 5% difference in ΔH°_c between liquid and gaseous methanol as compared to 0.7% for isooctane. This difference could be recovered by using exhaust gases to evaporate the liquid fuel.

Evaporating methanol in air adiabatically will cool the mixture. Using quadratic heat capacities for methanol in dry air, complete adiabatic evaporation of a stoichiometric mixture at 25°C would give a theoretical temperature drop of 142°K. (Assuming dry air = 21% O_2, 78% N_2, 1% Ar, and for methanol $c_p = 24.03 + 0.065T - 5.43 \times 10^{-6}T^2$ kJ/kg mol K.) Some previous estimates are 122°K (100) and 125°K (32). This temperature drop is 7.9 times the estimated drop of 18°K for isooctane. (91) Practically only a part of this will be realized, if for no other reason than that methanol will freeze at -97.6°C, and that moisture in the air will condense and freeze on the carburator at 0°C.

Since liquid methanol has only 45% of the specific heat of combustion that isooctane has, it will require 8.5 times as much heat to evaporate sufficient methanol to deliver the same heating value as isooctane in liquid fuel. Providing sufficient heat to vaporize the methanol has been one of the largest problems when methanol is used in internal combustion engines, particularly if they were designed for use with hydrocarbon fuels. Effective recovery of exhaust heat could possibly turn this to an advantage by improving the overall thermal efficiency.

If the methanol-air charge was allowed to cool the twenty-five degrees to 0°C, this would increase its energy density 9.2%. A proportional amount of evaporation would cool an isooctane charge 3.2°K and increase its energy density 1.1%. When combined with the greater energy density of 4.9%, fuel atomization with a 25°K cooling through adiabatic evaporation would give methanol a 13% greater energy density in the fuel charge than isooctane. This greater power available from an engine is one of the primary incentives for using methanol in race cars. (along with a higher octane rating)

VAPOR PRESSURE - The highly polar nature of methanol causes a relatively low vapor pressure and high boiling point compared to molecules of similar molecular weight. Isooctane by comparison with 3.57 times the molecular weight has a boiling point 34°C higher, and thus a much lower vapor pressure.

Most gasolines by comparison have a wide mixture of molecular weight compounds and thus boiling points. The lower molecular weight compounds provide sufficiently high vapor pressures to permit ready ignition under ambient conditions. Methanol's vapor pressure is considerably lower than these, which causes considerable difficulty in cold starts, making them impractical below 10°C (50°F). Using a low vapor pressure compound blended with the methanol, or as dual fuel is being considered for cold starts. Alternatively a small amount of methanol could be vaporized and ignited by electricity and used to provide the initial heat to vaporize sufficient methanol to start the car. The corollary of this is that there should be little problem with vapor lock using pure methanol in hot climates.

The opposite problem arises when methanol is mixed with gasoline. The polar and non-polar liquids do not mix well, and a very high nonlinear increase in the vapor pressure is observed rather than the decrease expected from calculating the vapor pressures proportional to the mole fractions. (275) This phenomena would require that some of the low molecular weight fraction of gasoline be removed if blends with methanol are used to prevent vapor lock in warm weather. This could be counterproductive to the goal of extending petroleum supplies but must be viewed in light of overall fuel requirements.

Methanol's low vapor pressure is beneficial from a safety viewpoint in that it reduces the chances of fire if spilled. Methanol's flame point or temperature at which a flame will ignite the liquid is 10 or 11°C while those for gasolines are usually below -20°C. (100)

SOLUBILITY - The polar nature of methanol makes it infinitely soluble in water while it still has enough hydrocarbon nature to be readily soluble in dry gasoline also. However, small amounts of water (~0.1%) and/or low temperature will cause phase separation. This would cause an engine to stall because it could not run on the denser methanol-water phase if it was adjusted for gasoline or a low methanol-gasoline blend. This poses a problem in distribution if methanol-gasoline blends were to be used as most conventional gasoline distribution systems and fuel grade methanol contain significant amounts of water and must be dried out. Small quantities of water in straight methanol can easily be tolerated in IC engines or boilers and are beneficial in reducing NO_x emissions.

The use of methanol in turbines, however, will require that stringent measures be taken to keep the methanol dry and free of salts. Methanol could easily be contaminated with salts dissolved in water which would cause severe corrosion problems during high temperature operating conditions in turbines. (158)

The specifics of the three component solubility phase diagram of methanol-gasoline-water are dependent on the particular distribution of the numerous aliphatic and aromatic compounds that gasoline is composed of. Numerous general phase diagrams have been published.

(14,16,100) Detailed studies are in progress to evaluate the particular effect of each compound, and to evaluate the possibilities of enhancing the water tolerance of methanol gasoline blends by various additives. (11, 290,328) Higher molecular weight alcohols are under serious considerations as they can easily be produced during the manufacture of technical grade methanol or "Methyl Fuel®". (270)

CORROSION AND SWELLING - The reactive polar hydroxyl group in methanol is noticeably different from petroleum hydrocarbons in its corrosion of metals. This is compounded when water and salts are dissolved in the fuel. Severe corrosion is noticed with zinc, lead and magnesium. (54, 100, 290) Aluminum and copper are also attacked more than with hydrocarbons. These materials need to be coated or lined with inert or resistant coatings. Dry methanol could be stored in conventional steel tanks without too much difficulty. However, even here the conventional rust preventative paints and metal coatings need to be re-examined for compatibility or for possible galvanic action.

Methanol is a strong solvent, and noticeably swells or softens many of the plastics or rubbers that are often used as gaskets or floats in conventional hydrocarbon fuel systems. Polyamides and methacrylate are affected and "Viton" floats in particular can swell up to 50%. (54, 75) Polyethylene and polyacetal seals appear to be compatable with methanol.

KNOCK RESISTANCE - The combustion properties of a fuel are of considerable importance in its use in IC engines. Methanol has a higher octane rating than isooctane, but its unusual burning characteristics have led to reported values of the Road Octane Number ranging from 106 to 114. Apparently high combustion pressure pulses are confused with knocking by conventional pressure sensors on CFR test engines. (100, 173) Different operating conditions and non-homogeneous fuel mixtures may also account for some of the variations. Experiments that take these factors into account estimate the RON of methanol is about 110 or higher as compared with 106 for many previous measurements. (157, 173, 161) The Motor Octane Number (MON) has been similarly estimated at from 87.4 to 94.6. (100, 193) Isooctane is taken as a standard and assigned RON and a MON = 100. The motor octane ratings appear to be more indicative of fuel performance in recent engines. A lucid discussion of octane ratings and their value is given by Benson. (24)

These high octane ratings permit considerably higher compression ratios using methanol than using gasoline thus permitting greater efficiency and power. Compression ratios up to 16:1 have been reportedly used in racing cars using methanol, (216) where it is considered a preferred fuel. The tendency of conventional fuels to knock have been reduced considerably in experiments in which the combustion chamber is altered to introduce swirl into the fuel charge. These effects and the corresponding benefits of higher compression ratios may apply to methanol also. Engine speed, compression ratio and temperature also affect the knocking tendency. (88) Using electronic controls and continuously variable flywheel transmissions to run an engine under optimum conditions will further enhance the overall efficiency.

When methanol is blended with gasoline in small quantities, it increases the octane rating considerably more than the proportion added. (161, 223) This effect depends on the octane ratings of the base gasoline with the Blending Road Octane rating varying from 118 to 148 for 15% methanol. The average variation is shown in Figure 1, but this should be used with caution as there is a large variation in the values using different gasolines with the same octane ratings. (173)

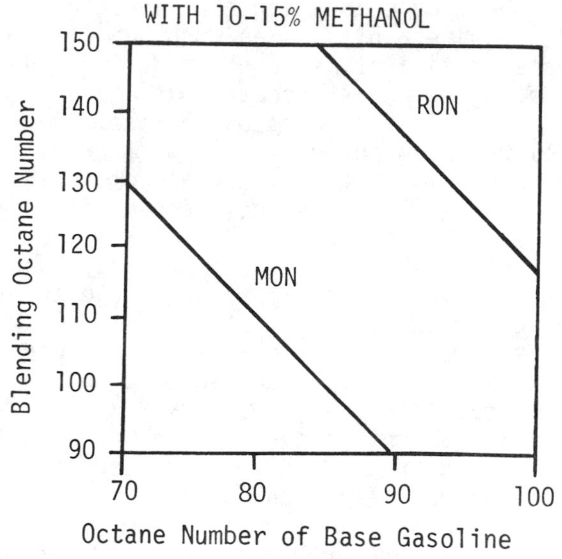

Fig. 1 - Blending octane number of methanol (173)

IGNITIBILITY - The high resistance to knock or spontaneous detonation of methanol is desired for use in the spark ignition Otto engine, but is the opposite of the spontaneous compression ignition desired for the Diesel Engine. The Cetane Number is used as a measure of the ignitibility of fuels in Diesel engines, with n-heptane having a CN = 56. Diesel fuels typically have Cetane numbers of 45-55. The Cetane rating of methanol was so low (<10) that it could not be measured directly. Extrapolation of tests using additives to pure fuel gave a CN = 3 for pure methanol, or CN = 2 for methanol + 10% H_2O. (100) Pure methanol is thus not suitable for use in conventional diesel engines. However, it can be used in conjunction with another fuel that has good compression ignition properties.

Another measure of ignitibility is the lowest temperature at which the fuel-air mixture ignites spontaneously. This depends on the operating conditions, with values from 400°C to 500°C being reported. Addition of water systematically increased the ignition temperature from 478°C for pure methanol to 495°C for 28% H_2O in methanol. (100)

Methanol has a correspondingly long ignition delay. This time between maximum compression and spontaneous ignition is dependent on the operating conditions. Theoretical estimates at a compression ratio of 6:1 and a fuel-air equivalence ratio of 0:77 gave an ignition delay of 26.6 msec for methanol compared with 2.2 msec for regular gasoline. (100) Methanol does have a comparatively low hot plate ignition temperature. (100) Hot combustion particles are usually not present to cause pre-ignition but the combustion chamber must be specifically designed to avoid hot surfaces, particularly in the spark plugs.

DISSOCIATION - As a correlary to the synthesis reaction, methanol dissociates into carbon monoxide and hydrogen as the temperature is raised. The equilibrium fraction of dissociation is shown in Figure 2 as a function of pressure.(Graph adapted from Reed 1976 ref351)For instance at equilibrium over 80% of the methanol is dissociated at 200°C and 10 atm. For this reason the combustion properties of methanol are to a large extent similar to mixture of carbon monoxide and hydrogen. The comparative properties have been studied in depth by Pefley et al. (205-207)

The major difference in properties is caused by the heat of dissociation, which for gaseous methanol is 90.56 kJ/mol at 25°C and one atmosphere pressure. This is 13.4% of the lower heat of combustion. For liquid methanol, the heat of dissociation (which includes evaporation) is 128.51 kJ/mol which is 20.1% of the lower heat of combustion. This suggests the possibility of recovering a significant amount of exhaust heat to evaporate and dissociate methanol thus increasing the efficiency.

If liquid methanol were injected into a hot container, evaporation, dissociation and a rise in temperature could increase the pressure enough so that the high pressure dissociated fuel could be introduced into the combustion chamber near the top of the compression cycle. This use of exhaust heat to pump up the gas would eliminate some of the wasted pumping work in the engine. Details of the energy recoverable by this means will require analysis of the kinetics of dissociation and the rates at which heat can be taken from the exhaust gases and added to the fuel.

FLAME SPEED AND MISSFIRE LIMITS - Hydrogen has a high flame speed that is considerably higher than petroleum fuels. This is reflected in the high flame speed of methanol which is

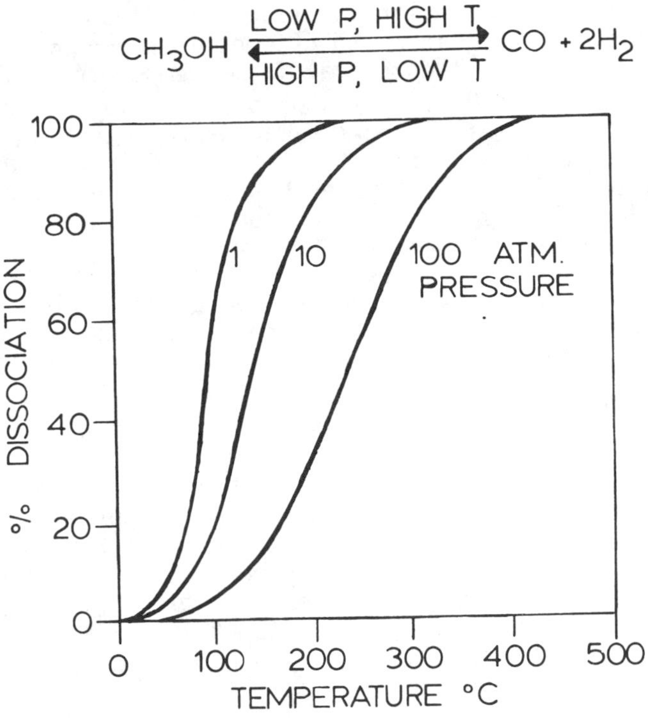

Fig. 2 - Methanol dissociation equilibria

higher than isooctane below a fuel-air equivalence ratio* of 1.3 and particularly in very lean fuel-air ratios. (100) Recent measurements of the time required to burn 1%, 10% and 90% of the mass charge shows that methanol burns significantly faster than indolene particularly in lean region. (339,340)See also(333)

Methanol has correspondingly wider missfire limits than does gasoline. The lean missfire limit or point where erratic combustion begins is typically 0.2 equivalence ratio units leaner for methanol than that of isooctane, going down to about 0.6 depending on the operating condition. (88) This allows a methanol engine to be run fairly lean with the corresponding benefits of higher efficiency and lower emissions compared to gasoline.

FLAME TEMPERATURE - Without dissociation, the adiabatic flame temperature could be calculated relatively easily using the heat of combustion to the gases and accurate heat capacities. The high temperatures in flames cause significant amounts of dissociation in the product gases. This increases the number of moles of gases in the mixture as well as absorbing considerable quantities of energy and reduces the flame temperature. The resulting nonlinear equations require iterative methods to solve. (288) Some of the reactions such as the formation of NO are comparatively slow, and must be included in accurate calculations.

* Fuel-air equivalence ratio Φ = (actual fuel/air ratio)/(stoichiometric fuel/air ratio)

Using equilibrium reaction values, isooctane burningin a stoichiometric amount of oxygen has a theoretical flame temperature of 3082°K with 22.12 moles of products as compared with 17 moles of undissociated products from 13.5 moles of reactants at atmospheric pressure. (288) Methanol (liquid) burning stoichiometrically in oxygen by comparison has an adiabatic flame temperature of 2890°K with 3.48 moles of dissociated products as compared to 3.0 moles of undissociated products from 2.5 moles of reactants. This is a 39% increase over the reactants as compared with 64% for isooctane. See also (335).

Diluting the gases with nitrogen by burning in air reduces the flame temperature and adds several more products and reactions to the calculations. The flame temperature of methanol in air has been estimated at 2194°K (3490°F), (86) or 250°K below that of isooctane. (31) Calculations of the adiabatic flame temperature as a function of the fuel-air equivalence ratio at a compression ratio of 8 are shown in Figure 3.*

* John Smith and David Kittelson, University of Minnesota, 1976, unpublished, using NASA's Flame Program. Equilibria data from Pefley et al.(207)

LUMINOSITY AND PARTICULATE FORMULATION - Methanol contains no carbon-carbon bonds and therefore it cannot form any unoxidized carbon particles or precursors to soot particles. (113) Higher molecular compounds in a methanol fuel could contribute to such products. Pure methanol, therefore, produces no soot and has a bluish flame with low luminosity that is nearly invisible in daylight. Radiation transfer is consequently lower with a methanol flame than petroleum flames.

This is an advantage where energy is to be extracted mechanically as in a turbine or internal combustion engine and where radiation energy is lost. Experiments indicate that ten percent less heat is lost to the engine coolant using methanol as compared to gasoline in an IC engine. This would be due to both lower radiation and convection losses due to a less luminous flame and lower temperatures. Lower radiation is a disadvantage in boilers or lamps where highly luminous flames are desired. However, the absence of soot deposits would retain good conductive heat transfer through the boiler shell. Furthermore, methanol flames will remove soot deposits left by previous fossil fuels. (83)

Numerous publications give further tables and comparisons of the properties of methanol

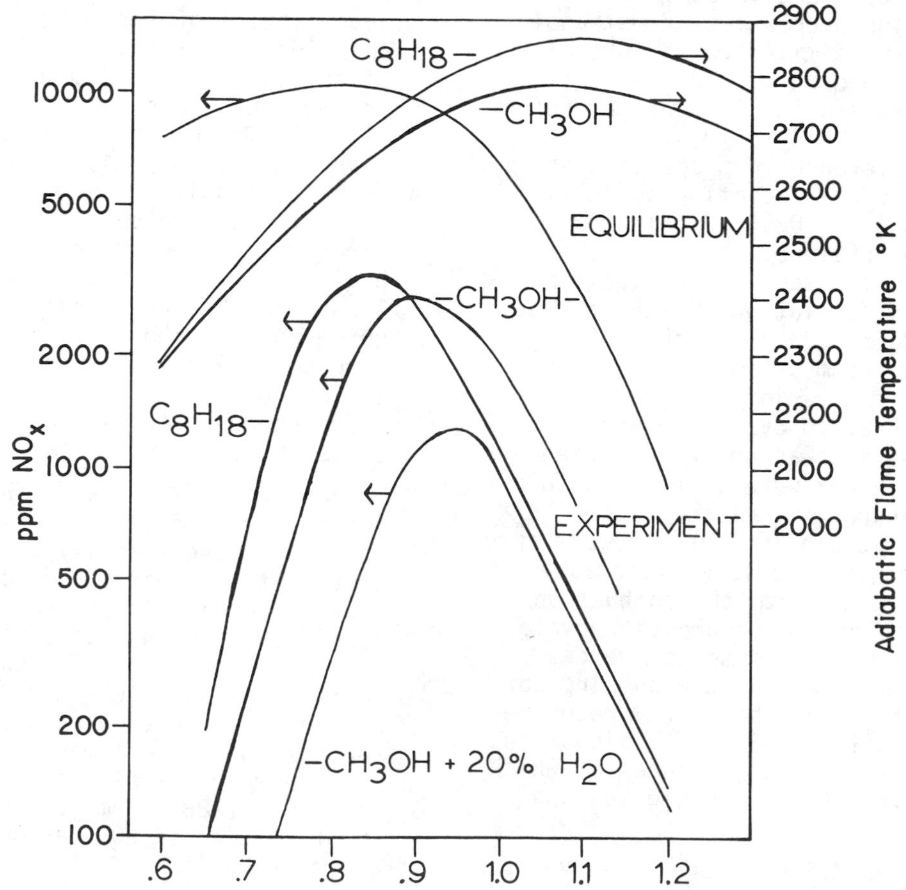

Fig. 3 - Flame temperature and NO_x emissions

with other fuels. (14,16,50,100,160,167,175,202, 241) Hydrogen has been advocated by many as the future fuel, but the very heavy storage containers required are a major disadvantage. (16) Ethanol is already in use in countries such as Brazil. (290) Hydrocarbon fuels from coal, shale oil, tar sands or biomass are the other major contenders.

EMISSIONS OF COMBUSTION PRODUCTS

NO_x EMISSIONS - Nitric oxide is formed from reaction of atomic oxygen or nitrogen with molecules N_2 or O_2.

$$O + N_2 = NO + N$$

$$N + O_2 = NO + O$$

These reactions are both very slow, with half lives on the same order as the expansion stroke in an engine. The formation of NO is thus governed by the kinetics rather than equilibrium considerations, and as a result has a very strong exponential temperature dependence. Experiments demonstrate this exponential decrease in NO_x emissions with temperature (and thus with equivalence ratio) in the lean region as the flame is diluted and cooled with air. (193) (See Figure 3) The equilibrium concentration by comparison has some variation with temperature, but this is small compared to the three orders of magnitude reduction obtainable by varying the equivalence ratio. Near the equivalence point the availability of oxygen becomes of major importance.

The lower flame temperature of a methanol flame results in substantially lower NO_x emissions than with isooctane below $\phi = 0.9$. Numerous researchers have advocated operating in this lean region with methanol to reduce emissions rather than resorting to catalytic methods. Some experimental results for instance show that the emissions for methanol could be decreased from a peak of 3000 ppm down to well below 100 ppm. (193) This compared with 230 ppm for isooctane at the lean limit. The peaks for methanol and isooctane were separated by about .05 equivalence units. This resulted in methanol giving half the NO_x emissions of isooctane in the lean region, though it actually had higher emissions in the region above an equivalence ratio of 0.9. Similar results were observed in another recent study. (340) (See Figure 3)

An unexpected result of recent experiments is that NO_x emissions reach a maximum as the compression ratio is increased, and then decrease at higher compression ratios. (32,233, 290) This unusual effect may permit higher compression ratios and efficiencies with reduced emissions.

Injecting water or adding it to the fuel lowers the peak flame temperatures and consequently the NO_x emissions. Mixing 20% water in, for instance, reduced NO_x emissions by a factor of 3 at the peak and by a factor of 10 at equivalence ratios around 0.7. (193) Decreasing the manifold pressure caused similar sharp reductions in NO_x emissions. The striking factor in these measurements is that with optimum timing, NO_x levels of 200 to 300 ppm were obtained at the optimum efficiency conditions. This corresponds to emissions of 0.6 to 0.8 g/mile for a vehicle averaging 8 mpg. on methanol. The 1978 federal standard of 0.4 g/mile of NO_x emissions could be met with an average of 140 ppm of NO_x which was easily attainable in the test engine.

Detailed comprehensive computer models are being developed at the Lawrence Livermore Laboratories to accurately model combustion and predict the amount of NO_x formed. (240) They have studied the evolution of the liquid fuel spray after atomization. They included time dependent effects, multi-dimensional geometry and the effects of gas swirl and piston motion to better analyze the stratified charge engine. They also coupled very detailed comprehensive chemical kinetics with a one dimensional transport model and are attempting to incorporate turbulence into the model.

UNBURNED FUEL AND ALDEHYDE EMISSIONS - The organic emissions from an engine using pure methanol are primarily methanol and aldehydes. These are technically not hydrocarbons, though they are often called that when comparing with the hydrocarbon emissions from gasoline and diesel fuel. Recent authors prefer to distinguish these as unburned fuel (UBF) and aldehydes. (136) Measurements of these emissions unearthed wide variations in sensitivity between many conventional detectors. Operating conditions often had to be altered to avoid factors such as condensing methanol on the walls or dissolving it in condensed water. Even when these factors were recognized, reported data on unburned fuel emissions have varied from several times those of gasoline during cold starts to an order of magnitude less in single cylinder engines. (32,54,136) Similarly aldehyde emissions have generally been higher than those of gasoline though some lower emissions have been measured.

In order to identify the factors affecting these emissions, studies were conducted at General Motors on how these emissions varied with fuel preparation and the temperatures involved. (136) They doubled the vaporization time for the fuel injector and added a 1 kW heater to account for the large heat of vaporization of methanol. With their standard arrangement, the UBF emissions were four times those of gasoline. Improved fuel preparation reduced these to 80% to 90%, eliminating any difference between fuels. Raising the air temperature 100°C lowered the UBF emissions 22% while raising the coolant temperatures 37°C reduced them 65%. (See Figure 4)

The aldeyde emissions were an order of magnitude above those of gasoline in the standard intake system. These were reduced 75 to 90% with the improved mixture preparation though still about four times higher than those of gasoline. These reductions in aldehyde emissions closely followed these of the unburned fuel, and were assumed to come from the partial oxidation of the UBF. They also noted a 70% reduction in CO emissions for methanol and 50% for gasoline through improved mixture preparation so that they both became comparable. The NO_x emissions on the other hand increased 20% and 15% respectively, probably due to higher flame temperatures.

Similar emission results were obtained by Volkswagon researchers who made efforts to heat the intake manifold with exhaust gases and improve the fuel-air spray atomizations. (32) All the emissions are thus closely tied to how well the methanol is vaporized and mixed with the air. (See Figure 4)

Increasing the compression ratio from 9.7 to 14 reduced the aldehyde emissions by half to levels comparable to those of gasoline. Adding 10% water reduced the aldehyde emissions by 40% and the NO_x by 50%. (32)

CARBON MONOXIDE EMISSIONS - These depend primarily on the air-fuel ratio. They are very low in the lean region and rise rapidly as the fuel-air equivalence ratio is enriched beyond 1.0. Experiments in single cylinder engines indicate that carbon monoxide emissions from methanol range from higher than to an order of magnitude lower than those from gasoline. (55, 205, 206) Further tests showed that CO emissions are a strong function of the fuel preparations, and correlate with the unburned fuel emissions. (136) Improving fuel preparation reduced CO emissions 70% for methanol and 50% for gasoline, resulting in comparable emissions for both. The CO emissions using the improved mixture preparation are shown in Figure 4. Rich operating during cold starts give high emissions which may require catalytic oxidation. These emissions are often higher than those of gasoline.

Elimination of the lead emissions will be one significant effect of using methanol directly or in blends as would conversion to unleaded gasoline. The carcinogenic and toxic properties of the aromatic compounds in gasoline need to be compared with the toxicity of methanol and aldehyde emissions. In a detailed analysis of the polynuclear aromatic emissions, Volkswagon researchers measured a total of less than 5% as many for methanol as for gasoline. (177) This reduction will need to be compared with possible increases in the fairly toxic aldehyde emissions from methanol. Both will vary considerably with the particular gasoline

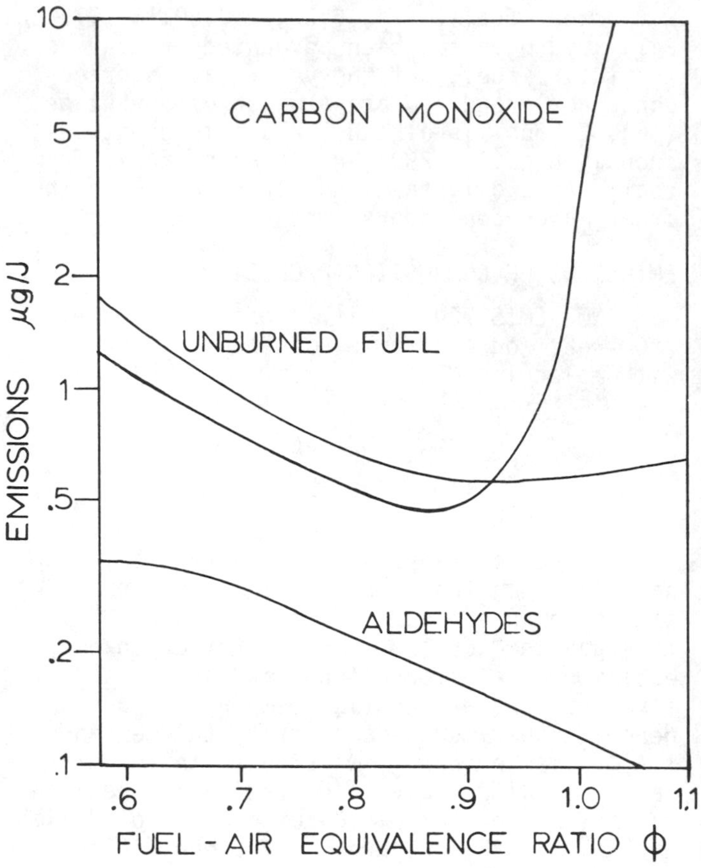

Fig. 4 - Methanol exhaust emissions

Table 4 - Boling Points and Vapor Pressures

Fuel	Mol. Wt.	Boiling Pt.	Vapor Pressure*
Methanol	32.04	64.5°C	37 kPa
Isooctane	114.23	99.3°C	15.5
Normal Gasoline (Summer)	~98	32-186°C	70
Normal Gasoline (Winter)			90

(From Ref. 100) * @ 37.8°C (100°F)

constituents and the engine operating conditions.

Particulates and sulfur compounds are totally absent from the exhaust gases of methanol flames. The formation of sulfuric acid mists by catalytic converters from sulfur compounds in gasolines by comparison has been recognized as a potentially significant health hazard and is now being studied.

Blends of methanol in gasoline result in changes in emissions proportional to the differences between the emissions of straight methanol and gasoline. (173) Varying engines and combustion conditions have resulted in wide variations reported by different researchers. (54) Systematic studies are being conducted now to quantify these figures. (10-12, 54, 290,323)

The toxicity of emissions and the formation of smog by automotive and industrial emissions have been the strongest impetus to study, regulate and reduce combustion emissions. Aldehydes are considered much more toxic and reactive in the photochemical smog forming reactions than methanol is. (74) The combined toxicity and reactivity of the entire combination of emissions will have to be assessed to compare methanol and gasoline, particularly as the Federal emission regulations were set for petroleum fuels. More study needs to be done on the combined effects on emissions of increasing the compression ratio, improving fuel atomization and vaporization, and injection or mixing water with methanol.

COMBUSTION APPLICATIONS

BOILERS - The electricity utility industry's use of clean fuels for intermediate and peak-load shaving electricity generation is increasing rapidly. The availability of natural gas is decreasing, thus increasing the demand for clean liquid fuels. By 1973 this demand had grown ten-fold in five years to 35,000 tonnes/day (280,000 bbl/day). This was expected to exceed 125,000 tonnes/day by 1980 and equal the automotive use by 1985 (885,000 tonnes/day). (91,166)

Seeing this market potential, Vulcan Cincinnati, Inc. in 1971 ran a small scale demonstration of the combustion of methanol in a boiler test stand at the Coen Company of Burlingame, California. (83) NO_x emissions were a quarter and a tenth of natural gas and #6 oil respectively. They then arranged a large scale demonstration in 1972 in a 49 MW commercial utility boiler at the A. B. Paterson Station of the New Orleans Public Service Inc. The test was also sponsored by two other U.S. power companies and monitored by 24 other international companies. (57,267) For the test Y shaped burner nozzles were used to enable complete combustion, and a centrifugal pump was installed as the fuel oil pumps could not handle the low viscosity, low lubricity of methanol. Otherwise no modifications were required for the multifuel boilers.

During the test no particulates were observed (methanol has no carbon-carbon bonds that could form soot), but instead the soot deposits in the furnace from previous operation on petroleum fuels were burned off by the methanol allowing higher heat transfer rates. The methanol flame is less luminous than that of oil or natural gas and has a lower flame temperature and thus less radiative and more convective heat transfer. This resulted in a 3% lower boiler efficiency than natural gas at equal loads. (86) No sulfur compounds were emitted since sulfur must be removed in manufacturing before the methanol synthesis process. They also noted that the carbon monoxide formed with methanol was less than that with oil and gas, and that aldehydes and acids were generally less than 1 and 10 ppm respectively and thus considered "insignificant" with proper combustions conditions. They concluded that methanol would be an excellent boiler fuel and its uses depend only on cost and availability. (84-86)

Plans have been made to build methanol plants in the Middle East to convert otherwise flared natural gas, and a number of utilities considered agreements to purchase this methanol. (49) Offshore Petroleum of Houston has begun looking for customers for its methanol to be made from offshore natural gas. (58) It appears therefore that utilities may be among the first to use methanol on a large scale and that this may come from natural gas in remote regions.

It will also be possible to use methanol for home or industrial furnaces. It has been recommended though that the entire fuel system be redesigned to accommodate the problems of corrosion, and allow for the special toxic and explosive hazards. Preheating the air and installing flame arrestors may also be advisable. (183)

TURBINES - General Electric Company has run tests in a combustion chamber that indicate that methanol is an excellent gas turbine fuel. (86,158) They anticipate a 6% gain in power output with pure methanol and a 12% gain with a mixture of 20% water in methanol over No. 2 distillate oil due to larger mass flows. Thermal efficiency would have increased 2% and 0% respectively. Nitrogen oxides were reduced 40% due to lower flame temperatures. Carbon monoxide emissions on the other hand increased three fold because of lower oxidation rates.

Turbines would have to be redesigned to make the best use of methanol's properties, though conventional turbines could probably be used. Modifications would include explosion proofing to accommodate methanol's lower combustion limit. Fuel flow rates would approximately double and additional means of lubrication would have to be provided to accommodate methanol's low lubricity.

Gas turbines are particularly susceptible to hot corrosion from sulfur and metals in the fuel or air. The sodium content of the fuel should be kept below 1 ppm. This is a potential problem with methanol since it readily mixes with water which could be contaminated with sodium salts. Proper precautions would have to be taken in transportation and distribution. The total absence of metallic elements and the lower flame temperature will result in lower corrosion and deposit rates. Three to four times longer operation between overhauls are expected as a result. (158)

Full scale tests of methanol have been run in an 18 MW gas turbine at Florida Power Corp., Bayboro site by United Technologies, Inc. (171)

The researchers found NO_x emissions from methanol were 74% less than with No. 2 oil while CO emissions were 30 ppm higher. These emissions were below the proposed federal standards for NO_x over the entire power range and for CO above 17 MW_e. They concluded that methanol appeared to be an ideal fuel for gas turbines as far as low emissions, excellent performance, relative ease of handling and storage, and a minimum of changes required for the engine. Supply and cost appeared to be the governing factors on the use of methanol would be introduced. Methanol looks like a promising fuel for the small automotive turbines under development if and when they are used.

CONVENTIONAL OTTO ENGINES - The use of alcohols in engines was considered even in the early days of the automobile. (43,114) Alcohols were used in Europe during the 30's and 40's. Recent tests of methanol in cars has caused a lot of interest. (206,224)

Straight Methanol - Detailed comprehensive experiments in single cylinder engines were performed by Ebersole and Manning to determine the operating ranges, efficiency, power, fuel consumption and emissions of methanol versus isooctane. (87,88) They conducted their experiments at two engine speeds, and limited their tests to a compression ratio of 7.5:1 to obtain knock-free operation under all conditions. At this low compression ratio they found maximum engine output was about equal for both fuels while methanol had 2.15 times the specific fuel consumption of isooctane.

The efficiency of an ideal adiabatic Carnot cycle is given by

$$\eta = 1 - (\frac{1}{CR})^{\gamma} - 1 = \frac{T_2 - T_1}{T_2}$$

CR = Compression Ratio
γ = Cp/Cv = Ratio of heat capacities

Isooctane has a slightly larger γ as it has more dissociated products and more nitrogen at the higher temperature. At the same compression ratio this would indicate an advantage of one percent over methanol. However, engines can operate on methanol at compression ratios of up to 14 or 16 compared to current gasoline engines at CR = 8.2. Methanol at a CR = 14 would ideally have at least a 15% advantage.

Further detailed studies of pure methanol in single cylinder engines have been conducted by the research departments of Volkswagon and Exxon Corporations. (29-33,173,193) Volkswagon experiments indicate that methanol gives between 5 and 12% (averaging 8%) improvement in thermal efficiency over gasoline at a compression ration of 8.2 at 2000 RPM. It has correspondingly 12% more power. Increasing the compression ratio to 14 improved the power a further 9% for methanol at a fuel to air equivalence ratio of 1.0. At an equivalence ratio of 0.89, friction losses were more pronounced giving a maximum power output at CR = 12. Both compression ratios gave thermal efficiencies above 40% as compared to 27% for gasoline in a single cylinder engine at CR = 8.2.

Experiments at Exxon highlighted the advantages of lean operation and lower pollutants as two additional significant advantages methanol has over gasoline. (193) Efficiency improves as the engine is operated lean with an optimum between fuel air equivalence ratios of 0.7 to 0.8. This lean operation provided efficiency advantages over isooctane ranging from 2% at high speeds and manifold presssures to 22% at lows speeds and manifold pressures.

The wide combustion limits of methanol permit smooth operation in the lean region while reducing the flame temperature and the NO_x emissions. They estimated that by leaning the fuel air mixture, emissions of the order of 100 ppm NO_x needed to meet the 1978 Federal Standards might easily be achieved under operating conditions while only sacrificing 1-2% in efficiency. If spark timing was not optimized this would still give only a 4% efficiency deficit. Operating under corresponding conditions with isooctane are not possible because the narrower missfire limits give ragged operation. The researchers concluded that 0-9% less loss in efficiency may be obtainable using methanol over gasoline in an Otto engine to reach NO_x emission limits. Using catalytic emission controls, relying on purely thermal emissions controls would give an 18-25% greater energy debit for gasoline. (193)

The weight of methanol may amount to 5% of the vehicle's total weight and thus increase the fuel consumption possibly by 2%. On the other hand, the greater power available in a methanol fueled engine would permit a smaller engine which would offset the increase in weight. The greater efficiency would also reduce the amount of fuel required for the same range. These improvements were estimated at 1% and 3% at CR = 8.2 and 12. (173) Some of the estimated energy advantages of methanol over gasoline are summarized in Table 5. Initial results using four cylinder engines confirm the above treads with some variations in magnitudes. (32,177)

In choosing synthetic fuels, the overall system efficiencies, economics and environmental aspects should be considered. The comparative efficiencies of synthesizing methanol from coal versus synthetic gasoline as well as the increased transportation costs must be combined with the greater operating efficiencies and pollution control advantages.

The system efficiencies for both fuels can be significantly improved by buffering the energy demand through energy storage. Hybrid combinations of heat engines and batteries or flywheels would allow an engine to be sized slightly over the average demand and operate at its optimum efficiency. Installing a flywheel transmission in a Pinto for instance improved its city fuel economy by over 50%. Further improvements were expected to increase this to over 100% with similar reductions in the total emissions. (105)

The stoichiometric air to fuel ratio (by volume) of methanol is 6.45 as compared to 15.1 for gasoline. With the same efficiencies a vehicle designed to run on pure methanol would need approximately double the fuel injection rate used for gasoline. In a carburetor either the jet or the venturi size could be changed to accommodate this. Vehicles using fuel injection systems could also be converted relatively simply. Conceivably the could be designed to run on either methanol, gasoline, or an intermediate blend with a simple mechanical or electronic adjustment. This could provide multifuel capability that would be valuable in the initial stages of setting up a methanol distribution system. One such vehicle is already in use. (114)

Providing sufficient heat to vaporize methanol, and to allow for cold starting requires more involved techniques. Methanol has a heat of vaporization of 1167 kJ/kg as compared with 300-400 kJ/kg for gasolines and thus requires three and a half times the heat to vaporize equal masses. Since approximately twice as much is needed for the same energy value, seven times as much heat must be recycled in a carburetted methanol engine as in one using gasoline. Conventional approaches to heating the intake manifolds by the exhaust gases are being suggested and experimented with. (100) Of possible major importance will be the use of heat pipes to conduct much larger amounts of heat from the exhaust gases to the air and fuel. At least two major programs are currently being directed at heat pipes for these automotive applications. (125,127) Catalysts may be another route to rapid dissociation.

This recovery of exhaust heat could possibly improve the thermal efficiency up to 20% if the methanol were dissociated under pressure. Advocates of dissociating methanol have shown improved unburned fuel and hydrocarbon emissions in the lean region, comparable CO emissions and higher NO_x emissions for the dissociated fuel as compared with methanol. (205,207) Ultrasonic vibration equipment should significantly improve the fuel atomization and combustion. (100) Further experimentation incorporating heat pipes and improved fuel atomization and heating need to be conducted to clarify the efficiency as well as emission benefits.

To be compatible with methanol, engines will need to be designed with adequate lubrication and cooling to accommodate methanol's low lubricacity and the absence of lead additives. All fuel distribution components will need to be inert in methanol or else protected. Spark plugs with good heat dissipation characteristics are needed to prevent preignition from a hot tip.

Table 5: Percentage Efficiency Advantages of Methanol over Gasoline

	CR=8.2[a]	CR=12[a]	CR=10[b]	CR=12[b]
Compression Ratio	0	+16	8-10	16-20
Efficiency	+8	+8		
Part Load or Lean Operation	+2	+2	10	10
Power Surplus	+1	+3		
Energy Density	-2	-2		
	9	25	18-20	26-30
			Thermal	Catalytic
Pollution Control			18-25	0-9
Detriments		total	36-45	26-39

a) Volkswagon (173) b) Exxon (193)

Blends with Gasoline - The possibility of mixing methanol with gasoline has received considerable attention, with vigorous debate between its advocates and critics. The primary advantages are the large octane boosting properties of methanol and the potential of extending the diminishing gasoline supplies in a form that is compatible with the existing vehicles without any alterations. (220) The major difficulties are corrosion problems, phase separation with water, vapor lock, poor drivability, cost, and toxicity. (275)

To quantify these factors, fleet tests of methanol-gasoline blends are being conducted by Volkswagon Research and Development Co. (Germany), the Swedish Methanol Development Co. (Sweden), Bartlesville Energy Research Center (ERDA, USA) and the University of Santa Clara with the City of Santa Clara (USA). (173, 290,10,207) [Ethanol-gasoline blends are being tested by Fiat Co. (Italy) and the University of Nebraska (USA), and are in use in Brazil. (290)] A number of other researchers and groups are testing methanol-gasoline blends in engines and in vehicles. (6,53,55, 100, 162,194,224,290).

The high Blending Octane ratings of methanol would make it an effective replacement for conventional lead additives. The Blending Road Octane value of methanol is typically 130, ranging from 150 in 83 RON gasoline to 116 in 110 RON gasoline as well as varying with the gasoline composition. The Blending Motor Octane rating is typically 100 and varies similarly. (See Fig. 1) (173) The blends give improved efficiencies and reduced emissions as expected from the mixing effect and approximately equal to gasoline under equivalently lean conditions.

Since methanol requires approximately double flow rates to apply the same energy rates, blends with gasoline using the same flow rate effectly lean out the fuel mixture. The optimum efficiency is obtained at a lean fuel air equivalence ratio of about 0.9. Adding methanol to older cars that were adjusted richer improves the fuel economy. Use of methanol blends in recent model cars as been criticized for reducing the efficiency because the engines ran leaner than optimum. (274-276)

Most vehicles could run directly on blends of up to 15% to 20% methanol in gasoline. (290) If all vehicles were fueled by these blends, they could be easily adjusted for optimum efficiency within pollution limits. Poor drivability during cold starts or in cold weather has been the major complaint of drivers using the blends. The vapor pressure may need to be controlled more carefully by varying the composition of the gasoline base between winter and summer to avoid vapor lock and cold start problems.

Chemical grade methanol is relatively pure with 0.05% or less water. Technical or fuel grade methanol however usually has the higher molecular weight alcohol fraction (containing high percentages of water) mixed back in, so that the final product may have 0.5% water. Conventional distribution systems may add again as much water. (290)

One of the potential problems is that of phase separation with water or temperature. Most gasoline distribution systems have water in the bottom of the tanks. Therefore, blending the fuel at the pump has usually been recommended. Water must be excluded or else other compounds added to the mixture to improve the water tolerance. Detailed solubility studies are currently being conducted to quantify this problem. (11, 290) With a proper fuel tank cover, water induced separation has not been a significant proplem in vehicle tests though occasional temperature induced phase separation has been noticed. (290)

A second factor is the large nonideal increase in the vapor pressure of the mixture. This could cause vapor lock problems in hot regions in vehicles that are not designed to accommodate it. (This is not a problem with pure methanol with its low vapor pressure.) Some of the lighter molecular weight fraction of the gasoline may have to be removed during the summer to counteract this increased vapor pressure. This could eliminate the economic advantages of blending in methanol.

Thirdly, corrosion problems will arise from automotive components not designed to handle methanol. Zinc and aluminum parts, lead plating (tern plating) and plastic parts especially VITON have been subject to corrosion or swelling. Fuel filters may take care of some of the problem.

Currently, the high cost of methanol makes most blends more expensive than gasoline. In the near future, these economics may be reversed, e.g. Mobil expects to spend 5¢/gal. to convert methanol to gasoline. (59) The prospects of a large scale production of synthetic methanol from coal or remote gas sources coupled with the need to stretch our rapidly dwindling petroleum supplies have been the primary incentive to consider these blends.

The electronic controls and microcomputers that are now being developed by the automotive industry will be one of the major factors in the development and introduction of alternatively fueled engines. These controls were first introduced to the mass automotive market in the 1975 model year in the U.S., to help optimize engine performance and keep pollutants below the legal limits. (231) The electronic control unit will monitor all the relevant parameters of pressure, temperature, engine speed fuel and air flows, and exhaust concentrations and can directly control the fuel and air flow.

Current control units are programmed and adjusted for a single fuel. They are being adapted to methanol engines. (69) Future control units will be made more versatile and monitor more sensors. They could be made to handle variable fuels or multiple fuels. Thus an engine could use either gasoline or methanol, or any blend of the two. The unit could also be programmed to inject water or use water in the fuel to reduce the flame temperature and emissions.

The problem of phase separation of methanol and gasoline due to water or temperature could be bypassed by using separate tanks for each, or by emulsifying the mixture. An imaginative solution that makes use of the phase separation phenomena is to specifically introduce sufficient water to the methanol gasoline mixture to induce separation, and then have a separate delivery system for each phase. (68, 290) The control unit could then choose from either or both phases and adjust the fuel and air flows accordingly.

STRATIFIED CHARGE ENGINES - The stratified charge engine (SCE) is receiving considerable attention because of the potential improvements in fuel economy and operation obtainable with it. However severe reductions of these gains often occur in attempts to reduce the NO_x emissions to low levels. Operating on methanol should give many of the benefits of reduced emissions previously noted. Experimental evidence is needed to confirm these expectations.

Lean operation in the SCE gives fuel economy improvements of the order of 28% to 38% over an Otto engine. (262) Operating on methanol could possible improve this another 10%. Improvements of 16-23% may be possible from the higher compression ratio compared with a conventional engine at CR=8. When combined with a 0-1% advantage from less pollution control, a stratified charge engine operating on methanol may have an estimated 44-71% better fuel efficiency for the system over the conventional gasoline at CR=8. For this reason, Vantine, et. al., have proposed that the methanol stratified charge engine be developed as the future engine. The rotary engine with charge stratification may also be of interest because of the low radiative losses, and clean characteristics of methanol.

DIESEL ENGINES - Compression ignition engines cannot be run directly on methanol as they require a fuel with a low auto ignition temperature (high cetane number). By the definition of high octane rating, methanol has a very low cetane rating of about 3. (100) Heptane by comparison has a cetane rating of 56 while diesel fuels are rated at 45 to 55. However dual fuel operation is possible and has been suggested. (21,100) In this mode of operation, an ordinary diesel fuel is injected in sufficient quantities to provide ignition, and methanol is used for the rest of the charge. Experiments using dual fuels in diesel engines at the University of Minnesota showed major reductions in the emissions of carbon particulates or smoke, and of the intake manifold temperature. (21) Efficiency and unburned fuel emissions were essentially unchanged.

When used under external combustion conditions with Stirling engines, or steam engines, methanol does not apparently have any specific advantages but would compare with fuel use in boilers.

INTRODUCTION OF METHANOL - Availability, distribution and conversion will be the major hindrances to the introduction of methanol as a fuel. Availability will depend largely on the economic and political climates or legislative action. Manufacturing facilities will probably be built when the costs of conventional petroleum fuels rise high enough to make synthetic fuels competitive. These pressures of rising demand on decreasing supply will occur either through the current depletion of the finite natural resources, or by political or legislative action such as the O.P.E.C cartel and its embargo in 1973 or the situations of South Africa or wartime Germany. The unpredictability of politically imposed prices may require government financial guarantees before industry will venture very far into the synthetic fuels market in the immediate future.

The inertia of the automotive industry to changes will require clearly demonstrated superiority and availability of the fuel along with very strong economic incentives, legislative measures or a major change in the public's values or buying habits. A classic example to the point is the current efforts of Mobil Oil Company to develop catalysts and methods to convert methanol to gasoline. (185,187) They are actively proposing to set up facilities in New Zealand to convert the abundant natural gas there to methanol and then to gasoline.(59) Conversion of methanol to gasoline increases the cost while decreasing the available energy. Furthermore an engine running on gasoline is less efficient than one optimized for methanol. (173,240) This demonstrates the preoccupation of initial cost in sales rather than lifecycle costs. In the short term, it is apparently easier to hide the increased operating costs and energy consumption in the future and maintain conventional engines rather than develop more efficient engines, and convert to alternative synthetic fuels which would increase the highly visible initial costs.

The actual technical and economic barriers of conversion and fuel distribution may not be large compared to the life cycle benefits to society. The pollution control measures that have been legislated in the U.S. resulted in the addition of extensive pollution control systems. These consequently required distribution of unleaded gasoline throughout the nation which demonstrates that major changes in engine design and fuel distribution can be made given sufficient incentives. Smaller, more efficient vehicles are being designed and built in response to the rising demand and legislative stipulations for lower fuel consumption.

The initial conversion of vehicles to use methanol will probably be easiest on vehicle fleets that have their own fuel distributing facilities. Supply on this small scale would be easy to guarantee. Conversion of vehicles and the distribution facilities would be localized and the corrosion and environmental hazards could be closely monitored. The city of Seattle set an example in its study of converting its municipal wastes to methanol or ammonia. (183) Forty-five percent of the estimated capacity of the proposed plant would be sufficient to supply all the city and county vehicle fleets as well as the taxis. This amounts to 6.2% of the city's vehicular miles. Similar convenient situations would occur where vehicles are confined to a certain region from geographical or political reasons.

As its supply increases methanol could be blended with gasoline. Vehicles may need minor adjustments and be checked for fuel compatibility. Adjustments to the base gasoline compositions and strict controls of the water content of the fuel may be required to minimize phase separation, vapor lock and cold starting problems. In California, legislation may be passed requiring the blends in the near future. (116,168)

For vehicles in general to be converted to pure methanol, enough distribution points will have to be set up in the region to permit some semblance of the freedom the public is used to. Alternatively the vehicles could be converted to multifuel capability, either mechanically or electronically, in which case there would be no difficulty travelling to areas lacking methanol distribution facilities.

With national petroleum resources being exhausted in the next couple decades, a synthetic fuel industry will be established to make up the shortages in fuels. This could begin to provide sufficient quantities of methanol to seriously consider large scale conversion. The conversion technology could also be well developed by then if intensive development and testing is begun now.

ELECTROCHEMICAL OXIDATION: FUEL CELLS

Many researchers and companies have made considerable efforts to develop fuel cells for generating electricity because of the potentially high conversion efficiencies possible. The fuel cell efficiency is limited primarily by local side reactions and losses, and not by the Carnot cycle conversion efficiency. Small modular units can be joined together to provide capacities ranging from small commercial sizes (e.g. 20-200 kW) up to utility sizes (25 MW) with working efficiencies expected to be above 35% to 40%. (192,213) These high efficiencies remain even at comparatively low power levels. Fuel cells will be highly reliable, automated and require little maintenace. These factors make fuel cells ideal for remote or dispersed electricitv generating applications. Methanol is considered an excellent fuel which can be used either directly or indirectly in fuel cells.

INDIRECT USE: THE HYDROGEN AIR FUEL CELL - The space program inspired the development of a highly reliable, light weight hydrogen/oxygen fuel cell by General Electric Co. (using acid ion-exchange electrolytes) and Pratt & Whitney Aircraft (using alkaline electrolytes). (192, 202,213,217) Commercial development of the H_2/O_2 fuel cell was continued by Pratt & Whitney (now United Technologies Corp.) supported by a consortium of 26 natural gas companies in their TARGET (Team to Advance Research for Gas Energy Transformation) program. The natural gas used is reformed to hydrogen, and fed to the fuel cell along with air as the oxidant. Twelve commercial units in the 20 to 200kW range are expected to be introduced after 1978. United Technologies Corp. has also begun a second program together with nine utilities to develop a 26 MW phosphoric acid fuel cell power generator for utility applications suitable for communities of 20,000 people. Initial delivery of the $5 million units is expected by 1980.

The acid fuel cells are hindered by the requirements for expensive, scarce noble metal catalysts. Palladium and ruthenium have been mixed with the platinum in an effort to improve its activity and reduce the costs. (63,120,210, 238,251) Efforts were also directed towards using a minimum of catalyst on a support which is often nickel. (41,63,251,252)

The search for inexpensive noble metal catalysts has uncovered tungsten carbide as a potential candidate. (22,40,100,131) The AEG and Telefunken Companies (Germany) are pursuing this catalyst for the direct as well as indirect methanol fuel cell.

Alkaline (KOH) electrolytes have been widely used in many fuel cell programs. They have the advantage in that silver or carbon is a suitable oxygen selective cathode, thus avoiding the problems of direct oxidation of the methanol. (100,184,265) They have the major disadvantage however that the carbon dioxide formed on oxidation combines with the electrolyte giving potassium carbonate.

$$CH_3OH + 2KOH + 3/2\ O_2 = K_2CO_3 + 3H_2O$$

The electrolyte must therefore be replenished, limiting the alkaline fuel cell to low power applications. The US Army is continuing its development of a 1.5 kW alkaline fuel cell. (100,109,110,282) The Shell Oil Company (in England) has made a number of studies on methanol fuel cells and is continuing basic research on catalysts. (100, 147,278-280) as are Britain's Electrical Research Association (Leatherhead, Surrey) and Battell Institute e.v. (Frankfurt/Main, Germany). (147) VARTA, Centre d'Etude de l'Energie Nucleare and Bosch are other organizations working on direct methanol fuel cells, while NASA, ERDA, General Electric, Union Carbide, Monsanto (US) and Hitachi Ltd. (Japan) are working on indirect fuel cells. (100,147,202)

Fuel cells have been considered for automotive propulsion, with demonstration vehicles assembled by General Motors, Institut Francais du Petrole, Siemans A.G. and K.V. Kordesch.(147 202) They are continuing their research on H_2/O_2 and indirect methanol fuel cells, but major commercial programs are still in the distant future. Large weight/power ratios and the high cost and limited availability of catalysts are the primary restrictions to automotive use.

Methanol is one of the preferred fuels for fuel cells because of the ease in which it can be catalytically converted to synthesis gas at low temperatures (250-350°C) with no carbon formed. (100) Sulfur removal to prevent poisoning the catalysts will also be minimal as these must be removed before the methanol is synthesized. (249) The use of conventional liquid storage is a major advantage as compared with pressurized or cyrogenic storage of gases. Methanol is also available from coal, municipal wastes and other sources as compared to the limited access utilities will have to clean liquid fossil fuels, in the near future.

DIRECT USE: THE METHANOL FUEL CELL - A number of fuel cells can utilize hydrocarbon fuels directly. Methanol has an advantage over other hydrocarbons in being miscible in the polar electrolytes used. (184) Low temperature (<100°C) low pressure fuel cells have received the most attention. Esso Research and Engineering Corporations (now EXXON) conducted considerable efforts to develop an acid electrolyte for methanol-air fuel cells. (61-63,128-130,184 200,253-255)

$$\begin{array}{ll} \text{ANODE} & CH_3OH + H_2O \rightarrow CO_2 + 6H^+ + 6e^- \\ \text{CATHODE} & 6H^+ + 6e^- + \tfrac{3}{2}O_2 \rightarrow 3H_2O \\ \hline & CH_3OH + \tfrac{3}{2}O_2 \rightarrow CO_2 + 2H_2O \end{array}$$

These cells have the advantage that the carbon dioxide formed is readily discharged from the cell. They have the disadvantage that the highly soluble methanol diffuses to the cathode and is oxidized directly by the oxidant. Catalysts selective only to oxygen that can tolerate the acid media have apparently not yet been found. Efforts were therefore directed towards installing permeable membranes to restrict the diffusion of the methanol to the cathode. Sulfuric acid was chosen over phosphoric acid as the electrolyte because of the smaller degree of polarization and side product formation. (63)

Meanwhile the French Alsthom Company apparentlydeveloped a superior cell design which provided better power to volume ratios in fairly nuetral electrolytes (pH 9-10). (100,269) In 1970 they joined Exxon Enterprises Inc. in a multimillion dollar effort to develop a commercial methanol-air fuel cell for small scale (up to 10 kW) dispersed and mobile applications. (39,147,268) With a 100 man force (compared to 1000 for United Technologies Inc.) this is the only other major fuel cell program moving toward commercialization.

HIGH TEMPERATURE FUEL CELLS - Moving to temperatures between 600-750°C, hydrocarbon fuels can be used directly with inexpensive catalysts, at greater efficiencies than at lower temperatures. (202) The Electric Power Research Institute (EPRI), for instance, is specifically developing the molten carbonate system for second generation utility fuel cells. (249) These are particularily suited for sustained applications where weight is no problem. Current performance goals are \$200/kW capital cost for a fuel cell operating 40,000 hours at 0.8v dc. They are also studying solid oxide electrolytes which operate above 1000°C. These cells must be very thin to minimize resistive loses, and thus are very susceptible to thermal and mechanical shocks. When developed, these systems will probably be used for stationary continuous applications under constant temperature conditions. Neither system seems suitable to mobile applications.

Further details on fuel cells in general, and methanol fuel cells in particular can be found in references (20,26,34,65,79,101,119,36 138,184,196,201,215,287).

METHANOL: A FEEDSTOCK FOR FOOD

Direct consumption of methanol is toxic to most animals and plants. However a number of species of single cell algae bacteria and yeasts grow well on a feedstock of methanol and inorganic nutrients. These products have a very high protein content, and they are currently being used to replace milk or soybeans in calf feeds. (48,232) Thus methanol can be considered an indirect metabolic fuel via the manufacture of single cell protein.

Full scale plants are being built in Europe to grow single cell protein (SCP) on methanol. (286) This development could have significant impact in the future on the food crisis. Agriculture in much of the world and particularly the arid regions is limited by the availability of water. Large scale solar energy plants in the large deserts of the Sahara, Australia, northern India and southwestern United States or in the tropical oceans could use atmospheric carbon dioxide and water to produce methanol. This in turn could be used as the feedstock for the single cell protein. The overall conversion efficiency could possibly be considerably higher than that of conventional agriculture.

THE HAZARDS OF METHANOL

As a volatile, strongly polar alcohol, methanol has numerous biological environmental physical and chemical hazards. Methanol is a major industrial chemical and solvent and has long been connected with ethanol as a denaturant. With the numerous proposed large scale uses of methanol, especially as a fuel, the hazards are receiving more serious attention.

BIOLOGICAL HAZARDS - Methanol is toxic and regulated by the Federal Hazardous Substances Act. (300) It must be labeled with "DANGER," "POISON," and the skull and crossbones symbol. The label must also bear the statement "Cannot be made nonpoisoneous." The statement of hazard must include "Vapor harmful" and "May be fatal or cause blindness if swallowed."

A detailed review of the biohazards of methanol has recently been made by H. S. Posner. (319)

Methanol exhibits immediate as well as long range accumulative effects. Initially methanol has a mild transient narcotic inebriating effect. Then there is a latent period varying in length, from less than an hour to three days after which the toxic symptoms begin to appear. These symptoms of methanol poisoning may include "weakness, dizziness, headache, sensation of heat, nausea, abdominal pain followed by vomitint, dyspnea, acidosis, visual disturbances, convulsions, coma, and death". (319) The presence and severity are strongly dependent both on the exposure and the individual's personal tolerance. Methanol damages the central nervous system, having the most obvious effect on the optic nerve, causing a strong sensitivity to light and temporary or permanent visual impairment or blindness. Muscle rigidity, tremulousness and spacticity have also been noted. In addition, progressive degenerative damage occurs in the kidneys, liver, heart and other organs. Methanol is eliminated very slowly from the body. Many of the toxic effects are believed to be due to the metabolites of methanol such as formaldehyde or formic acid, which are far more toxic than methanol itself.

Ingestion - Direct ingestion brings about the most rapid response, with 50-100 ml (2-4oz.) usually being a fatal dose, though 25-50 ml (1-2 oz.) has often been fatal if not treated immediately. Individual tolerance varies widely. Strict measures must be taken to prevent its oral consumption. Siphoning fuel by mouth should be strictly avoided. All references to "alcohol" in its distribution and discussion should be avoided wherever possible, to avoid its confusion with ethanol.

Inhalation - High concentrations of methanol vapors can develop acute posioning after brief exposures. A concentration 1,000 ppm will cause irritation to the eyes and mucous membranes while 5,000 ppm can cause stupor or sleepiness. (306) The odor is barely detectable at 2,000 ppm. (318) One to two hours exposure to 50,000 ppm will result in deep narcosis and possibly death. Tolerance levels have been conservatively estimated by the Advisory Center on Toxicology on the basis of preventing the accumulation of methanol and its metabolites in the body. (301) A value of 200 ppm was estimated as the upper tolerance limit for a steady exposure eight hours per day for a single 40 hour week. Other tolerance values are given in Table 6.

Dermal Contact - The immediate effects of methanol on the skin are the removal of grease and drying the skin, typical of other solvents. However, methanol can also be absorbed through the skin and cause the toxic and lethal effects described previously purely via this route.

Exposure to the Eyes - The immediate effects of methanol are similar to other strong solvents, and it should be washed out promptly. Whether by direct contact, or inhaled or ingested, methanol will cause blurring of vision, an extreme sensitivity to light (photophobia) and inflamation (conjuctivitis). Severe exposure can destroy the optic nerve leading to blindness, and cause definite eye lesions. Sometimes the visual symptoms may clear up initially, but return to cause blindness.

Treatment - First aid should be given immediately, and then a physician called as soon as possible. Patients who have inhaled the vapors should be removed from the area immediately, and artificial respiration administered if breathing has stopped. Clothing wet with methanol should be removed immediately and the skin washed. If liquid methanol has contacted the eyes, they should be washed continuously with water for fifteen minutes while being kept open; the patient should then be taken to an opthalmologist at once. If methanol has been ingested, the patient should be induced to vomit and the stomach emptied as soon as possible. Possible treatment is: "ANTIDOTE - If swallowed: Give a tablespoonful of salt in a glass of warm water and repeat until vomit fluid is clear. Give two teaspoons of baking soda in a glass of water. Have patient lie down and keep warm. Cover eyes to exclude light." (306)

Medical treatment by hemodialysis along with the administration of ethanol and bicarbonate as necessary appears to be preferred to peritoneal dialysis on the basis of being more rapid and having fewer residual complications. Fixed-bed charcoal cannisters are also being considered in conjunction with the hemodialysis. (319)

ENVIRONMENTAL HAZARDS - Berger discusses the environmental hazards in the use of methanol in comparison with gasoline. (304) Primary considerations are the air pollution caused by combustions and the potential problems of large scale spills. Secondary aspects would be the biological hazards associated with manufacture and distribution as discussed previously.

Methanol burns with a lower flame temperature, reducing the nitrogen oxides formed during combustion in air. Methanol can be mixed with water and burned at lean stoichiometric air fuel ratios to further reduce the nitrogen oxides formed far below that possible with gasoline.

Carbon monoxide and hydrocarbon emissions are reduced while aldehyde emissions are increased somewhat. However, the overall reactivity of the emissions is probably still considerably lower than that of gasoline. Methanol would produce no sulphur oxides nor sulphuric acid mists because of the strict desulfurization required in its manufacture. It produces no soot or other particulates when burned. Adding methanol to gasoline as an octane booster would also eliminate some of the lead emission. (See discussion of the use of methanol as a fuel.)

The high polynuclear aromatic content of gasoline causes its vapors to be more carcinogenic than those of methanol even though its threshold limit value is currently rated higher (500 ppm versus 200 ppm). A high benzene content in gasoline can also give it a toxicity rating on ingestion similar to that of methanol* Otherwise methanol is considered more toxic on ingestion than gasoline, though few cases of gasoline ingestion have been encountered. (304, 320) The toxicology of gasoline is reviewed by the American Petroleum Institute. (303)

*NOTE: OSHA has recently set an emergency standard of 1 ppm for airborn benzene over an 8-hour day because of its carcinogenic nature.

Actual tolerance to various forms of administration in laboratory tests are summarized in the Registry of Toxic Effects of Chemical Substances. (316) The lowest toxic dose recorded for humans on ingestion was 100 mg/kg. while the lowest lethal dose was 304 mg/kg. The lowest concentration of vapors for which toxic effects were observed on inhalation was 300 ppm.

Methanol is probably toxic to most plants, animals, and aquatic organisms. Methanol is soluble in water and so could not be mopped up as oil spills are, but it would be easily diluted with large quantities of water. Some bacteria thrive on methanol, which is actively being considered as a feedstock in growing single cell protein. Berger suggests the use of the bacterium Pseudomonal fluorescens to consume spilled methanol. (304)

PHYSICAL AND CHEMICAL HAZARDS - The most obvious hazard is that of explosion or combustion. The flammability limits are 7.3 to 36% by volume in air at STP. (311) Vapor saturated air at 20°C (68°F) contains 13% methanol putting it in the explosive region. Air saturated with gasoline is too rich to explode. (15) Dehn discusses the use of inerts to reduce the flammability limits over the liquid surface, and gives a good review of previous work. (307) Flame arrestors and inert gas blankets may be required. (299)

Care must also be taken to keep oxidizing agents separate from methanol as it will react vigorously with them. The chemical reactions of methanol are discussed by Monick. (315) Because of its high vapor pressure methanol will have to be enclosed to prevent evaporation, but a tightly sealed container must not be exposed to heat or to the sun to avoid rupture. A methanol fire should be fought with CO_2 or dry chemical or foam extinguishers or with a water spray (rather than a jet to prevent spreading the flame) if these are unavailable.

Noticeable corrosion has occurred with methanol in contact with ternplated (lead plated) fuel tanks, magnesium, copper, lead, zinc and aluminum parts, and some synthetic gaskets. These need to be avoided or protected. The water present in conventional pertroleum handling facilities would have to be removed, as this would cause dilution, contamination and rust problems. Otherwise methanol can be stored in regular steel tanks.

The Manufacturing Chemists Association describes appropriate procedures to handle and transfer volatile fuels. (311-314)

Further details and references on the hazards and handling of methanol are available in papers or texts by American Industrial Hygiene Association, (302) Deichman & Gerarde, (308) Gleason et. al., (309) Koivusala, (310) Patty et. al., (318) and Sax (320).

SUMMARY

Historically methanol has been one of the major alternative synthesis fuels particularily when petroleum fuels were scarce. It has also been used as one of the preferred premium fuels for race cars. During the last decade, the problems of pollution and the impending depletion of petroleum fuels have spurred studies of synthetic fuels. Methanol is emerging from these studies as potentially a more efficient fuel overall, capable of meeting the most stringent federal standards with little reduction in conversion efficiency in vehicles.

Tests show that methanol is well suited for combustion in boilers and turbines in utility electricity generation. This may be the first large scale application of methanol as a fuel. Methanol's high performance combustion properties make it capable of producing significantly more power at higher efficiencies in high compression internal combustion engines than petroleum fuels in conventional IC engines. This is enhanced further in the stratified charge engine. Methanol's high octane ratings make it unsuitable for direct use in diesel engines, but could be used in a dual fuel mode where a conventional diesel fuel is used to ignite the fuel. A low specific heat value and consequently larger volumes and weights for methanol is an inherent disadvantage for methanol compared to petroleum fuels. This is partially offset by greater efficiencies in operation.

High performance has been accompanied by substantial reductions in emissions since the lower flame temperature of methanol results in much lower NO_x emissions primarily in the lean region. Aldehyde emissions have been noticeably higher. However the magnitude is a very strong function of the fuel vaporization and combustion conditions and correlates closely with unburned fuel emissions. No particulates or sulfur emissions are formed. Corrosion of unprotected parts designed for use with petroleum fuels is a particular problem to be addressed.

Recent technological advances in microcomputers, stratified charge combustion, heat pipes, fuel injection and atomization offer significant potentials in converting to an efficient versatile methanol or multifuel economy, with considerably higher fuel efficiency and lower emissions. Blends of methanol in gasoline also appear to be a possible means of introducing methanol into the market. Corrosion, water free distribution, economics and careful handling procedures in light of its toxicity are the main problems to be overcome. Large scale use of methanol is currently limited by manufacturing capacity and economics, but these should ease as the synthetic fuel industry becomes established. The overall efficiencies and costs from synthesis through are needed to properly evaluate alternative fuels.

Next to hydrogen, methanol appears to be one of the strongest contenders as a fuel for electrochemical oxidation in fuel cells. A large program is underway to develop methanol-air fuel cells, with numerous other smaller research programs continuing. Success of these programs could provide a highly efficient, pollution free means of generating electricity.

Single cell protein is being manufactured using methanol as a feedstock. This could provide a major alternative source of protein in the near future.

The toxicity of methanol will require more stringent handling and labeling procedures. Further precautions must also be taken to prevent explosions and corrosion.

Methanol appears to be one of the more promising alternate synthetic fuels providing high conversion efficiencies, with low pollution under a wide variety of applications.

This paper is adapted from two sections of a more comprehensive review of methanol written as M.S. papers by the author the fall of 1976 at the University of Minnesota. The other sections deal with the synthesis and economics of methanol produced from conventional as well as renewable resources. The entire review was kindly published by ERDA and is available from the National Technical Information Center, Springfield, Virginia as publication # NP-21727.

BIBLIOGRAPHY: FUEL USES AND HAZARDS OF METHANOL

This bibliography is representative of recent publications on methanol and related synthetic fuels and topics primarily through the early part of 1976 with some recent additions. A good introduction to European publications can be found in the bibliography in reference 100. The names and addresses of many researchers active in the field are reproduced in the list of participants attending the Engineering Foundation Conference in 1974 (91) and the Swedish Methanol Development Co. in 1976 (290).

I wish to thank the numerous authors who have provided reprints of their papers from which much of this bibliography was compiled. Information on corrections and omissions would be very welcome as well as future papers pertaining to methanol to update these bibliographies.

REFERENCES

1. H. G. Adelman, R. S. Devoto, and D. G. Andrews, "Reduction of Automobile Exhaust Emissions with Methyl Alcohol as Fuel." Stanford University, Contract EHSH 71-010 for EPA, 1971.

2. H. G. Adelman, G. D. Andrews, and R. S. Devoto (Stanford University, Stanford, CA), "Exhaust Emissions from a Methanol-Fueled Automobile." Soc. Auto. Eng. Paper No. 720693, San Francisco, August 21-24, 1972, 16 pp.

3. H. Adelman, "Characterization and Research Investigation of Methanol and Methyl Fuels in Automobile Engines." Automotive Power Systems contractors coordination meeting, Ann Arbor. ERDA-64, 6 May 1975, pp. 345-346, Energy Res. Abs. 1-51222.

4. R. R. Adt, Jr., R. D. Deopker, and L. E. Poteat, "Methanol-Gasoline Fuels for Automotive Transportation: A Review", prepared for US Environmental Protection Agency, Alternative Automotive Power Systems Division, Ann Arbor, Michigan, November 1974.

5. R. R. Adt, "Methanol-Gasoline Fuels for Automobile Transportation", Auto. Power Systems contractors coordination meeting, Ann Arbor ERDA-64, 6 May 1975, pp. 344-345, Energy Res. Abs. 1-5121.

6. R. R. Adt, Jr., K. A. Chester, J. Pappos, and M. R. Swain, "Methanol-Gasoline Blends: Performance and Emissions," Am. Inst. Chem. Eng. Mtg., Boston, Mass., Sept. 1975.

7. R. R. Adt, Jr., Hebb Greenwell and M. R. Swain, "The Hydrogen and Methanol Fueled-Air Breathing Automobile Engine." Hydrogen Theme Conference #2. Miami, Florida, 1975, 12 pp.

8. J. R. Allsup, "Gas from Coal as an Automotive Fuel." Soc. Auto. Eng. Paper No. 730802, 1973, Bureau of Mines, US Dept. of the Interior.

9. J. R. Allsup and R. D. Fleming (Bartlesville Energy Research Center, Bartlesville, OK), "Emission Characteristics of a Prime Mover for Hybrid Vehicle Use." Bureau of Mines BM-RI-7988, 1974, 37 pp.

10. J. R. Allsup (Bartlesville Energy Research Center) "Methanol Gasoline Blends as Automotive Fuel" for Auto. Eng. Paper No. 750763 September 8-11, 1975, Milwaukee, Wisc.

11. J. R. Allsup, "Methanol/Water/Fuel Composition, Temperature Phase Diagrams", Bartlesville Energy Research Center, ERDA, Bartlesville, OK, Summer, 1976.

12. J. R. Allsup, "Engine Work with Methanol Blends in Multicylinder Vehicles", Bartlesville Energy Research Center, ERDA, OK, Summer 1976.

13. Alsthom: Laboratoire de Recherche et de Developpement, 91-Massy, France. Operation Pile A Combustible Alsthom-Jersey, P.M. Domenjoud, manager. (Now Alsthom-Exxon)

14. American Petroleum Institute, Committee for Air and Water Conservation, Task Force EF-12, "Use of Alcohol in Motor Gasoline-A Review" API Publication No. 4082, August 1971.

15. American Petroleum Institute, Committee on Mobile Source Emissions, Task Force EF-18, "Alcohols: A Technical Assessment of Their Application as Fuels," API No. 4261, July, 1976.

16. Carl J. Anderson, Beverly Berger, Jack Carlson, William Crothers, David Gregg, John

Grens, and Alan Pasternack (Lawrence Livermore Labs) "LLL Contribution to the AEC Methanol Report: Fuel Utilization and Environmental Impact." University of California, Livermore, California, UCID-16442, January 18, 1974, prepared for US AEC, Contract W-7405-Eng-48.

17. A. J. Andreatch and R. Feinland, "Continuous Trace Hydrocarbon Analysis by Flame Ionization." Analytical Chemistry, Vol. 32, (1960) p. 1021.

18. H. A. Ashby, (Environmental Protection Agency, Ann Arbor, MI) "Emissions from the Methanol Fueled Stanford University Gremlim." NTIS PB-218 420, August 1971.

19. G. S. Bahn, "Theoretical Nitric Oxide Production Incidental to Autoignition and Combustion of Several Fuels Homogeneously Dispersed in Air Under Some Typical Hypersonic Flight Conditions." LTV Aerospace Corp., Hampton, Virginia, NASA-CR-2455, 1974.

20. Bernard S. Baker, ed., "Hydrocarbon Fuel Cell Technology, A Symposium." New York, Academic Press, 1965. 150th Nat'l Mtg., Am. Chem. Soc. Div. Fuel Chem. Atlantic City.

21. K. D. Barnes, D. B. Kittelson, and T. E. Murphy (University of Minnesota) "Effect of Alcohols as Supplemental Fuel for Turbocharged Diesel Engines." SAE Paper No. 750469, February 24-28, 1975.

22. L. Baudendistel, H. Bohm, J. Heffler, G. Louis, and F. A. Pohl (A.E.G. Telefunken Frankfurt-Niederrad FRG) "Fuel Cell Battery with Non-Noble Metal Electrodes and acid Electrolyte." 7th Intersoc. Energy Conv. Eng. Conf. Paper No. 729004, 1972, p. 20.

23. P. Bedard, "When is a carburetor not a carburetor?" Car and Driver, December 1974.

24. J. Benson, (General Motors) "What good are Octanes?" ChemTech. Vol. 5, January 1976, pp. 16-22.

25. Beverly J. Berger, "Environmental Aspects of Methanol As Vehicular Fuel. Health and Environmental Effects." Lawrence Livermore Lab UCRL-76076 September 25, 1974, 12 pp., 1974 Engineering Foundation Conference. "Methanol as an Alternate Fuel." New England College, Henniker, NH, July 7-12, 1974.

26. Carl Berger, ed.,"Handbook of Fuel Cell Technology, Prentice-Hall,"New Jersey, 1968, 607 pp.

27. J. E. Berger (Shell Oil Company, Houston, Texas) "Alcohols in Gasoline" Statement to the Subcommittee on Priorities and Economy in Government, Joint Economic Committee, May 22, 1974.

28. Jerry E. Berger (Shell Oil Co.) "Some Fuel for Thought." Ecolibrium Winter, 1974, Vol. 3, No. 4, p. 4 (4).

29. W. E. Bernhardt, "Kinetics of Nitric Oxide Formation in Internal Combustion Engines." Paper C 149/71, Conference on Air Pollution Control in Transport Engines (Inst. Mech. Engr) Solihull, England, November 1971.

30. W. Bernhardt, W. Behrens, P. Heidemeyer, W. Geffers, "Ermittlung Polyzyblischer aromatishcher Kohlenwasserstoffe im Automobilabgas in Abhangigkeit von Motorkonzept und Fahrzustand." VW-Forschungsbericht No. F2-74/75, June 1974.

31. W. E. Bernhardt, W. Lee, (Volkswagenwerk AG) "Combustion of Methyl Alcohol in Spark-Ignition Engines." 15th Int'l Symp. on Combustion, Tokyo, Japan, Paper No. 136, August 15-31, 1974.

32. Winfried E. Bernhardt, (Volkswagenwerk AG)"Engine Performance and Exhaust Emission Characteristics from a Methanol-Fueled Automobile." Paper at 1975 GMR Symposium "Future Automotive Fuels-Prospects, Performance, and Perspective" General Motors Technical Center, Warren, Michigan, October 6-7, 1975.

33. Ing. W. Bernhardt, (Volkswagenwerk AG) "Methanol as an Automotive Fuel-Problems and Expectations," Symp. "Methanol as a Fuel", Royal Swedish Acad. Eng. Sci., Stockholm, March 23, 1976.

34. R. E. Biddick and D. L. Douglas, "An Alkaline Methanol-Air Primary Battery System." Am. Chem. Soc. Div. Fuel Chem. Proc. Vol. 9, No. 3, Pt. 1, April 1965, pp. 113-124.

35. R. Bignell and G. Humphrey, "LNG or LCF" Gas Tech '74, Amsterdam, October 1974.

36. H. Binder and G. Sandstede, "Research Trends in Batteries and Fuel Cells with Regard to Future Energy Supplies" (Ger.) Chemie Ingenieur Technik. Vol. 47, No. 2, pp. 51-55, January 1975.

37. I. Bjerle (University of Lund, Sweden) "Different Processes for the Production of Synthesis Gas," Symp. "Methanol as a Fuel", Royal Swedish Acad. Eng. Sciences, Stockholm, March 23, 1976.

38. J. O. M. Bockris and T. Srinivasan,"Fuel Cells—Their Electrochemistry." McGraw-Hill, 1969.

39. J. O. Bockris and R. A. Fredlein, "Electrochemistry for Ecologists." Plenum Pub., 1973.

40. H. Böhm and K. Maass, (Telefunken AG, Frankfurt am Main, West Germany) "Methanol/Air Acidic Fuel Cell System." 9th Intersoc. Energy Conv. Eng. Conf. San Francisco, California, August 26-30, 1974, Proc. p. 836-840.

41. D. B. Boies and A. Dravnieks, "New High-Performance Methanol Fuel Cell Electrodes." Electrochem. Tech. Vol. 2, 1964, p. 351.

42. J. A. Bolt and D. H. Holkeboer, "Lean Fuel/Air Mixture for High Compression Spark-Ignited Engines." SAE Summer Meeting Paper #380D, 1961.

43. J. A. Bolt, "A Survey of Alcohol as a Motor Fuel," in Soc. Auto. Eng. SP-254, June 1964. "Alcohols and Hydrocarbons as Motor Fuels."

44. R. L. Bradow (Environmental Protection Agency) Statement before the Subcommittee on

Priorities and Economy in Government Joint Economic Committe. May 21, 1974.

45. P. Breisacher and R. Nichols, "Review of Alcohols as a Reciprocating Engine Fuel" The Combustion Institute (Western States Section) 1973 Fall Meeting. October 29-30.

46. P. Breisacher and R. Nichols (Aerospace Corp., El Segundo, CA) "Fuel Modification: Methanol Instead of Lead as the Octane Booster for Gasoline." Central States Section of Combustion Inst., Spring Meeting, Madison, Wisconsin, March 26-27, 1974.

47. N. D. Brinkman, N. E. Gallopoulos and M. W. Jackson (General Motors Research Labs.) "Exhaust Emissions, Fuel Economy, and Driveability of Vehicles Fueled with Alcohol Gasoline Blends." Soc. Auto. Engr. Paper #750120, Detroit, February 24-28, 1975, 27 pp.

48. N. D. Brinkman, (General Motors Research Labs., Warren, Mi.) "Vehicle Evaluation of Neat Methanol-Compromises Among Exhaust Emissions, Fuel Economy and Driveability." NATO/CCMS Int'l Symp. on Automotive Propulsion Systems, April 19, 1977, Arlington, Va.

49. Donald P. Burke, "CW Report: Methanol." Chemical Week, September 24,1975, pp. 32-42.

50. Cameron Engineers, Inc. (Denver Co.) "Synthetic Fuels Data Handbook." T. A. Hendrickson (Compiler), 1975.

51. Grady Stanley Canada, "High Pressure Combustion of Liquid Fuels." PhD Thesis, Pennsylvania State University, 1974, Dissert. Abs. Vol. 35, p. 5907-B, Xerox #75-10,789.

52. Edward J. Canton, S. S. Lestz and W. E. Meyer, (Pennsylvania State University) "Lean Combustion of Methanol-Gasoline Blends in a Single Cyliner SI Engine." Soc. Auto. Eng. Fuels and Lubricants Mtg. TX.Paper #750698, June 3-5, 1975.

53. Edward J. Canton (Pennsylvania State University) "Lean Combustion of Methanol Gasoline Blends in a SI Engine." Center for Air Environmental Studies CAES #394-75, May 1975, 83 pp.

54. Philip E. Cassady, "The Use of Methanol as a Motor Vehicle Fuel." Mathematical Sciences Northwest Inc. Seattle, Wash., MSNW Report #74-243-2 Prepared for the City of Seattle, Dept. of Lighting, Seattle, Washington, September 1974, 111 pp., 43 ref.

55. P. E. Cassady (Math. Sciences N.W. Inc. Bellevue) "The Use of Methanol as a Motor Fuel." 169th Am. Chem. Soc. Nat'l. Mtg., Philadelphia, April 6-11, 1975, 43 ref.

56. R. R. Cecil (ESSO Research and Engineering Co., Linden, New Jersey) "Exxon Experience with Alcohols in Motor Gasoline." Statement to Subcommittee on Priorities and Economy in Government, Joint Economic Committee, May 21, 1974.

57. Chemical Week, "Two Votes for Methanol." Vol. 112, March 14, 1973, p. 73.

58. Chemical Week, "Offshore Petroleum (Houston) Readying Methanol Plant Plans off Gulf Coast." Nov. 12, 1975, p. 9.

59. Chemical Week, "Mobil May Build Gasoline-from-Methanol Plant." May 26, 1976, p. 32.

60. Chemical Week, "Is Alcohol Next Candidate for Fuel Pumps?" Vol. 114, No. 5, January 30, 1974, p. 33.

61. G. Ciprios, "Methanol Fuel Cell Battery." Proc. 20th Am. Power Sources Conf. 1966, pp. 46-49.

62. G. Ciprios, "Methanol-Air Fuel Cell Battery." Intersoc. Energy Conv. Eng. Conf. No. 669001, 1966, pp. 9-14.

63. G. Ciprios, "Recent Developments in Methanol-Air Fuel Cells." Intersoc. Energy Conv. Eng. Conf. No. 679038, 1967, p. 357.

64. C. F. Clark (Stanford Research Institute, CA) "Transportation and Distribution of Methanol" Symp. "Methanol as a Fuel." Royal Swedish Acad. of Eng. Sciences, Stockholm, March 23, 1976.

65. G. Cohn (Englehard Industries Inc., Newark, N.J.) "Direct Hydrocarbon Fuel Cell." Proc. 17th Ann. Power Sources Conf. May 21-23, 1963, p. 96. Contract No. DA-36-039 SC-90691.

66. J. M. Colucci (General Motors Res. Labs) "Methanol/Gasoline Blends-Automotive Manufactures Viewpoint." Paper @ 1974 Engineering Foundation Conference, "Methanol as an Alternate Fuel," Henniker, N.H., July 7-12, 1974.

67. E. S. Corner and A. R. Cuningham, "Value of High Octane Number Unleaded Gasolines in the U.S." Am. Chem. Soc. Mtg. Los Angeles, CA, Div. of Water, Air and Waste Chemistry, March 1971.

68. W. T. Crothers and C. J. Anderson, "A Practical Approach to the Introduction of Alternate Automotive Fuels." Calif. Univ., Lawrence Livermore Lab. UCRL-51779, March 14, 1975, 23 pp.

69. W. T. Crothers (Lawrence Livermore Labs) "The Use of Methanol in Transportation." UCID-16528 Rev. 1, February 3, 1975 (original July 1, 1974) USAEC Contract W-7405-Eng-48, 35 pp.

70. J. C. Davis, "Can Methanol Fuel Contend." Chem. Eng., June 25, 1975, pp. 48-50.

71. R. R. Davison and W. D. Harris, "Methyl Alcohol as Motor Fuel." Texas Engineering Expt. Sta. Tech. Bull. 74-2, April 1974.

72. Harry Davitian (Cornell University) "Energy Carriers in Space Conditioning and Automotive Applications: A Comparison of Hydrogen, Methane, Methanol, and Electricity." 9th Intersoc. Energy Conv. Eng. Conf. Paper No. 749037, August 26, 1974.

73. Davy Powergas Ltd. London, "Methanol."

74. G. V. Day, "The Prospects for Synthetic Fuels in the U.S." IEEE (UK) Int. Conf. "Energy, Europe, and the 1980's," 1975.

75. R. C. Deskin and L. Conforti (E. I. DuPont de Nemours and Co., Wilmington, De.) "Fluid Resistance of Viton." DuPont Viton Bulletin No. 15, July 1965.

76. E. Dickson and E. E. Hughes, "Impacts of Synthetic Liquid Fuel Development for the Automotive Market." Automotive Power Systems Contractors Coordination Meeting, ERDA-64, May 6, 1975, pp. 369-385. Energy Res. Abs. 1-5124.

77. W. A. Dietz, "Response Factors for Gas Chromatographic Analysis." J. of Gas. Chromatography, Vol. 5, 1967, p. 68.

78. B. Dimitriades and T. C. Wesson, "Reactivities of Exhaust Aldehydes." Bureau of Mines R.I. 7527, May 1971.

79. Andrew Drawnieks and D. B. Bois, "Methanol Fuel Cells with Dissolved Oxidents." Am. Chem. Soc. Div. Fuel Chem., Vol. 7, No. 4, pp. 223-233, September 1963.

81. Dennis H. Eastland (Davy Powergas, Inc.) "Fuel and Energy Uses of Methanol." Paper presented to the University of Pittsburgh School of Engineering; 2nd Ann. Symp. Coal Gasification and Liquefication, August 5, 1975.

82. A. G. Dixon, A. C. Houston, and J. K. Johnson (Shell Research Ltd.England) "An Automatic Generator for the Production of Pure Hydrogen from Methanol." 7th Intersoc. Energy Conv. Eng. Conf. Paper No. 729161, 1972, pp.1084-90.

83. R. W. Duhl and T. O. Wentworth (Vulcan-Cincinnati, Inc.) "Methyl Fuel from Remote Gas Sources." Am. Instit. Chem. Eng. Soc. Calif. Section 11th Annual Mtg., April 16, 1974, Los Angeles.

84. R. W. Duhl (Vulcan Cincinnati, Inc.) "Methanol, A Boiler Fuel Alternative." Am. Inst. Chem. Eng., 8th Annual Mtg., Boston, Mass., Sept. 7-10, 1975.

85. R. W. Duhl, "Methanol as a Boiler Fuel." submitted for publication, Chem. Eng. Prog. February 1976.

86. R. W. Duhl (Vulcan Cincinnati, Inc.) and J. W. Boylan (A. M. Kinney, Inc.) "Use of Methanol as a Boiler Fuel." IV A - Symposium Swedish Academy of Engineering Sciences, Stockholm, Sweden, March 23, 1976.

87. G. D. Ebersole, "Power Fuel Consumption, and Exhaust Emission Characteristics of an Internal Combustion Engine Using Isooctane and Methanol." Ph.D. Thesis, The University of Tulsa, OK, 1971.

88. G. D. Ebersole (Phillips Petroleum Co.) and F. S. Manning (U. of Tulsa) "Engine Performance and Exhaust Emissions: Methanol versus Isooctane." Soc. Auto. Eng. Nat'l. West Coast Mtg., San Francisco, Calif., Soc. Auto. Eng. Paper 720692, August 21-24, 1972.

89. G. Egloff, "Motor Fuel Economy of Europe." Ind. and Eng. Chem., Vol. 30, No. 10, 1938, pp. 1091-1104. 24 ref.

90. Energy and Pipelines and Systems, "Methanol as Gas Substitutes." June 1974, Vol. 1, No. 6, pp. 55-57.

91. Engineering Foundation, New York, N.Y. 1974 Engineering Foundation Conference on "Methanol as an Alternate Fuel.""Vol. I--Conference Report," 70 pp., "Vol. II--Reprints of Papers," 450 pp., New England College, Henniker, N.H., July 7-12, 1974.

92. Environmental Protection Agency, Washington, D.C., "Fact Sheet--Use of Alcohol as a Motor Fuel." FS-12b, EPA/OMSAPC/May 1974.

93. Environmental Protection Agency, " Proc. Symposium-Workshop on Alternative Fuels." Ann Arbor, Mich., October 15, 1974, 368 pp.

94. Environmental Protection Agency, "Proc. of the Solvent Reactivity Conference." Ecological Research Series, EPA-65013-74-010, November 1974, 34 pp., PB 238-296/AS.

95. N. D. Esau (Amoco Oil Co., Whiting, In.) "Automotive Fuel System Deterioration Effects Due to the Use of Fuel Composed of 90% Amoco Unleaded plus 10% Ethanol or Methanol." Internal Communication, June 19, 1974.

96. Exxon Enterprises, Inc., New York, N.Y., Alsthom-Exxon Joint Fuel Cell Activity. Attn. C. E. Heath, Chairman,or Exxon Research and Engineering Co., Corporate and Government Research, Linden, N.J., 07036, Attn. John M. Longo (Formerly Jersey Enterprises, N.Y. Div. of Standard Oil Col. of N.J.).

97. Exxon Research Co., Attn. Eugene Elzinga, "Applications of Alsthom-Exxon Alkaline Fuel Cells to Utility Power Generation." EPRI Project 584-1, November 1976.

98. L. S. Ettre and H. N. Claudy, "Hydrogen Flame Ionization Detector." Chemistry in Canada." Vol. 34, September 1960.

99. E. Faltermayer, "The Clean Synthetic Fuel that's Already Here." Fortune, Vol. XCII, No. 3, September 1975, pp. 147-154.

100. Federal Ministry for Research and Technology, Bonn. "On the Trail of New Fuels--Alternative Fuels for Motor Vehicles." 578 pp. Translated (May-June 1975) from the German "Neun Kraftstoffen auf der Spur-Alternative Kraftstoffe für Kraftfahrzeuge." Bundesministerium für Forschung und Technologie, Bonn, 1974, pp. 1-282, Gersbach & Sohn Verlag, Munchen, UCRL-Trans-10879; 2974, 395 pp., 155 ref. ERDA Res. Abs. 1 (1), 0412, 1976.

101. Arnold Fickett, "Fuel Cells: Versatile Power Generators." EPRI Journal, April 1976, pp. 14-19.

102. R. A. Findlay (Phillips Petroleum Co., Bartlesville, OK) "Methanol as Engine Fuel." Fi-141-74, July 19, 1974.

103. R. E. Fitch and J. D. Kilgroe (Consolidated Engineering Tech. Corp. Mountain View, CA) "Investigation of a Substitute Fuel to Control Automotive Air Pollution." Final Report CETEC-Report No. 01800-FR, February 1970, for the National Air Pollution Control Agency NTIS, No. PB 194 688.

104. R. D. Fleming and T. W. Chamberlain, "Methanol as Automotive Fuel; Part I: Straight Methanol." Soc. Auto. Eng., Paper No. 750121, February 24-28, 1975.

105. A. A. Frank (University of Wisconsin) "Fuel and Emission Characteristics of a Flywheel-Heat Engine Vehicle." 1975 Flywheel Tech. Symp., Univ. of Calif., Berkeley, Calif., Nov. 10, 1975, NTIS No. CONF-75113 Contract DOT-05-30112.

106. R. Ganeshan (Selas of America, Nederland) "Methanol as Fuel-Cheaper than LNG." Oil and Gas J., July 24, 1972, pp. 61-62.

107. David Garrett and T. O. Wentworth, "Methyl-Fuel--A New Clean Source of Energy." Am. Chem. Soc. Div. Fuel Chem., 1973 Ann. Mtg., Paper No. 9, August 27, 1973.

108. The German Tribune, "VW and Mercedes run Smoothly on Meths." No. 623-21, March 1974, p. 9, Transl. from Der Tagesspiegel, February 24, 1974.

109. Edward A. Gillis (U.S. Army Mobility Equipment. R. & D Center) "1.5 KW Open Cycle Hydrocarbon-Air Fuel Cell." 7th Intersoc. Energy Conv. Eng. Conf. Paper No. 729165, 1972.

110. E. A. Gillis (U.S. Army, Ft. Belvior) "Methanol Fuel for Fuels Cells." 1974 Eng. Foundation Conf., "Methanol as an Alternate Fuel." Henniker, N.H., July 7-12, 1974.

111. J. C. Gillis et al (Institute of Gas Technology, Chicago, IL) "Synthetic Fuels for Automotive Transportation." presented at Spring Meeting of Combustion Inst., Madison, Wis., March 26, 1974.

112. J. C. Gillis, J. B. Pangorn, and K. C. Vyas, "The Technical and Economic Feasibility of Some Alternative Fuels for Automotive Transportation." Institute of Gas Technology, Chicago, IL., 19th Intersoc. Energy Conv. Eng. Conf., Newark, N.J., Paper No. 759128, August 22, 1975, pp. 856-862,

113. S. R. Gollahalli and T. A. Brzustowski (U. of Waterloo, Ontario) "Flame Characteristics in the Wake of a Burning Methanol Drop." Combustion and Flame, Vol. 24, 1975, pp. 273-275.

114. C. H. Gonnermann, J. S. Moore, and P. W. McCallum (Mueller Assoc., Inc., Baltimore, Md.) "Fueling Automotive Internal Combustion Engines with Methanol--Historical Development and Current State of the Art." 10th Intersoc. Energy Conv. Eng. Conf., Newark, Paper No. 759127, August 22, 1975, pp. 849-855, Contract E.P.A. No. 68-03-0388.

115. Bill Green, California Legislature Assembly Bill No. 3255 supported by document: "Methanol-Gasoline Blends, A Summary of Experiences, University of Santa Clara, 1968-1974."

116. Bill Green and Keene, Assembly Bill No. 443, California Legislature, 1975-1976 Regular Session. January 9, 1975 (re. adding methanol to gasoline).

117. D. P. Gregory and R. F. Defour, "Utilization of Synthetic Fuels Other than Hydrogen." Inst. Gas Tech. Project 8936 for Oak Ridge Nat'l Lab, May 1972.

118. D. P. Gregory and R. Rosenberg (Institute of Gas Technology, Chicago, IL) "Synthetic Fuels for Transportation and National Energy Needs." Soc. Auto. Eng. Pub. SP-383, "Energy and the Automobile." July 1973, pp. 37-45.

119. P. G. Grimes, B. Fiedler, and J. Adam (Allis Chalmers Mfg. Co., Milwaukee, Wis.) "Preliminary Study of the Alkaline Methanol-Hydrogen Peroxide Fuel Cell." 13 pp.

120. Th. Gugenberger, M. Jung and A. Winsel, "Hydrazine-Methanol Catalysts." Industrial Colloquium, Battele -Frankfurt, 1971.

121. G. Hagey and A. J. Parker, Jr., "Technical and Economic Criteria for the Selection of Alternative Fuels for Personal Automotive Transportation." 9th Intersoc. Energy Conv. Eng. Conf., San Francisco, CA, August 26, 1974.

122. G. Hagey and A. J. Parker, Jr., (Environmental Protection Agency and R. P. Mueller & Assoc.) "Status and Summary of In-Progress Research Activities on Alternative Fuels." 2nd Symp. on Low Pollution Power Systems Development, NATO/CCMS, Dusseldorf, Germany, November 4-8, 1974, Conf-741151, pp. 356-367.

123. William Hampton and Nicholas Iammartino, "Will Autos go Alcoholic?" Chemical Engineering, Vol. 82, No. 15, July 21, 1975, pp. 58-61.

124. Philip Handler (National Academy of Sciences, EPA, Washington, D.C.) "Report by the Committee on Motor Vehicle Emissions." February 15, 1973.

125. John L. Harned (General Motors Corp.) "Heat Pipe Early Fuel Evaporation." Soc. Auto. Eng., Paper No. 760565, St. Louis, MO., June 7-10, 1976.

126. J. A. Harrington and R. M. Pilot, "Combustion and Emission Characteristics of Methanol." Soc. Auto. Eng., Paper 750420, February 1975.

127. G. A. Harrow, W. D. Mills, A. Thomas, and I. C. Finlay (Thorton Research Center. Shell Research LTD., U. K. and Nat'l. Eng. Lab., Scotland) "The Vapine--A Practical System for Producing Homogeneous Gasoline-Air Mixtures." Soc. Auto. Eng., Paper No. 760564, St. Louis, MO., June 7-10, 1976.

128. C. E. Heath (Esso R & E Co.) "The Methanol-Air Fuel Cell." Proc. 17th An. Power Sources Conf., 1963, pp. 96-97.

129. C. E. Heath (Esso R & E Co.) "Methanol Fuel Cells." Proc. 18th An. Power Sources Conf., May 19-21, 1964, pp. 33-36.

130. C. E. Heath, "Fuel Cells." Science, Vol. 180, No. 4086, Letter, pp. 543-544, May 11, 1973.

131. Carl Heinz, Harold Bohm, and Franz A. Poh (NASA Washington, D.C.) "Cost Effectiveness of Tungsten Carbine-Carbon Fuel Cells." Translation into English from Wiss. Ber. AEG-Telefunken FRG, Vol. 46, No. 3-4, 1973, pp. 109-116, by Scientific Translation Serv., Santa Barbara, NASA-TI-F-15748, July 1974, 31 pp., NTIS, No. N74-28543.

132. H. Heitland, "Energy Workshop-Report on the Use of Methanol in Volkswagons." Eng. Foundations Conf., "Methanol as an Alternate Fuel." Henniker, June 1974.

133. Herbert Heitland, Winfried Bernhardt, and Lee Wenpo (Volkswagenwerk AG) "Comparative Results on Methanol and Gasoline Fueled Passenger Cars." 2nd Symposium on Low Pollution Power Systems Developments (NATO/CCMS) November 4-8, 1974, Dusseldorf, Germany, pp. 387-395.

134. J. Hellbach and W. Bernhardt, "Mögliche Alternativ-Kraftstoffe fur Verbrennungsmotoren."

Volkswagen Research Report F2-74/3, February 1974.

135. Samuel Sanders Hetrick, "The Effects of Oxy-Hydrocarbon Fuel on Exhaust from Spark Ignition Engines." The Pennsylvania State University, College of Engineering, Ph.D. Thesis, PRL-5-67, 1967.

136. D. L. Hilden and F. B. Parks (General Motors) "A Single Cylinder Engine Study of Methanol-Emphasis on Organic Emissions." Soc. Auto. Eng. No. 760378, February 23-27, 1976, Detroit, Mich., Gen. Mot. Res. Pub. GMR-2072 R.F & L-596.

137. O. Hirao (U. of Tokyo, Japan) "Studies of Methanol as a Fuel for Automobile Engines in Japan." Symp. "Methanol as a Fuel," Royal Swedish Acad. Eng. Sci., Stockholm, March 23, 1976.

138. J. P. Hoare, "Oxygen Overvoltage Measurements on Bright Platinum in Acid Solutions, IV Methanol Solutions." J. Elect. Soc. Vol. 113, 1966, pp. 846-851.

139. J. A. Hoess and R. C. Stahman, "Unconventional Thermal, Mechanical, and Nuclear Low-Pollution-Potential Power Sources for Urban Vehicles." Soc. Auto. Eng., Paper No. 690231, January 13-17, 1969, Detroit.

140. Phillip Ronald Hooker, "Transportation of Crude Oil and Natural Gas from the Arctic as a Cold Dispersion of Oil in Methanol." Stanford University, Petroleum Engineering, Ph.D. Thesis, 1975, Diss. Abs. Vol. 35, No. 5, p. 2429-B, Xerox No. 75-25, 546; 150 pp.

141. M. Hubbard, J. J. Bonilla, K. W. Randall, and J. D. Powell (U. of Calif., Davis and Stanford University) "Closed Loop Control of Lean Fuel-Air Ratios Using a Temperature Compensated Zirconia Oxygen Sensor." Soc. Auto. Eng. Paper No. 760287, Detroit, February 23-27, 1976.

142. R. W. Hurn and T. W. Chamberlain (Bartlesville Energy Research Center) "Fuels and Emissions--Update and Outlook, 1974." Soc. Auto. Eng., Paper No. 740694, Milwaukee, September 9-12, 1974.

143. R. W. Hurn (Bureau of Mines, Bartlesville, OK) "Alternative Fuels--Methanol." 2nd NATO-CCMS Symp. Low Pollution Power Systems Development, Dusseldorf, Germany, November 4-9, 1974, pp. 375-387.

144. R. Hurn, "Experimental Work with Methanol/Gasoline Blends: A Status Report." Automotive Power Systems Contractors coordination meeting, ERDA-64, May 6, 1975, pp. 356-368, ERDA Energy Res. Abs. 1-5123.

145. R. W. Hurn, J. R. Allsup, and B. H. Eccleston, "Characteristics of Methanol as Internal Combustion Fuel." 10th Intersoc. Energy Conv. Eng. Conf., Newark, Delaware, August 17-22, 1975.

146. Hydrocarbon Processing, "Report Complete on 'Methyl Fuel' Merits." April 1973, p. 17. (Re: Methanol in Industrial Boiler).

147. Nicholas R. Iammartino, "Fuel Cells Fact and Fiction." Chemical Engineering, Vol. 81, No. 11, May 27, 1974, pp. 62, 64.

148. Imperial Chemical Industries Ltd., England, "New Protein." August 1973, 20 pp.

149. J. C. Ingamells, "Fuel Economy and Cold Start Driveability with some Recent-Model Cars." Soc. Auto. Eng., Fuels and Lubricants Meeting, Paper 740522, Chicago, June 17-21, 1974.

150. J. C. Ingamells and R. H. Lindquist, (Chevron Research Laboratory, Richmond, Cal.) "Methanol as a Motor Fuel." Chemtech, 1974.

151. J. C. Ingamells and R. H. Lindquist, "Methanol as a Motor Fuel." 1974 Engineering Foundation Conference "Methanol as an Alternate Fuel," Henniker, New Hampshire, July 1974.

152. J. C. Ingamells and R. H. Lindquist, "Methanol as a Motor Fuel or a Gasoline Blending Component." Soc. Auto. Eng. Paper No. 750 123, Detroit, February 24-28, 1975, Chevron Research Co., Richmond, Calif.

153. N. Iwai, T. Tsuruga, S. Kobayashi, and H. Sudo (Japanese Automobile Research Inst., Tokyo, Japan) "Automotive Engine for Methanol-Water Mixture." 2nd NATO-CCMS Symp. on Low Pollution Power Systems Development, Dusseldorf, FRG, November 4-8, 1974, pp. 395-403.

154. F. Jaarsma, "Impact of Future Fuels on Military Aero-Engines." National Aerospace Lab., Amsterdam. Advisory Group for Aerospace Research and Dev.; Neuilly Sur Seine, France, 1974, Ann. Mtg., Paris, France, September 26, 1974.

155. R. G. Jackson and R. M. Tillman (Continental Oil Co., Ponca City, OK) "Automotive Uses of Methanol Fuel." 1974 Eng. Foundation Conf. "Methanol as an Alternate Fuel," Henniker, N.H., July 7-12, 1974.

156. R. G. Jackson (Continental Oil Co.) "Role of Methanol as a Clean Fuel." Soc. Auto. Eng. Paper No. 740642 @ Mid-Continent Sect. Mtg., October 27, 1973.

157. R. G. Jackson and R. M. Tillman, "Potential for Methanol as an Automotive Fuel." Soc. Auto. Eng. Paper No. 750118.

158. P. M. Jarvis (General Electric Co., Gas Turbine Products Div.) 1974 Eng. Foundation Conf. "Methanol as an Alternative Fuel." July 8, 1974.

159. J. E. Johnson, "The Storage and Transportation of Synthetic Fuels: A Report to the Synthetic Fuels Panel." Oak Ridge Nat'l. Lab. ORNL-TM-4307, September 1972.

160. R. T. Johnson (U. of Missouri, Rolla) "Energy and Synthetic Fuels for Transportation: A Summary." Soc. Auto. Eng. Paper No. 740599, August 12-16, 1974, 55 ref.

161. R. T. Johnson (U. of Missouri, Rolla) "Evaluation of Methyl Alcohol as a Vehicle Fuel Extender." Dept. Transportation Contract DOT-OS-40104, August 1975, 166 pp., NTIS No. PB-251108.

162. R. T. Johnson and R. K. Riley (U. of Missouri, Rolla) "Single Cylinder Spark Ignition Engine Study of the Octane, Emissions, and Fuel Economy Characteristics of Methanol-Gasoline-Blends." Soc. Auto. Eng., Detroit Congress, SAE Paper No. 760377, February 23-27, 1976.

163. R. T. Johnson, R. K. Riley, and M. D. Dalen (U. of Missouri, Rolla) "Performance of Methanol-Gasoline Blends in a Stratified Charge Engine Vehicle." Soc. Auto. Eng., Paper No. 760546, @ St. Louis, MO, June 7-10, 1976.

164. R. W. Johnston, J. G. Neuman, and D. Agarwal (General Motors Corp.) "Programmable Energy Ignition System for Engine Optimization." Soc. Auto. Eng., Paper No. 750348, Detroit, February 24-28, 1975.

165. F. H. Kant (Exxon Res. & Eng. Co.) "Feasibility Study of Alternative Automotive Fuels." Alternative Automotive Power Systems Coordination Meeting, Ann Arbor, Michigan, May 13-16, 1974.

166. F. H. Kant (Exxon Res. & Eng. Co.) "Feasibility of Alternative Automotive Fuels." 1974 Eng. Foundation Conf., "Methanol as an Alternate Fuel," Henniker, N.H., July 7-12, 1974.

167. F. H. Kant, R. P. Cahn, A. R. Cunningham, M. H. Farmer, and W. Herbst, "Feasibility Study of Alternative Fuels for Automotive Transportation. Vol. I, Executive Summary; Vol. II, Technical Section; Vol. III, Appendices." by Exxon Research and Engineering Co., Linden, N.J., for Environ. Protect. Agency.EPA 460/3-74-009-A, Band C., June 1974, Contract No. 68-01-2112, NTIS No. PB-235582.

168. Kapiloff, Assembly Bill No. 662, Calif. Legisl., 1975-76 session, February 3, 1975 (re adding Methanol to Gasoline).

169. F. L. Kester, "On-Board Hydrogen Storage Method by Methanol-Steam Reforming Method." Automotive Power Systems Contractors Coordination Meeting, Ann Arbor, ERDA-64, May 6, 1975, pp. 323-328, ERDA Energy Res. Abs. 1-5118.

170. Thomas F. Kirkwood and Allen D. Lee, "A Generalized Model for Comparing Automobile Design Approaches to Improved Fuel Economy." RAND Corp. Santa Monica, Calif., January 1975, R-15 62-NSF, 142 pp., NSF/RA/N-75-046, NTIS No. PB-244 385/1PSK.

171. R. D. Klapatch, "Gas Turbine Emissions and Performance on Methanol Fuel," ASME-IEEE Joint Power Generation Conf., Portland, Oregon, September 28-October 1, 1975.

172. W. R. Knox (Monsanto Co.) "Methanol Fuel-Long-Range Implication for Petrochemicals." 1974 Eng. Foundation Conf. "Methanol as an Alternate Fuel," Henniker, N.H., July 7-12, 1974.

173. A. Koening, W. Lee, and W. Bernhardt, (Volkswagenwerk,AG, Ger.) "Technical and Economical Aspects of Methanol as an Automotive Fuel." Soc. Auto. Eng., Paper No. 760545, 1976.

174. Stanley S. Kurpit, "1.5 and 3 KW Indirect Methanol-Air Fuel Cell Power Plants." 10th Intersoc. Energy Conv. Eng. Conf. Paper No. 759036, 1975.

175. D. E. Lapedes, M. G. Hinton, and J. Meltzer (Aerospace Corp, El Segundo, Calif.) "Current Status of Alternative Automotive Power Systems and Fuels. Vol. I. Executive Summary; Vol. II. Alt. Auto Engines; Vol. III. Alternative Nonpetroleum-Based Fuels; Vol. IV. Electric and Hybrid Power Systems,"prepared by the Aerospace Corporation for the Environmental Protection Agency. Alternative Automotive Power Systems Divs., Aerospace Report No. ATR-74 (7325) -1, Vol. I-IV EPA No. EPA-460/3-74-013-a (V.1), 51 pp.; -b (V.2), 463 pp.; -c (V/3), 380 pp., July 1974, Contract No. 68-01-0417.

176. C. W. LaPointe and W. L. Schultz (Ford Motor Co.) "Comparison of Emission Indexes within a Turbine Combustor operated on Diesel Fuel or Methanol." Soc. Auto. Eng. Paper No. 730669 presented at National Powerplant Meeting, Chicago, June 18-22, 1973.

177. Wenpo Lee and Winfried Geffers (Volkswagenwerk AG) "Engine Performance and Exhaust Emission Characteristics of Spark-Ignition Engines Burning Methanol and Methanol-Gasoline Mixtures." Am. Inst. Ch. Eng. 90th Mtg., Boston, Paper No. 31d, September 9, 1975.

178. Richard Lewis, "Methanol and Ethanol-Short History, Current Production, Future and Available Literature." Center for Studies of the Physical Environment, Univ. of Minnesota, February 27, 1974.

179. H. J. Liebhafsky and E. Cairns, "Fuel Cells and Fuel Batteries." John Wiley and Sons, New York, N.Y., 1968, p. 452.

180. J. W. Lincoln, "Methanol and Other Ways Around the Gas Pump." Garden Way Press, Vermont, May 27, 1976.

181. R. H. Lindquist and J. C. Ingamells (Chevron Research Co., San Francisco) "Methanol from a Fuel Suppliers Viewpoint." 1974 Eng. Foundation Conf., "Methanol as an Alternate Fuel," Henniker, N.H., July 7-12, 1974.

182. Marine Engineering/Log, "Clean Energy by Conventional Ship." Vol. 78, No. 10, September 1973, pp. 112-118.

183. Mathematical Sciences Northwest, Inc., Seattle, Wa.,"Feasibility Study: Conversion of Solid Waste to Methanol or Ammonia." MSNW 74-243-1, Sect. IV "Technical Assessment of Conversion Product Uses," September 6, 1974, 109 pp., 51 ref., prepared for the City of Seattle, Dept. of Lighting, Seattle, Washington.

184. J. M. Matsen, and D. G. Levine (Esso R & E, Co., Linden, N.H.) "Cathode Performance in the Methanol-Air-Fuel Cell." J. Electrochemical Tech., Vol. 5, No. 5-6, May-June 1967, p. 266.

185. John P. McCullough (Mobil Chemical Co.) "Converting Coal into Gasoline." First Burton W. Logue Memorial Lecture, Am. Chem. Soc., Tulsa, OK, May 20, 1976.

186. W. J. McLean (Cornell Univ., Ithaca, N.Y.) "Alternative Automotive Fuels: Some Prospects and Problems." Joint Tech. Mtg. Combustion Inst., San Antonio, TX, Conf-750458-11, April 21, 1975.

187. S. L. Meisel, J. P. McCullough, C. H. Lechthaler, and P. B. Weisz, "Gasoline from Methanol in One Step." Chemtech. February 1976, Vol. 6, pp. 86-89.

188. Louis Meites, "Electrochemical Data." Vol. I, Part IA, Wiley, New York, 1974.

189. R. H. Meyer (Northeast Utilities) "Methanol for Utility Use." 1974 Engineering Foundation Conf. "Methanol as an Alternate Fuel." Henniker, N.H., July 7-12, 1974.

190. Ronald G. Minet, "Synthetic Fuel. Technological Solution to Air Pollution Problems." Chem. Eng. World., Vol. 9, No. 4, 1974, pp. 63-68.

191. John A. Monick, "Alcohols: Their Chemistry, Properties, and Manufacture." Reinhold Book Co., New York, 1968.

192. Carle C. Morrill, "Fuel Cell for Improved Electrical Power Supply." Pratt and Whitney Aircraft, E. Hartford, Conn., 1973, 6 pp.

193. W. J. Most and J. P. Longwell, (Exxon Res. & Eng. Co.) "Single-Cylinder Engine Evaluation of Methanol--Improved Energy Economy and Reduced NO_x." Soc. Auto. Eng., Paper No. 750119, February 24-28, 1975.

194. W. J. Most, "Methanol and Methanol-Gasoline Blends as Automotive Fuels." at "Combustion of Alternate Fuels and Combustion of Coals," Combustion Institute, Central States Section, April 5, 1976, Columbus, Ohio.

195. A. W. Nash and D. A. Howes, "Principles of Motor Fuel Preparation and Applications." Vol. I, 2nd Ed., John W. Wiley and Sons, N.Y., 1938, pp. 461-561, 200 ref.

196. National Technical Information Service, "Fuel Cells." NTI Search, No. COM-74-11533 (270 abstracts covers 1969-September, 1974).

197. S. J. Neustadtl, "The Methanol Alternative Now . . . or Never?" Technology Review, March/April 1974, p. 61.

198. J. S. Ninomiya, A. Golovoy, and S. S. Labana (Ford Motor Co., Dearborn, MI) "Effect of Methanol on Exhaust Composition of a Fuel Containing Toluene, n-Heptane, and Isooctane." Air Pollution Control Assoc., Vol. 20, No. 5, May 1970.

199. Oil & Gas J., "HNG Finds Ready U.S. Market for its Saudi Methyl Fuel." Oct. 22, 1973, p. 36.

200. E. H. Okrent and B. L. Tarmy (Esso R & E Co.) "Methanol-Air Fuel Cell." Chem. Eng. Prog., Vol. 62, No. 5, May 1977, pp. 83-84.

201. N. J. Palmer, B. Lieberman, and M. A. Vertes, "A Comparison Between External and Internal Reforming Methanol Fuel Cell Systems." Am. Chem. Soc., Div. Fuel Chem., Vol. 9, No. 3, pt. 1, pp. 135-154, April 1965.

202. J. Pangorn and J. Gillis (Institute of Gas Technology, Chicago, IL) "Alternative Fuels for Automotive Transportation--A Feasibility Study, Vol. I - Executive Summary; Vol. II - Technical Section; Vol. III - Appendices," July 1974 for the Environmental Protection Agency, EPA-460/3-74-012-a, b, and c, 26 pp., 265 pp., and 105 pp.

203. A. Pasternak, "Methyl Alcohol: A Potential Fuel for Transportation." Lawrence Livermore Lab. Rpt., UCRL-76293, 1974.

204. T. H. Paulsen, "Methyl-Fuel Project Serves as Attractive Petro-chemical Base." Oil & Gas J., Vol. 71, No. 40, Oct. 1, 1973, pp. 68-69.

205. R. K. Pefley, M. A. Saad, M. A. Sweeney, J. D. Kilgroe, and R. E. Fitch (Univ. of Santa Clara, CA) "Study of Decomposed Methanol as a Low Emission Fuel--Final Report." Office of Air Programs, Environmental Protection Agency, NTIS No. PB-202 732, April 30, 1971. Contract EHS-70-118.

206. R. K. Pefley, M. A. Saad, M. A. Sweeney, and J. D. Kilgroe (Univ. of Santa Clara, CA) "Performance and Emission Characteristics Using Blends of Methanol and Dissociated Methanol as an Automotive Fuel." 6th Intersoc. Energy Conv. Eng. Conf., SAE Paper No. 719008, Aug, 3-5, 1971, Proc., pp. 36-46.

207. R. K. Pefley (Univ. of Santa Clara, CA) "Methanol/Gasoline Blends--University Viewpoint." Paper @ 1974 Engineering Foundation Conference, "Methanol as an Alternate Fuel," Henniker, N.H., July 7-12, 1974.

208. John Perry, Jr., "Lower Power Methanol-Air Battery." D.A. Project 1S7-62705-AH-94,ECOM-4213, April 1974, NTIS No. AD-779183, 24 pp.

209. G. Persson (Nat'l Environmental Protection Board, Sweden) "Environmental Consequences." Symp. "Methanol as a Fuel," Royal Swedish Acad. Eng. Sci., Stockholm, March 23, 1976.

210. O. A. Petry, B. I. Podlovchenko, A. N. Frumkin, and Hira Lal, "The Behavior of Platinized-Platinum and Platinum-Ruthenium Electrodes in Methanol Solutions." J. Electroanal. Chem., Vol. 10, 1965, p. 253.

211. F. Pischinger, "Methanol as a Fuel for Vehicle Engines." Ver. DTSCH, Ing. Ber., Vol. 224, 1974, pp. 59-66 (German).

212. E. Plassmann (Technischer Überwachungs-Verein Rheinland, e.V., Köln FRG) "Possible Alternate Fuels for Internal Combustion Engines with Regard to LPG and LNG." 2nd NATO-CCMS Symp. on Low Pollution Power Systems Development, Dusseldorf, FRG Nov. 4-8, 1974, pp. 366-375.

213. William H. Podolny, "Fuel Cell Powerplants for Rural Electrification." United Aircraft Corp., Pratt & Whitney Aircraft Div., Dec. 1971, 8 pp.

214. R. I. Pollack, "Environmental Aspects of Methanol as Vehicular Fuel: Air Quality Effects." Lawrence Livermore Labs, UCRL-76064, CONF-740727-1, @ 1974 Engineering Foundation Conference "Methanol as an Alternate Fuel," Henniker, N.H., July 8, 1974.

215. D. Pouli and J. R. Huff, "The Anodic Oxidation of Methanol on Platinum in Alkaline Solution." Am. Chem. Soc. Div. Fuel Chem., Vol. 9, No. 3, Pt. 1, April 1965, pp. 89-102, or Hydrocarbon Fuel Cell Technology, Academic Press, N.Y., 1965, pp. 103-119.

216. T. Powell (Hofstra University) "Racing Experiences with Methanol and Ethanol-Based Motor-Fuel Blends." SAE No. 750124, Feb. 24-28, 1975 @ Detroit.

217. Pratt & Whitney Aircraft, "Advanced Technology Fuel Cell Program-Interim Report." Jan. 1974, EPRI 114.

218. A. A. Quader, "Lean Combustion and the Misfire Limit in Spark Ignition Engines." Soc. Auto. Eng., International Automobile and Manufacturing Meeting, Toronto, Paper No. 741055, Oct. 1974.

219. S. L. Quick and G. D. Kittredge. (Nat'l Air Pollution Control Admin., Ann Arbor, MI) "Control of Vehicular Air Pollution through Modifications to Conventional Power Plants and Their Fuels." Proc. 2nd Clean Air Cong., NAPCA, 1970, pp. 631-639.

220. T. B. Reed and R. M. Lerner (Lincoln Laboratory, MIT, Lexington, Mass.) "Methanol: A Versatile Fuel for Immediate Use." Science, Vol. 182, Dec. 28, 1973, pp. 1299-1304.

221. T. B. Reed and R. M. Lerner (Lincoln Lab., MIT, Lexington, MA): Methanol Information Sheet, No. 1, January 1974; Methanol Information Sheet, No. 2, March 1975; Methanol Information Sheet, No. 3, April 7, 1975; Methanol Information Sheet, No. 4, January 1976.

222. Thomas B. Reed, "Methanol for Fuel: A Bibliography on the Production and Use of Alcohols as Fuel." M.I.T. Energy Lab., Methanol Div. July 1, 1974, 90 ref.

223. T. B. Reed, "Advantages of Neat and Blended Operation of Methanol Fuel in Vehicles." 1974 Eng. Foundation Conf., "Methanol as an Alternate Fuel," Henniker, N.H., July 7-12, 1976.

224. T. B. Reed, R. M. Lerner, E. D. Hinkley, and R. E. Fahey (Lincoln Lab., M.I.T.) "Improved Performance of Internal Combustion Engines Using 5-30% Methanol in Gasoline," 9th Intersoc. Energy Conv. Eng. Cong., San Francisco, Paper No. 749104, Aug. 26-30, 1974, pp. 952-955.

225. T. B. Reed (M.I.T. Energy Lab, Cambridge) "Comparison of Methanol and Methanol-Blends." 1974.

226. T. B. Reed (M.I.T., Cambridge) "Use of Alcohols and Other Synthetic Fuels in Europe from 1930-1950." 1975.

227. T. B. Reed, "The Role of Methanol in Various World Energy Systems." Symp. "Methanol as a Fuel," Royal Swedish Acad. Sci., Stockholm, March 23, 1976.

228. T. Remmets, "Future of Methanol as an Engine Propellant and as Fuel." (Ger.) Erdol Und Kohle, Vol. 29, No. 2, Feb. 1976, pp. 69-74.

229. P. W. Reynolds (Agricultural Div.,I.C.I.) "Protein from Natural Gas; Pseudomonas Methylatropha." Energy World, May 16, 1975, pp. 8-10.

230. L. Riekert, "Energy Conversion by Chemical Processes." Chemie Ingenieur, Vol. 47, No. 2, January 1975, pp. 48-50 (Ger.).

231. J. G. Rivard (Bendix Electronic Co.) "Electronic Control Unit for Production Electronic Fuel Injection Systems." Soc. Auto. Eng., S.A.E. 760242, Detroit, Feb. 23-27, 1976.

232. M. Rosenzweig and S. Ushio, "Protein from Methanol." Chemical Engineering, Jan. 6, 1974, pp. 62-63.

233. Royal Swedish Academy of Sciences & Swedish Methanol Development Co., "Methanol as a Fuel Proc. Symp., Stockholm, Sweden, March 23, 1976." IVA-Meddelande No. 195.

234. T. W. Ryan, III, S. S. Lestz, and W. E. Meyer, "Extension of the Lean Misfire Limit and Reduction of Exhaust Emissions of an S.I. Engine by Modification of the Injection and Intake Systems." Soc. Auto. Eng. Paper #740105, Detroit, February 1974.

235. E. Sawicki, T. R. Hauser, T. W. Stanley, and W. Elbert, "The 3-Methyl--2-Benzothiazolene Hydrazone Test." Anal. Chem., Vol. 33, No. 1, Jan. 1961.

236. A. W. Scarratt, "The Carburation of Alcohol." Soc. Auto. Eng. Trans., 1921.

237. M. Schaffrath, "Methanol as Engine Fuel." Erdol Und Kohle, Vol. 29, No. 2, Feb. 1976, pp. 64-68 (Ger.).

238. H. Schmidt and W. Vielstich, "Einflusz von Edelmetall-Mischkatalysatoren auf die anodische Oxydation von Methanol and Formiat." Z. Analyt. Chem., Vol. 224, 1967, p. 84 (Ger.).

239. Matthias Schwarzmann, "Methanol: A Raw Material for Synthesis and an Energy Source." 1975, 21 pp., (UCRL-Trans-10908), Transl. from Chem.-Ing.-Tech., Vol. 47, No. 2, 1975, pp. 56-61.

240. A. L. Shrier, "Methanol Fuel." Am. Chem. Soc. Symp., "The Role of Technology in the Energy Crisis." Atlantic City, N. J., September 8-13, 1974, ACS Div. of Petro. Chem., Preprint, Vol. 19, No. 3, August 1974.

241. Society of Automotive Engineers, New York, "Alcohols and Hydrocarbons as Motor Fuels." Soc. Auto. Eng. Special Publication SP-254, June 1964.

242. P. Soedjanto and F. W. Schaffert, "Transporting Gas-LNG vs. Methanol." Oil and Gas Journal, June 11, 1973, pp.88-92.

243. P. J. Soukup, (Amoco Oil Co., Whiting, Indiana) "Methanol in Gasoline." August 11, 1970.

244. E. S. Starkman, F. M. Strange and T. J. Dahm, "Flame Speeds and Pressure Rise Rates in Spark Ignition Engines." Soc. Auto. Eng. Publication 83V-1, July 1959, S. A. E. International West Coast Meeting, Vancouver, B.C., August 1959.

245. E. S. Starkman, G. K. Newhall and R. D. Sutton, "Comparative Performance of Alcohol and Hydrocarbon Fuels." Soc. Auto. Eng. Special Publication "Alcohols and Hydrocarbons as Motor Fuels." SP-254, June, 1964.

246. E. S. Starkman, R. F. Sawyer, R. Carr, G. Johnson and L. Muzio (Univ. of California, Berkeley) "Alternative Fuels for Control of Engine Emission." Air Pollution Control Assoc., Vol. 20, No. 2, February, 1970, pp. 87-92.

247. M. Steinberg, F. J. Salzano, M. Beller and B. Manowitz, "Methanol as a Fuel in the

Urban Energy Economy and Possible Source of Supply." Brookhaven Nat'l Lab., BNL 17800, April 1973, 19 pp.

248. R. J. Stettler, (General Motors) "Initial Evaluation of Coal Derived Liquid Fuels in a Low-Emission Turbine Combustor." Conf.: "Combustion of Alternate Fuels and Combustion of Coal." The Combustion Institute, Central States Section, Columbus, Ohio, April 5-7,1976.

249. R. P. Stickles, E. Interess, G. C. Sweeney, P. E. Mawn and J. M. Parry, "Assessment of Fuels for Power Generation by Electric Utility Fuel Cells." Prepared by Arthur D. Little Inc., Cambridge, Mass., for Electric Power Research Institute, Palo Also, Ca., EPRI 318, October 1975, 310 pp.

250. J. Stone et. al., (Mitre Corp., McLean, Va.) "Survey of Alcohol Fuel Technology--Interim Report." M74-61, Rev. 1, July 1974.

251. Christopher L. Sylwan, (Royal Institute of Technology, Stockholm,Sweden) "Methanol Feul Cell Electrodes Consisting of Platinized Nickel Matrices." Energy Conversion, Vol. 15, No.3/4, 1976, pp.137-141.

252. H. Tamura and C. Iwakura, "Studies on the Methanol Fuel Cell: A New Electrode Catalyst for the Anodic Oxidation of Methanol." Technol. Rep. Osaka Univ., Vol. 17, 1967, p.549.

253. B. L. Tarmy, (Esso R. & E. Co.) "Methanol Fuel Cells." Proc. 16th Ann. Power Sources Conf., 1962, pp. 29-31, contract D. O. D. No. DA-36-039 SC-89156.

254. B. L. Tarmy and G. Ciprios, "The Methanol Fuel Cell Battery." Eng. Develop. Energy Conv., ASME 1965, p. 272.

255. B. L. Tarmy, (Esso R. & E. Co.) "Methanol Fuel Cell Batter." Proc. 19th Ann. Power Sources Conf., May 18-20, 1965, pp. 41-43.

256. Texaco Research Center (Beacon, N.Y.) "Evaluation of Methanol as a Component of Motor Fuels."

257. R. M. Tillman, O. L. R. Spilman and J. M. Beach, "Potential for Methanol as an Automotive Fuel." S.A.E. 750118, February 24-27, 1975, Detroit.

258. R. E. Train (Environmental Protection Agency) Statement before the Subcommittee on Priorities and Economy in Government, Joint Economic Committee, May 21, 1974.

259. Arthur E. Uhl, (Bechtel Inc.) "Fuel Energy Systems: Conversion and Transport Efficiencies." 9th Intersoc. Energy Conv. Eng. Conf. Paper No. 749084, August 26-30, 1974, NTIS No. A75-10554.

260. U.S.A.E.C., Division of Reactor Development and Technology, Synthetic Fuels Panel, "Hydrogen and Other Synthetic Fuels: A Summary of the Work of the Synthetic Fuels Panel." Report No. TID-26136, Sept. 1972, NTIS #PB-224482.

261. U.S. Bureau of Mines, "Measurement of Emissions from Engines Fueled with Methanol or Methanol/Gasoline Mixtures." Proc. Technical Discussions held in Denver, August 9, 1974.

262. H. C. Vantine, J. Chang, B. Rubin and C. Westbrook, "The Methanol Engine: A Transportation Strategy for the Post-Petroleum Era." Lawrence Livermore Lab., UCIR-961, November 26, 1975, 58 pp.

263. D. Vendil, "Methanol as an Energy Source." Kemisk Tidskrift (Chem. Journal) Vol. 88, No. 3, March 1976, p. 30-35 (Swedish).

264. D. Vendil, (Swedish Methanol Development Co., Sweden) "Methanol as a Fuel--A Swedish National Interest." Symposium on "Meth anol as a Fuel." Royal Swedish Acadamy of Engineers, Stockholm, March 25, 1976.

265. W. Vielstich, "Alcohol Air Fuel Cells--Development and Application." Am. Chem. Soc. Div. Fuel Chem., Vol. 9, No. 3, Pt. 1, April 1965, pp. 60-73.

266. Vulcan Cincinnati, Inc., "Methyl Fuel Firing Test." Restricted Report of Private Test, December 1971.

267. Vulcan Cincinnati, Inc., "Methyl Fuel Combustion Test, Vol. I and II." Report of Test at A. B. Paterson Plant, restricted to the sponsors, Dec. 15, 1972, 1000 pp.

268. Wall Street Journal, "Fuel Cell Development Set by French Firm, Jersey Standard Unit." December 1970.

269. B. Warszawski, B. Verger and J. C. Dumas, "Alsthom Fuel Cells for Marine and Submarine Applications." Marine Technology Society Journal, Vol. 5, No. 1, Jan/Feb, 1971, pp.28-41.

270. T. O. Wentworth, "Methyl Fuel Could Provide a Motor Fuel." Chem. Eng. News, September 17, 1973.

271. T. O. Wentworth, (Vulcan Cincinnati Co. Cincinnati, Ohio) "Outlook Bright for Methyl-Fuel." Environ. Sci. Techn., Vol. 7, No. 11, Nov. 1973, pp. 1002-1003.

272. E. C. Wenzel, (Emission Free Fuels, Inc.) "Water /Alcohol Solutions in Internal Combustion Engines." March, 1974.

273. J. R. White, C. N. Rowe and W. J. Koehl (Mobil R. and D. Corp.) "Physico-Chemical Properties of Methanol Related to Fuel Use." 1974 Eng. Foundation Conf., "Methanol as an Alternate Fuel," Henniker, N. H., July 7-12, 1974.

274. Eric E. Wigg and R. S. Lunt, "Methanol as a Gasoline Extender-Fuel Economy, Emissions, and High Temperature Driveability." Soc. Auto. Eng., S.A.E. paper No. 741008, Toronto, October 1974.

275. E. E. Wigg, (Exxon Res. & Eng. Co., Linden, N. J.) "Methanol as a Gasoline Extender: A Critique." Science, Vol. 186, No. 4166, November 29, 1974, pp. 785-790.

276. Eriç E. Wigg and Robert S. Lunt, "Methanol-Gasoline Blends: How Promising Are They?" Automotive Engr., Dec. 1974, pp.38-42.

277. R. C. Wilhoit and B. J. Zwolinski, "Physical and Thermodynamic Properties of Aliphatic Alcohols." J. Phys. Chem. Ref. Data Vol. 2, 1973, Suppl. No. 1, pp. 1-40 to 1-54.

278. K. R. Williams, M. B. Andrew and F. Jones, "Some Aspects of the Design and Operation of Dissolved Methanol Fuel Cells," Hydrocarbon Fuel Cell Technol. Symp. Am. Chem. Soc., Div. Fuel Chem., Vol 9, No. 3 Pt. 1, April 1965, pp. 125-134.

279. K. R. Williams, "Methanol Fuel Cells." Advances in Science, Vol. 22, No. 105, 1966, pp. 617-622, (A Review).

280. K. R. Williams and A. G. Dixon (Shell Research Ltd. Chester, England) "Hydrocarbon and Methanol Fuel Cell Power Systems." Performance Forecast Selec.Static Energy Convers. Devices, 29th Meeting, AGARD Propul. Energy Panel, Air Force Aero Propul. Lab., 1967, pp. 634-651.

281. C. Winter, et. al., "Energy Imports, LNG vs. MeOH." Chemical Eng., Vol. 80, No. 112, November 12, 1973, pp. 233-238.

282. J. E. Wynn, (U.S. Army Electronics Command, Ft. Monmouth) "Methanol Oxygen Fuel Cells." 24th Power Sources Symp., Atlantic City, from Technical Report ECOM-02387-F, February 1968, pp.198-207.

283. Tamechika Yamamoto (Catalysts and Chem. Ind. Co., Japan) "Environmental Pollution and Systematization of Chemical Techniques--Use of Methanol as a Fuel." Chemical Economy and Eng. Rev., Vol. 4, December 1972, pp.52-57.

284. Tamechika Yamamoto, "Synthesis and Utilization of Methanol in Natural Gas Producing Areas (for Food Production)." Chemical Economy & Eng. Rev., Vol. 5, 1973, p. 7.

285. J. Yasuda (Mitsubishi Gas Chemical Co., Japan) "Production of Methanol Fuel from Synthesis Gas." Symp. "Methanol as a Fuel," Royal Swedish Academy of Engineering Sciences, Stockholm, March 23, 1976.

286. R. J. Young (Imperial Chenical Industries, Ltd., Agricultural Div.) "Fermentation Protein from Methanol." Am. Chem. Soc., 169th Nat'l Meeting, Paper 82, April 6-11, 1975.

287. Anon., "Fuel Cell." Transl. from French Patent1,371,815, UCRL-Trans-10882, June 1974, 11 pp.

288. A. G. Gaydon and H. G. Wolfhard, "Flames: Their Structure, Radiation and Temperature." Chapman and Hall 3rd Ed., 1970,p.304.

289. F. D. Rossini et. al., "Selected Values of Physical and Thermodynamic Properties of Hydrocarbons and Related Compounds." Am. Petroleum Inst., Carnegie Press, 1953.

290. Swedish Methanol Development Co. (Stockholm, Sweden) Proc. Seminar "Methanol as a Fuel." Vol. I-Seminar Report,36 pp., Vol II-Seminar Papers, 118 pp. March 21-22,24, 1976, Dag Vendil, Chairman.

291. B. Baratz, R. Ouellette, W. Park and B. Stokes (Mitre Corp.) "Survey of Alcohol Fuel Technology. Vol. I." M74-61-Vol. 1, November 1975, 143 pp. NTIS No. PB-256007/6WE.

292. B. Stokes and W. Park (Mitre Corp.) "Survey of Alcohol Fuel Technology. Vol. II." M74-61-Vol. II, November 1975, 68 pp. PB 256008/4WE.

293. R. R. Adt, Jr., R. D. Doepker, L. E. Poteat, K. C. Chester, C. N. Kurucz, J. M. Pappas and M. R. Swain, "Characterization of Methanol/Gasoline Blends as Automotive Fuel--Performance and Emissions Characteristics." Prepared for U.S. Environmental Protection Agency, Alternative Automotive Power Systems Division, Ann Arbor, Mi., 1976.

294. American National Standards Inst. Inc., "Metric Practice." August 19, 1976, ANSI Z 210.1-1976, ASTM E 380-76, IEEE Std. 268-1976, Pub. Inst. Elec. Electronic Eng. Inc., New York, New York.

295. G. S. Canada and G. M. Faeth, "Combustion of liquid Fuels in a flowing Combustion Gas Environment at High Pressures." Proc. 15th Int'l Symp. on Combustion, Tokyo, Japan, August 25, 1974, pp. 419-428.

296. A. W. Crowley, J. P. Keubrich, M. A. Roberts, W. S. Koehl, W. L. Waschner and W. T. Wotring, "Methanol-Gasoline Blends Performance in Laboratory Tests and in Vehicles," Inter-Industry Emission Control Program -2 (IIEC-2) Progress Report No. 1, Warrendale Pa., Soc. Auto. Engr. Inc., January 1975.

297. R. D. Gleming and T. W. Chamberlain, (Bureau of Mines, Bartlesville, Ok.) "Methanol as Automotive Fuel. Part I. Straight Methanol." Bureau of Mines CONF-750264-1, 1976, 24 pp., Auto. Engr. Congr & Expo., Detroit, February 24, 1975.

298. David L. Hagen, "Methanol: Its Synthesis, Use as a Fuel, Economics,and Hazards." M.Sc. Papers, University of Minnesota, December 1976, 180 pp. 609 ref., Published by U.S. Energy Research and Development Administration, Report No. NP-21727.

299. J. D. Rogers, Jr., "Ethanol and Methanol as Automotive Fuels." E.I. du Pont de Nemours & Co., Inc., Petroleum Chemicals Division, Report No. P813-3, November 1973.

300. Federal Hazardous Substances Act, U.S. Public Law 86-613,74 Stat. 372-81.

301. Advisory Center on Toxicology, "Report on Methanol" in "Toxicity Evaluation of potentially Hazardous Materials, Part III." 1959, pp. 82-103. Washington, D.C. National Research Council-National Academy of Sciences.

302. American Industrial Hygiene Association, "Methanol Hygiene Guide."

303. American Petroleum Institute "API Toxicological Review: Gasoline 1st Edition." New York, New York, 1967.

304. B. J. Berger (Lawrence Livermore Labs.) "Environmental Aspects of Methanol as Vehicular Fuel: Health and Environmental Effects." UCRL-76076.

305. Chemical Rubber Company, "Handbook of Analytical Toxicology." 1975.

306. Chemical Solvents Corp., "Synthetic Methanol." 1960.

307. J. T. Dehn, "Flammable Limits Over Liquid Surfaces." Combustion of Flame, Vol. 24, 1975, pp. 231-238.

308. William B. Deichmann and Horace W. Gerarde, "Symptomatology and Therapy of Toxicological Emergencies." Academic Press, New York and London, 1964.

309. M. N. Gleason, R. E. Gosselin, H. C. Hodge and R. P. Smith, "Clinical Toxicology of Commercial Poisons." pp. 155-158, Williams and Williams Co., Baltimore, 1969.

310. M. Koivusalo, "Methanol" in "International Encyclopedia of Pharmacology and Therapeutics." Vol. 2, "Alcohols and Derivatives." Sec. 20, pp. 465-505 Pergamon Press, New York, 1970.

311. Manufacturing Chemists Association (Washington, D. C.) "Properties and Essential Information for Safe Handling and Use of Methanol." Chemical Safety Data Sheet SD-22, 1970.

312. Ibid. "Recommended Practices for Bulk Loading and Unloading Flammable Liquid Chemicals to and from Tank Trucks." Technical Bulletin TC-8, 1975.

313. Ibid., "Recommended Safe Practices and Procedures--Flammable Liquids: Storage and Handling of Drum Lots and Smaller Quantities." Safety Guide SG-3, 1960.

314. Ibid., "MCA CHEM-CARD--Transportation, Emergency Guide: Methanol." CC-69, 1965.

315. John A. Monick, "Alcohols, Their Chemistry, Properties and Manufacture", Reinhold Book Company, New York, 1968.

316. National Institute for Occupational Safety and Health, Public Health Service Center for Disease Control, "Registry of Toxic Effects of Chemical Substances." U.S. Dept. H.E.W.1975.

317. National Safety Council -- "Data Sheet on Methanol."

318. Frank A. Patty, David W. Fassett and Don D. Irish, eds., "Industrial Hygiene and Toxicology." Vol. II, 2nd Revised Edition, "Methanol." p. 1409, John Wiley & Sons, New York, 1973, 2377 pp.

319. Herbert S. Posner, "Biohazards of Methanol in Proposed New Uses." J. Toxicology and Environmental Health, Vol. I 1975, pp. 153-171, National Institute of Environmental Health Sciences, Research Triangle Park, N.C., (118 ref.).

320. N. I. Sax, "Dangerous Properties of Industrial Materials." 4th ed., pp. 908-981, Van Nostrand Reinhold Co., New York, 1975.

321. R. R. Adt, Jr., K. C. Chester, C. N. Kurucz, J. M. Pappas and M. R. Swain, "The Effects of up to 30 Volume Percent Addition per. se. on the Basic Performance and Exhaust Emission Characteristics of a Carbureted Spark Ignition Engine." Int'l Symp. Auto. Tech. & Automation, Rome, September 27-October 1, 1976.

322. R. R. Adt, Jr., K. A. Chester and C. K. Wiesner, (Univ. of Miami, Coral Gables) and J. M. Pappas and M. R. Swain, (Hawthorne Research & Testing Inc., Coral Gables, Florida) "The Effect of Blending Methanol With Gasoline on Geometric Distribution With 5% Lean Mixtures." NATO/CCMS 4th Int'l Symp. Auto. Propulsion Systems, April 19, 1977, Arlington, Va.

323. J. R. Allsup, (Bartlesville Energy Research Center) "Experimental Results Using Methanol and Methanol/Gasoline Blends as Automotive Engine Fuel." BERC/RI-76/15, January 1977, 87 p.

324. W. E. Bernhardt, A. Koenig, W. Lee and H. Menrad (Volkswagenwerk A.G. Wolfsburg, Fed. Republic Germany) "Recent Progress in Automotive Propulsion Systems, NATO/CCMS 4th Int'l Symp. on Auto Propulsion Systems, April 19, 1977, Arlington, Va.

325. M. C. Branch, K. Wolfe and N. Ishikawa (Univ. of California, Berkeley) "Combustion of Methanol and Methanol Blends in a Stratified Charge Engine." Proc.11th Intersoc. Energy Conf. pp. 115-121, September 12, 1976, ERA 2: 17691.

326. Norman D. Brinkman (General Motors Research Laboratories) "Effect of Compression Ratio on Exhaust Emissions and Performance of a Methanol-Fuelled Single-Cylinder Engine." Submitted to Soc. Auto. Eng. Passenger Car Mtg., Detroit, Mi., September 26, 1977.

327. L. S. Caretto, "Other Engines, Other Fuels: An Overview." Soc. Auto. Eng. Paper 760608, San Francisco, August 9-12, 1976, 13 pp.

328. B. H. Eccleston and F. W. Cox, (Bartlesville Energy Research Center, Bartlesville, Ok.) "Physical Properties of Gasoline/Methanol Mixtures." BERC/RI-76/12, January 1977, 79 pp.

329. R. D. Doepker, R. R. Adt, Jr., K. A. Chester, T. L. Helmers and C. K. Wiesner (Univ. of Miami, Coral Cables, Fla.) "The Effect of Methanol Addition to Gasoline on Total and Individual Hydrocarbons Methanol and Formaldehyde Emissions from a Carburetted Spark Ignition Engine." NATO/CCMS 4th Int'l Symp. Auto. Propulsion Systems, April 21, 1977, Arlington, Va.

330. Energy Research and Development Administration, Highway Vehicle Systems Contractors Coordination Meeting: Ann Arbor, Mi., May 4, 1976. ERDA-76-136. E. E. Ecklund, (ERDA), "Overview of Alternative Fuels Utilization Program," pp. 254-260. R. Pefley, (U. Santa Clara), "Summary of Methanol Mobile Power Research Investigations," pp. 281-291. R. R.Adt, Jr., (Univ. Miami), "Characterization of Methanol/Gasoline Blends as Automotive Fuel." pp. 291-293. J. Allsup, (Bartlesville Energy Research Center), "Characterization of Methanol as an Automotive Fuel." pp. 294-297.

331. Energy Research and Development Administration, "Alternative Fuels Utilization Report." No. 1, April 1977, Published by Div. Trans. Energy Conserv., Alternative Fuels Utilization Branch, Washington, D.C., Edited

and Distributed by Mueller Associated, Inc., 1900 Sulfur Spring Rd., Baltimore, Md. 21227.

332. J. H. Freeman (The Sun Co., Inc.) "Alcohols--A Technical Assessment of Their Applications as Fuels." Submitted to Soc. Auto. Eng. Passenger Car Mtg., Detroit, Mi, September 26, 1977.

333. N. A. Henein, T. Singh, J. Rozanski and P. Husak (Wayne State Univ., Detroit), "Flame Speeds, Performance and Emissions with Methanol Indolene Blends." NATO/CCMS 4th Int'l Symp. Auto. Propulsion Syst., Low Pollution Power Systems Develop, Washington D.C., April 18-22, 1977.

334. T. Hirano, M. Kinoshita (Ibaraki Univ., Japan), "Gas Velocity and Temperature Profiles of a Diffusion Flame Stabilized in the Stream Over Liquid Fuel." Proc. 15th Int'l. Symp. on Combustion, Tokyo, August 25, 1974, pp.379-387. Combustion Institute, Pittsburgh, 1975.

335. John Houseman and D. J. Cerini, (Jet Propulsion Lab., Pasadena, Ca.), "Onboard Hydrogen Generation for Automobiles." Proc. 11th Intersoc. Energy Conv. Eng. Conf. Paper No. 769001, State Line, Nv., September 12-17, 1976, pp. 6-16.

336. R. W. Hurn, (ERDA, Bartlesville Research Cntr) "Alternative Fuels--The Outlook and Options Within the Next Decade." NATO/CCMS 4th Int'l Symp. Auto.Propulsion Systems, April 19, 1977.

337. N. Ishikawa and M. C. Branch (Univ. California, Berkeley), "Experimental Determination of the quenching Distance of Methanol and Iso-Octane/Methanol Blends." Conf. on Combustion of Coal, Oil Shale, and Tar Sands, Combustion Problems related to the Enhancement of Energy Efficiency, Salt Lake City, Paper 22, April 19, 1976, 22 pp., ERA 2:12697

338. W. Lee, A. Koenig and W. Bernhardt (Volkswagenwerk A.G., Wolfsburg, Germany) "Potential for Methanol-Gasoline Blends as Automotive Fuels," Proc. 11th Intersoc. Energy Conv. Eng. Conf. pp. 105-114, September 12,1976.

339. J. A. LoRusso, "Combustion and Emissions Characteristics of Methanol, Methanol-Water, and Methanol-Gasoline Blends in a Spark-Ignition Engine." M.S. Thesis, Mass. Inst. Tech., May 1976.

340. J. A. LoRusso, R. J. Tabaczynski (Mass. Inst. Tech.), "Combustion and Emissions Characteristics of Methanol, Methanol-Water, and Gasoline-Methanol Blends in a Spark Ignition Engine." Proc. 11th Intersoc. Energy Conv. Eng. Conf., Vol. I, pp. 122-132, September 12, 1976, ERA 2:17689.

341. H. Menrad, W. Lee and W. Bernhardt, (Volkswagenwerk A,G,) "Development of a Pure Methanol Fuel Car." Submitted to Soc. Auto.Eng. Passenger Car Mtg., Detroit, Mi., September 26, 1977.

342. W. J. Most and E. E. Wiggs, "Methanol adn Methanol-Gasoline Blends an Automotive Fuels." presented @ The Combustion Institute, Central States Section, Spring Meeting, April 5-6, 1976, Columbus, Ohio.

343. M. Noguchi, T. Bunda, M. Suniyoshi (Toyota Motor Co., Ltd.), J. Kageyama (Fuji Electric Co., Ltd.), S. Yamaguchi (Nippon Soken Inc.), "A Study on Reformed Fuel for an Automotive Gasoline Engine." NATO/CCMS 4th Int'l Symp. Auto.Propulsion Syst. Low Pollution Power Syst. Devel., Arlington, Va., Session 6, April 18-22, 1977.

344. A. J. Parker, Jr., (Mueller Assoc. Inc. Baltimore, Md.), "Alcohols and Gaseous Fuels from Biomass," NATO/CCMS 4th Int'l Symp. Auto. Propulsion Systems, April 19, 1977, Arlington, VA.

345. Peter Sunn Pederson and Klaus Bro (Purdue Univ.) "Alternative Diesel Engine Fuels: An Experimental Investigation of Methanol, Ethanol, Methane and Ammonia in a d.i. Diesel Engine with Pilot Injection." Submitted Soc. Auto. Eng. Passenger Car Mtg., Detroit, Mi., September 26, 1977.

346. R. K. Pefley, et. al., "Characterization and Research Investigation of Methanol and Methyl Fuels in Automobile Engines." University of Santa Clara, California, Report ME 76-2, October 1976.

347. L. E. Poteat, (University of Miami), "Compatibility of Automotive Materials with Methanol/Gasoline Blends." NATO/CCMS 4th Int'l Symp. Auto Prop. Syst., April 19, 1977, Arlington, Va.

348. R. Schmidt, "Method of and apparatus for Improved Methanol Operation of Combustion Systems." U.S. Patent 3,986,350, October 19, 1976, 8 pp.

349. J. Van der Weide and P. Tiedema (TNO, Delft), "Alternative Fuels With Regard to LPG and Methanol." NATO/CCMS 4th Int'l Symposium Auto. Propulsion Systems., April 19, 1977, Arlington, VA.

350. Washington Center for Metropolitan Studies, Washington, D.C., Coordinator Conference on "Capturing the Sun Through Bioconversion." Washington, D.C., March 10, 1976, T. B. Reed, "When the Oil Runs Out: A Survey of our Primary Energy Sources and the Fuel We Can Make from them." Proc. pp. 336-388; "Liquid Fuels: Workshop No. 6." pp. 389-392; "Speech by D. Vendil (Swedish Methanol Development Co., Stockholm)." pp.393-402; H. Heitland, (Drive Train Research, Englewood Cliffs) "Volkswagen Alternative Fuel Programs." pp. 403-416; R. W. Hurn, (ERDA) "Properties and Characteristics of Gasoline/Methanol Fuel." pp. 425-436.

351. Reed, T.B.,"net Efficiencies of Methanol production from Gas, Coal, Waste or Wood,"Am. Chem. Soc., Div. Fuel Chem., New York, Vol 21, No. 2 Paper No.16, April 4-9,1976.

Table 6 - Estimated Tolerance Values for Methanol

Duration	PPM
Single but not repeated exposure	
1 hour	1,000
8 hours	500
24 hours	200
5 x 8 hour working days	200
168 hours	50
30 days	10
60 days	5
90 days	3
Single or repeated exposures	
1 hour out of every 24 hours	500
1 x 2 hr or 2 x 1 hr every 24 hours	200

Source: Advisory Center on Toxicology,(301) cited by Posner.(319)

APPENDIX: PROPERTIES OF METHANOL CH_3OH

Molecular Weight 32.042 g mol^{-1}

Synonyms: Carbinol, Colonial Spirit, Columbian Spirit, Methyl Alcohol, Methyl Hydroxide, Monohydroxymethane, Pyroxylic Spirit, Wood Alcohol, Wood Naptha, Wood Spirit. French - Alcool Methylique; Italian - Alcool Metilico, Metanolo; German - Methylalkohol; Polish - Metylowry Alkohol.

Temperature	Refractive Index	Density	Vapor Pressure	
C		kg m^{-3}	k Pa	mm Hg.
-20		0.8287		
-10		0.8194		
0	1.3361	0.8100		
10	1.33224	0.8007		
15	1.33034	0.7960	9.8856	74.15
20	1.32840	0.79131	13.0120	97.60
25	1.32652	0.78664	16.9575	127.19
30	1.32457	0.78196	21.8832	209.88
40	1.3207	0.7726	35.4677	266.03
50	1.3169	0.7633	55.6106	417.11
60		0.7546	84.6032	634.58
70		0.7448		
80		0.7347		
90		0.7242		
100		0.7132		

Density: Francis Equation $\rho_{SL} = A - BT - C/(E-T)$

Temp. Range	A	B x 10^3	C	E
-20→ 50°C	0.84638	0.9321	423.28	11641
60→100°C	0.86867	0.6111	17.267	283.08

Vapor Pressure: P

$$\ln (P/kPa) = 15.76129944 - 2.845920984 \times 10^3\ KT^{-1} - 3.743415457 \times 10^5\ K^2T^{-2} + 2.188669828 \times 10^7\ K^3T^{-3}$$

From 288.15→337.65°K Accuracy: T ± 0.002 K, P ± 1 Pa

Boiling Point: T_b = 337.664K ± .002K., 64.514°C

Melting Point (=Triple Point): T_f =-97.56 ± 0.02°C

Critical Temperature: T_c = 239.43°C, 512.58K

Critical Pressure: P_c = 8096kPa (79.9 atm)

Critical Density: ρ_c = 0.272 gcm^{-3}

Heat of Fusion: ΔH = 32.13 ± .05 kJ mol^{-1}

Heat of Vaporization: @64.70°C, 760 mm Hg ΔH_v = 34.48±.04kJmol^{-1}

Heat of Vaporization: @25.0°C, 125.45mmHg ΔH_v = 37.40±.15kJmol^{-1}

THERMODYNAMIC FUNCTIONS OF IDEAL METHANOL GAS AT ONE ATMOSPHERE:

Temperature	Entropy S	Heat Capacity Cp	Heat of Formation	Gibbs Energy of Formation
K	J $K^{-1}mol^{-1}$	J $K^{-1}mol^{-1}$	ΔH_f° kJmol^{-1}	ΔG_f° kJ mol^{-1}
0	0	0	-190.21	-190.21
273.15	235.94	42.47	-200.16	-165.60
298.15	239.70	43.89	-201.08	-162.42
300	239.99	44.02	-201.17	-162.17
400	253.59	51.42	-204.72	-161.17
500	265.73	59.50	-207.90	-134.22
600	277.48	67.03	-210.58	-119.24
700	288.32	73.12	-212.80	-103.85
800	298.57	79.66	-214.60	-88.12
900	308.24	84.89	-216.10	-72.22
1000	317.44	89.45	-217.19	-56.19

STANDARD STATE PROPERTIES: @25°C	LIQUID	GAS
Heat of Combustion ΔH_c° kJ mol^{-1} to $H_2O(l)$	-726.13±.4	-764.08±.4
Heat of Combustion ΔH_c° kJ mol^{-1} to $H_2O(g)$	-638.11±.4	-676.05±.4
Heat of Formation ΔH_f° kJ mol^{-1}	-239,03±.4	-201.08±.4
Entropy S° J K^{-1} mol^{-1}	127.24±.2	239.70±.2
Gibbs Energy of Formation ΔG_f° kJ mol^{-1}	-166.82±.4	-162.42±.4
Heat Capacity C_p° J $K^{-1}mol^{-1}$	81.17±.1	43.89±.1
Heat of Dissociation ΔH_d° kJ mol^{-1}	128.51±.4	90.56±.4

Conversion factor of 4.1840 J/cal. assumed throughout.
Sources: Wilhoit, R. C. and Zwolinski, B. J., (Thermo. Res. Ctr., Texas A & M Univ.) "Physical and Thermodynamic Properties of Aliphatic Alcohols," J. Phys. and Chem. Ref. Data, Vol. 2, 1973, Supplement No. 1, 1973, pp. 1-40 to 1-54.
Gibbard, H. F. and Creek, J.L., "Vapor Pressure of Methanol from 288.15 to 337.66K," J. Chem. Eng. Data, Vol. 19, No. 4, 1974, p. 308-310.

Alternative Diesel Engine Fuels: An Experimental Investigation of Methanol, Ethanol, Methane and Ammonia in a D.I. Diesel Engine with Pilot Injection *

Klaus Bro and Peter Sunn Pedersen
The Tech. Univ. of Denmark

FROM A PRODUCTION POINT OF VIEW it has been found (1)[x] that the most promosing alternatives to gasoline and diesel fuel (made from crude oil, oil shales or coal) are Methanol, Ethanol, Methane, Hydrogen and Ammonia, all with superiour octane ratings which make these fuels potentially very suitable for spark ignition engines. During the last few years a very considerable amount of work has been done on S.I. engine applications of these fuels.

The amount of work on diesel engine applications reported in the literature is, however, much more modest and in most cases the pollution aspects are not considered. Thus the purpose of the present work has been to provide a general knowledge of the suitability of the mentioned alternative fuels for diesel engines and to identify topics where further research is needed.

x Numbers in parantheses designate References at end of paper

As the self ignition quality of the five mentioned fuels is very poor (as a high octane rating equals a low cetane rating) an independent source of ignition must be provided. For diesel engine applications the most obvious method is to inject a small amount of the ordinary diesel fuel ("pilot fuel") through the normal injection system. So the engine will have two fuel systems, one for the pilot fuel (which makes "pure diesel operation" possible) and one for the alternative (main) fuel. In this way it is possible to run a diesel engine with a high octane fuel as the main energy source.

A number of investigations regarding dual fuel operation of diesel engines using pilot injection as the source of ignition are reported in the literature. During W.W. II biogas from municipal waste was used with good results (2). In the 1950's Alberstein et al (3) added small amounts of volatile fuels to the intake air of a diesel engine in order to increase the power

*Paper 770794 presented at the Passenger Car Meeting, Detroit, Michigan, September 1977.

ABSTRACT

The results of an experimental investigation of Methanol, Ethanol, Methane and Ammonia as primary fuels for a high speed direct injection diesel engine are described. The fuels were added to the intake air and ignited by injection of a small amount of pilot diesel fuel (30 percent on energy basis).

All of the four fuels were found applicable with Methane as the most suitable and Ammonia as the least suitable. Experimental data are presented regarding engine power output, efficiency, smoke and gaseous emissions, and the different types of combustion observed during the experiments are discussed.

SAE/PT-80/19/$02.50

output and to reduce the smoke ("Fumigation") but the purpose was not to use the volatile fuel as the main energy source. Similar experiments were performed by Havemann et al (4), (5), in order to investigate the effects on smoke, power output and ignition delay of adding Ethanol with heavy fuel oil as the main fuel - and more recently, by Hirako and Ohta (6), who added Methanol and Ethanol to the intake air.

Lowe and Brandham (7) and Whitehouse (2) have investigated the use of Methane (Natural Gas) in large engines, determining the effects on engine efficiency of the ratio between the amounts of pilot fuel and main fuel (i.e. Methane). Tesarek (8) investigated the use of Natural Gas in a small high speed diesel engine and replaced up to 80 percent (on an energy basis) of the diesel fuel by Natural Gas with the main purpose being to reduce smoke emissions. Karim et al (9), (10), (11) have made comprehensive investigations of combustion phenomena in dual fuel engines, for example regarding the effects of intake charge temperature and mixture strength on the occurrence of knock. Most of the work deals with Methane but the use of Hydrogen has also been investigated.

Gray et al (12) have investigated the use of Ammonia in a diesel engine, but they were only able to substitute a rather modest amount of diesel fuel by Ammonia. Starkman et al (13) used Ammonia in a diesel engine but a spark plug was used to provide the ignition.

In the following an experimental investigation of the effects on power output, fuel economy and emissions of the amount of pilot fuel (at constant amount of the alternative main fuel) and the amount of the alternative fuel (at constant amount of pilot fuel) is described.

APPARATUS

The engine used for the investigation was a single cylinder, water cooled, direct injection 4-stroke diesel research engine of 0.09 m bore and 0.10 m stroke coupled to a Zölner eddy current dynamometer. A cross-section of the engine, which features variable compression ratio (from 4.5:1 to 17.7:1) is shown in figure 1, and the

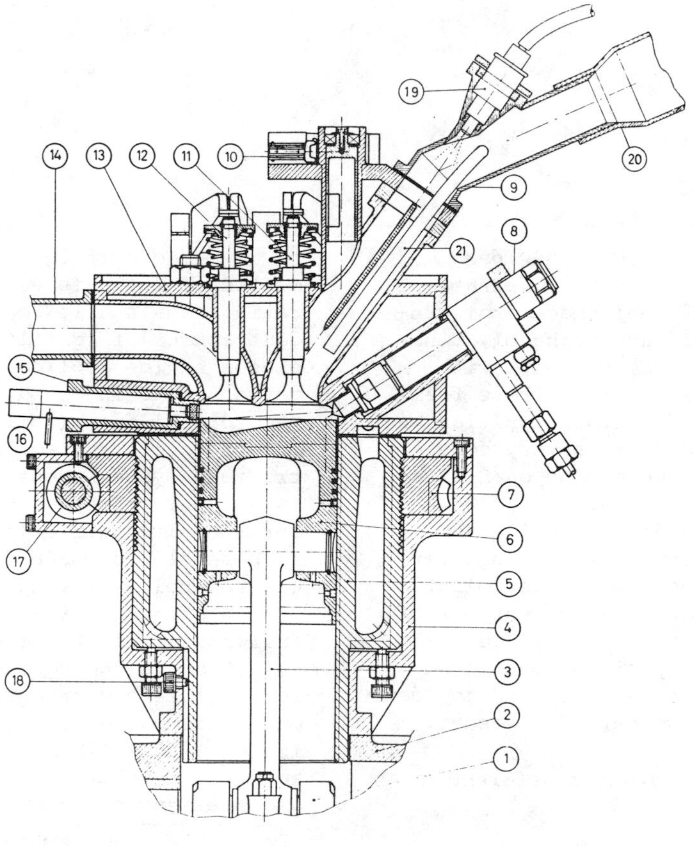

Fig. 1 - Cross-section of cylinder unit of the research diesel engine. Numbers refer to: 8) Pilot fuel injection nozzle, 9) Intake pipe 11) Intake valve, 12) Exhaust valve, 14) Exhaust pipe, 19) Electronic fuel injection nozzle for the liquid fuel, 20) Electric heater. 21) Supply tube for gaseous fuels

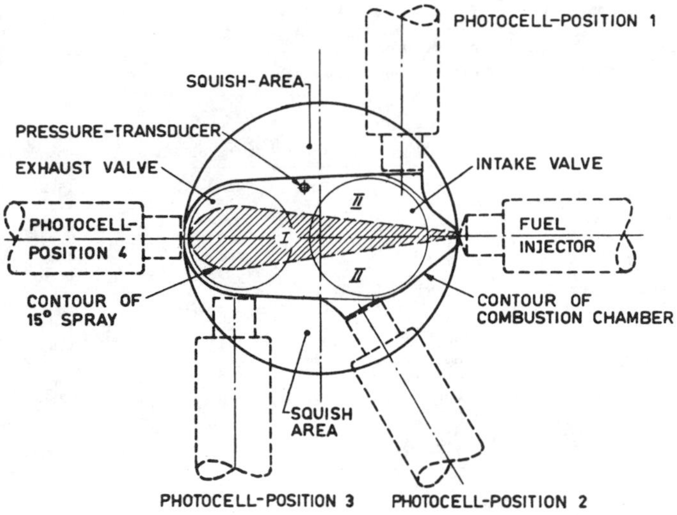

Fig. 2 - The combustion chamber, seen from below, showing the contour of a fuel spray with 15 degrees cone angle, and the two different combustion zones: I: Pilot fuel spray and main fuel combustion zone, II: Main fuel combustion only

flat, oblong combustion chamber with a minimum of air swirl is shown in figure 2. A detailed description of the engine, which was developed at the Technical University of Denmark, is given in (15).

The air drawn in by the engine was passed through a measuring orifice, a large air tank (in order to dampen the pressure pulsations from the engine) and a 2 kW electric heater element (position 2 in figure 1), which is able to preheat the intake air up to about 300 degrees C (approx. 570 degrees F) before mixing with the alternative fuel. As no valve overlap was used all of the intake air was trapped in the cylinder.

The gaseous alternative fuels Ammonia and Methane (from high pressure tanks) were added continuously to the intake air through a tube (position 21 in figure 1) ending very close to the intake valve. The amount of gaseous fuel added was determined by using a ROTA-flowmeter, a pressure gauge and a thermocouple.

The liquid alternative fuels Methanol and Ethanol were injected into the intake air using a Bosch electronic fuel injection nozzle with intermittent spray positioned in the intake pipe close to the cylinder head (position 19 in figure 1). The fuel supply pressure from the electric fuel pump was maintained constant at 2 bars and the amount of fuel supplied was controlled by a simple electronic circuit, by which the duration of the injection period could be varied. The amount of fuel injected was calculated from the time used for emptying a calibrated bottle.

The pilot diesel fuel was injected through a single hole fuel nozzle (position 8 in figure 1). The injection pressure was determined by using a piezoelectric pressure transducer, while the injector needle lift was established by a variable inductance type transducer. The amount of pilot fuel was calculated from the time used for emptying a calibrated bottle.

The combustion chamber pressure was indicated by a piezoelectric pressure transducer and the start of combustion was determined from the differentiated signal dP/dt.

The exhaust gas was analyzed for carbon monoxide CO, carbon dioxdide CO_2 and hydrocarbons HC (C_6) using NDIR-analyzers, for total hydrocarbons HC (C_1) by using a FID-analyzer and for nitrogen oxides NO and NO_x (= NO + NO_2) by using a chemiluminescence analyzer. A Hartridge smokemeter was used to determine smoke emissions.

EXPERIMENTAL PROCEDURE

A complete experimental investigation of the effects of all major engine variables on power, efficiency and emissions is too comprehensive even if only a single fuel is considered. It was therefore decided to define a reference condition for the engine based on a preliminary investigation and to perform the two parameter variations considered to be the most important: a variation of the amount of pilot fuel (with a constant amount of the alternative main fuel) and a variation of the amount of the alternative main fuel (with a constant amount of pilot fuel).

Table 1 - Specification of the Reference Condition used in both Series of the Experiments. The Engine Friction Loss at 1500 RPM is approximately 0.75 kW.

x : Diesel Fuel Equivalent

	Units	Diesel	Methanol	Ethanol	Methane	Ammonia
Engine speed	RPM	1500	1500	1500	1500	1500
Power output	kW	1.61	1.61	1.61	1.61	1.61
Pilot fuel consumption	kg/hour	-	0.200	0.202	0.245	0.242
Main fuel consumption	kg/hour	0.893	1.031	0.775	0.396	1.594
Specific fuel consumption[x]	g/kWh	554	401	408	463	572
Total efficiency	percent	14.7	20.3	19.9	17.6	14.2
Intake air consumption	kg/hour	22.1	23.2	21.8	20.1	20.3
Air excess ratio for main fuel	-	1.65	3.51	3.13	2.97	2.10
Total air excess ratio	-	1.65	2.41	2.18	1.92	1.52
Intake air temperature	deg. C	120	121	120	118	120
Ignition delay of pilot fuel	deg. CA	21	28	26	24	32
Start of injection	deg. CA b TDC	34	32	34	36	34
Injection pressure	bar	125	125	125	125	125
Max. cylinder pressure	bar	48	48	51	38	44
Exhaust gas temperature	deg. C	398	299	322	340	376
Smoke density (Hartridge)	% Hartr.	24	6	4	0	2

Table 1 specifies the reference-condition chosen. During the preliminary investigation it was found that 30 percent (on an energy basis) was the smallest amount of pilot fuel consistent with a reasonable efficient combustion in the present engine. In order to obtain a good atomization of the pilot fuel a high injection pressure and a small nozzle orifice diameter is desirable, but in order to limit the amount of pilot fuel to 30 percent an opening pressure for the nozzle of only 125 bars has to be used. The smallest nozzle available was used (orifice diameter 0.25 mm). The start of pilot fuel injection for each alternative fuel was chosen such as to optimize the engine efficiency.

As can be seen from table 1 no reference condition was defined for Hydrogen. This is due to the fact that in the present engine it was only possible to replace about 10 percent of the diesel pilot fuel with Hydrogen. When larger amounts of Hydrogen were added, the ignition stability of the pilot fuel was affected seriously, rapidly causing very frequent ignition failures and finally stopping the engine. Several times attempts were made to use larger amounts of Hydrogen, but only the above mentioned results were found. This observation differs from the findings of Karim and Klat (10), who were able to obtain satisfactory combustion within certain (although not too wide) limits of the Hydrogen/air-ratio. At too lean mixtures Karim and Klat found ignition failures, while too rich mixtures gave severe knocking. In the present engine ignition was not affected at very lean mixtures, but ignition failures (and no knocking) were found at richer mixtures. Due to this Hydrogen was not investigated further.

EXPERIMENTAL RESULTS

COMBUSTION PHENOMENA - Figures 3 to 6 show the cylinder pressure P and the differentiated pressure signal dP/dt versus time for Methanol, Ethanol, Methane and Ammonia respectively in the reference condition. As can be seen from the figures, there are pronounced differences in the course of combustion between the two alcohols and the two gases. The dP/dt-signals in figures 3 and 4 show two maxima, indicating that the combustion of the two alcohols is delayed compared to the combustion of the pilot fuel.

In principle three different types of combustion are possible, the one occurring depending on the air/fuel ratio for the alternative fuel:

(1) Single Combustion - Combustion only takes place inside the pilot fuel spray area (area I in figure 2) and only the pilot and probably the alternative fuel inside the pilot fuel spray area burns. This leads to decreased efficiency and increased emission of unburnt fuel, the latter mainly coming from area II (figure 2).

(2) Consecutive Combustion - The first combustion phase corresponds to "single combustion". The resulting increase in pressure and temperature in the combustion chamber is thought to make it possible for a flame to progress into the mixture of air and alternative fuel outside the pilot fuel spray area (area II in figure 2). The combustion of this mixture forms the second combustion phase which is clearly distinguishable in the cylinder pressure diagram and very pronounced in the dP/dt-signal (see figures 3 and 4).

(3) Simultaneous Combustion - Once the pilot fuel ignites combustion starts to take place both inside and outside the pilot fuel spray area. The pilot fuel and the alternative fuel burn in a single combustion phase.

The type of combustion found in a specific case depends on the ignition limits of the alternative fuel, the air/alternative fuel-ratio and the amount of pilot fuel. Further, during the preliminary experiments aimed at defining an appropriate reference condition, the intake air temperature was found to have a strong influence on the type of combustion occurring. In particular it was found that the Methanol

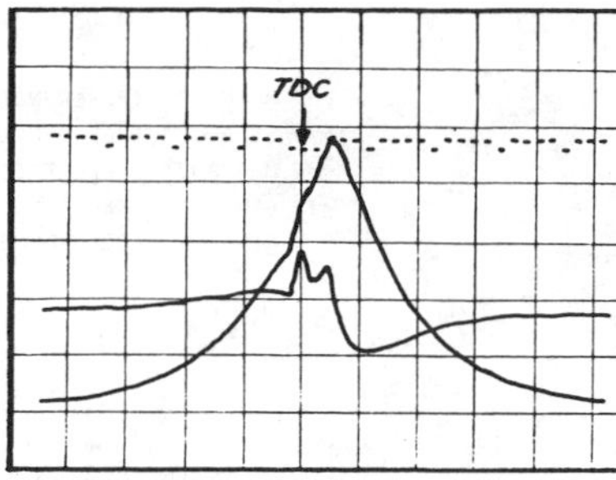

Fig. 3 - Time history of cylinder pressure (P) and the time-derived (dP/dt) in the reference condition (see table 1) for Methanol. Dots indicate degrees crank angle with 4 degrees CA between dots. Pressure scale 10 bars/division, time scale 2 msec/division

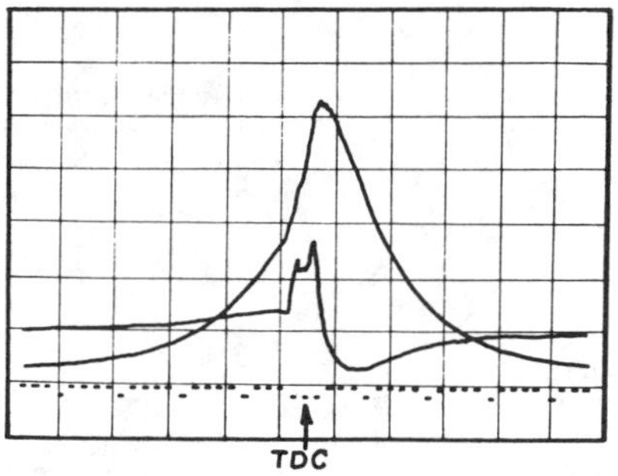

Fig. 4 - Time history of cylinder pressure (P) and the time derived (dP/dt) in the reference condition (see table 1) for Ethonal. Dots indicate degrees crank angle with 4 degrees CA between dots. Pressure scale 10 bars/division, time scale 2 msec/division

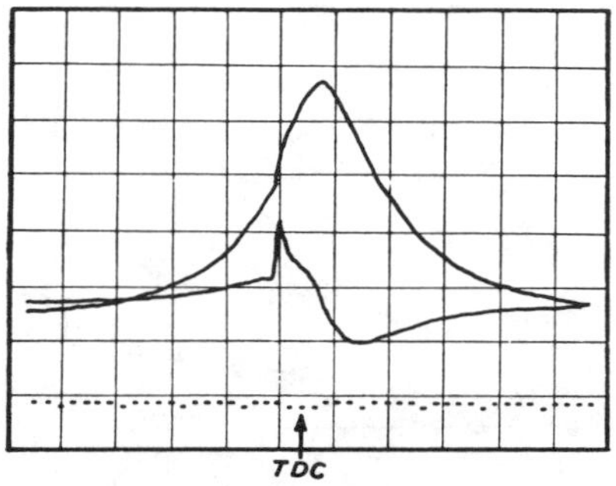

Fig. 5 - Time history of cylinder pressure (P) and the time derived (dP/dt) in the reference condition (see table 1) for Methane. Dots indicate degrees crank angle with 4 degrees CA between dots. Pressure scale 10 bars/division, time scale 2 msec/division

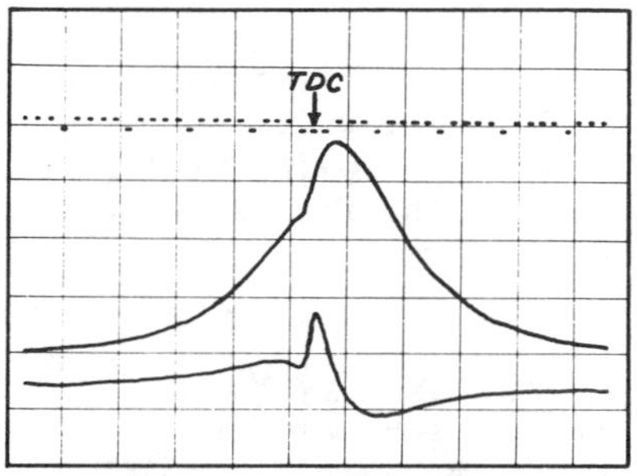

Fig. 6 - Time history of cylinder pressure (P) and the time-derived (dP/dt) in the reference condition (see table 1) for Ammonia. Dots indicate degrees crank angle with 4 degrees CA between dots. Pressure scale 10 bars/division, time scale 2 msec/division

combustion was very sensitive to changes in intake air temperature and that intake air temperatures below 120 degrees C (249 degrees F) could only be applied with considerable difficulty. It was also found that small changes in the pilot fuel injection system (like deposits on the nozzle) could cause the combustion to change from the consecutive type into single combustion with resulting loss in power and efficiency and increased emission of unburned fuel.

EFFECTS OF AMOUNT OF PILOT FUEL - Figures 8-17 present the main results of the experimental investigation of the effects of the amount of pilot fuel on power, efficiency, ignition delay, maximum cylinder pressure and exhaust emissions. All results are presented as functions of the total air excess ratio λ_{tot} (for definitions, see Appendix A). The amount of main fuel is kept constant so that the part of the total energy input supplied by the pilot fuel varies with λ_{tot} as shown in figure 7. For comparison a variation of the amount of fuel in pure diesel operation was performed and the results are shown together with the results from the dual-fuel operation.

As can be seen from figures 7, 8 and 9 Methanol requires at least 30 percent of the energy input to be pilot fuel, while the three other fuels can operate on lower amounts of pilot fuel. However the amount of pilot fuel (on energy basis) required for achieving the optimum engine efficiency for each fuel is 35 percent for Methanol, 34 percent for Ethanol, 43 percent for Methane and 35 percent for Ammonia and the best total efficiency is obtained with Methanol.

The pronounced decrease in efficiency and power output for Methanol and Ethanol near the total air excess ratio 2.5 can be referred to a change in the type of combustion from the consecutive type found in the reference condition

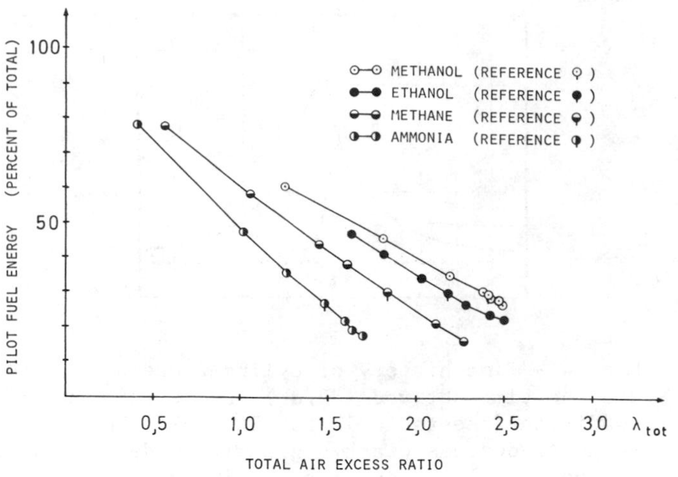

Fig. 7 - The percentage of the total energy input, supplied by the pilot (diesel) fuel as a function of the total air excess ratio λ_{tot} (see table 1). This figure applies to figures 8 to 17 only, which show the results of varying the amount of pilot fuel with constant amount of each of the four alternative fuels

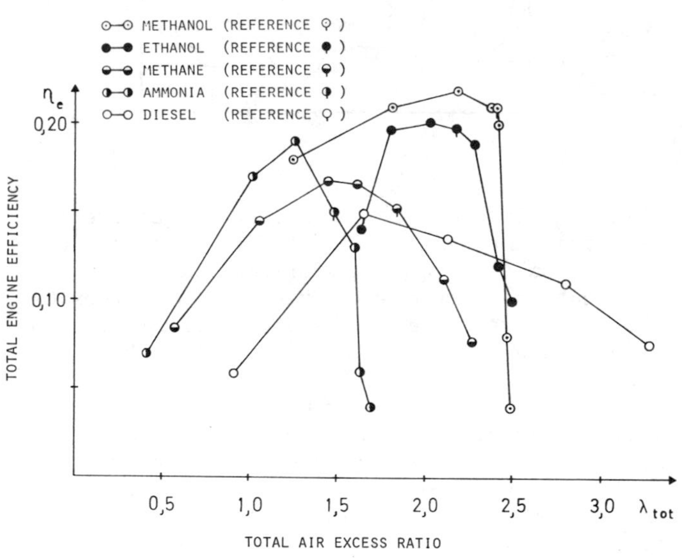

Fig. 9 - Variation of total engine efficiency η_e with amount of pilot fuel

to the single combustion type, resulting in a sharp increase in the emission of unburned fuel (figures 14 and 15).

Ammonia shows a similar character but naturally no increase in the HC-emission is found as Ammonia is not a hydrocarbon. However figures 16 and 17 show that the emission of unburned Ammonia is very considerable near the total excess air ratio 1.5. Figures 16 and 17 show NO and NO_x emission, but it must be remembered that

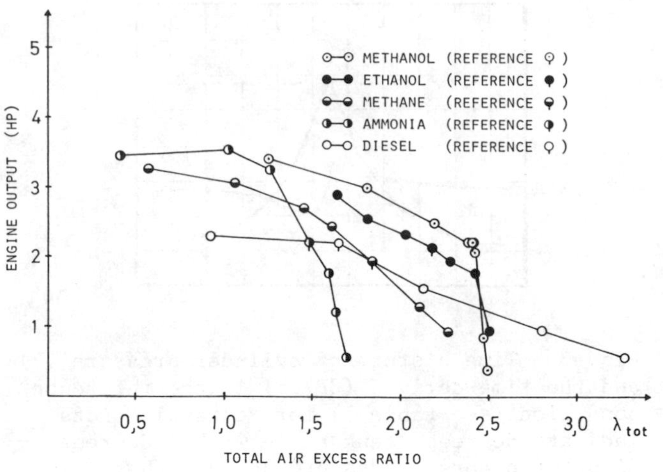

Fig. 8 - Variation of engine output with amount of pilot fuel

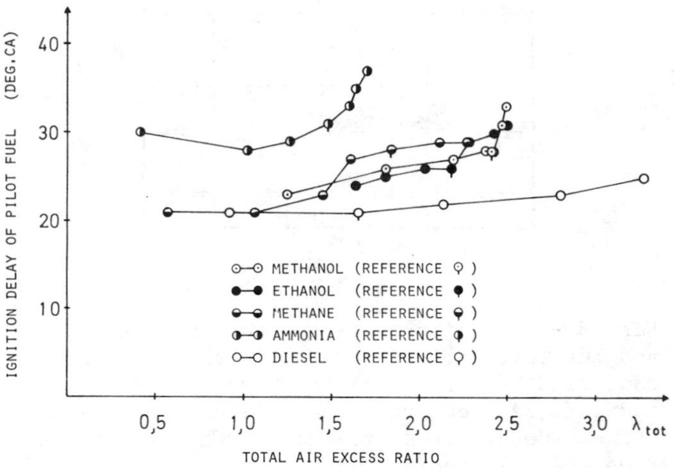

Fig. 10 - Variation of ignition delay of the pilot fuel (in degrees crank angle) with amount of pilot fuel

a chemiluminescence analyzer is sensitive to Ammonia when analyzing for NO_x (due to the converter) and not, when analyzing for NO. Thus the vast difference between NO_x and NO-emission around the total air excess ratio 1.5 is due to a strong emission of unburned Ammonia, indicating that the combustion of Ammonia only takes place in or near the pilot fuel spray area (I in figure 2). The combustion of Ammonia in general is slow and it is very difficult to determine whether it is of the single combustion type or of the consecutive type, but the former seems most likely.

As shown in figure 10, all of the four fuels influence the ignition process of the pilot diesel fuel, with Ammonia giving the strongest increase in ignition delay. There

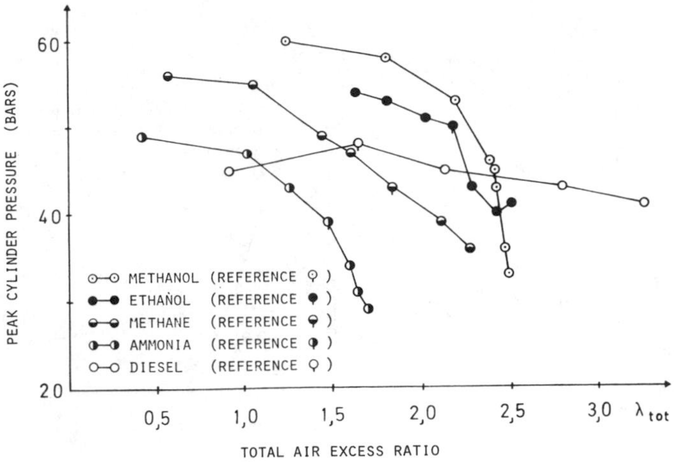

Fig. 11 - Variation of peak cylinder pressure with amount of pilot fuel

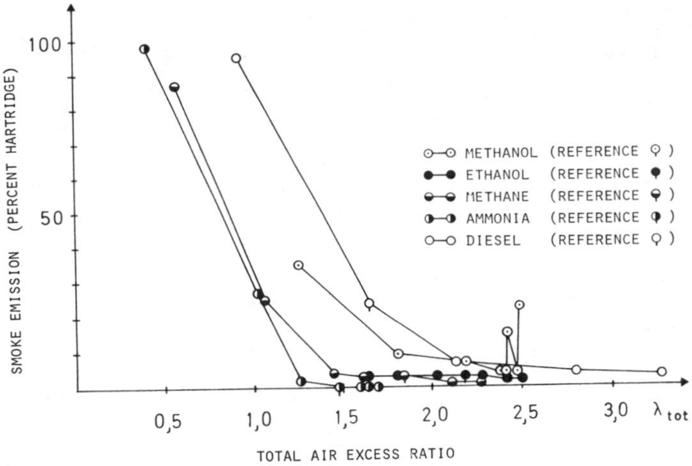

Fig. 12 - Variation of smoke emission, measured by a Hartridge smokemeter, with amount of pilot fuel

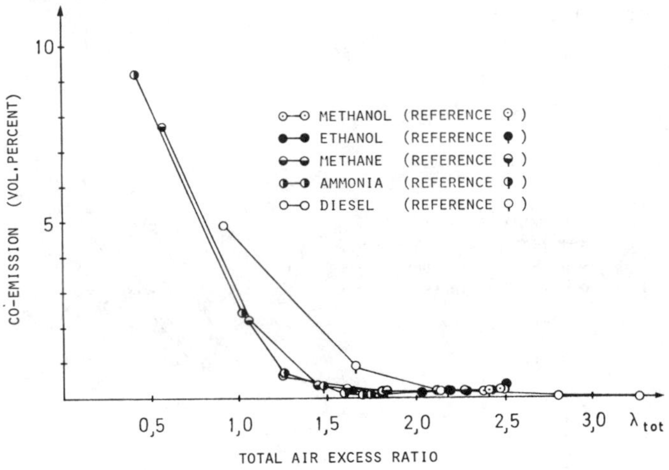

Fig. 13 - Variation of CO-emission with amount of pilot fuel

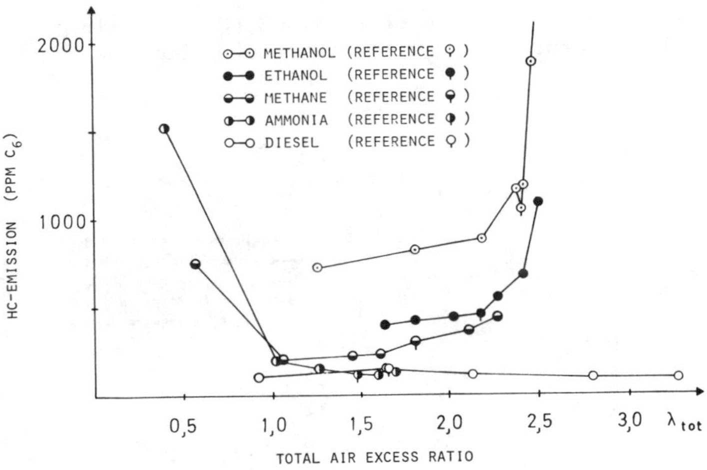

Fig. 14 - Variation of HC (C_6)-emission, (NDIR-measurements) with amount of pilot fuel

is a general tendency towards less influence at larger amounts of pilot fuel (low total air excess ratio) which might be due partly to a higher temperature level at the higher engine load and partly to a relative decrease in the concentration of the low cetane rated main fuels. The latter possibility seems most likely as the variation in ignition delay with engine load for pure diesel operation is much less than with any of the four alternative fuels.

The maximum combustion pressures for dual-fuel operation are generally higher than for pure diesel operation as shown in figure 11, but the increase is modest and is not thought to prohibit the use of any of the four fuels.

Except at rich mixtures, the amount of pilot fuel has no effect on smoke emission (figure 12), which is reduced considerably compared to pure diesel operation. The same is true regarding the CO-emission at larger amounts of pilot fuel, while the CO-emission at low amounts of pilot fuel is about equal to the values for pure diesel operation.

The emission of unburned fuel (figures 14 and 15 for Methanol, Ethanol and Methane, figure 16 and 17 for Ammonia) in dual-fuel operation is considerably higher than in pure diesel operation and in general is improved at larger amounts of pilot fuel due to improved combustion of the alternative fuel outside the pilot fuel spray area as discussed earlier.

The NO and NO_x-emission at dual-fuel operation is not very much influenced by the amount of pilot fuel except for the well-known dependence on the total air excess ratio, giving

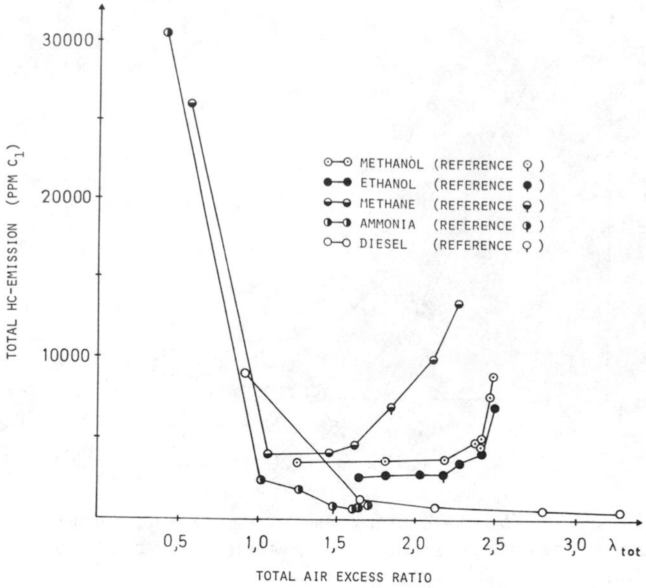

Fig. 15 - Variation of total HC (C_1)-emission (FID-measurements) with amount of pilot fuel

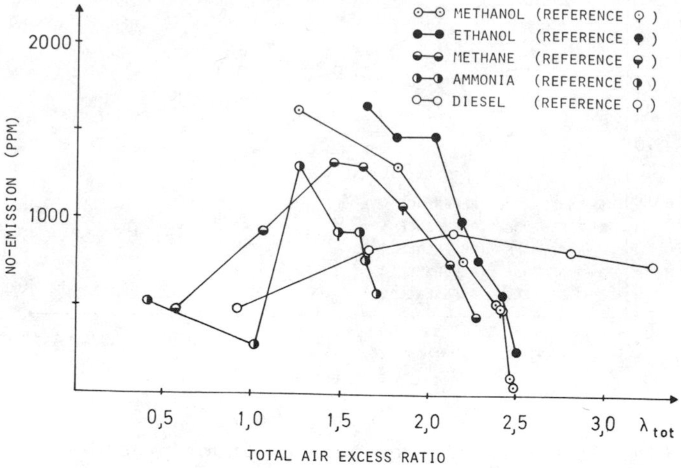

Fig. 16 - Variation of NO-emission with amount of pilot fuel

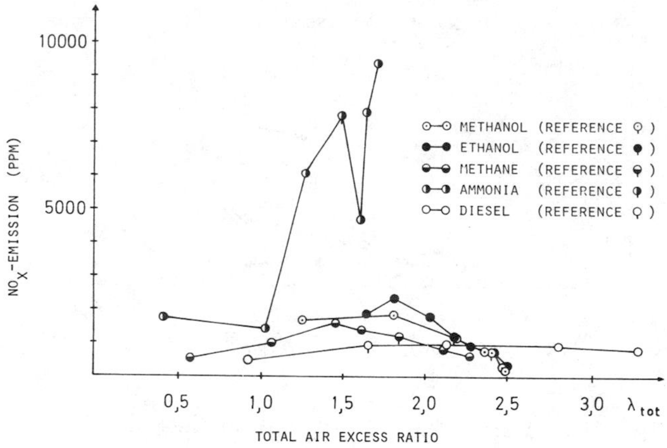

Fig. 17 - Variation of NO_x-emission with amount of pilot fuel

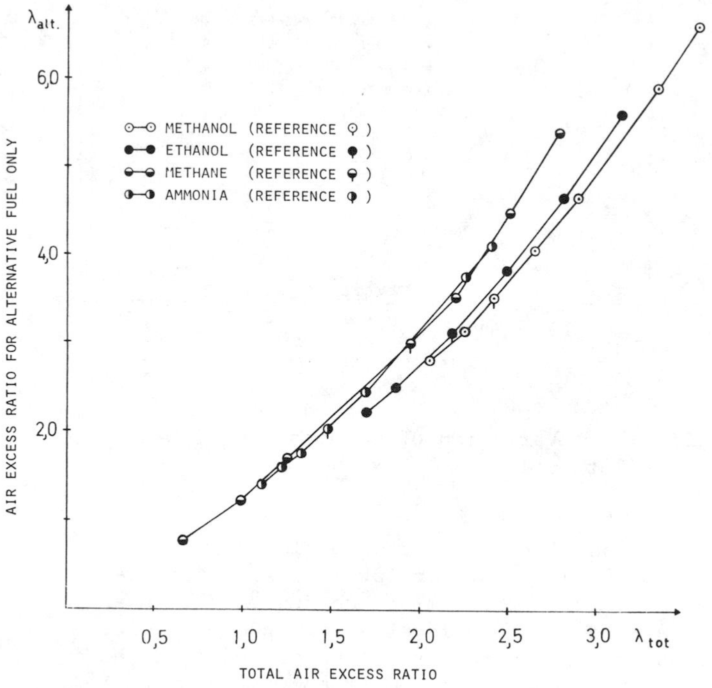

Fig. 18 - Relationship between the total air excess ratio λ_{tot} and the air excess ratio λ_{alt} for the alternative (main) fuel only, at a constant amount of pilot fuel (see table 1). This figure applies to figures 19 to 28 only, which show the results of varying the amount of each of the four alternative fuels with constant amount of pilot fuel

the maximum emission strength at a mixture somewhat leaner than stoichiometric.

EFFECTS OF AMOUNT OF ALTERNATIVE FUEL - Figures 19-28 present the main results of the effects of amount of (main) alternative fuel on power, efficiency, ignition delay, maximum cylinder pressure and exhaust emissions. Like before the results are presented as functions of the total air excess ratio λ_{tot}, but a key to translate this into air excess ratio for the alternative fuel only λ_{alt} is given in figure 18.

Figures 19 and 20 in general show that the engine output and the total efficiency increases

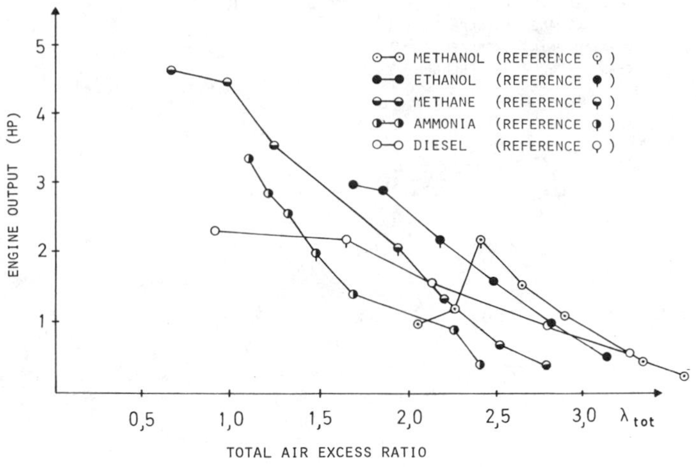

Fig. 19 - Variation of engine output with amount of alternative fuel

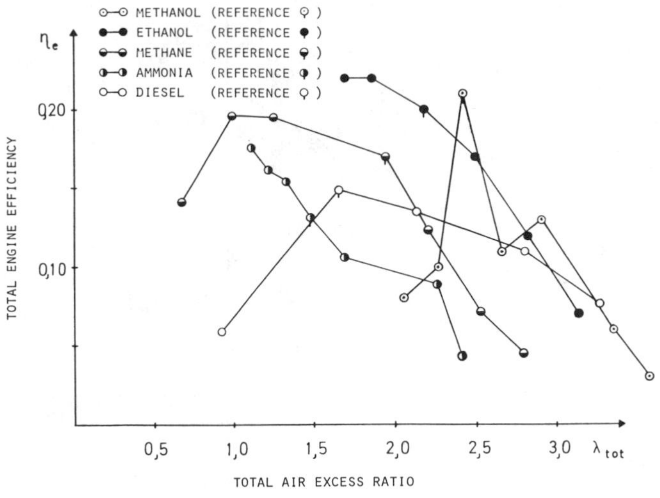

Fig. 20 - Variation of total engine efficiency $_e$ with amount of alternative fuel

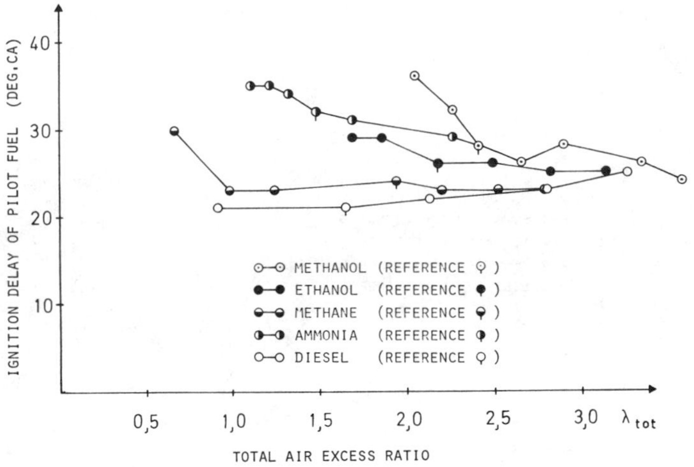

Fig. 21 - Variation of ignition delay of the pilot fuel (in degrees crank angle) with amount of alternative fuel

as the concentration of the alternative fuel increases. The increase in efficiency is mainly due to an increase in mechanical efficiency, resulting from the higher indicated mean effective pressure. Methanol however behaves in a different manner, as both engine output and efficiency decrease considerably when the total air excess ratio λ_{tot} becomes less than 2.5, corresponding to an air excess ratio for the Methanol of about 3.5.

From figure 21 it can be seen that although the ignition delay of the pilot fuel increases at a higher rate when lowering λ_{tot} below 2.5 than at higher λ_{tot}-values, the ignition delay itself is not larger than when using Ammonia at λ_{tot} between 1.0 and 1.5, in which case the total efficiency is much better. So the decrease in power and efficiency for Methanol at λ_{tot} lower than 2.5 is mainly not due to late ignition but, as can be seen in figures 25 and 26, due to a change in the combustion type of the Methanol (into "single combustion") indicated by the increase in the emission of unburned fuel and a change in the dP/dt-signal. So in this engine Methanol will burn in very lean mixtures (air excess ratio about 3 and leaner), while in spark ignition engines Methanol will only burn at mixtures richer than an air excess ratio of about 1.5 (17). No fully satisfactory explanation was found to this behaviour, but part of the explanation could be that inlet air temperature

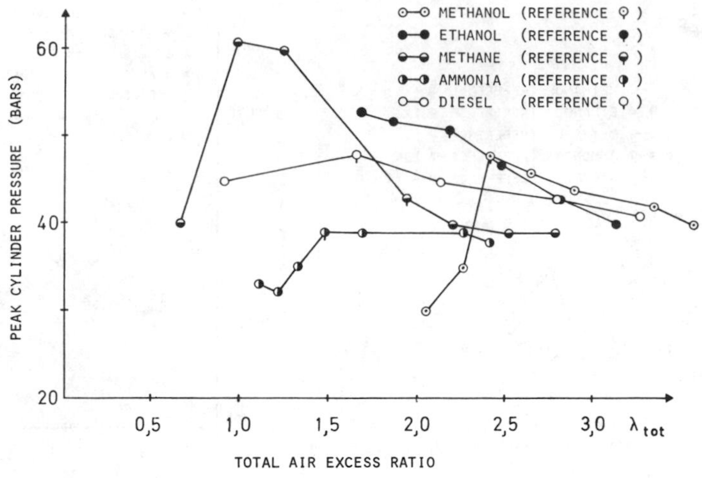

Fig. 22 - Variation of peak cylinder pressure with amount of alternative fuel

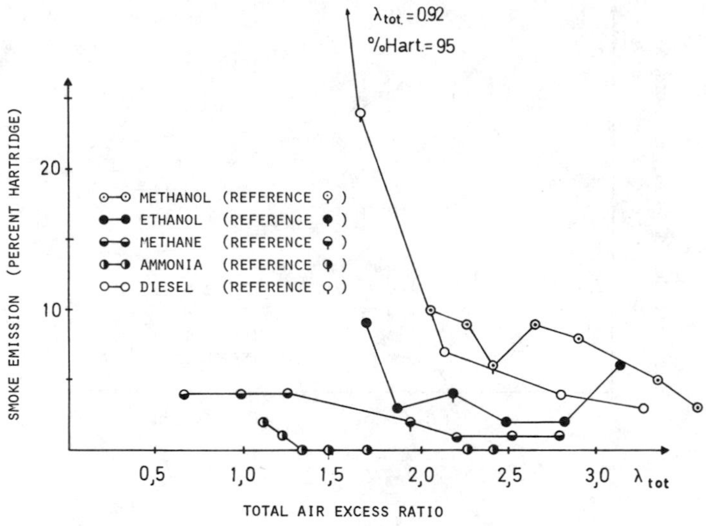

Fig. 23 - Variation of smoke emission, measured by a Hartridge smokemeter, with amount of alternative fuel

and not inlet mixture temperature was held constant. At larger amounts of Methanol the mixture temperature decreases due to Methanol's high latent heat of vaporization and this might cause excess cooling and poor mixing.

Figure 21 shows that the ignition process of the pilot fuel is influenced by all of the four alternative fuels, causing increases in the ignition delay up to about 50 percent when using Methanol and Ammonia in "rich" mixtures, while Methane only causes marginal increases except at mixtures richer than stoichiometric.

The maximum combustion pressures are slightly higher when using the two alcohols than in pure diesel operation, but lower when using Ammonia. The largest increase in maximum pressure however is found near a stoichiometric mixture when using Methane, in which case the engine output is approximately doubled. In general the change in combustion peak pressure is considered to be modest and is not thought to prohibit the use of any of the alternative fuels.

Methane and Ammonia act as quite efficient smoke-suppressors, allowing virtually smoke-free operation even at rich mixtures as can be seen from figure 23. The two alcohols only slightly influence the smoke-emission with Me-

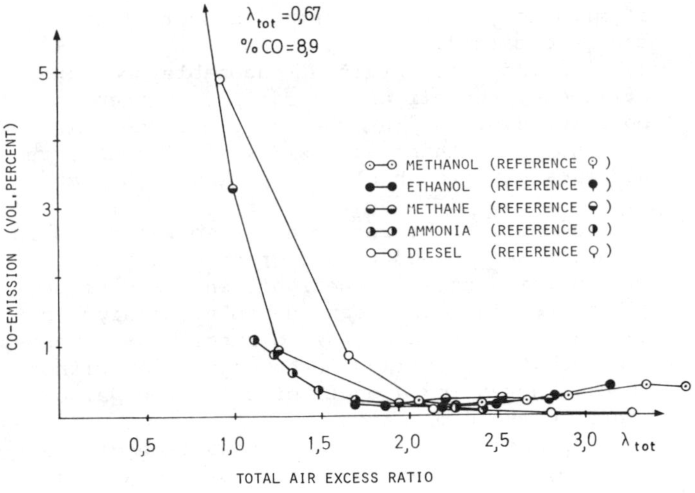

Fig. 24 - Variation of CO-emission with amount of alternative fuel

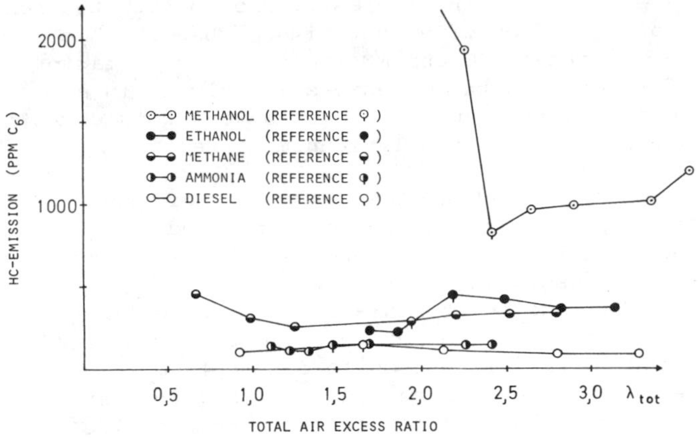

Fig. 25 - Variation of HC (C_6)-emission (NDIR-measurements) with amount of alternative fuel

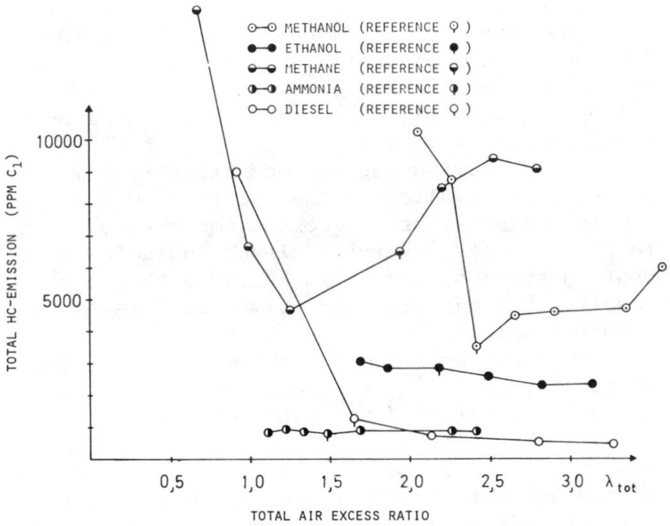

Fig. 26 - Variation of total HC (C_1)-emission (FID-measurements) with amount of alternative fuel

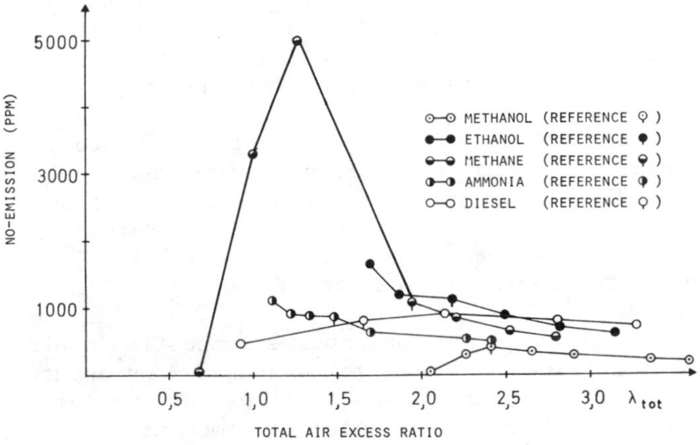

Fig. 27 - Variation of NO-emission with amount of alternative fuel

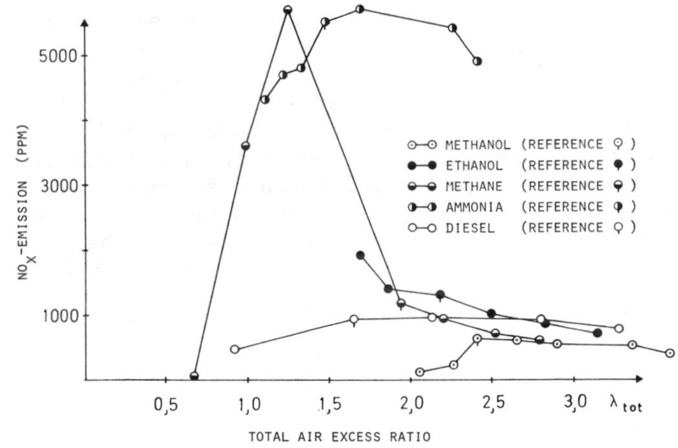

Fig. 28 - Variation of NO_x-emission with amount of alternative fuel

thanol giving a slight increase and Ethanol a slight decrease.

The CO-emission is largely unchanged except at richer mixtures where a certain decrease is found (figure 24).

When using any of the four fuels a considerable increase in the emission of unburned fuel is found as can be seen in figures 25 and 26 for Methanol, Ethanol and Methane and figures 27 and 28 for Ammonia.

The emission of Nitrogen oxides when using Ethanol and Ammonia does not differ very much from the emission in pure diesel operation as can be seen in figures 27 and 28. Due to the cooling effect of evaporating Methanol, this fuel decrease NO and NO_x-emissions by a factor of about 2. Methane behaves largely as in a spark ignition engine giving a very considerable emission at a slightly lean mixture.

OTHER OBSERVATIONS - A number of other observations of engine operating conditions and fuel interactions were made, some of which are discussed in the following.

Methanol - It has already been mentioned that intake air temperatures below 120 degrees C caused the combustion to change from the consecutive type to single combustion. Further it was found that air temperatures between 120 and 150 degrees C made the consecutive combustion more stable but at higher temperatures (about 200 degrees C) heavy knock occurred.

In order to control the power output of a dual-fuel engine the ideal would be to supply a constant amount of pilot fuel, corresponding to the indicated power demand at idle and to increase power simply by adding an appropriate amount of the alternative (main) fuel to the unthrottled intake air. This is not possible with any of the four fuels investigated here if a reasonable engine efficiency (compared to pure diesel operation) and low emissions of unburned fuel are to be maintained. For Methanol, the air excess ratio must be between 3 and 6 to do so as can be seen from figures 18, 20, 25 and 26.

Ethanol - The consecutive combustion could be maintained with air temperatures down about 90 degrees C and in general it was no problem to obtain a stable consecutive combustion. At intake air temperatures above 160 degrees C heavy knock occurred and at intake air temperatures about 200 degrees C self-ignition of the Ethanol occurred during the compression stroke.

In order to obtain a reasonable engine efficiency the air excess ratio for Ethanol must be lower than about 5 as can be seen from figures 18 and 20.

Methane - It was possible to run the engine with intake air temperatures down to about 30 degrees C, but with a certain loss in efficiency due to a weak second period of the consecutive combustion. A distinct consecutive combustion was seen with intake air temperatures above 130 degrees C and at higher temperatures the second period becomes very fast - at 200 degrees C giving a higher rate of pressure rise than the first period. Despite this high intake air temperature no knocking occurred.

The ignition delay of the pilot fuel was nearly unaffected except at very rich mistures where the oxygen concentration is considerably decreased. For this reason it is believed that the amount of pilot fuel can be decreased considerably below the values used in the present engine, provided that a sufficiently good atomization of the very small amount of fuel can be obtained.

In order to obtain a reasonable good engine efficiency the air excess ratio of Methane should be below 3.5, but in order to limit the emission of unburned Methane and oxides of nitrogen, the air excess ratio should not be lower than about 2.

Ammonia - The intake air temperature could be decreased to about 60 degrees C without ignition failures, but the ignition delay of the pilot fuel increased considerably. At high intake air temperatures the combustion was improved and at 300 degrees C the engine ran without knock. Increasing amounts of pilot fuel also improved the combustion.

In order to obtain a reasonable engine efficiency the air excess ratio for Ammonia should be below 2, but even in that case there is a strong emission of unburned Ammonia (figures 27 and 28) which represents a major problem to engine applications of Ammonia.

CONCLUSION

The investigation showed that all of the four mentioned alternative fuels can be used as fuels for the high speed direct injection diesel engine with pilot injected diesel fuel as the source of ignition.

The following conclusions are strictly speaking only valid for the engine used for the experiments, as the combustion system differs from most commercial diesel engine combustion systems, but it is believed that the results qualitatively are valid for direct injected diesel engines in general.

METHANE was found to be the most suitable alternative fuel. It was possible to run the engine at total air/fuel-ratios below stoichiometric, thus increasing the maximum output of the engine considerably. The ignition delay of the pilot injected diesel fuel was only slightly affected except at very rich mixtures.

AMMONIA was found to be the least suitable alternative fuel, mainly because of a strong emission of unburnt Ammonia (Odour). Ammonia also gave the strongest increase in the ignition delay of the pilot injected diesel fuel and the smallest increase in maximum power output and efficiency. The latter effects are thought to be due mainly to a very slow combustion of the Ammonia.

METHANOL and ETHANOL showed the same type of behaviour, Ethanol being the most suitable. Methanol was found to be very sensitive to operating conditions. Small changes in the amount of pilot fuel or the amount of Methanol could cause large changes in the power output and the same was found regarding the pilot fuel injection system. A satisfactory explanation of this sensitivity was not found.

EMISSIONS - The use of any of the four alternative fuels reduced the emission of smoke from the engine and increased the emission of unburnt fuel (hydrocarbons and/or Ammonia). The emission of carbon monoxide was largely unchanged for all of the four fuels. The emission of nitrogen oxides increased moderately using Ethanol and Ammonia, while Methane gave a nearly 5-fold increase near the stoichiometric mixture and Methanol gave a somewhat lower emission.

NEED FOR FURTHER RESEARCH - The present investigation has only dealt with the effects of a few parameters. There is a general need for further research regarding the effects of the same parameters using other engines and regarding other parameters not dealt with in the present work. The different behaviour of Hydrogen in the present engine and in the engine used by Karim and Klat (10) might indicate that some engine design variables are important. The present work has shown that Methanol is very sensitive to a number of variables and the causes for this sensitivity should be investigated further.

ACKNOWLEDGEMENTS

The authors wish to express their thanks to Statens teknisk-videnskabelige Forskningsråd (The Danish Council for Scientific and Technical Research), who sponsored this investigation.

REFERENCES

1. H.E.C. Andersen, "Brændstoffer fra alternative energikilder - grundlæggende krav og fremstillingsmuligheder" (Engine Fuels from Alternative Energy Sources for Present Combustion Engines - Basic Demands and Production Possibilities). Technical University of Denmark, Laboratory for Energetics, Report RE 75-12, 1976

2. N.D. Whitehouse, "Advances in British Dual-Fuel and Gas Engines". Diesel Engineers and Users Association, Publication 353, 1973

3. A. Alberstein, W.B. Swim and P.H. Schweitzer, "Fumigation Kills Smoke". SAE-Transactions, Vol. 66 (1958), p 574

4. H.A. Havemann, M.R.K. Rao, A. Natarajan and T.L. Narasimhan, "Alcohol with Normal Diesel Fuels - Part I". Gas and Oil Power, January 1955, p 15

5. H.A. Havemann, M.R.K. Rao, A. Natarajan and T.L. Narasimham, "Alcohol with Normal Diesel Fuels - Part II". Gas and Oil Power, February 1955, p 45

6. Y. Hirako and M. Ohta, "Effect of Lean Pre-Mixture on the Combustion in Diesel Engines". Bulletin of Japan Society of Mechanical Engineers, Vol. 16 (1973), No. 101, p 1750

7. W. Lowe and P.T. Brandham, "Development and Application of Medium Speed Gas Burning Engines". Proceedings of the Institution of Mechanical Engineers, Vol. 186 (1972), No. 6

8. H. Tesarek, "Investigations Concerning the Employment Possibilities of the Diesel-Gas Process for Reducing Exhaust Emissions, Especially Soot (Particulate Matters)". SAE-paper 750158, 1975

9. G.A. Karim, S.R. Klat and N.P.W. Moore, "Knock in Dual-Fuel Engines". Proceedings of the Institution of Mechanical Engineers, Vol. 181 (1967), Pt. 1, No. 20 p 453

10. G.A. Karim and S.R. Klat, "Experimental and Analytical Studies of Hydrogen as a Fuel in Compression Ignition Engines". ASME-paper 75-DGP-19, 1975

11. G.A. Karim, "Combustion in Dual-Fuel Engines - A Status Report". Preprint A1. 8th International Congress on Combustion Engines, CIMAC, Brussels, May 1968

12. T.J. Gray, E. Dimitrof, N.T. Meckel and R.D. Quillian Jr., "Ammonia fuel - Engine Compatibility and Combustion". SAE-paper 660156, 1966

13. E.S. Starkmann, G.E. James and H.K. Newhall, "Ammonia as a Diesel Fuel: Theory and Application". SAE-paper 670946, 1967

14. T.J. Pearsall and C.G. Garabedian, "Combustion of Anhydrous Ammonia in Diesel Engines". SAE-paper 670947, 1967

15. P. Sunn Pedersen, "A Diesel Research Engine". Technical University of Denmark, Laboratory for Energetics, Report RE 74-11, 1974

16. K. Bro, "Forsøg med Metanol, etanol, metan, brint og ammoniak med pilotindsprøjtning i en forsøgsdieselmotor" (An Experimental Investigation of Methanol, Ethanol, Methane, Hydrogen and Ammonia as Fuels in a Research Diesel Engine with Pilot Injection). Technical University of Denmark, Laboratory for Energetics, Report RE 74-14, 1974

17. "Neuen Kraftstoffen auf der Spur". Bundesministerium für Forschung und Technologie, Bonn 1974

18. H.A. Steiger and J.A. Smit, "Development and Practical Application of the Large-Bore Direct-Drive Dual-Fuel Engine as Propulsion Unit for LNG-Carriers". ASME-paper 75-DGP-1, 1975

APPENDIX A: DEFINITIONS

Air Excess Ratio λ - In general λ is defined as

$$\lambda = \frac{\text{Air/Fuel-Ratio}}{\text{Stoichiometric Air/Fuel-Ratio}} = \frac{1}{\Phi} \quad \text{(A-1)}$$

where Φ is the equivalence ratio. In the present context both total air excess ratio λ_{tot} and air excess ratio for the alternative fuel only λ_{alt} are used. The definitions are:

$$\lambda_{tot} = \frac{M_a}{(A/F)_{s,mf} \cdot M_{mf} + (A/F)_{s,pf} \cdot M_{pf}} \quad \text{(A-2)}$$

$$\lambda_{alt} = \frac{M_a}{(A/F)_{s,mf} \cdot M_{mf}} \qquad \text{(A-3)}$$

where:

M_a	mass of cylinder air charge per cycle
M_{mf}	mass of main fuel per cycle
M_{pf}	mass of pilot fuel per cycle
$(A/F)_{s,mf}$	stoichiometric Air/Fuel-Ratio for main fuel
$(A/F)_{s,pf}$	stoichiometric Air/Fuel-Ratio for pilot fuel

The stoichiometric Air/Fuel-Ratios are given in Table A 1.

Total Engine Efficiency η_e - The following is used for η_e :

$$\eta_e = \frac{E}{\dot{M}_{mf} \cdot H_{mf} + \dot{M}_{pf} \cdot H_{pf}} \qquad \text{(A-4)}$$

where:

E	engine power output to brake (kW)
$\dot{M}_{mf}$	mass flow of main fuel to engine (kg/s)
$\dot{M}_{pf}$	mass flow of pilot fuel to engine (kg/s)
H_{mf}	net heat of combustion for main fuel (kJ/kg)
H_{pf}	net heat of combustion for pilot fuel (kJ/kg)

The net heat of combustion values used are listed in table A 1. As can be seen the net heat of combustion for Methanol and Ethanol are based on data for the liquid, while the values used for Methane and Ammonia are based on data for the gases. This difference corresponds to the conditions during the experiments, which are also believed to be the most realistic for future applications. The energy required for the evaporation of the Ammonia and for compensating the cooling due to the expansion for both Methane and Ammonia is not taken into account as this was taken from the surroundings. Under conditions where this is not possible (very low ambient temperatures), the energy required can be taken from waste heat in the engine cooling water or exhaust gases.

The energy consumption of the electric intake air preheater is not included in the energy input to the engine in the calculation of the total engine efficiency, as this energy could also have been supplied by the waste heat in the exhaust gases. For convenience this was not done during the experiments.

Ignition Delay - The ignition delay of the pilot fuel is defined as the time from the start of needle lift to the first visible deviation from the motored trace of the differentiated cylinder pressure signal dP/dt.

Table A 2 - Fuel Specifications. L indicates liquid fuel, G indicates gas

		Net Heat of Combustion	Stoichiometric Air/Fuel-Ratio
		kJ/kg	kg/kg
Diesel Fuel	(L)	44300	15.0
Methanol	(L)	19674	6.4
Ethanol	(L)	26790	9.0
Methane	(G)	55150	17.2
Ammonia	(G)	18585	6.1

Three-Way Conversion Catalysts on Vehicles Fueled with Ethanol-Gasoline Mixtures*

J. J. Mooney, J. G. Hansel and K. R. Burns
Engelhard Industries Div.
Engelhard Minerals & Chemicals Corp.
Menlo Park, N. J.

Current worldwide requirements for new energy has renewed interest in alcohol as an extender to replace a certain fraction of gasoline as an automotive fuel. There is particular interest in fermentation-derived ethanol or coal-derived methanol as blending agents in motor gasoline. Brazil, as well as other countries including some farming areas of the United States, has encouraged the use or proposed the use of alcohols in gasoline blends(1)*. The use of alcohols in motor fuel has been discussed, and in some instances, used in limited quantities for nearly fifty years(2,3). Early use was often in an era when motor fuel quality and environmental considerations were not as important as they are today. Additionally, the high octane properties of alcohol may be of greater interest than they have been historically.

*Numbers in parentheses designate references listed at the end of the paper.

Fundamentally a problem has existed using various blends of alcohol and gasoline because of differences in stoichiometric air/fuel ratios of the two fuels. A different calibration of the carburetor is required for various ratio blends of the two fuels to maintain equivalent exhaust emission and driveability characteristics; which, on a realistic basis, has not been practical. With the introduction of the three-way conversion (TWC) catalyst/closed-loop feedback air/fuel metering control system, which is designed to control the air/fuel mixture, a potential solution to this fundamental problem appeared to exist. With this consideration, an investigation was conducted on such a system to characterize system response with various fuel blends and to determine the inherent capability of the system to compensate the air/fuel mixture in view of changed fuel stoichiometry.

*Paper 790428 presented at the Congress and Exposition, Detroit, Michigan, February 1979.

ABSTRACT

Three-way conversion catalysts systems which provide control of three emission components — hydrocarbons, carbon monoxide, and oxides of nitrogen — generally require the use of closed-loop feedback control of the air/fuel metering system to provide a near stoichiometric exhaust gas composition. As an alternate fuel such as ethyl-alcohol is added to gasoline, the stoichiometric air/fuel ratio is altered. Conventional air/fuel metering systems do not compensate for this change. This paper studies the ability of the Volvo closed-loop control fuel injection system, now in use on vehicles in the United States, to maintain the stoichiometric air/fuel mixture needed for optimum emission control and to compensate for the change in fuel stoichiometry encountered in gasoline/alternate fuel mixtures. The system was found to have the ability to completely compensate for gasoline/ethyl alcohol mixtures up to 30% ethyl alcohol with no loss in emission control. Gasoline/methyl alcohol blends were also tested successfully.

There is definite potential for the three-way conversion catalyst/closed-loop systems to provide our nation with flexibility in fuel formulation not otherwise found in conventional fuel metering systems. With proper planning and engineering, non-petroleum derived fuels can be blended with gasoline without requiring engine adjustments or compromising emission controls.

SAE/PT-80/19/$02.50

INTRODUCTORY DISCUSSION

Limited current use of alcohol/gasoline blends is generally accomplished without efforts to compensate for the significant difference in stoichiometric air/fuel ratios of the fuel components. For example, commercial unleaded gasoline has a stoichiometric air/fuel ratio of about 14.6, whereas, ethanol has a stoichiometric air/fuel ratio of 9.0. As may be seen in Figure 1 (see Appendix B for details), the stoichiometric air/fuel ratio of an ethanol/gasoline blend varies linearly between these values. Moreover, a conventional gasoline carburetor, or fuel injection system, will tend to follow a path similar to the Fuel Metering Characteristic Line in Figure 1 as the percentage of ethanol is increased from 0% (100% gasoline) in an ethanol/gasoline blend. As a result, conventional carburetors that are manufactured for normal gasoline use, near a stoichiometric air/fuel ratio of 14.6, would provide an air/fuel ratio of perhaps 14.3 (from Figure 1) if operated on a 30% ethanol/gasoline blend. 14.3 air/fuel ratio is considerably leaner than the 12.8 stoichiometric air/fuel ratio for 30% ethanol (from Figure 1) and could require carburetor adjustment in order to maintain equivalent performance.[4] Conversely, if a carburetor is adjusted at near the stoichiometric point for 30% ethanol in gasoline, and then the vehicle is operated on 100% gasoline, the resulting richer mixture will produce substantially increased CO and HC emissions — and will be wasteful of energy.

Figure 1

ETHANOL/GASOLINE STOICHIOMETRY

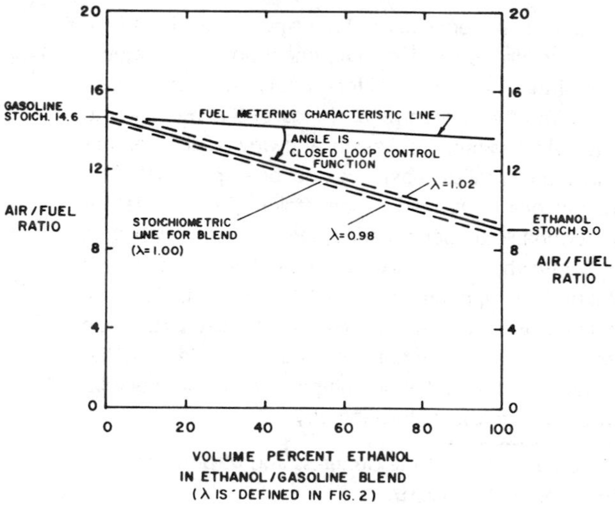

Among the recognized potential problem areas associated with the use of these alcohol/gasoline blends are[1,4,5,6]

- Volatility — leading to cold and hot starting difficulties.
- Mixture distribution within the engine.
- Water pick-up and component separation.
- Corrosion and chemical attack of fuel system and engine system components.
- Stoichiometric differences between alcohol and gasoline with the associated exhaust emission considerations.

It is the last item — the differences in air/fuel ratio requirements with consideration of exhaust emission control — to which this paper is specifically directed.

When considering systems capable of controlling HC and CO emissions from a vehicle fueled with ethanol/gasoline or methanol/gasoline blends, several existing systems (designed for gasoline) may be examined, as in Tables I and II. The last system in Tables I and II, namely, the development of closed-loop fuel metering systems for use with a three-way conversion (TWC) catalysts has — from the tests discussed in this paper — provided a means for achieving low exhaust emissions and automatically adjusting the fuel injection system for the differences in stoichiometry among a variety of fuel blend components.

TABLE I

FOUR EXISTING SYSTEMS (FOR GASOLINE) FOR POTENTIAL CONVERSION TO OPERATION ON ALCOHOL/GASOLINE BLENDS WITH CONTROL OF HC, CO AND NOx EMISSIONS.

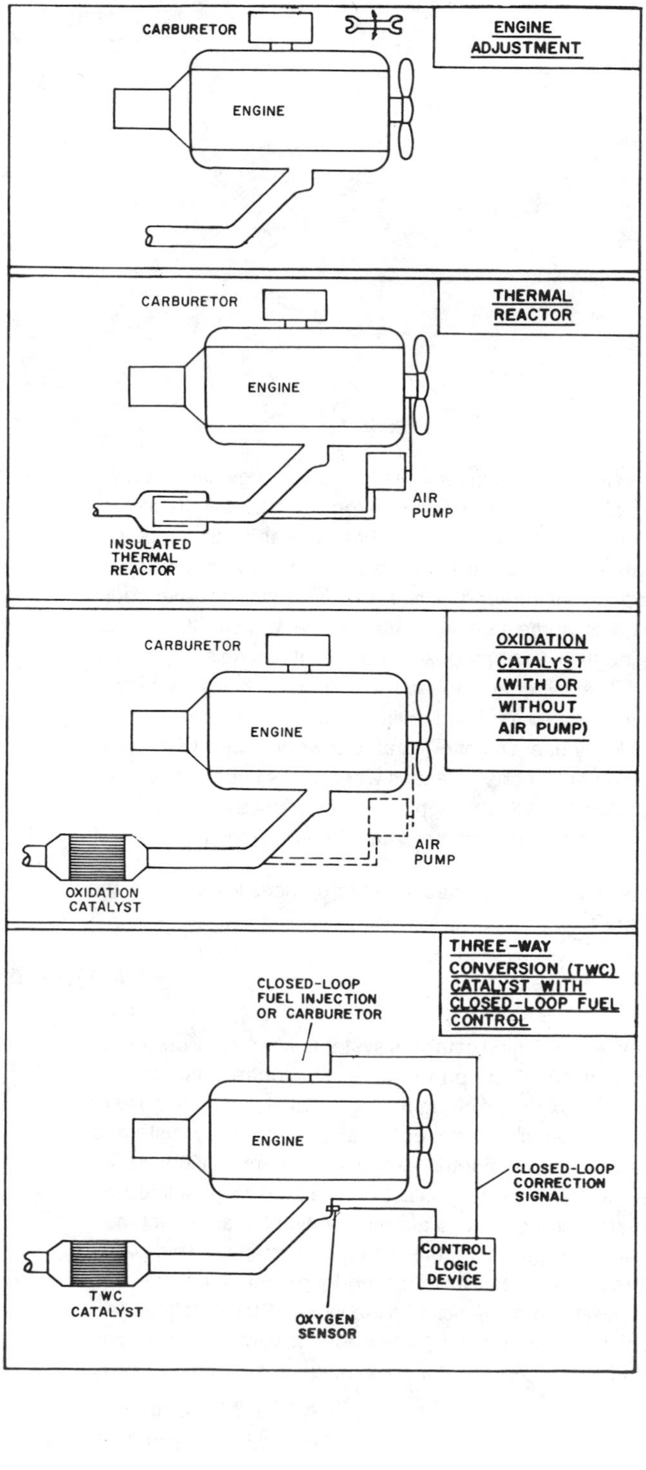

TABLE II

Advantages and Disadvantages of Systems Described in Table I When Considered for Operation on Alcohol/Gasoline Blends with Control of HC, CO and NOx Emissions

System	Advantages	Disadvantages
Engine Adjustment	• Inexpensive	• Each change in alcohol/gasoline blend would require a further adjustment or possible recalibration of the carburetor in order to minimize HC and CO emissions. • Control of HC and NO_x emissions would require retarding ignition timing with associated fuel consumption penalties. • Further NO_x control must be achieved separately (Exhaust Gas Recirculation).
Thermal Reactor	• Reactor material technology is known.	• Air pump required if vehicle was set for alcohol/gasoline blend and subsequently operated on gasoline (engine runs rich). • High exhaust temperatures necessary for thermal reactor operation requires rich air/fuel mixtures or retarded ignition timing with associated fuel consumption penalties. • If operated on alcohol/gasoline blend without carburetor adjustment (lean mixture), thermal reactor might not function. [Thermal reactors generally require thermal energy from reactor combustion — from excess fuel (rich mixtures) — in order to function for HC and CO control]. • NO_x control must be achieved separately.
Oxidation Catalyst	• Proven technology • HC and CO control with lean mixtures without air pump. • Catalyst will function at low exhaust temperatures. • Engine scheduled maintenance cost reduced by 50 to 75% as a result of required system reliability and use of unleaded fuel.	• Air pump required for rich mixtures. • Probable fuel consumption penalty — when rich mixtures result as alcohol blend approaches zero. • NO_x control msut be achieved separately.
Three-Way Conversion (TWC) Catalyst with Closed-Loop Fuel Control	• Proven technology. • Automatic compensation for high altitude driving or for changes in fuel. • Can be used for wide ranges of ethanol/gasoline and methanol/gasoline mixtures. • System also reduces NO_x emissions as well as HC and CO. • No fuel consumption penalty (EPA predicts 8% gain over comparable oxidation catalyst system). • Engine scheduled maintenance cost reduced by 50 to 75% as a result of required system reliability and use of unleaded fuel.	• Not completely engineered for alcohol/gasoline mixtures.

Through this new technology, as indicated in this paper, fuel blends containing up to at least 30 percent ethanol may be utilized with such a system while maintaining low exhaust emissions and satisfactory driveability. It would be expected, therefore, that a vehicle adjusted for optimal system performance on a 30% ethanol/gasoline mix would enjoy similar capability of perhaps $\pm$30%, i.e., 0% ethanol and 60% ethanol. Existing systems may have to undergo materials or construction changes in order to handle the potentially reactive properties of alcohols.

The flexibility of the TWC closed-loop system, to allow use of a variety of fuels while maintaining effective exhaust emission control, could be a very important element in long term future strategy for minimizing oil imports. Considering that the life of a vehicle is approximately 10 years, then 10 years after introduction of the TWC closed-loop system, substantially the entire vehicle population of a country could utilize fuels with a wide range in properties and specifications. For certain regions in the world where alcohol or other fuel components could be utilized as gasoline extenders, such system flexibility might be incorporated with an overall improvement of energy and environmental considerations.

THREE WAY CONVERSION (TWC) CATALYST STOICHIOMETRIC REQUIREMENTS — A TWC catalyst provides simultaneous removal of HC, CO and NO_x from engine exhaust gas[7,8,9]. High percentages — often 80 percent or higher — of all three pollutants may be removed in this manner if the exhaust gas composition remains, on the average, within a narrow band of air (oxygen)/fuel ratio, as may be seen in Figure 2. Although the air/fuel ratios may vary considerably on an instantaneous basis, the average A/F ratio within each interval of a few seconds duration must lie within a band of about 1 or 2 percentage points on either side of the stoichiometric air/fuel ratio for the fuel in use in order for the TWC catalyst to effectively remove the HC, CO and NO_x. For example, if the stoichiometric air/fuel ratio for unleaded gasoline is 14.6, the required band of control may be approximated by the values 14.45 to 14.75 air/fuel ratio. Although this requirement may appear to be formidable, the technology is developed, is commercially available, and has been used by at least five vehicle manufacturers in recent years. One such vehicle is a 4-cylinder Volvo[8] sold primarily in California which uses an Engelhard TWC catalyst and a R. Bosch closed-loop fuel injection system with similar closed-loop systems being employed or contemplated by most automobile manufacturers for the United States market including the Ford Pinto.[10] The successful control of A/F ratio hinges on the closed-loop control system; that is, a means by which the oxygen content of the exhaust is continuously monitored and the air/fuel ratio adjusted to maintain the stoichiometric exhaust gas composition necessary for the TWC catalyst to function. Additionally, both fuel economy and engine performance are near maximum when the air/fuel ratio is held at stoichiometric. The oxygen content of the exhaust, which is related to air/fuel ratio, is determined with an oxygen sensor[8,11] placed in the exit of the exhaust manifold (see Table I). This solid electrolyte cell produces a millivolt signal of 0 to 500 mv when slightly lean and 500 to 800 mv when slightly rich as shown in Figure 3. The response of the oxygen sensor is very rapid, detecting changes in the exhaust gas within a few milliseconds and responding with a very steep voltage characteristic at the stoichiometric point $\lambda = 1.00$* — which is independent of fuel composition.

Figure 2

THREE-WAY CATALYST CONVERSION FUNCTION. HC, CO & NO_x CONVERSION EFFICIENCY AS A FUNCTION OF LAMBDA (λ)

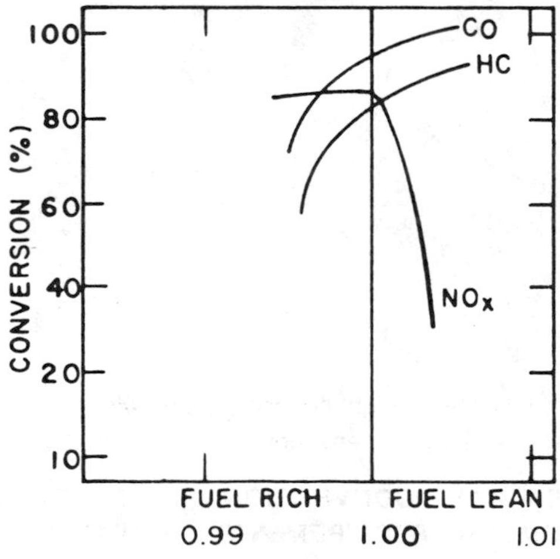

$$\lambda = \frac{\text{TOTAL } O_2 \text{ IN EXHAUST}}{\text{STOICHIOMETRIC } O_2}$$

It is this steeply rising characteristic that permits a relatively simple electronic control circuit (or logic device) to provide correction signals to a fuel injection device or carburetor, and thus maintain accurate closed-loop control of air/fuel ratio in these production vehicles. The lower schematic in Table 1 indicates a logic device which compares the sensor output voltage with a predetermined value or set point, and then provides the correction signal.

*λ is defined in Figure 2.

Figure 3

OXYGEN SENSOR MILLIVOLT OUTPUT AS A FUNCTION OF EXHAUST GAS COMPOSITION (λ)

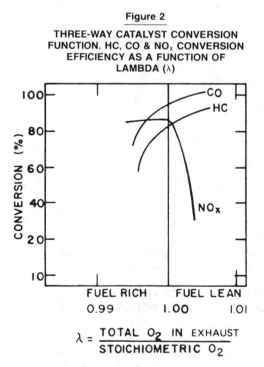

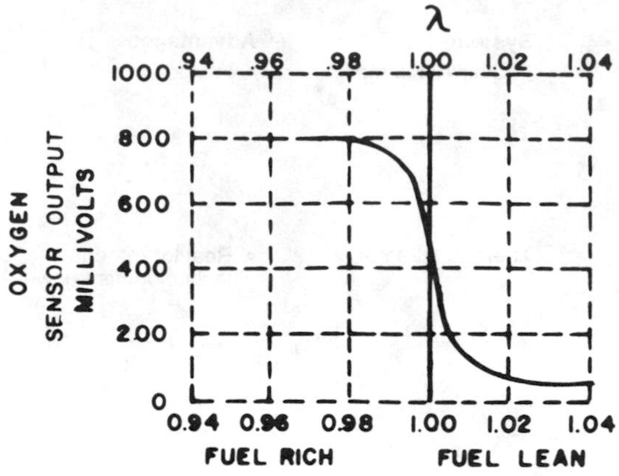

As noted earlier, the production system automatically compensates for changes in air density when operating at varying altitudes — and thus meets this requirement as regulated in the United States[12]. It is anticipated that this TWC catalyst closed-loop control system will be in major use in the United States beginning with vehicle model years 1980 and 1981[9].

Can this system also compensate for fuel compositions which vary widely in type and chemical composition? With the above as background, a test program was initiated at Engelhard in which several ethanol/gasoline, a few methanol/gasoline blends, and other mixtures containing diesel oil and water were used for testing in a 1977 prototype Volvo TWC catalyst vehicle, which was engineered for gasoline use only. The purpose of the subject tests was to examine the emissions and vehicle driveability to determine the ability of existing closed-loop control systems to maintain a stoichiometric air/fuel ratio with the test blends. It should be emphasized that no attempt was made to adjust the spark timing or to otherwise reoptimize the closed-loop control and TWC catalyst for the use of ethanol/gasoline blends. In fact, the system on the Volvo test vehicle is designed to compensate for tendencies of the air/fuel ratio to go rich with increasing altitudes. On the other hand, a system designed for alcohol/gasoline blends should be designed to compensate for air/fuel ratio in just the opposite direction (lean).

DESCRIPTION OF TESTS

VEHICLE SYSTEM — A Volvo TWC catalyst test car with a closed-loop control fuel injection was used in the program. The vehicle was a prototype of the 1977/78 Volvo vehicle currently sold in California.[8] The specific vehicle specifications are listed below:

Vehicle	VOLVO Model 244, 4-Speed Manual Transmission

Engine	VOLVO B21, 4-Cylinder 2.1 Liters displacement. No exhaust gas recycle.
Fuel-Metering System	R. BOSCH K-Jetronic Fuel Injection.
Closed-Loop Control System	R. BOSCH Oxygen Sensor (λ-Sond) and Logic Device. [Fuel-Metering and Closed-Loop Control Systems are Integral]
Dynamometer Inertia Weight	3000 Pounds (1350 kg.) 13.2 Road HP at 50 MPH (includes 10% added HP for air-conditioning)
TWC Catalyst	ENGELHARD PTX-516, TWC-16, (102 cubic inches, catalyst volume).

Both the Engelhard catalyst as well as the R. Bosch oxygen sensor had been used for less than 300 miles of operation at the time these alcohol/gasoline blend tests were begun. As noted earlier — and repeated here for emphasis — no attempt was made to reoptimize the engine or control system parameters for the alcohol/gasoline blends.

ETHANOL (AND METHANOL)/GASOLINE BLEND COMPONENTS — The gasoline was an unleaded fuel very similar to that specified by the United States Government, Environmental Protection Agency, for vehicle emissions testing[12]. The ethanol was a commercial denatured product and the methanol was of reagent quality. The significant properties of the fuel components are listed below:

Ethanol	
Ethyl Alcohol	97%
Ethyl Acetate	1%
Methyl Iso-Butyl Ketone	1%
Aviation Gasoline	1%
Methanol	
Reagent Grade	99.9%
Unleaded Gasoline	
Research Octane	91.8
API Gravity 60/60°F	64
Vapor Pressure (100°F)	8.4 psi
Pb	0.023 gm/gal.
Sulfur	300 ppm
Phosphorous	0.5 ppm
Distillation ASTM D-86	
IBP	99°F
50%	212°F
EP	400°F

OTHER FUEL BLENDS WITH GASOLINE — Additional testing was conducted with other fuel compounds blended with gasoline. Various fuel mixtures containing gasoline/ethanol were formed using diesel fuel and water. The primary purpose of this testing was to assess system performance with a broader range of fuel characteristics, including contaminants.

TEST PLAN — Exhaust emissions tests were conducted on the Volvo TWC Catalyst test car with the following ethanol/gasoline blends (measured on a volume basis):

0% Ethanol	100% Gasoline	Baseblend
10% Ethanol	90% Gasoline	
20% Ethanol	80% Gasoline	
30% Ethanol	70% Gasoline	
40% Ethanol	60% Gasoline	
50% Ethanol	50% Gasoline	

The United States Federal Test Procedure was used to determine exhaust emissions[12]. In addition to the above, a series of brief dynamic driving cycles were run on each of the above blends, and in addition 60% ethanol/40% gasoline was evaluated for the purpose of examining more closely the characteristics of the closed-loop control system. Test data included recordings of the output of the oxygen sensor as explained in Appendix A. Exhaust emissions tests were also conducted with the following methanol/gasoline blends (measured on a volume basis):

0% Methanol	100% Gasoline	Baseblend
10% Methanol	90% Gasoline	
20% Methanol	80% Gasoline	

RESULTS

EMISSION TESTS — The United States Federal Test Procedure (FTP) emissions results are listed in Table III along with some brief comments about overall vehicle driveability during the emissions tests on the chassis dynamometer. The HC, CO and NO_x emissions (gm/mi) are also plotted in Figure 4. The individual sample bag emissions of HC, CO and NO_x (gm/bag) for each of the three bags of the FTP tests are plotted in Figures 5, 6 and 7.

TABLE III

Volvo Model 244 Vehicle
R. Bosch Lambda Sond
Engelhard TWC-16 Catalyst
PTX-516 Converter
Fresh Catalyst
Fresh Oxygen Sensor

1975 Federal Emissions Test Procedure
Results

HC gm/mi	CO gm/mi	NO_x gm/mi	Blend	Remarks
Baseline, No Catalyst				
0. 93	13.68	2.47	100% Gasoline	Baseline. Dummy Converter
Ethanol/Gasoline Blends				
0.14	3.19	0.11	100% Gasoline	Baseblend
0.31	3.90	0.11	10% Eth./90% Gas.	One False Start Bag 3
0.24	2.77	0.14	20% Eth./80% Gas.	
0.11	2.61	0.07	30% Eth./70% Gas.	
0.14	0.78	1.77	40% Eth./60% Gas.	Some Loss of Drivability.
0.22	0.90	1.55	50% Eth./50% Gas.	Poor Drivability During Heavy Accelerations.
Methanol/Gasoline Blends				
0.14	3.19	0.11	100% Gasoline	Baseblend
0.16	3.20	0.09	10% Meth./90% Gas.	
0.15	3.20	0.05	20% Meth./80% Gas.	

Figure 4

1975 FEDERAL TEST PROCEDURE HC, CO & NO_x EMISSIONS (gm/mi) AS A FUNCTION OF VOLUME PERCENT ETHANOL IN ETHANOL/GASOLINE BLEND.

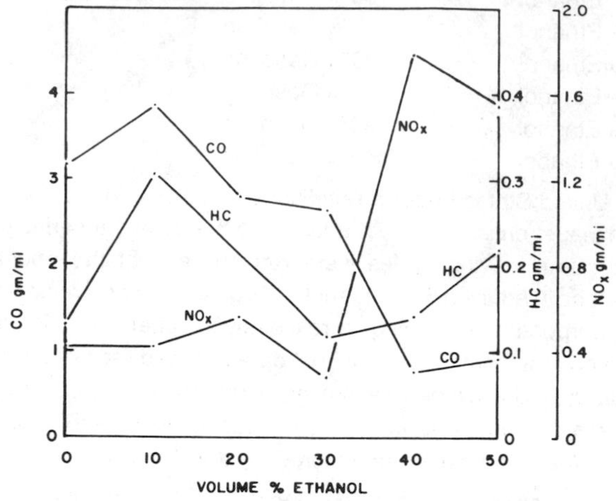

Figure 5

1975 FEDERAL TEST PROCEDURE EMISSIONS OF HC (gm/bag) FOR BAGS 1, 2 & 3 AS A FUNCTION OF VOLUME PERCENT ETHANOL IN ETHANOL/ GASOLINE BLEND.

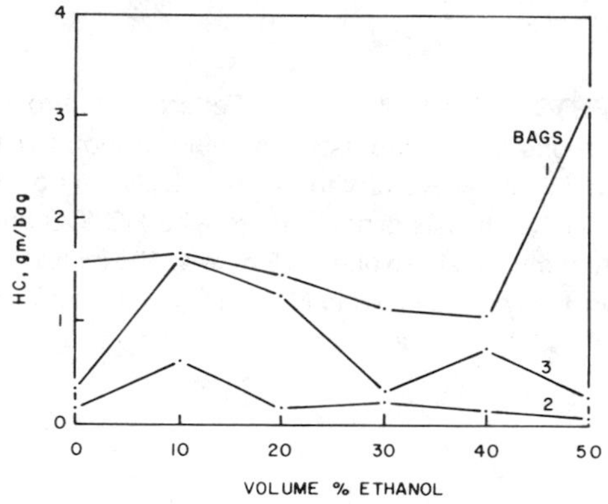

Figure 6

1975 FEDERAL TEST PROCEDURE EMISSIONS OF CO (gm/bag) FOR BAGS 1, 2 & 3 AS A FUNCTION OF VOLUME PERCENT ETHANOL IN ETHANOL/ GASOLINE BLEND.

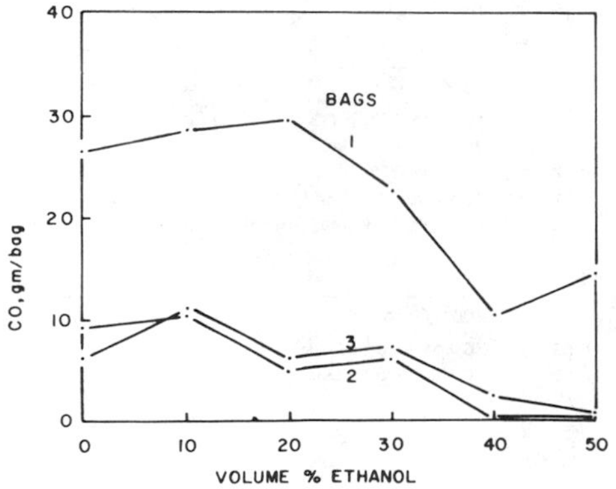

Figure 7

1975 FEDERAL TEST PROCEDURE EMISSIONS OF NO_x (gm/bag) FOR BAGS 1, 2 & 3 AS A FUNCTION OF VOLUME PERCENT ETHANOL IN ETHANOL/ GASOLINE BLEND.

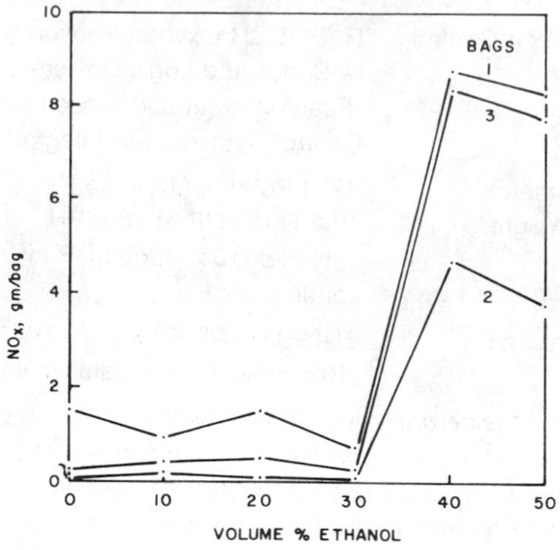

The most significant aspects of the FTP emission tests are the very low vehicle emissions and the fact that these low emissions were maintained with 30% ethanol in the fuel blend. HC/CO/NO_x levels were in the vicinity of 0.2/3.0/0.1 gm/mi with baseline (no catalyst) emissions of 0.93/13.68/2.47 gm/mi. As compared to 30% ethanol, the FTP emission tests at 40% and 50% ethanol indicate an increase in NO_x and decrease in CO emissions. This observation, concerning CO and NO_x emissions, is consistent with a drift in control of the closed-loop air/fuel ratio control towards a slightly leaner mixture. Consider the fuel metering characteristics described in Figure 1. At 40% ethanol/gasoline blend, the required closed-loop control stoichiometric air/fuel ratio for TWC catalyst activity is 12.2, whereas the air/fuel ratio without closed-loop control would probably be located near the Fuel Metering Characteristic Line at about 14.0. This difference poses a significant challenge to the control system which is probably periodically reaching the limits of its compensation control range and during dynamic driving with more than 30% ethanol it is likely that the closed-loop system (as set) cannot always maintain a stoichiometric ($\lambda = 1.00$) air/fuel ratio to maintain maximum TWC catalyst conversion efficiency for HC, CO and NO_x. As a result, the average air/fuel ratio moves slightly lean, and according to Figure 2 the NO_x conversion efficiency should markedly decrease while CO and HC conversion efficiencies increase.

The data in Table III and Figures 4 through 7 confirm the expected trends for CO and NO_x; however, HC emissions do not continue to decrease as anticipated. The explanation for this HC behavior is relatively complex and relates directly to vehicle driveability (hesitation, etc.) which may be partially overcome by engine adjustments.

Appendix A gives a further explanation of the results obtained during this investigation as related to measurements of the oxygen sensor output. Appendix B is a theoretical discussion of the effect of the ethanol (in an ethanol/gasoline blend) on the fuel metering system characteristics (open and closed-loop).

Limited testing was also conducted with methanol as a gasoline extender. The data presented in Table III demonstrates that blends of up to 20% methanol had no influence on emission results compared with the baseblend case. Although further testing would be required to define the operational limits of methanol blends with gasoline, these results demonstrate that the TWC catalyst closed-loop system is also compatible with varying methanol/gasoline blends in the same manner as described above for ethanol/gasoline blends.

Subsequent to the above tests, additional tests were conducted using other components blended with gasoline as listed in Table IV. Because this testing was conducted approximately nine months later with different humidity and temperature conditions the baseline (without catalyst) emissions are somewhat different and cannot be compared exactly with the first series of tests.

TABLE IV

Volvo Model 244 Vehicle
R. Bosch Lambda Sond
Engelhard TWC-16 Catalyst
PTX-516 Converter
Fresh Catalyst
Fresh Oxygen Sensor

1975 Federal Emissions Test Procedures
Results

HC gm/mi	CO* gm/mi	NO_x gm/mi	Blend	Research Octane Number	Fuel Viscosity (Centipoise)
0.32	5.74	0.08	100% Gasoline	92.9	0.41
Without Catalyst					
1.22	17.99	2.70	100% Gasoline	92.9	0.41
0.33	3.55	0.10	70% Gasoline 23% Ethanol 7% Diesel Fuel	98.6	0.57
Without Catalyst					
1.39	15.87	2.34	Same as Above	98.6	0.57
0.38	5.76	0.09	70% Gasoline 25% Ethanol 2.5% Diesel Fuel 2.5% Water	N.A.	N.A.
0.30	3.83	0.16	70% Gasoline 30% Ethanol	102.6	0.57

*The higher CO emissions vs. those shown in Table III are attributable to lower ambient temperatures during the preconditioning soak period.

Again, the flexibility of the system to maintain proper emission control for various fuel blends was demonstrated; there are, however, some variations of the emission results which might be attributable to different fuel viscosities as evidenced by the test results presented in Table IV; however, the effect of viscosity is not clear. In theory, the closed loop control system should automatically compensate for differences caused by small increases of fuel viscosity and maintain constant exhaust emission levels.

Although a thorough study of the potential octane benefits of blending alcohol with gasoline was not contained in this investigation, Research Octane Number measurements were made for some fuel blends presented in Table IV. It is significant that the addition of 30% ethanol improved the RON of unleaded gasoline from 92.9 to 102.6. Furthermore, as shown in Table IV, ethanol appears to more than offset the known poor octane characteristics of fuels such as diesel oil.

FUTURE IMPACT

These studies show that the Volvo closed-loop system has the inherent capability for compensating for gasoline/alternate fuel mixtures. The Volvo system was designed for 100% gasoline. In order to expand and optimize the system for alternate fuel utilization, design and engineering programs are required on the materials of the fuel wetted parts and the control range for optimum compensation. Additional work for automatic ignition timing changes or control would also be helpful. Closed loop carburetion was not tested in this study, however, it would be expected that closed loop carburetion might be a more formidable task than with fuel injection.

If the above mentioned work is undertaken and incorporated into production systems as they are introduced in the early 1980's, it will provide us with valuable flexability in fuel utilization in future years. The system with alternate fuel utilization capability also provides the oil companies with greater flexability of fuel formulation compared to current specifications. The net result would make us less dependent on petroleum derived fuels and capable of utilizing domestically produced coal based or renewable resource based fuels. The fractions of these fuels could be increased or decreased seasonally as fuel source availability fluctuates — all of this without adversely affecting auto exhaust emissions.

CONCLUSIONS

The conclusions from this study as to the effect of alcohol/gasoline fuel blends on the operation and exhaust emissions of a TWC catalyst closed-loop control vehicle are as follows:

- With an emissions/vehicle system designed for gasoline only as a fuel, ethanol/gasoline fuel blends containing at least 30% ethanol may be utilized with excellent vehicle operation and low HC, CO and NO_x exhaust emissions — as measured on a chassis dynamometer using United States (Federal) emissions test procedure. To accomplish this, the closed-loop control system automatically provided a stoichiometric air/fuel ratio of 12.8 for the 30% ethanol/gasoline blend and a 14.6 air/fuel ratio for 100% gasoline.
- Generally good control of HC and CO emissions was obtained with up to a 50% ethanol/gasoline blend, but vehicle driveability was degraded in some modes of dynamic driving.
- Methanol/gasoline fuel blends containing up to 20% methanol provided excellent vehicle operation and low HC, CO and NO_x exhaust emissions. Methanol concentrations above 20% were not evaluated.
- It is anticipated that this vehicle when optimized for performance and emission control on a 30% ethanol/gasoline fuel would enjoy perhaps a $\pm$ 30% range of control; i.e., 0/100% to 60/40% ethanol/gasoline mixture.
- Although not demonstrated for blends other than mentioned in this report, it appears possible to design a feedback air/fuel compensation system which could maintain stoichiometry with fuel blends ranging from 100% gasoline to 100% methanol or ethanol.
- Compensations for changes in fuel stoichiometry with ethanol/diesel oil/water/gasoline mixtures was found.

This work has demonstrated that a basic system exists which can operate on a variety of fuels while maintaining low exhaust emissions.

Questions concerning the compatibility of alcohol/gasoline blends with current vehicle fuel system materials, as well as other potential handling problems with these blends, have not been addressed in this paper — but can probably be overcome with materials/engineering modifications.

For the future we can take advantage of the demonstrated flexibility of this system to utilize fuels of various properties. In order to realize these advantages, immediate consideration should be given to planning and properly engineering such a system.

ACKNOWLEDGEMENTS

We would like to express appreciation to Volvo AB, Gothenburg, Sweden and, in particular, to Dr. Stephen Wallman for his cooperation in providing the test vehicle and technical assistance which made this investigation possible.

REFERENCES

1. G. L. Witzenberg, "Alcohol Fuel Getting Attractive", *Wards Auto World,* Volume 13, July, 1977.
2. D. A. Howes, "The Use of Synthetic Methanol as a Motor Fuel", Jour. Inst. Petr. Tech., 19, 301 (1933).
3. W. J. W. Pleeth, "Alcohol, A Fuel for Internal Combustion Engines", Chapman and Hall, London (1949).
4. G. Publow and L. Grinberg, "Performance of Late Model Cars with Gasoline-Methanol Fuel", SAE Paper No. 780948 (1978).
5. E. E. Wigg and R. S. Lunt, "Methanol as a Gasoline Extender — Fuel Economy, Emissions and High Temperature Drivability", SAE Paper No. 741008 (1974).
6. H. Menrad, W. Lee and W. Bernhardt, "Development of a Pure Methanol Fuel Car", SAE Paper No. 770790 (1977).
7. J. J. Mooney, C. E. Thompson and J. C. Dettling, "Three-Way Conversion Catalysts — Part of the New Emission Control System", SAE Paper No. 770365 (1977).
8. G. T. Engh and S. Wallman, "Development of the Volvo Lambda-Sond System", SAE Paper No. 770295 (1977).
9. "An Update on Three-Way Catalysts" *Automotive Engineering,* Vol. 85, August 1977.
10. "Ford's 2.3 Liter Four Adopts Three-Way Approach", *Automotive Engineering,* Oct. 1977.
11. R. Zechnall, G. Baumann and H. Eisele, "Closed-Loop Exhaust Emission Control System With Electronic Fuel Injection", SAE Paper 730566 (1973).
12. *Federal Register,* Vol. 42, No. 124, June 28, 1977.

APPENDIX A

OXYGEN SENSOR OUTPUT SIGNALS

Since the oxygen sensor is an electrolytic cell, and the output voltage is sensitive to the air/fuel ratio, the measurement of this signal can provide considerable information about the behavior of the closed-loop control system — and especially with the ethanol/gasoline blends. The response time of the sensor is very fast, and if the output voltage is recorded on a high speed (fast response time) recorder, relatively brief engine events may be studied in some detail. For this work a Hewlett-Packard Model 7402A Oscillographic Recorder was utilized.

A brief portion of the Federal Test Procedure (FTP) driving cycle between 640 seconds and 680 seconds — shown on the bottom of Figures 8 and 9 — was chosen as a driving cycle which could be driven accurately and repetitively with the driver's aid* as a guide. The acceleration rate was moderate. For each of the ethanol/gasoline blends — 0%, 10%, 20%, 30%, 40%, 50% and 60%** ethanol — the vehicle was driven over this driving cycle while the oxygen sensor output was recorded, as well as the tailpipe emissions (after the catalyst) of HC, CO and NO_x. Figure 8 provides a composite of the recorded sensor output voltage at 0%, 30%, 50% and 60% ethanol. The corresponding after-catalyst CO and NO_x vehicle emissions on a concentration basis are shown in Figure 9. In each case the horizontal axis is time which increases from right to left and is the same scale for all portions of each of the two figures. Maximum vehicle speed was 26.5 MPH.

The sensor voltage traces in Figure 8 have a scale of 0 to 1000 millivolts (mv) [or 1.0 volt]. For discussion purposes the $\lambda = 1.00$ stoichiometric point is about 550 mv (from Figure 3). Lower voltages (down to 0 mv) represent lean air/fuel ratios and voltages above 500 mv represent rich air/fuel ratios***The brief upsets in the voltage traces labeled as 1-2 and 2-3 at 0% ethanol, etc., represent transmission shift points of the vehicle. The CO and NO_x emissions shown in Figure 9 are on a scale of zero to 1000 PPM for CO and zero to 500 PPM for NO_x. It should be emphasized that these emissions are measured on a concentration basis.

The sensor voltage trace for zero percent ethanol provides a basecase with which to compare the other sensor voltage traces. Note that the normal oscillations cover the range of about 50 to 800 mv. The mean voltage may be considered to be very close to about 550 mv (or about $\lambda = 1.00$). At 30% ethanol, the voltage trace basically still resembles that of 0% ethanol, except that the maximum voltage of the oscillation is reduced to 700 mv, and the minimum is also reduced somewhat. The NO_x emissions have increased slightly but are still very low. The increase in CO emissions — particularly the appearance of the two spikes — cannot readily be explained. The FTP emissions data in Table III, as well as the Bag 2 CO plot in Figure 6, do not indicate any increase in CO mass emissions at 30% ethanol. The overall HC, CO and NO_x emissions are still very low. At 50% ethanol in Figure 8 the oscillations indicate that the air/fuel ratio was in the rich area (above 500 mv) on the average less of the time than was the case at 0% and 30% ethanol. Thus, although the system is still controlling, the overall stoichiometry has moved slightly more lean than was the case at 30% ethanol. Figure 9 shows that CO emissions are low, but that NO_x emissions have increased. The most dramatic changes in sensor output voltage as well as CO and NO_x emission concentrations occur at 60% ethanol. At this condition the sensor voltage indicates that the air/fuel ratio was lean for almost all of the accelerations — meaning that the closed-loop control system had reached the limit of compensation control and was no longer maintaining an average air/fuel ratio at stoichiometric ($\lambda = 1.00$) for optimum TWC catalyst function. Note in Figure 9 that NO_x emissions increased. However, CO emissions as may be anticipated from Figure 3, have essentially been reduced to zero.

With the vehicle fully warmed-up, excellent vehicle driveability on the chassis dynamometer was maintained up to and including 40% ethanol. At 50% ethanol a very slight hesitation could be detected during the initial part of the acceleration. At 60% ethanol some vehicle hesitation was apparent throughout the acceleration. Recall that in the FTP tests in Table III some loss of driveability was noticed at 40% ethanol and at 50% ethanol poor driveability was observed in heavy accelerations.

---

* Speed vs time trace utilized by the vehicle driver to follow the Federal Test Procedure driving cycle.

**60% ethanol blend was added in order to further study the limits of this particular closed-loop control.

***The quasi-sinusoidal oscillations of the oxygen sensor output voltage (rich-lean-rich-lean, etc.) such as indicated in Figure 8 at 0% ethanol, are a normal function of the closed-loop control system[8,9]. This oscillation, which has a quasi-period of less than about one second, is the means by which the logic device attempts to maintain the exhaust gas oxygen sensor output voltage near the set point of about 500 mv, so that the TWC catalyst will function.

Figure 8

OXYGEN SENSOR OUTPUT AS A FUNCTION OF VOLUME PERCENT ETHANOL IN ETHANOL/GASOLINE BLEND FOR PORTION OF FEDERAL TEST PROCEDURE (FTP) DRIVING CYCLE, BAG 2,

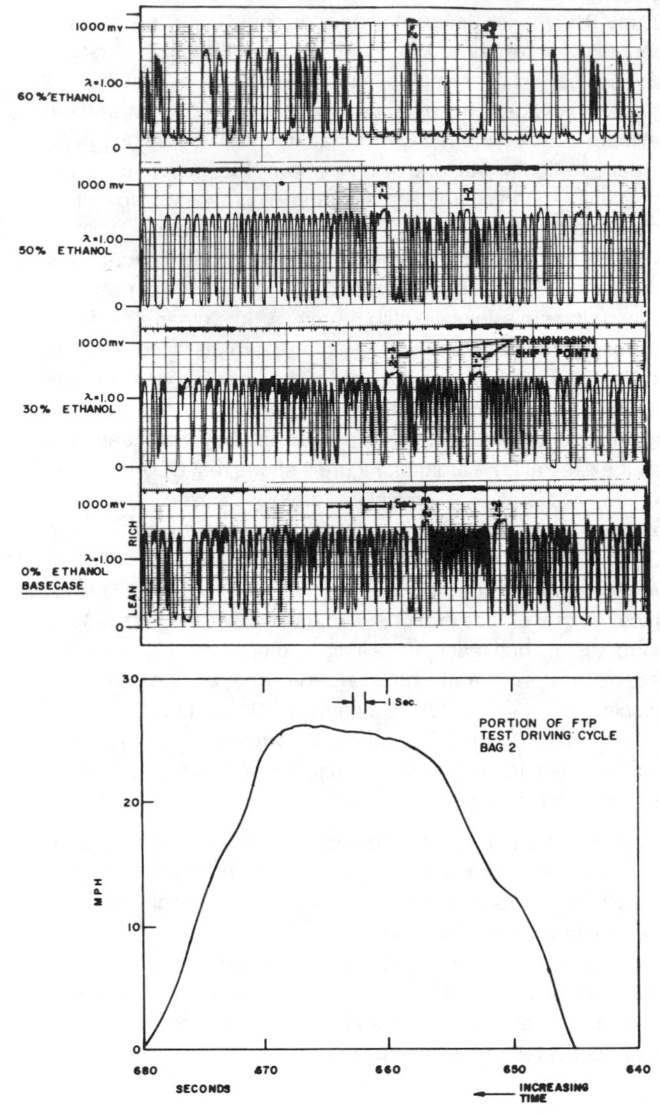

Figure 9

CO AND NO_X EMISSIONS AS A FUNCTION OF VOLUME PERCENT ETHANOL IN ETHANOL/GASOLINE BLEND FOR PORTION OF FEDERAL TEST PROCEDURE (FTP) DRIVING CYCLE, BAG 2.

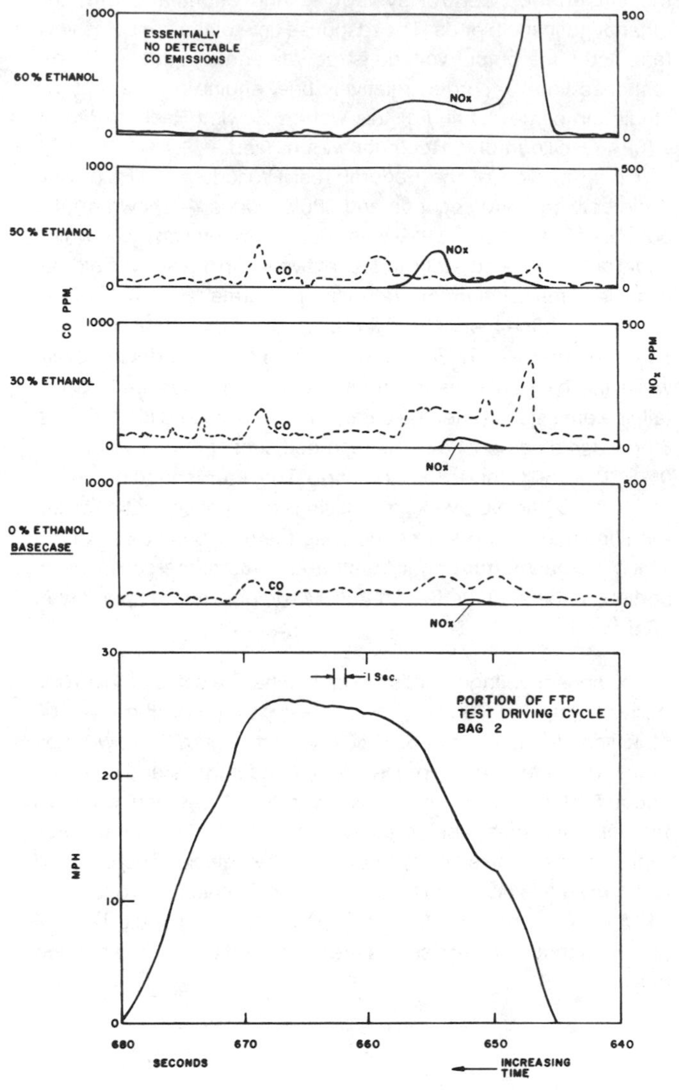

APPENDIX B

STOICHIOMETRIC CONSIDERATIONS, ETHANOL/GASOLINE BLENDS

The stoichiometric air/fuel ratio, specific weights and viscosities of the ethanol/gasoline blend components are shown below.

Figure 1 represents the air/fuel ratio of any blend between 0% ethanol and 100% ethanol. The left axis is 100% gasoline; the right axis is 100% ethanol.

$\lambda = 1.00$ represents the stoichiometric air/fuel ratio. Note that the stoichiometric air/fuel ratio for any intermediate blend varies linearly with the percentage of ethanol. The lines $\lambda = 0.98$ and $\lambda = 1.02$ represent the far outer limits of the average air/fuel ratio during dynamic driving of a closed-loop control TWC catalyst vehicle. The relative closeness of these two lines is evidence of the type of in-use closed-loop control required for good TWC catalyst performance with gasoline.

The line in Figure 1, labeled Fuel Metering Characteristic Line (FMCL), represents, as a first approximation, the behavior of a conventional open-loop gasoline metering system (carburetor or fuel injection) which was designed for gasoline only (at the stoichiometric air/fuel ratio) and then utilized for ethanol/gasoline blends. The approximation is based on the theory that a conventional, open-loop fuel metering system will meter air and fuel according to their volumes and not their weight, and that the viscosity of an ethanol/gasoline blend is the same as that of gasoline alone. With these approximations the FMCL is generated. This line would be horizontal if the density of the two fuel components were equal. Because the density of ethanol is 6% greater than that of gasoline, the FMCL slopes downward from Left to Right. [It can be shown that since the viscosity of ethanol is substantially greater than that of the gasoline components (see Table above) the FMCL may move upward slightly toward a level position.]

The significant conclusion to be drawn from the above approximations is that the closed-loop control system, in order to maintain a stoichiometric air/fuel ratio with the ethanol/gasoline blend, must, in effect, angle the FMCL down to the Stoichiometric Line. The angle through which the swing is made is labeled the Closed Loop Control Function. As anticipated, the higher the percentage ethanol in the blend, the greater the Closed Loop Control Function must be in order to maintain the stoichiometric air/fuel ratio for the TWC catalyst to function.

Component	Stoichiometric Air/Fuel Ratio	Specific Weight (20°C)	Viscosity (20°C)
100% Commercial Unleaded Gasoline	14.6	Approx. 6.17 lb/gal.	Approx. 0.2 to 0.7 Centipoise for Components
100% Ethanol	9.0	6.57 lb/gal.	1.2 Centipoise

*λ is defined in Figure 2.

Practicality of Alcohols as Motor Fuel*

T. O. Wagner
Amoco Oil Co.
Chicago, IL

D. S. Gray, B. Y. Zarah, and A. A. Kozinski
Amoco Oil Co.
Naperville, IL

ALCOHOLS HAVE BEEN USED as motor fuel on a limited basis almost since the automobile was invented. Methanol has been and still is used as fuel for racing cars with specially designed engines where high power is wanted regardless of fuel consumption or cost. It also has been used at low concentrations (ca 0.1 per cent) in gasoline to prevent fuel line freeze-ups in winter. Ethanol was occasionally sold as a blend with gasoline when it was surplus during the 1930s. Isopropyl alcohol has been used at low concentrations (ca 1 per cent) in gasoline as a carburetor anti-icing agent. Tertiary butyl alcohol, a relatively low-cost byproduct from chemical manufacture, is presently used as concentrations up to 7 per cent as an octane-improving agent

*Paper 790429 presented at the Congress and Exposition, Detroit, Michigan, February 1979.

ABSTRACT

Several alcohols have been or are being used as motor fuel in limited applications where their unique properties provide a specific benefit. Recently, proposals have been made to manufacture methyl and ethyl alcohols from nonpetroleum feedstocks for use as a primary energy source. Such a program, it is claimed, would reduce U.S. dependence on foreign oil; manufacturing ethanol would also stabilize prices of farm commodities.

An economical and abundant feedstock for methanol manufacture is North Dakota lignite. In a plant consuming lignite and self-sufficient in utilities, energy in the methanol would represent about 45 per cent of the energy in the feedstock. The cost of energy from the methanol produced would be about twice that from gasoline at 1979 prices. Liquefaction of lignite or coal to a synthetic crude with subsequent refining by conventional processes would provide motor fuel at somewhat higher energy efficiency and lower cost than by methanol synthesis. Municipal solid wastes could also be used as feedstock for methanol manufacture, but the cost would be even higher than with lignite as feedstock.

The lowest cost and most abundant feedstock for ethanol manufacture is corn. With current farming and ethanol manufacturing practices, which use mainly petroleum-based fuels, energy in the ethanol plus that in the byproducts is only about 55 per cent of the energy in the petroleum fuels consumed. If the ethanol plant used only nonpetroleum fuels such as coal or agricultural wastes for process energy, the efficiency based on just the petroleum fuels used in farming and collection of wastes would range from 130 to 160 per cent; overall efficiency would remain at 55 per cent or less. Energy in the byproducts represents about 40 per cent of the energy output, which greatly complicates evaluation of the efficiency of the plant. In any event, the cost of energy from the ethanol produced would be at least four times that from gasoline. The high cost results from large capital charges, large operating costs and, especially, large raw material costs.

While both methanol and ethanol have characteristics that complicate distribution and detract from product quality, poor energy yield and high cost are the most serious deterrents to commercialization of either alcohol. Methanol is definitely preferred, but neither represents a practical alternative unless the price of petroleum-based fuels rises markedly relative to that of the alcohols.

SAE/PT-80/19/$02.50

for unleaded gasoline in special refining situations; a related compound, methyl tertiary butyl ether, is used similarly in Europe.

Recently, it has been proposed that methanol or ethanol be produced from "renewable" raw materials--anything but petroleum gases or liquids--and be used as motor fuels. Of the alcohols, only these two can be produced practically with current technology from nonpetroleum raw materials. Using methanol or ethanol as motor fuel, it is claimed, would free the United States from dependence on foreign oil. Using ethanol would also stabilize prices of farm commodities. While the motives of an alcohol program appear noble, it would have unprecedented economic and technological implications which, we fear, have not been fully examined.

If commercialization of alcohol fuels comes about, it will likely be the result of government rule rather than the rational action of a free market, and the government will undoubtedly define the strategy under which commercialization takes place. However, responsibility for translating the general, and likely, vague and conflicting, provisions of the strategy into practice would fall on refiner/marketers. They would have to arrange for a reliable supply of the alcohol, transport it to a blending facility, blend it into gasoline without violating industry and legal specifications for motor fuel, and distribute the alcohol-containing fuel to service stations. They also would have to face the wrath of vehicle manufacturers, the motoring public, service station operators, government regulators, legislators, and news media if prices are too high, supply is uneven, or product quality is improper.

For any new fuel to be practical, the cost of manufacturing it must be competitive with cost of alternative fuels, the energy yield from the raw materials consumed must be reasonable, and it must be possible to distribute the fuel practically. Of course, the fuel must also meet the quality requirements of the equipment in which it will be used. Our paper will discuss the cost, energy yield, distribution, and product quality of methanol and ethanol as motor fuels from the practical viewpoint of the refiner/marketer.

PROPERTIES OF ALCOHOLS

The alcohols all have a common feature. Their molecular structure includes an OH, or hydroxy radical, which gives them certain characteristics: high latent heat of vaporization and high solubility in water. These water-like characteristics are most apparent in the alcohols of low molecular weight, methanol and ethanol, because the OH radical predominates over their short hydrocarbon chains; they are least apparent in the alcohols of high molecular weight, tertiary butyl or heavier alcohols, because their longer hydrocarbon chains predominate over the OH radical. These characteristics can be advantages or disadvantages, depending on what function the alcohol is to serve. Certain properties of the alcohols of interest, which will be discussed throughout the paper, are shown in Table I.

The various alcohols have been used mainly because their unique properties provided some specific benefit that outweighed other disabilities. The wide flammability limits and high latent heat of methanol make it attractive as a racing fuel. The partitioning of isopropyl alcohol between water and gasoline and its good freeze-point depression made it an effective carburetor anti-icing agent. The high Motor octane number and modest water affinity of tertiary butyl alcohol make it attractive as an octane-blending component. Ethyl alcohol is not particularly suited as a motor fuel, but was blended with gasoline when it was surplus for lack of a better use. However, none of the alcohols has ever been used as a primary energy source on a wide scale because they were all too expensive and generally had other disabilities relative to gasoline.

CURRENT SUPPLY OF GASOLINE, METHANOL, AND ETHANOL

Current U.S. consumption of motor gasoline

Table I
Properties of Alcohols and Gasoline

Compound	Chemical Formula	Molecular Weight	% Oxygen by weight	Net Heat of Combustion, MBTU/gallon	Stiochiometric A/F Ratio, lbs. air/lb. fuel	Latent Heat of Vaporization, MBTU/gallon
Methanol	CH_3OH	32	50	57	6.4	3.3
Ethanol	C_2H_5OH	46	35	76	9.0	2.6
Isopropyl alcohol	C_3H_7OH	60	27	86	10.3	2.1
Tertiary butyl alcohol	C_4H_9OH	74	22	93	11.1	1.7
Gasoline (typical)	C_8H_{15}	111	0	115	14.6	0.8

is about 7,000,000 barrels per calendar day (7 MMBPCD)*. As a result of the federal motor vehicle economy standards, gasoline demand is generally projected to be relatively stable over the next five years, and then begin to decline. About half of the crude oil now used to make gasoline and other products is produced domestically, and the other half is imported. The price of domestic crude ranges from about $5 to $16 per barrel depending on the applicable federal price regulation. The price of foreign crude is about $16 per barrel, landed in the United States. The composite price of crude is equalized among refiners through the federal "Entitlements" program, and is now about $14 per barrel. The resulting wholesale price of gasoline presently ranges from $0.40 to $0.50 per gallon, depending on location, terms of delivery, and grade. (At the time this paper was written, early in 1979, the wholesale gasoline market was unstable because of disturbances in Iranian crude oil production.) The lower heat of combustion of gasoline is about 115 MBTU/gallon, so cost of energy from gasoline is about $4/MMBTU.

Present U.S. methanol production is about 85 MBPCD, and the largest plant produces about 18 MBPCD. Most of the current supply is consumed as an industrial chemical, and there was about 10 per cent unused capacity in 1978(1).** Nearly all methanol currently is made from natural gas, and the wholesale price is about $0.45 per gallon. The lower heat of combustion of methanol is 56 MBTU/gallon, so the cost of energy from methanol is about $8/MMBTU.

Present U.S. ethanol production is about 33 MBPCD(2). Of this, about 25 MBPCD is made in six plants by hydration of ethylene derived from petroleum, and 8 MBPCD is made in hundreds of small plants by fermentation of various agricultural materials--mainly wheat, corn, and sorghum. Ethanol made from petroleum is used solely as an industrial chemical. Federal law permits only ethanol made by fermentation to be used in drugs, cosmetics, and beverages, and essentially all the fermentation ethanol is consumed in these products. The small amounts of gasoline-ethanol blends that have been sold recently on a "demonstration" basis in several Midwest states have included both petroleum and fermentation ethanol. There is presently little unused capacity for producing either type of ethanol. Irrespective of source, the current wholesale price of ethanol ranges from $1.20 to $1.40 per gallon, depending on the location, terms of delivery, purity, and how it is denatured (made unfit to drink). The lower heat of combustion of ethanol is 76 MBTU/gallon, so the cost of energy from ethanol is about $17/MMBTU.

Since current production of methanol is only 1 per cent of gasoline demand, and current total production of ethanol is only 1/2 per cent of gasoline demand (and even most of that is made from petroleum), it is apparent that completely new alcohol industries would have to be developed if alcohols are to supplement or replace petroleum as a motor fuel. It is also apparent that using either of the alcohols as motor fuel at their present prices would have severe inflationary effects.

MANUFACTURE OF METHANOL

Rather than use natural gas to make methanol, lignite or coal, municipal solid wastes, waste biomass, or specifically-grown biomass have been suggested as methanol feedstocks(3, 4). Lignite is an inexpensive, low-grade coal particularly suited for methanol synthesis. Proposed plants would have capacities around 120,000 barrels per stream day (120 BMPSD)* or seven to eight times that of the largest present-day plant. Due to the large quantities of feedstocks needed, the plants would likely be located near the raw material source.

PROCESSES - A schematic diagram of a plant for producing methanol from lignite or coal is shown in Figure 1(5). The plant would likely be located in North Dakota, where lignite is abundant, and it would be self-sufficient in utilities. Run-of-mine lignite is crushed, dried, and pulverized in a preparation section. Some of the prepared lignite is used as plant fuel for steam and electric power generation. Pulverized lignite is fed to steam/oxygen-blown gasifiers (partial combustors) to produce a synthesis gas consisting of CO and H_2. Gasification also produces acidic contaminants such as H_2S and CO_2. All H_2S and most of the CO_2 are removed by scrubbing the synthesis gas with an amine solution. The H_2S is later recovered as elemental sulfur. Subsequently, part of the clean synthesis gas undergoes a CO-shift conversion to adjust the $H_2/CO/CO_2$ ratios to the necessary values. This conversion produces additional CO_2, which is partly removed by amine scrubbing, and then the shifted gas is mixed with the remainder of the synthesis gas. Methanol synthesis, via the low-pressure process with a Cu/Zn/Cr catalyst, is the final step. The

*The letter M is used to denote thousands in this paper.

**Numbers in parentheses designate References at end of paper.

*Capacity per Stream day and per Calendar day differ by the plant operating factor, the fraction of the calendar days it is "on stream." Alcohol plants would be expected to have operating factors of about 90 per cent.

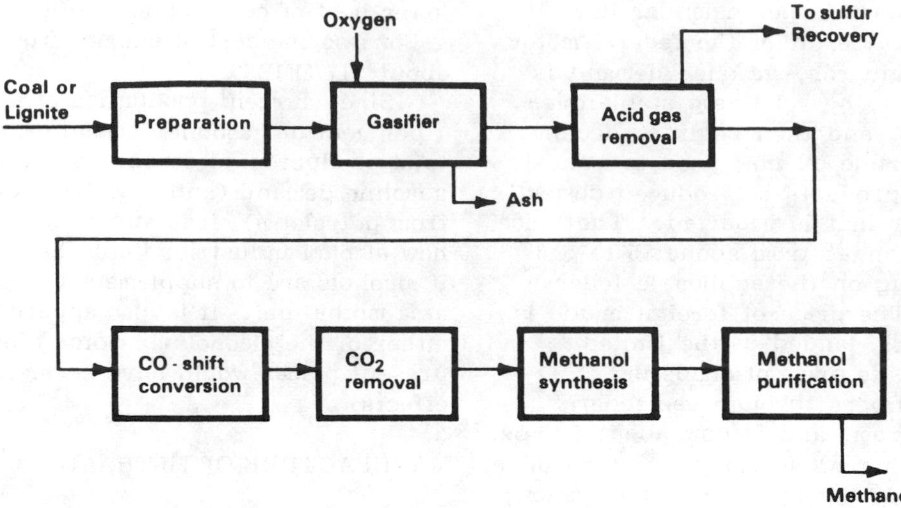

Fig. 1 - Process schematic for methanol production from coal or lignite

raw methanol is then purified to remove water. Currently, aldehydes and higher alcohols are also removed after methanol synthesis, but this may not be necessary or even desirable for methanol to be used as motor fuel.

The cost of disposing of municipal solid wastes is increasing as more stringent regulations are imposed on landfill and open burning. Recovery of some energy from these wastes could partly offset the high disposal cost, and at the same time provide a cleaner method of disposal. One way to recover energy is to convert the wastes to methanol. A city with a population of 3 million would typically generate wastes for a methanol plant of about 11 MBPCD capacity. Because motor fuel demand for a city of this size would be in the order of 80 MBPCD, the methanol produced could serve as a supplement to gasoline, but it would not be adequate to serve as a complete replacement.

Conversion of wastes to synthesis gas is still in its infancy, and to date it has been carried out only in small demonstration units that have been burdened with severe operating problems. For all methanol feedstocks, the difference in processing lies in the conversion of the feedstock to synthesis gas. Figure 2 shows a schematic diagram for conversion of municipal wastes to methanol(5). The slagging Purox vertical-shaft furnace gasifier, developed by Union Carbide, is used conceptually because none has been commercialized. The wastes are

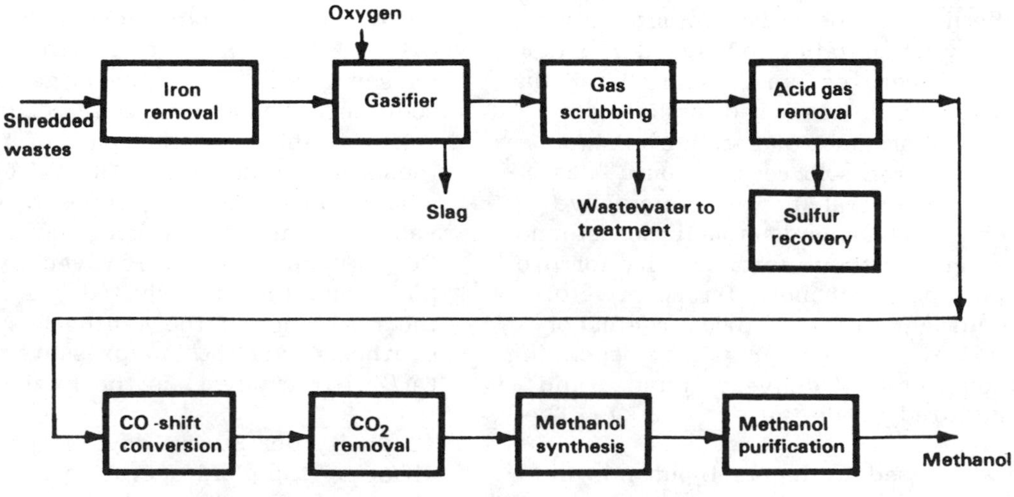

Fig. 2 - Process schematic for methanol production from municipal solid wastes

first shredded and then passed under a magnet to remove ferrous material. The iron-free wastes are then gasified with oxygen; any remaining ash, glass, or metal residues are fused into a slag at the bottom of the gasifier where temperatures reach about 3,000°F. The slag is discharged from the gasifier into a quench bath and later disposed. The product synthesis gas is cleaned by water scrubbing and other means to remove any particulates, entrained oils, H_2S and CO_2. CO-shift conversion for $H_2/CO/CO_2$ ratio adjustment, methanol synthesis, and methanol purification are accomplished in a manner similar to that for lignite feed.

Biomass broadly includes cellulosic agricultural, forestry, and marine products and wastes, and represents a potentially significant indigenous source of energy. Currently, about 320 MM tons/year of agricultural residues are left on fields after grains are harvested, and forestry residues amount to about 120 MM tons/year(6). The total energy in these wastes is equivalent to about 2 MMM bbl. of crude oil, or about 30 per cent of the U.S. annual consumption. Only a small fraction of residues are collected now and, of course, not all is recoverable. Additional materials could be available through culture of "woody" plants specifically as a methanol feedstock(7, 8, 9).

Some success in developing wood gasification processes has been achieved by Union Carbide and Battelle Columbus Laboratories. The Union Carbide Purox process employs a vertical-shaft furnace gasifier in which wood chips are partially oxidized to produce a synthesis gas of CO and H_2 in a two-step process. (The same gasifier as described conceptually for processing municipal wastes.) Oxygen is used to burn the wood chips in the lower part of the furnace, and the heat of combustion then partially gasifies the chips in the upper part. The Battelle process employs a fluid-bed gasifier with hot sand circulated to provide gasification heat. The hot sand is heated in a combustor that burns wood char. Circulating the sand eliminates the need for an oxygen plant, and could reduce energy requirements.

With both processes, clean-up of the synthesis gas and subsequent methanol synthesis would be the same as with a lignite feedstock. These processes generate 1 to 2 BTU of low-grade waste heat for each BTU of methanol synthesized, which means, of course, that they consume large amounts of feedstock per unit of methanol produced.

COSTS - Capital investments for methanol plants thought to be of optimum size(5) using lignite or municipal solid wastes as feedstocks are itemized in Table II. We assumed the lignite plant to be located in North Dakota and the waste plant to be located in Chicago. The lignite plant to produce 120 MBPSD of methanol would require a total investment of $1.8 MMM or $15 M per daily barrel of methanol. It would use three pounds of lignite to produce one pound of methanol. The waste plant to produce 12 MBPSD of methanol would cost $338 MM, or $28 M per daily barrel of methanol. It would use about five pounds of wastes to produce one pound of methanol. For comparison, a new, large (ca 200 MBPSD) modern refinery making gasoline and distillate fuels would require an investment of $3 M to $4 M per daily barrel of crude run.

Costs at the plant gates for producing methanol from the two raw materials are shown in Table III. Annualized capital charges amount to about 20 per cent of the investment. A credit of $27 is allowed per ton of sulfur recovered from the lignite, and we assumed that the waste plant would be paid $5 per ton for wastes it accepts. Methanol from the lignite plant would cost $0.35/gallon and methanol from the waste plant would

Table II
Investments for Methanol Plants

Feedstock: 96 MM pounds/day Lignite
Product: 120 MBPD methanol (33 MM pounds/day)

Item	1978 Costs, $MM
Coal preparation	57
Gasifiers and gas cleaning	307
Oxygen plant	278
Gas compression, acid gas removal, and CO-shift conversion	378
CO_2 removal from shifted gas	81
Sulfur plant	23
Methanol synthesis and purification	160
Storage and utilities	287
Royalty and startup	106
Working capital	89
Total investment	1,766
$M/BPSD methanol	15

Feedstock: 15 MM pounds/day Wastes
Product: 12 MBPD methanol (3.3 MM pounds/day)

Item	1978 Costs, $MM
Waste preparation and iron removal	38
Gasifiers	108
Oxygen plant	28
Gas cleanup and sulfur recovery	3
CO-shift conversion	11
Gas compression	37
CO_2 removal from shifted gas	10
Methanol synthesis and purification	31
Methanol storage and utilities	36
Royalty	6
Startup	15
Working capital	15
Total investment	338
$M/BPSD methanol	28

Table III
Annual Operating Costs for METHANOL Plants

Feedstock: 96 MM pounds/day Lignite
Product: 120 MBPSD methanol (33 MM pounds/day)

Item	1978 Costs, $MM
Capital charges*	349
Taxes and insurance	20
Maintenance	50
Operating labor	9
Supervision	6
Overhead	35
Lignite ($6/ton)	94
Catalysts and chemicals	10
Water	6
Sulfur credit ($27/ton)	(1)
Total	578
$/gallon methanol	$0.35

Feedstock: 15 MM pounds/day Wastes
Product: 12 MBPSD methanol (3.3 MM pounds/day)

Item	1978 Costs, $MM
Capital charge*	67
Taxes and insurance	5
Maintenance	11
Operating labor	2
Supervision	1
Overhead	7
Purchased power	1
Purchased water	1
Catalysts and chemicals	1
Municipal waste credit ($5/ton)	(13)
Sulfur credit ($27/ton)	nil
Total	83
$/gallon methanol	$0.50

*Based on 12% discounted cash flow, 50% tax rate, and 13 sum-of-years-digit-depreciation period.

cost $0.50/gallon. The cost of methanol from the lignite plant does not include transportation to market which, for example, would be about 5 cents/gallon to Chicago.

It is significant that the estimated cost of methanol from the lignite plant is less than the current market price, which indicates that a new lignite plant could compete with currently-operating plants using natural gas. However, for equal energy content, the cost of methanol from the lignite plant still is much higher than the cost of gasoline. Methanol at even lower cost could be produced from natural gas now being flared to the atmosphere in the Middle East or Mexico. Of course, using foreign gas as raw material would not help the United States achieve energy self-sufficiency.

ENERGY BALANCE - The energy consumption and production of methanol plants using lignite and municipal solid wastes as feedstocks are shown in Table IV. A lignite plant that is self-sufficient in utility requirements has a thermal efficiency of 45 per cent. A waste plant that purchases electric power but that is self-sufficient otherwise has a thermal efficiency of 40 per cent(5).

These efficiencies have significance only in comparison to the efficiencies with which lignite and wastes can be used for other purposes. Lignite can be burned as boiler fuel with an efficiency of about 80 per cent. This, of course, does not produce a liquid fuel. Coal or lignite can be converted into a synthetic crude with an efficiency of 60 or 70 per cent (and perhaps at somewhat lower cost) from which it can be refined into gasoline and distillates with an efficiency of at least 90 per cent. The only alternative way wastes presently can be used for energy production is to burn them as boiler fuel. Technology to burn such low-quality fuels is not well developed, but it probably could be done with an efficiency of around 60 per cent. Thus, in either event, converting lignite, coal, or wastes into a liquid form to make them easier to transport and burn involves substantial energy penalties. Further, converting lignite or coal to methanol may not be as economical and efficient a route as liquefaction.

Table IV
Energy Balance for Methanol Plants

Lignite	MBTU/Gallon Methanol
Consumption	
Lignite	-127
Total consumption	-127
Production	
Methanol	+57
Sulfur	+0.3
Total production	+57
Net gain (loss)	(70)
Efficiency--45 per cent	

Muncipal Solid Wastes	MBTU/Gallon Methanol
Consumption	
Wastes	-141
Purchased power	-1
Total consumption	-142
Production	
Methanol	+57
Sulfur	+0.2
Total production	+57
Net gain (loss)	(85)
Efficiency--40 per cent	

MANUFACTURE OF ETHANOL

Ethanol can be manufactured from any feedstock containing carbohydrates using well-developed, commercial processes. Suitable feedstocks include corn, wheat, sorghum, sugar beets, sugar cane, potatoes, and other grains(10). In the United States, corn would probably be the preferred feedstock because it is cheapest and most abundant. (Its growers are also the most active politically.) Sugar cane has technical advantages as a feedstock, but due to climatic restrictions, it can be grown only in areas in certain states--Texas, Louisiana, Florida, and Hawaii--already largely used for sugar cane production.

Ethanol can also be manufactured from feedstocks containing cellulose using highly complicated processing that has not been commercialized. The cellulose is first hydrolyzed by acid treatment or enzymatic action, and then processed in a manner similar to that for simple sugars. A processing scheme recently announced for manufacturing ethanol from forestry and similar wastes using enzymatic conversion is claimed to be competitive with current fermentation processes for manufacturing ethanol(11).

PROCESSES - Production of ethanol from grains such as corn is shown schematically in Figure 3(12). The grain is first ground and cooked with water to convert the starch to sugar with the enzyme amylase. The sugar is then fermented with yeast to produce raw ethanol and a high-protein (ca 30 per cent) material commonly known as distillers' dried grain (DDG). The raw ethanol is distilled to remove impurities such as higher alcohols and to remove most of the water. Ethanol forms an azeotrope with 5 per cent water, and the last step in producing anhydrous ethanol is an extractive distillation with benzene. The DDG is dried and recovered for sale as a cattle feed. The optimum capacity of an ethanol fermentation plant is small because of difficulties in collecting and storing raw materials, and typically would be about 1.4 MBPSD. The plant would consume about 23 M bushels of corn per day, which is the production from about 90 M acres, or 140 square miles of farmland. Because the feedstock would become available only once each year, the plant would need to have storage facilities for at least 8 MM bushels of corn.

Production of ethanol from sugar cane requires only simple, well-established processes since the fermentable sugar is obtained directly from the sugar cane. The cane is first cut and ground, and the cane juice is extracted by maceration. After clarification by filtration and concentration through evaportation, the juice is fermented with yeast to yield raw ethanol. A series of distillation steps, including a final extractive distillation with benzene, are used to obtain anhydrous ethanol. Sugarbeets could be an alternative feedstock, but would be more expensive than sugar cane.

COSTS - Investments and operating costs for ethanol plants of 1.4 MBPSD capacity have been reported by several investigators, with somewhat different results. Estimates we think reasonable for a plant using corn as feedstock are shown in Table V. We believe that the initial investment plus startup costs and working capital would be about \$44 MM in 1979 dollars. This is equivalent to \$30 M per daily barrel of ethanol. Annual operating costs for the production of anhydrous ethanol are shown in Table VI. The total cost per gallon of ethanol is \$1.88, but it is assumed that the DDG can be made as a byproduct and sold at \$120/ton. (This price for DDG is higher per pound than the price assumed for corn.) DDG sales yield a credit of 43¢/gallon and reduce the net ethanol cost to \$1.45/gallon, which is a little higher than the current market price of ethanol made by fermentation. Current prices likely do not reflect the full costs of capital required by a new plant.

In addition to DDG, the byproducts from a

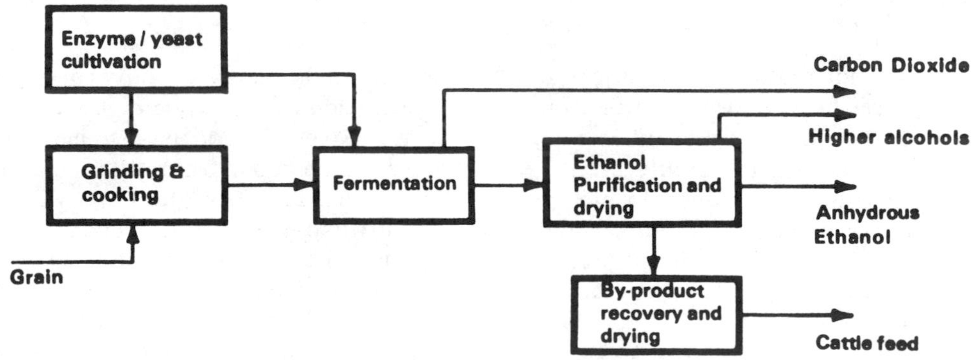

Fig. 3 - Process schematic for ethanol production from grain by fermentation

Table V
Investments for Ethanol Plant

Feedstock: 23 M bushels/day corn
Product: 1.4 MBPSD ethanol

Item		1978 Costs, $MM
Plant site and preparation		3.0
Office, laboratory, and other buildings		3.2
Steam plant		11.7
Water plant		0.4
Alcohol plant		18.1
Grinding and cooking	6.0	
Fermentation	4.2	
Alcohol recovery and drying	4.4	
By-product recovery and drying	3.5	
Alcohol storage		2.7
Startup		1.2
Working capital		3.6
Total		43.9
$M/BPSD ethanol		31

Table VI
Operating Costs for Ethanol Plant

Feedstock: 23 M bushels/day corn
Product: 1.4 MBPSD ethanol

Item		$M/Yr.
Capital charges* (same basis as for methanol)		8,700
Taxes and insurance		600
Maintenance		1,200
Labor		600
Supervision		200
Overhead		1,300
Utilities, alcohol plant		4,500
Steam	4,400	
Cooling water	20	
Electric power	80	
Utilities, DDG plant		2,900
Steam	2,000	
Cooling water	100	
Electric power	50	
Fuel		750
Corn		17,500
5.8 MM bushels/year at $2.60/bushel	15,000	
1.9 MM bushels/year at $1.30/bushel	2,500	
Benzene		3
Total		37,500
$/gallon ethanol		1.88
DDG credit (72 M tons/year at $120/ton)		(8,600)
Net costs		28,900
$/gallon ethanol		1.45

corn-fermentation ethanol plant include a mixture of higher alcohols, commonly known as fusel oil, and carbon dioxide. The quantity of oil is very small and is usually consumed internally as a fuel. The carbon dioxide is produced in significant quantities and represents about 30 per cent by weight of the corn feed to the plant. However, in determining the production costs of ethanol, no credit was given for CO_2 because revenue from its sales would not cover compression and distribution costs if the supply were greatly expanded.

Our estimate of the cost of ethanol from a new plant is based on several highly favorable assumptions that may not be valid for the long term. First, in contrast to methanol manufacture, capital charges are not the major element of total cost; corn is. "Fresh" corn was assumed to cost $2.60/bushel, and "distressed" corn $1.30/bushel, giving a composite cost of about $2.30/bushel(12). It is very doubtful if the operator of an ethanol plant could secure long-term contracts for a large supply of corn at a price of $2.30 per bushel in 1979 dollars. Although growers may have to sell corn for that price in some years, they definitely would want a higher price on a long-term arrangement, especially if large amounts of grain are being used in an ethanol program. Just recently, the Illinois Farmers Union demanded that the Carter administration raise the "target" price of corn to $3.92/bushel. Further, there really is no established market in distressed corn. Its availability is sporadic, and in many cases, it is simply worked off in fresh corn at the price of fresh corn. Since one bushel of corn produces about 2.6 gallons of ethanol, this component of ethanol cost is simply 40 per cent of the per bushel cost of corn, and any escalation in cost of corn causes proportional (and very significant) escalation of ethanol cost. Also, we assumed that all DDG can be sold at the present market price of $120/ton as cattle feed. This probably would not be true if manufacture of ethanol was expanded. About seven pounds of DDG are produced per gallon of ethanol. The current market for DDG is about 0.3 MM tons/year but manufacture of 10 per cent of U.S. motor fuel as ethanol would yield over 37 MM tons/year of DDG. The additional DDG would probably displace soybean meal as feed, and disrupt both the DDG and soybean markets. Thus, large scale manufacture of ethanol would undoubtedly incur higher raw material costs and lower byproduct credits than we assumed.

Energy consumption and production in manufacture of ethanol by corn fermentation is shown in Table VII(13). The alcohol section alone consumes 108 MBTU/gallon of ethanol produced. The ethanol plus the higher alcohols have an energy content of 77 MBTU/gallon, so the efficiency of this section is only 71 per cent. This means that 1.4 BTU are consumed for every BTU produced. Overall thermal efficiency of the complete plant is even lower. When the energy used in growing corn and the energy used in drying the DDG are included, the efficiency is only 56 per cent. This means that 1.8 BTU are consumed for every BTU produced as ethanol, other alcohols, and DDG. There is no option but to dry the DDG and use it as feed or fuel because of the problems in disposing of it other ways.

The significance of the energy efficiency of a corn-fermenting ethanol plant is somewhat more complex than that of a methanol plant. Nearly all ethanol plants now in operation burn natural gas or oil for process energy. Also, gasoline or diesel fuel is used exclusively for farming to grow the corn, except for some irrigation that uses natural gas. Thus, manufacture of ethanol as now practiced consumes 95 MBTU of energy from natural gas or petroleum per gallon of ethanol produced <u>over and above the energy gained when the ethanol and byproduct DDG are burned</u>.

This analysis probably overstates the efficiency of ethanol production, because cattle feed equivalent to the DDG can be obtained from soybeans using farming energy of less than 1/4 the energy credited to DDG.

Of course, a logical question is whether anything can be done to make ethanol manufacture

Table VII
Energy Balance for Ethanol Plant

		MBTU/Gallon Ethanol	
Farming Consumption			
Seed corn, herbicides, and insecticides		2	
Fertilizer		20	
Cultivation and irrigation		21	
Drying and transportation		3	
Subtotal			46
Alcohol Plant Consumption			
Grinding and cooking		24	
Fermentation		1	
Beer still		40	
Distillation		29	
Dehydration		14	
Subtotal			108
By-Product Plant Consumption			
Concentration		43	
Drying		20	
Subtotal			63
Production			
Ethanol		76	
DDG		45	
Higher alcohols		1	
Subtotal			122
Net gain (loss)			(95)
Efficiency	56%		

more efficient in energy. Since the energy content of the products cannot be changed, either the energy consumption must be reduced, or a lower-valued form of energy must be used. Both farming and ethanol manufacture are industries that have been under strong economic pressure for years, and since energy represents a significant element of cost in both industries, there have been continued and effective efforts to reduce energy consumption. Further reductions will not be easy to make and are not likely to be large. There remains the option of using different energy sources, such as wastes or coal. While this would not improve the total energy balance, it might make its consequences less important.

Proposals have been made to burn the field residues from corn, called "stover," for process energy in the ethanol plant. Dried stover yields about 160 MBTU/gallon of ethanol when burned, but energy of about 10 MBTU/gallon would be consumed in collecting, transporting, and drying the stover(14). Although technology to burn low-energy wastes like stover in small, dispersed plants is only poorly developed, potentially the stover could supply all but 11 MBTU of the 171 MBTU/gallon process energy required. Energy from petroleum would still be needed for farming, 46 MBTU/gallon of ethanol, and to collect the stover, 11 MBTU/gallon. If the deficit in process energy was supplied by petroleum, then the energy efficiency of the ethanol operation, based on just the liquid fuel produced and the petroleum consumed, is 76/(46+10+11) or 113 per cent. If the deficit in process energy is supplied by some source other than petroleum, the efficiency becomes 135 per cent. However, there is question whether use of all stover to supply process energy is realistic. Some of the stover would have to be left on the field to improve soil "tilth" and to control erosion. The total ethanol/DDG operation cannot be self-sustaining if more than 30 per cent is left behind. Further, it is significant that little or no stover is used as fuel now because it is not competitive in cost with coal or price-controlled natural gas or oil.

Another alternative in ethanol manufacture is to use lignite or coal as fuel for the ethanol plant. In this case, ethanol manufacture would amount largely to a mechanism for converting lignite or coal into liquid fuel, just like a methanol plant. If, say, lignite provided all process energy, the total input per gallon of ethanol would be 46 MBTU from petroleum for farming, and 171 MBTU from lignite for process energy. The energy efficiency of the ethanol operation, again based on the liquid fuel produced and the petroleum consumed, is 76/46 or 160 per cent.

Although using lignite for process energy improves the energy efficiency of the ethanol operation based on petroleum, the lignite itself is not used with high efficiency. After crediting petroleum for the ethanol produced, the lignite can be credited only for the byproducts, DDG, and higher alcohols, so the efficiency with which lignite is used is only 46/171 or 33 per cent. The lignite could be used for methanol synthesis at higher efficiency, at much lower cost, and without tying up the large amounts of land that ethanol manufacture requires.

Although cost and, in the United States, supply are serious disabilities, a major advantage for using sugar cane as feedstock for ethanol manufacture is that it has a large net energy production. The "bagasse" or sugar cane residue is available at the fermentation plant and can be used to supply all process energy requirements. As a result, conversion of sugar cane to ethanol, including farming, consumes only 25 MBTU of energy for every gallon of ethanol produced, so each gallon of ethanol produced represents a net gain of 51 MBTU.

DISTRIBUTION OF ALCOHOLS

Since current alcohol supply is small and already fully utilized, use of alcohol as motor fuel on anything beyond a demonstration scale would have to be based on a completely new industry. Regardless of how aggressive the alcohol strategy becomes, only limited volumes will be available for many years. Thus, alcohols would be used initially as fuels in specialized applications, such as stationary gas turbines for power generation, or as a supplement to motor fuel, such as Gasohol, in areas where the alcohol is locally available.

As is well known, methanol and ethanol have properties that make them difficult to distribute in bulk: high affinity for water, and the tendency to separate from hydrocarbons at low temperatures and/or in the presence of small amounts of water. There is no question but that both pure alcohols and gasoline/alcohol blends can be distributed successfully in a dedicated distribution system, if it is specially designed to exclude water and if adequate care is taken in operations. There are circumstances where distribution of alcohols under all these restrictions would be relatively easy, but not necessarily economic. For example, ethanol could be produced in a plant located near a corn supply in Nebraska, then transported a few miles by a dedicated truck transport (ca 8 M gallons) to a municipal power plant where it would serve as fuel for boilers or spark-ignition engines generating electric power or pumping water for the community. Or, it could serve as fuel for a local industry. Modest amounts of alcohols could be utilized this way.

However, distributing alcohols or alcohol/

gasoline blends on a large scale as motor fuel would present major problems. Excluding all water from currently-installed refinery, pipeline, terminal, bulk plant, and service-station tanks would be very costly, if not impossible. Fortunately, water is nearly insoluble in gasoline and normally causes no problem. If gasoline contacts water, as it might when a terminal or service station tank is agitated during filling, any water present may be temporarily suspended as droplets, but only an insignificant amount will be dissolved. After agitation ceases, the suspended water soon settles to the bottom. The gasoline draw-off point in the tank is located high enough to avoid the water, and the water bottoms are drained or pumped out periodically.

Solubility of alcohols in gasoline depends on the gasoline's temperature and water content. Ethanol is fully soluble in dry gasoline at high temperature, but methanol is not. Solubility of both alcohols decreases as temperature falls, and it decreases dramatically in the presence of water. Small amounts of water will extract the alcohol, and a two-phase system--gasoline and alcohol/water--results. This increases the amount of bottoms by the amount of alcohol in the gasoline, and raises the chances that bottoms will inadvertently be delivered to customers. Of course, with proper care, the alcohol/water phase can be detected and then drained and discarded, just as water can be drained from gasoline alone, but the costly alcohol would be lost as a result. Even if enough water is not present to cause phase separation, alcohol blended in gasoline will increase the amount of dissolved water that can be retained. This could increase corrosion rates of pipelines, tankage, delivery vehicles, and fuel systems of motor vehicles.

Sites of any alcohol plants would probably not coincide with refinery sites, so distributing alcohol/gasoline blends would require a system at least partially redundant to present gasoline distribution systems. An example of a fully-dedicated, separate, distribution system for ethanol is shown in Figure 4.

In the existing system for collecting crude oils and distributing motor fuels and distillates, most crude oil is gathered in the oil fields by pipelines. In some fields, wells are not grouped closely enough or volumes are too small for a pipeline to be economic, and the gathering is done by truck. The crude is collected in terminals and then distributed by large-diameter pipelines, ships, or barges to refineries. The refined products are distributed from the refineries by pipelines, ships, or barges to terminals. (There may be intermediate terminals that are not shown.) The pipelines and vessels handle a variety of products in succession--leaded and unleaded gasolines, jet fuel, diesel fuel, and heating oils. In rare cases a pipeline may handle both these "clean" products and crude oil. Deliveries from terminals may be made directly to service stations or industrial customers, or through smaller intervening terminals, called bulk plants. Nearly all deliveries from terminals are made by truck transports, which generally have capacities of about 8 M gallons in two to four compartments. Terminal tanks are usually dedicated to a single product, but truck transports often are not because it would severely complicate scheduling and raise costs. Service stations typically have one tank per grade of gasoline offered, but large stations may have multiple tanks. Except where size is restricted by law, the capacity of service station tanks usually is 6 M gallons or more, so that the total contents of a compartmented truck transport can

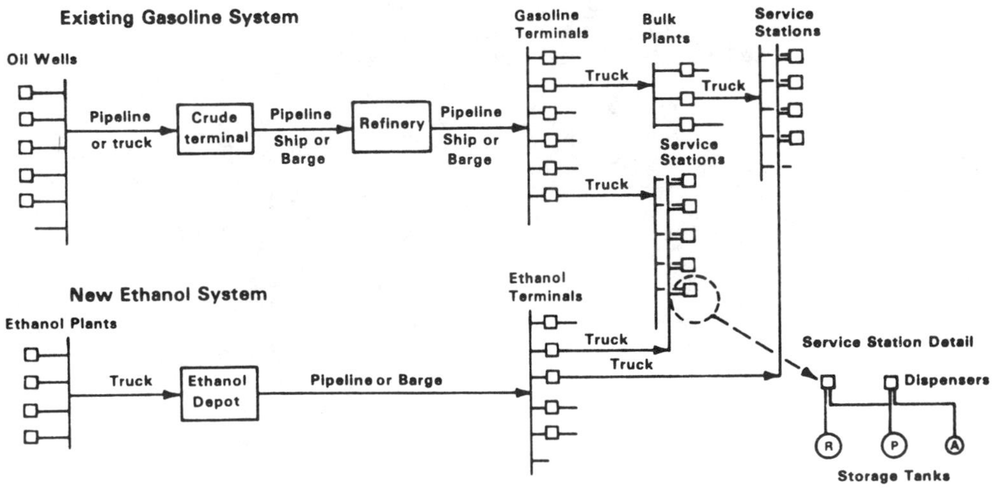

Fig. 4 - Distribution of gasohol with dedicated ethanol facilities

be accepted at a station each delivery.

In the alcohol system, ethanol would be gathered by truck transport from plants located near the supply of corn. Because a large area is required to grow the corn feedstock, the plants would probably be too far apart for a pipeline to be economic. The ethanol would be collected at terminals, then transported by small-diameter pipeline to ethanol distribution terminals. These terminals probably would be located in existing gasoline terminals where possible. From the terminals, ethanol would be delivered by compartmented truck transports to service stations or industrial customers. At service stations, the ethanol would be stored in its own tank and be added to the gasoline via blending dispensers as vehicles are serviced. If the alcohol terminals are not located in the gasoline terminals, separate, dedicated truck transports would be required for alcohol delivery.

With proper design and operating procedures, it should be possible to distribute Gasohol to vehicles without contamination using the system shown, but troubles could still occur if the vehicle tanks contained water. Of course, the additional facilities would raise costs. The added cost would depend on specific details of the system, and generalization is impossible. However, itemized costs for one specific case are shown in Table VIII. Here it is assumed that ethanol is produced at plants located through mid-Indiana and mid-Illinois and used to provide Gasohol in the Chicago metropolitan area. This represents a rather simple supply situation, because there is no other city of such large size located so near a potential ethanol supply. A system for supplying methanol as a motor fuel supplement would be similar except for siting of the methanol plant.

In the example, total motor fuel demand for the area is taken at 150 MBPCD, so ten plants of 1.5 MBPCD capacity are required to supply the ethanol. The ethanol distribution system is patterned after Figure 4. The total investment in the ethanol system, not including the ethanol plants, is $42 MM, about $3 M per daily barrel of ethanol, and annualized operating costs are $22 MM, or about 10 cents per gallon of ethanol dispensed.

Because ethanol would displace 10 per cent of the gasoline, it would offset part of the gasoline distribution cost. The most significant savings would be in truck deliveries from terminals to service stations. Applying credit for this saving might reduce the net additional ethanol distribution costs to 6 or 7 cents per gallon of ethanol, or slightly over 1/2 cent per gallon of total gasoline. Since gasoline demand is not expected to grow significantly in the future, the ethanol distribution facilities would not be a substitute for expanded gasoline facilities, and would represent entirely new capital.

Obviously, if alcohols are to be commercialized on a large scale, refiner/marketers would want to conduct adequate tests to determine the most economic distribution system that will suffice for each locality. Perhaps the ethanol could be blended at the terminal truck loading racks, avoiding separate tanks for ethanol and blending dispensers at service stations. However there would be large risks involving product liability in such tests, and they may have to be done with financial indemnification of the refiner/marketer by either the alcohol supplier or the sponsoring government agency.

PRODUCT QUALITY

When a refiner/marketer examines the prac-

Table VIII
Initial Capital Requirements and Annual
Operating Costs of an Alcohol Distribution System

Item	Initial Capital Requirement $MM	Direct Operating Costs $MM/Year
Alcohol plants' inventory	2.8	-
Transportation to terminals (rural, 50 miles average)	-	5.5
Terminal storage and inventory	2.2	0.1
Pipeline transport (8-inch diameter, 150 miles) and inventory	24.0	2.1
Product terminal and inventory	9.1	0.2
Truck transport to service stations (urban, 20 miles average)	-	6.0
Service station storage and blending pumps	3.5	0.2
Total	41.6	14.1
	Distribution Costs	
	$MM/Year	$/Gallon
Direct operating costs	14.1	0.07
Capital charges	8.3	0.03
Total	22.4	0.10

ticality of a component for motor fuel, a major concern is how it affects the ability to meet performance-related product specifications--mainly octane number and volatility. Concern also exists about any adverse effect it will have on durability of the fuel systems of customers' cars--fuel filter plugging, white metal corrosion, and elastomer deterioration. Components derived from petroleum may differ in octane number and volatility but they are usually similar otherwise. Components from nonpetroleum sources may differ in other ways. If the proposed fuel component is highly unusual, it may be necessary to consider exhaust and evaporative emissions, safety in handling, toxicity, and fuel mileage. Currently, methanol is prohibited as a motor-fuel component under Section 211(f) of the Clean Air Act, but prohibition of ethanol has been waived by EPA on the basis that it will not raise total vehicle emissions significantly because the volume currently used is very small. No unusual safety or toxicity problems are anticipated with methanol or ethanol.

Octane number and volatility are most important to a refiner/marketer because controlling them is costly, and they directly affect satisfaction of motorists. Research and Motor octane numbers of methanol, ethanol, and typical lead-free regular and premium gasolines are shown in Table IX. Also shown are octane numbers of 10 per cent blends of the alcohols in the lead-free regular gasoline. The pure alcohols have very high octane numbers, and they raise octane quality of the lead-free regular gasoline when they are blended into it. The gain in Research octane number is especially large. However, recent-model U.S. cars respond mainly to Motor octane number, and the gain in Motor octane number, while still good, is not spectacular.

It is frequently assumed that adding 10 per cent ethanol to lead-free regular gasoline makes it equal in octane quality to lead-free premium gasoline. This is not true. The example in Table IX, which is typical, shows that adding 10 per cent ethanol raises Motor octane number of the lead-free regular gasoline only from 83 to 85, short of the 87 octane number of the lead-free premium gasoline.

Even though 10 per cent ethanol does not make regular gasoline into premium, the improved octane number does have value. If the cost of raising octane number by conventional refinery processing is taken as 40 cents/octane-barrel, or 1 cent/octane-gallon, the 2-unit gain in Motor octane number obtained by blending 10 per cent ethanol into lead-free regular gasoline is worth 2 cents per gallon of gasoline, or 20 cents per gallon of ethanol. But, since ethanol costs about $1 per gallon more than gasoline, ethanol is not an economic octane-blending agent. Octane improvement costs would have to rise severalfold before a refiner would consider using ethanol for this purpose. On the other hand, methanol is an economical octane-blending agent and would be useful were it not for its low energy content and affinity for water. The Motor octane grain with methanol is about two units, the same as with ethanol, and since cost of methanol per gallon is about the same as cost of gasoline, methanol yields essentially zero-cost octane improvement. However, methanol has never been used for this purpose because the adverse effects it has on product quality are too serious.

Methanol and ethanol have markedly different volatility and combustion properties than gasoline, which affect carburetion requirements. Methanol and ethanol, being nearly pure compounds, have very narrow boiling ranges compared to gasoline, a mixture of many diverse compounds. The alcohols require several times as much heat to vaporize them, so more exhaust heat must be directed to the intake manifold. And because they are, in essence, already partly oxidized or burned, they need less air for combustion. While an engine can be designed specifically to operate on either straight methanol or ethanol, it cannot be designed practically to operate interchangeably on straight alcohols

Table IX
Octane Numbers of Alcohols and Gasolines

	Research	Motor
Methanol	112	91
Ethanol	111	92
Lead-free regular gasoline	92	83
Plus 10 per cent methanol	96 (132)	85 (103)
Plus 10 per cent ethanol	96 (132)	85 (103)
Lead-free premium gasoline	97	87

( )Apparent blending octane number of alcohol at concentration shown.

and gasoline.

When methanol and ethanol are added to gasoline, they alter the energy content, latent heat of vaporization and stoichiometric A/F ratio of the blend proportionately. Since they are quite different chemically from gasoline, they have large and curious effects on volatility of the blend. As shown in Figure 5, addition of either 10 per cent methanol or ethanol raises RVP and depresses front-end distillation temperatures.

The differences in volatility caused by blending the alcohols into gasoline affects vehicle performance in two ways. First, the higher heat of vaporization of the alcohol or alcohol/gasoline blend reduces vaporization in the intake system, which impairs warm-up performance. To restore performance, it would be necessary to increase manifold heat, which cannot be done practically without vehicle redesign. Second, the higher RVP and lower distillation temperatures (higher volatility) make vehicles more prone to vapor lock and stalling once they are warmed up. To compensate for this tendency, the RVP and distillation temperatures of the base blend before alcohol addition would have to be reduced. This requires that some of the most volatile components--butanes and pentanes--be withdrawn from the gasoline pool. These are relatively inexpensive, clean-burning components of gasoline and the only alternative use for them is as burner fuels.

Both methanol and ethanol are more active chemically than gasoline components, so they may cause novel problems in the fuel systems of customers vehicles. Attacks by alcohols on fuel tanks and blockage of fuel filters have been reported. In Amoco's test car fleet, several plastic or elastomeric fuel-system parts were seriously degraded during operation on blends of 10 per cent methanol or ethanol in gasoline. Redesign of the components could eliminate the problem in new cars, but it would still be troublesome during the transition unless existing cars were retrofitted. The liability and warranty implications from alcohol use are obvious to both the refiner and the vehicle manufacturer.

Alcohols would be expected to give poorer fuel mileage than gasoline because of their lower energy content(15). However, a mileage improvement of 6 per cent is frequently quoted, based on a Gasohol demonstration sponsored by the Nebraska Agricultural Industrial Products Utilization Commission despite the fact that the energy content of Gasohol is 3 per cent less than gasoline. This claim cannot be substantiated, because detailed results of the Nebraska demonstration have never been published. It is interesting to note that for Gasohol to yield 6 per cent better mileage, a vehicle would have to "see" ethanol as having an energy content of 184 MBTU/gallon rather than 76 MBTU/gallon as found by calorimetry. One carefully-controlled vehicle fleet test on methanol at 15 per cent in gasoline showed 5 per cent higher fuel consumption than with gasoline(16). No fully-reliable fleet data are available for ethanol gasoline blends, but one short-term test of Gasohol by

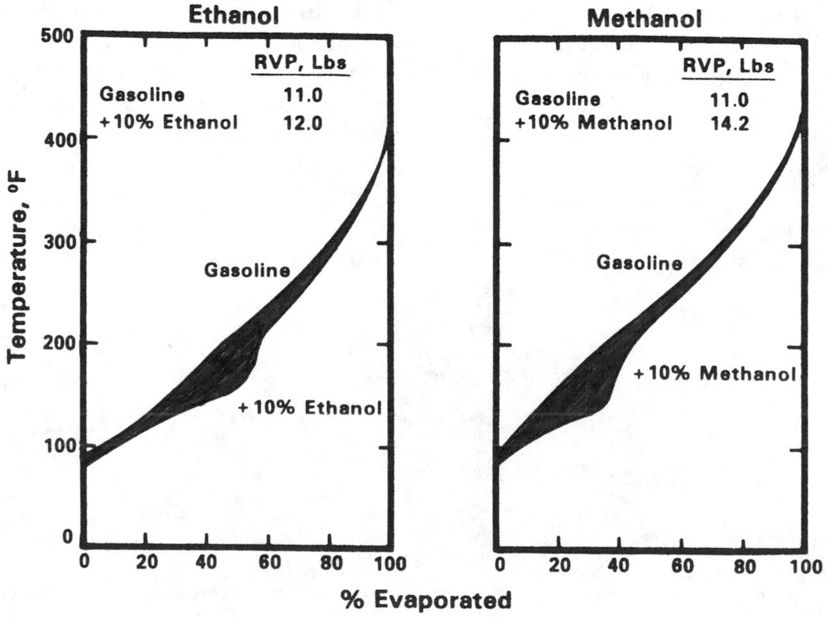

Fig. 5 - Effects of alcohols on vapor pressure and distillation

the Illinois Department of Administrative Services indicated 9 per cent higher fuel consumption than with gasoline(17).

Substituting alcohols for gasoline effectively leans carburetion unless some compensating adjustments are made. Thus, it is logical to expect that emissions of exhaust HC and CO might be reduced somewhat, although exhaust emissions of unburned alcohol and aldehydes and evaporative emissions may increase(18). Results of idle tests using exhaust analyzers of service-station quality have shown substantial reductions in HC and CO with Gasohol. Of course, similar benefits could be obtained with gasoline simply by adjusting the carburetor for a leaner mixture. FTP tests on catalyst-equipped cars using exhaust analyzers of laboratory quality have shown lower CO, higher NO_x, and mixed results on HC. Overall, use of alcohols as blends would probably have minor effect on vehicle emissions, unless vehicle owners found it necessary to retune engines to restore driveability. Returning could increase HC and CO emissions.

COMMERCIALIZATION OF ALCOHOLS

Commercialization of alcohols on any substantial scale would involve major new investments and would result in significantly higher product costs to motorists. An orienting example is shown in Table X. We assumed that the energy equivalent of 10 per cent of current motor fuel demand (700 MBPCD) is to be met by production of methanol from lignite or ethanol from corn. This would require production of 1,440 MBPCD of methanol and 1,060 MBPCD of ethanol.

The investment and product costs are those developed earlier. We assumed, rather arbitrarily, that additional transportation investments of $6 M/daily barrel would be needed with methanol and $3 M/daily barrel would be needed with ethanol. The added cost of transportation over that for gasoline was assumed to be $0.10/gallon for methanol and $0.05/gallon for ethanol. The total new investments required would be in the range of $30 to $35 MMM. This sum roughly equals the current annual expenditures of the domestic petroleum industry in all types of activities related to finding and producing petroleum.

Compared to gasoline at current market price, total costs at wholesale for the 700 MBPCD of gasoline replaced would be roughly double with methanol and five times as much with ethanol. Even if gasoline price is assumed to be as high as $0.55/gallon, as it might be if made solely from free-market or imported oil, the alcohols remain much more costly. Methanol is clearly the preferred alternative to ethanol. With methanol, transportation costs represent a relatively large fraction of total cost. This is partly because higher unit costs were assumed for methanol due to its longer distance from the major markets, and partly because a larger volume must be transported for equivalent energy. If the methanol could be used nearer the plants, and in a manner requiring only a simple distribution system, methanol would be more nearly competitive with gasoline.

Whether commercialization of alcohols would add to the nation's domestic supply of liquid fuels depends on how plant operation is structured. Methanol plants consuming lignite, coal, or biomass would presumably use no petroleum-based materials as feedstocks or as plant fuels, although small amounts of petroleum-based fuels would be consumed in mining, harvesting, and transporting raw materials and products. Otherwise, all fuel produced would represent a net addition to the liquid fuel supply.

Ethanol plants using oil as plant fuel would be large net consumers of petroleum. Those using gas as fuel would provide net additional liquid fuel at the expense of gas consumption.

Table X
Economics of Commercializing Alcohols

	New Capital Investments, $MMM	Product Costs at Wholesale $/Gallon		Total $MMM/Year	
Methanol for Lignite (1,440 MBPCD)					
Plant	22	0.35		7.7	
Transportation	9	0.15		3.3	
Total	31	0.50		11.0	
Ethanol from Corn (1,060 MBPCD)					
Plant	32	1.45		23.5	
Transportation	3	0.10		1.6	
Total	35	1.55		25.1	
Gasoline (700 MBPCD)					
Refinery	0	0.45	0.55*	4.8	5.9*
Transportation	0	0.05		0.5	
Total	0	0.50	0.60*	5.3	6.4*

*Estimated cost if gasoline was made entirely from foreign crude at $16/bbl.

Those using lignite, coal, or biomass would be net suppliers of liquid fuel, just like methanol plants, although at markedly higher cost.

However, with current technology, alcohol manufacturing practices and crude oil prices, the economic interests of the public would not be served by commercializing either methanol or ethanol.

THE REFINER/MARKETER'S VIEWPOINT

We want to consider now the incentives for a refiner/marketer to offer alcohol-based fuels. It is becoming obvious that the real cost--the cost in uninflated dollars--of domestic crude oil will have to rise in the future to sustain production, and that the present rate of production cannot be sustained indefinitely at reasonable cost. Possible alternatives are to import more crude, use alcohols as fuels, or produce synthetic crudes from shale, tar sands or coal, and then refine more-or-less conventional fuels and other products from them.

At the present time, the cost of all alcohols and, especially, ethanol is substantially higher than the cost of liquid fuels made from foreign crude or from domestic crude exempt for price restraints. Further, there are no indications that the cost of the alcohols can be reduced significantly. This means that there will be no economic incentive for a refiner/marketer to offer alcohol-based fuels until the real price of gasoline made from petroleum or synthetic crudes climbs to the equivalent price of the alcohols. The cost of synthetic crudes from shale, tar sands, or coal are very uncertain, but estimates have been quoted in the range of $20 to $30 per barrel in 1979 dollars for refinery grade crudes. At these prices, gasoline might cost $0.70 to $0.90 per gallon; methanol would begin to be competitive, but ethanol would not.

Provisions of the 1978 Energy Tax Act and recent actions of several state governments tend to distort the basic economics of alcohol use. The Tax Act exempts gasoline containing 10 per cent or more of alcohol produced from agricultural or forestry feedstocks from 4¢ per gallon of federal excise tax. This amounts to 40¢ per gallon of ethanol or $16.80 per barrel, more than the present cost of foreign crude landed in the United States. Also, several states exempt Gasohol from some local excise taxes; one state allows an exemption of more than 6¢ per gallon. With the federal exemption, the combined subsidy for alcohols may reach an incredible $42 per barrel! Since the real cost for adding 10 per cent ethanol to gasoline at current prices is about 10¢ per gallon, these exemptions put Gasohol roughly on an apparent break-even basis. Earlier legislation provided for low-cost loans for alcohol plants, which could reduce capital charges in ethanol manufacturing and further reduce apparent ethanol cost.

As a result of these measures, there are now localities where tax exemptions have made the cost to the refiner/marketer for alcohol-based fuels about equal to that of hydrocarbon fuels. Of course, the subsidy does nothing to reduce the real cost of the alcohols. It only shifts part of the cost to others who often do not know they are paying it. In the case of the alcohol subsidy, the cost must eventually show up as reduced road construction and maintenance, or higher taxes on nonexempt motor fuel or some other commodity. But even though the refiner/marketer is protected by this subsidy, it would not be a responsible action to offer alcohol fuels, especially ethanol, on anything but a test basis as long as the real cost is so high and the energy balance is so poor.

However, even if the real cost of alcohols matched that of petroleum fuels, there would still be little incentive for a refiner/marketer to use alcohol fuels. They provide modest octane improvement, but have adverse effects on volatility, and they are much more difficult to distribute through service stations. And while methanol could enhance total liquid fuel supply, ethanol would not unless use of petroleum fuels were completely prohibited in its manufacture.

REFERENCES

1. Anon., Chemical and Engineering News, January 22, 1979, p13.

2. Anon., Chemical and Engineering News, December 18, 1978, p10.

3. Hagen, D. L., "Methanol: Its Synthesis, Use as a Fuel, Economics, and Hazards," ERDA NP-21727, December, 1976. (MS thesis, University of Minnesota).

4. Barr, W. J., and Parker, F. A., "Sources of Alcohol Fuels for Vehicle Fleet Tests," ERDA CONS/2693, August, 1977.

5. Clark, F. C., and Ushiba, K. K., "Methanol Fuel: Its Manufacture and Utilization," Stanford Research Institute, March 1977.

6. Alich, J. A., and Witmer, J. G., Solar Energy, Vol. 19, p625.

7. Ernest, R. Kent, Proceedings of the Second Pacific Chemical Engineering Congress, 1977, Vol. I, p187.

8. Park, W. R., Second Annual Fuels from Biomass Symposium, Rensselaer Polytechnic Institute, June, 1978.

9. Hokenson, A. E., et al., "Chemicals from Wood Wastes," Raphael Katzen Associates, Cincinnati, December, 1975, NTIS No. PB-262-489.

10. David, M. L., et al., "Gasohol Eco-

nomic Feasibility Study" Report prepared under contract to the Energy Research and Development Center, University of Nebraska, in cooperation with the Nebraska Agricultural Products Industrial Utilization Commission, July, 1978.

11. Todisch, M. R., et al., Fuels and Chemicals from Biomass, Purdue Agricultural Experimental Station, Publication 7290.

12. Scheller, W. A., and Mohr, B. J., "Gasoline Does, Too, Mix with Alcohol," Chemtech, October, 1977, p616.

13. Scheller, W. A., and Mohr, B. J., "Net Energy Analysis of Ethanol Production," 171st National meeting of the American Chemical Society, New York, April, 1976, Vol. 21, No. 2, p29.

14. Inman, R. E., "The Cost of Plant Biomass Production," Biotech and Bio Engineering, Symposium No. 5, 1975, p43.

15. Brinkman, N. D., et al., "Exhaust Emissions, Fuel Economy and Driveability of Vehicles Fueled with Alcohol/Gasoline Blends," Society of Automotive Engineers, preprint 750120.

16. Koenig, A., et al., "Technical and Economical Aspects of Methonal as an Automotive Fuel," Society of Automotive Engineers, preprint 760545.

17. Demel, J. R., Illinois Department of Administrative Services, Press Release, May 8, 1978.

18. Anon., Analysis of Gasohol Fleet Data to Characterize the Impact of Gasohol on Tailpipe and Evaporative Emissions, U.S. Environmental Protection Agency, December, 1978.

Alcohols in Diesel Engines - A Review*

Henry Adelman
Univ. of Santa Clara

ALCOHOLS HAVE BEEN USED IN INTERNAL COMBUSTION ENGINES since their invention. The lower alcohols methanol and ethanol are known to be excellent fuels for spark ignition (SI) engines due to their high octane ratings, lean flammability limits, increased thermal efficiency and low exhaust emissions [1,2]*. In addition, alcohols have been shown to be good gas turbine (GT) fuels since they give lower exhaust emissions and longer engine life [3,4].

While SI and GT engines can use alcohol fuels with minimal modifications to their fuel delivery systems, the diesel engine has not been a good candidate for alcohols. In these compression ignition (CI) engines the extremely low cetane ratings (0-10) of the alcohols prevents their autoignition. Therefore, alcohols have been used in unmodified diesel engines only with the addition of large amounts (10%-20%) of ignition accelerators which are typically nitrate compounds [5-8]. These chemicals increase the cetane ratings of the alcohols by autoigniting at low temperatures and providing an ignition source for the fuel.[9]. Unfortunately, they are expensive and because they contain nitrogen, they increase NOx emissions dramatically. Only in one case has a diesel engine been made to operate on pure ethanol by using a higher compression ratio of 25:1 [5].

Since the alcohols have not been suitable CI engine fuels when used alone, their use in conjunction with diesel fuel has been extensively studied. Historically, alcohols have been added to the engine intake air (fumigation) since they do not mix well with diesel fuels[10]. However, even in this application alcohols can cause ignition problems.

Even though alcohols are not presently good CI engine fuels, experiments on their optimum use continue since some areas of the world need to produce and use fuels from their own biomass and coal resources instead of importing petroleum.

This paper will try to summarize the work done to date and highlight the areas which require further research. First, the important property differences between alcohols and traditional diesel fuels will be discussed followed by an examination of the emissions and thermal efficiency characteristics of alcohol fueled diesel engines.

FUEL PROPERTIES-STOICHIOMETRY

DIESEL FUEL- Diesel engines are not like conventional spark ignited engines which require a relatively homogeneous mixture of fuel and air that must be carefully mixed according to the fuel stoichiometry. Instead, compression ignition engines operate by injecting a liquid fuel direcly into the combustion chamber where upon

---

* Numbers in brackets designate References at the end of paper.

*Paper 790956 presented at the Fuels and Lubricants Meeting, Houston, Texas, October 1979.

ABSTRACT

Pure alcohols and alcohol-diesel fuel blends have been compared to diesel fuels in terms of engine power, efficiency and emissions. It is apparent that pure alcohols are poor diesel fuels as their properties are significantly different from the traditional fuels. These differences are examined and in some instances methods of offsetting the alcohols' deficiencies are assessed. Further areas of study are identified since it is possible that some regions of the world will need to use alcohols as diesel fuel extenders in the near future.

SAE/PT-80/19/$02.50

injection into the hot, compressed air, the fuel vaporizes, spontaneously ignites and burns locally as a diffusion flame. Thus, the engine is referred to as a compression ignition engine (C.I.). This local diffusive burning does not require that the whole combustion chamber be filled with a combustible mixture and therefore the fuel stoichiometry is not an important consideration except as it influences the fuel energy input. Stoichiometry is, however, a consideration in diesels in that an overall fuel lean mixture must always be maintained to avoid the exhaust emissions of large amounts of particulates, unburned fuel and carbon monoxide.

Diesel fuels, like gasoline, require about fifteen pounds of air per pound of fuel for a stoichiometric air-fuel ratio (A/F) of 15 to 1. Since a diesel engine does not have a throttle the amount of air inducted per cycle is fixed and power output is controlled by the amount of fuel injected. In operating over its load range a diesel engine may have A/F's from 20:1 for maximum power to 100:1 for idling; note that the overall A/f is always fuel lean.

ALCOHOLS- Methanol and ethanol have stoichiometric A/F's which are lower than diesel fuel. However, stoichiometric mixtures of diesel fuel and alcohols have roughly the same energy, so the ratio of the stoichiometric A/F for diesel fuel to that for alcohols is a measure of the increased mass of alcohol fuel which is required for the same power output. For example, methanol with a stoichiometric A/F of 6.4 to 1 requires roughly 2.3 times more fuel for a given energy input, while ethanol with a stoichiometric A/F of 9 to 1 requires about 1.7 times more fuel than diesel. In addition, the densities of methanol (0.79 gm/ml) and ethanol (0.78 g/ml) are lower than for diesel fuel (0.84 g/ml) requiring 2.5 times more volume of methanol and 1.8 times more volume of ethanol for the same energy as a unit volume of diesel fuel. These large increases in alcohol fuel volumes would require changes in the diesel fuel metering system to provide the full range of power output.

ALCOHOL-DIESEL FUEL BLENDS- Methanol and ethanol are not very soluble in diesel fuel so blends of these two fuels would usually take the form of emulsions. However, here the word blend will be used to describe any method of using the two fuels simultaneously as by inducting the alcohols in the intake air (fumigation) or by using a separate alcohol fuel injection system as well as emulsions. Since methanol and ethanol liquid fuels have only 40% and 56% of the volumetric energy of diesel fuel respectively, it would not be possible to deliver full power with an alcohol emulsion in an unmodified fuel system. However, it would be possible to provide more fuel energy with fumigation since the excess air could be utilized by the alcohol fuel. Hence, it is possible to increase the power output from diesels by use of alcohol diesel fuel blends.

FUEL PROPERTIES- VOLATILITY AND HEAT OF VAPORIZATION

DIESEL FUEL- Diesel fuel must evaporate and mix with air before igniting and burning. However, since the fuel does not have to vaporize and mix with air outside the engine, its volatility need not be as high as for spark ignition (SI) engine fuels. The volatility of diesel fuels, like gasolines, is measured by the distillation temperatures versus percent evaporated. For example, a typical diesel fuel consists of a mixture of hydrocarbon compounds which boil from 366°F (186°C) to 640°F (337°C) with 50% evaporation at 494°F (257°C) [11].

While the volatility of diesel engine fuel is not as critical as for SI engine fuels, the heat of vaporization can have an effect on the air-fuel mixture temperature when the fuel evaporates after injection. Diesel fuel has a low heat of vaporization. For example, hexadecane or cetane which is used as a reference fuel has a heat of vaporization of 97 Btu per pound. The vaporization of a stoichiometric mixture of cetane and air (A/F=15) will cause the mixture temperature to drop about 30°F (about 17°C). Since the ignition delay of a fuel sprayed into air is extremely sensitive to temperature the heat of vaporization will be an important comparison for alternate fuels.

ALCOHOLS- Methanol and ethanol have low vapor pressures at room temperature but very high heats of vaporization of 502 and 396 BTU per pound respectively. Thus the vaporization of a stoichiometric amount of methanol in air will lower the adiabatic mixture temperature by 356°F (198°C) while a stoichiometric mixture of ethanol will drop 200°F (111°C). These large reductions in mixture temperatures compared to diesel fuel are a major problem for alcohols since diesel engine operation relies on the rapid autoignition of the injected fuel when it mixes with air. Since the ignition delay, or time to the start of burning, is a strong function of temperature for all fuels, the vaporization and mixture cooling of alcohols will result in increases in their ignition delays. This issue will be discussed in a later section on cetane ratings.

ALCOHOL-DIESEL FUEL BLENDS- The high heats of vaporization of alcohols become less of a problem as the ratio of alcohol to diesel fuel decreases. In some cases up to 30% methanol has been used in fumigated diesels with no changes to the engine such as intake heating or higher compression ratio to compensate for the increased heat of vaporization of the alcohol [12]. In addition, 20% methanol or ethanol have been used in emulsions in unmodified engines [13]. The changes in volatility of alcohol emulsions is not known. However, it has been shown that with water-diesel fuel emulsions the fuel droplet contains water in its center. Heating this composite droplet in air causes the water to boil and shatter the droplet producing faster fuel vaporization [14]. Whether a similar process occurs with alcohol-diesel fuel emulsions is not known.

FUEL PROPERTIES- CETANE RATING

DIESEL FUEL- The cetane rating of a diesel engine fuel is a measure of its ability to autoignite quickly when it is injected into the compressed and heated air in the engine. Like octane ratings for SI engines, a high cetane number indicates a low propensity for diesel engine knock. Diesel fuels have typical cetane ratings of 40 to 60 while high octane fuels as gasoline which are difficult to autoignite have cetane numbers of about 10 to 20 indicating their poor suitability as diesel fuels. As a general rule, fuels that work well in diesels are poor for SI engines and vice versa.

ALCOHOLS- Pure alcohols such as methanol and ethanol have very low cetane numbers ranging from zero to about five [5] indicating difficulty in achieving their autoignition in diesel engines. As mentioned in the section on volatility, one reason for the alcohols' low cetane numbers is their high heats of vaporization which lowers mixture temperatures and increases ignition delays. This heat of vaporization effect can be counteracted by various methods which will increase mixture temperatures in the engine such as heated intake air and higher compression ratios. On the other hand methods are available to increase the alcohol fuel's cetane rating by adding compounds to the fuels which autoignite very easily even at low temperatures and provide an ignition source for the alcohols. These compounds which are typically nitrates are called ignition accelerators or improvers. However, they can be expensive and their toxicities in the raw fuel or as a component of the engine exhaust are not known.

ALCOHOL-DIESEL FUEL BLENDS- The addition of alcohols to diesel fuel in the form of emulsions or solutions may lower the cetane rating of the fuel. Some recent evidence shows that 20% methanol or ethanol emulsions with 5% surfactant lower the cetane rating of the diesel fuel from 45 to 30 and 43 respectively [13]. While the ethanol emulsion causes very little cetane degradation, a 20% ethanol solution showed a cetane number of only 33. Ignition delay for all the hybrid fuels was increased by about the same amount so cetane number may not be an adequate criterion for mixed alcohol-diesel fuel performance.

When the alcohol fuels are fumigated there is generally an increase in the ignition delay of the injected diesel fuel. The fact that this increased delay was due to mixture cooling, caused by the alcohol's high heat of vaporization, was demonstrated by inducting the fuel as liquid droplets and as vapor. When ethanol was inducted as a mist of fine droplets, ignition delay increased while vaporized ethanol induction caused no change or a slight decrease in ignition delay [15]. Other tests with fumigated ethanol indicate increased ignition delay with increasing amounts of alcohol [16]. The ignition delay increases could be reduced by raising the compression ratio or by adding the ignition accelerators nitromethane and aniline to the inducted ethanol instead of to the injected diesel fuel which is known to have little effect. Similar increases in ignition delay have been observed when methanol was carbureted in a diesel engine [6]. In general, the ignition delay increases with fumigated alcohols can be reduced or cancelled out by heating the intake air to counteract the alcohol's evaporative cooling effects. However, excessive mixture heating can cause the rapid autoignition of the alcohol-air mixture resulting in the same "knocking" phenomena as in spark ignition engines. Thus, engineers and designers of alcohol systems for diesels must recognize the paradoxical problem of their having a low cetane number if injected and a low octane number if fumigated in a heated intake mixture.

FUEL PROPERTIES- MATERIALS COMPATIBILITY AND WEAR

DIESEL FUEL- Diesel engine components and diesel fuel are, of course, made to be compatible. Since diesel fuel is a good lubricant, it is used to lubricate parts of the fuel injection pump. Heavy duty diesel engines are noted for their long service life implying a low wear rate of materials. However, the high lubricating oil contamination from the diesel combustion process requires frequent maintenance to achieve low engine wear rates.

ALCOHOLS- Pure alcohols have very poor lubricity and could cause accelerated wear of diesel fuel systems and engine components. Union Oil Company tests report high steel-on-steel rubbing wear with methanol but this could be controlled by the use of a lubricity additive designed for gas turbine fuels [17]. In some current tests with methanol, 1% castor oil was added to the alcohols to provide fuel pump lubrication [7]. Fuel system lubrication problems could be solved by the use of multi-fuel pumps which do not rely on the fuel for lubrication. However, the effects of alcohols on internal engine wear are not known at this time.

ALCOHOL-DIESEL FUEL BLENDS- The use of inducted, vaporized alcohol mixtures should not cause any wear or materials problems in diesel engines. The authors through limited experience have not found any fuel pump problems when using methanol mixed with diesel fuel or jet fuel in a gas turbine to form emulsions. More investigations are needed before it is known if any materials or wear problems will occur.

EXHAUST EMISSIONS-DIESEL FUEL

HYDROCARBONS- Similar to spark ignition (SI) engines, gaseous hydrocarbons or unburned fuel appears in the exhaust of diesel engines. The source of these emissions is also believed to be somewhat similar to that for SI engines being due to incomplete fuel combustion and flame quenching along with problems unique to the diesel as dripping or leakage of the fuel injectors. In general, diesel engines emit less HC than SI engines since the diffusion flame is surrounded by air and comes in contact with a smaller portion of the combustion chamber.

The quality of CI engine hydrocarbon emissions is as important as their quantity. One set of chemical analyses showed that 1% to 5% of diesel vehicle HC emissions are non-reactive methane while about 70% are reactive compounds (benzene, ethylene, propylene and toluene) [11]. Ethane and acetylene constituted the remainder of HC emissions. In comparison, a similar catalyst equipped gasoline fueled vehicle emitted about 30% to 50% methane and 40% reactive hydrocarbons. Polynuclear aromatic (PNA) compound emissions seemed to be greater for the diesel engine. Thus, the gaseous hydrocarbon emissions from diesel exhaust appeared to be more photochemically reactive, toxic and carcinogenic than that from a catalyst equipped gasoline vehicle. Oxidation catalysts have been tried on diesel engines but they are not as effective in HC control as for an SI engine because the diesel exhaust is colder and the particulate emissions would soon coat the active sites [18].

CARBON MONOXIDE- Carbon monoxide (CO) emissions from a diesel engine are generally very low due to the lean air-fuel ratios. Tests with similar diesel and catalyst equipped gasoline cars showed about the same CO emissions [11]. Since oxidation catalysts remove up to 90% of the CO from gasoline exhaust the diesel must have had inherently lower CO emissions.

OXIDES OF NITROGEN- Oxides of nitrogen (NOx) emissions from CI engines appear to be low due to the lean air-fuel ratios but they are not as low as they would be for a homogeneously charged engine operating at the same lean equivalence ratio [19]. This is undoubtedly due to the diffusive and droplet burning nature of the diesel flame where combustion occurs at close to stoichiometric conditions regardless of the amount of excess air. Consequently, A/F ratio has little effect on diesel NOx emissions while it has a major influence on SI emissions. NOx control schemes such as exhaust gas recirculation can be effective on diesel engines but other emissions such as HC and particulates will increase [20]. Present NOx catalysts cannot be used with diesels since they require a certain combination of oxygen, carbon monoxide and hydrogen which is only found in the exhaust of stoichiometric mixtures.

PARTICULATES- Particulate emissions from diesel engines are from 50 to 100 times those from gasoline engines [11]. Diesel particulates consists of liquid aerosols and solid carbonaceous particles resulting from the heterogeneous nature of the combustion process where fuel vaporizes and mixes with air in a non-uniform manner. Fuel rich mixture pockets burn and form gaseous carbon which then condenses into solid particulates [21]. Other gaseous HC compounds can then condense on these nuclei. PNA compounds have been observed on diesel particulates and the carcinogenicity of these particles is currently under investigation by the Environmental Protection Agency (EPA) [22]. The size of diesel particulates is also important since smaller particles can enter and remain in the respiratory system and may cause severe health problems.

OXYGENATES- Oxygenated species such as aldehydes and ketones are present in diesel exhaust as they are in gasoline exhaust. Tests with a diesel vehicle showed aldehyde emissions of about 0.12 gm/mi compared to 0.032 gm/mi for a similar catalyst equipped gasoline vehicle [11]. Since catalysts can remove about 80% to 90% of aldehydes from gasoline exhaust, a non-catalyst vehicle would be expected to have similar aldehyde emissions as a diesel fueled car.

ODORS- Diesel engine exhaust has a characteristic odor which is offensive to many people. Tests are being conducted to determine the compounds in diesel exhaust which contribute to this odor [23].

OTHER SPECIES- Since diesel fuels contain 0.1% to 0.5% sulfur, sulfur compounds can appear in diesel exhaust. The major sulfur compound in diesel exhaust is sulfur dioxide (SO_2) which can be converted to sulfuric acid during the exhaust and subsequent mixing with the atmosphere [24].

EXHAUST EMISSIONS-ALCOHOLS

Neat alcohols have not been tested extensively in diesel engines due to their extremely low cetane numbers. Ethanol has been used in one case with a high compression engine while pure methanol has required the use of spark ignition in a diesel type engine. Otherwise alcohols have been used in an impure form with the addition of large amounts of additives which decrease the ignition delays of these fuels. The limited emissions data from these tests is presented next.

HYDROCARBONS- The hydrocarbon or unburned fuel emissions from pure alcohol fueled diesel engines have not been reported. In one case, a direct injection diesel type engine was modified to use low cetane fuels by adding a spark ignition system [25]. When fueled with methanol, the HC emissions were the same as for a normal SI engine. Tests with ethanol in a high compression ratio diesel did not report emissions [5]. While no chemical analyses have been made, the HC composition of alcohol fueled diesel exhaust is expected to be similar to that of SI engine exhaust with unburned alcohols predominating.

Alcohols have been used in place of diesel fuel in some cases when ignition modifying compounds were added to the fuels. In one test, 16% hexyl nitrate was required for methanol operation in an unmodified engine. Unburned fuel emissions were two to ten times higher than for diesel fuel but no diagnostic tests were made to find the reason for these increases [8]. Also, no measurement was made of the exhaust hydrocarbon composition, thus the effects of the hexyl nitrate additive are not known. In Swedish tests with methanol, 20% of the ignition improver Cetonox was necessary to provide a cetane rating of 35 [7]. If the air was preheated to 70°C, this requirement was reduced to 12% additive. Hydrocarbon emissions were low but were not compared to those from diesel fuel. Note that any nitrogen-containing additives have the potential for forming dangerous compounds as hydrogen cyanide or ammonia in combustion with fuels containing hydrogen.

CARBON MONOXIDE- Carbon monoxide emissions data do not exist for pure alcohol fueled diesel engines. Tests with a diesel type engine modified to include a spark ignition system show about twice the emissions of CO for methanol as for gasoline but values were still very low [25]. Other tests in a diesel engine with methanol containing 16% hexyl nitrate also showed about twice the CO compared to diesel fuel [8]. However, all emissions were higher from this engine and no attempt was made to explain or optimize its behavior. It should be noted that similar increases in CO were observed in gas turbines using methanol and ethanol instead of fuel oil but these higher emissions were traced to inadequate fuel injection system performance using alcohols [4]. When the fuel injection system was changed to produce smaller alcohol droplets, CO emissions were reduced to levels below those for fuel oils, a trend which had been predicted by computer modeling [3]. Since alcohols produce the same CO as hydrocarbon fuels in spark ignition engines, the higher comparative values seen in diesel engines may be due to differences in fuel injection characteristics with alcohols. Fuel injection systems designed for alcohols should eliminate this problem.

OXIDES OF NITROGEN- The emissions from pure alcohol fueled diesel engines have not been reported. However, in other internal combustion engines such as spark ignition engines and gas turbines, the use of alcohols reduces NOx smissions significantly [1-4]. This trend was also apparent in a spark ignited version of a diesel engine where NOx was reduced by a factor of three by using methanol instead of gasoline [25].

Other alcohol fuel tests in unmodified engines required the use of ignition improving compounds. Since these chemicals contain nitrogen, it is probable that they will increase NOx emissions. This suspicion has been verified in one test where methanol containing 16% hexyl nitrate produced twice the NOx emissions as for diesel fuel [8]. If methanol can be used in its pure form in diesel engines much lower NOx emissions should result.

PARTICULATES- Pure methanol has not been observed to form soot during combustion. Furthermore, methanol does not contain any inorganic materials as sulfur, so particulate emissions from diesel engines are expected to be nonexistent. Ethanol also does not appear to form soot during combustion and no evidence of soot was seen in tests using 100% ethanol in a high compression ratio diesel [5]. In tests with 10% cyclohexanol nitrate in ethanol, the Bosch smoke number, which is an indication of particulate density, was close to zero for all load conditions [5]. If particulates are observed in alcohol diesel operation, they will probably be due to the combustion of any lubricating oil which reaches the combustion chamber. The ability to combust without forming particulates makes alcohols very attractive fuels for diesel engines since the proposed EPA standard of 0.6 gm/mi particulates for light-duty vehicles apparently cannot be met by existing domestic light duty diesel-fueled vehicles.

OXYGENATES- No measurements have been reported for oxygenated species in the exhaust of alcohol fueled diesel engines. However, in view of the increased levels of aldehydes when using methanol or ethanol in a spark ignition engine, higher aldehydes might be expected in alcohol diesel exhaust.

ODORS- The odor characteristics of diesel engines fueled with alcohols have not been reported. However, the authors have noted that spark ignition engine exhaust has a different odor with methanol than with gasoline which may be due to the presence of unburned methanol and formaldehyde.

OTHER SPECIES- Since alcohol fuels do not contain sulfur compounds, no sulfur oxide emissions are expected. However, the addition of nitrogen containing ignition acceleration additives to alcohols may cause the formation of some danerous hydrogen-carbon-nitrogen compounds as HCN. This issue must be resolved in laboratory testing.

EXHAUST EMISSIONS-ALCOHOL-DIESEL FUEL "BLENDS"

FUMIGATION- Mixed fuel operation with alcohols and diesel fuels can take several forms. Historically, alcohols have been introduced into diesel engines by mixing them with the intake air which is termed fumigation [26]. The injected diesel fuel then acts as a pilot ignition source for the homogeneous alcohol-air mixture. The resulting combustion is a composite of diesel diffusive burning and premixed alcohol combustion which may produce emissions patterns with characteristics of both diesel and spark ignition engines.

Hydrocarbons- When alcohols are introduced into the diesel intake air, hydrocarbon emissions generally increase with increasing amounts of alcohols [12,19]. Unfortunately no analyses were made of the chemical composition of the exhaust which may have uncovered the source of the increased emissions. However, it is known that homogeneously mixed spark ignition engines emit unburned fuel due to premixed flame quenching at the cylinder walls, so this mechanism would be expected to increase fumigated diesel emissions. In some instances the slow burning of the lean alcohol-air mixture may also result in higher unburned fuel emissions due to its bulk quenching during the engine's expansion stroke [16]. The theory that fumigated diesel HC emissions result from the same mechanisms as in SI engines is supported by their similar trends with equivalence ratios [19]. Thus, fumigated operation may always result in higher hydrocarbon emissions than pure diesel operation. However, if the chemical composition of the HC emissions is improved with alcohol addition, mass increases might be acceptable. Further, if particulates are significantly reduced, catalytic cleanup of the exhaust may be possible.

Carbon Monoxide- As long as there is an overall fuel lean mixture in an engine, CO emissions will be low. Tests with fumigated ethanol and methanol show that CO remains relatively fixed unless the amound of alcohol added to the air is high enough to cause an overall fuel rich mixture which will result in high emissions. [12,19].

Oxides of Nitrogen- NOx emissions from pure diesel fuel operation do not appear to be a strong function of equivalence ratio as they are in homogeneous spark ignition engines [19]. However, when homogeneous alcohol-air mixtures are inducted, NOx emissions show the same trends with equivalence ratio as for SI engines, peaking at slightly lean overall mixtures. Depending on the relative amounts of inducted alcohols and injected diesel fuel and the overall equivalence ratio, NOx can either increase or decrease with fumigation.

Particulates- Fumigation of alcohols generally reduces smoke emissions which have previously been used as a measure of particulate mass [12,19]. However, the correlation between visible smoke and particulate mass and size distribution has not been established. In addition, the chemical composition of the particulates must be determined indicating that more sophisticated particulate studies are needed. The effectiveness of inducted alcohols as smoke suppressors depends on the overall equivalence ratio and the relative amounts of alcohol and diesel fuels.

Oxygenates- No studies have been made of the effects of fumigation on oxygenates in diesel exhaust. However, because the use of alcohols in SI engines causes higher aldehyde emissions than with hydrocarbon fuels, these emissions must be examined.

Odors- Again, no information is available on the effects of alcohols on diesel exhaust odors.

Other Species- To the extent that alcohols contain no sulfur and produce little PNA, these emissions should decrease in proportion to the amount of energy supplied by the alcohols.

DUAL INJECTION- Other dual fueling methods include a separate injection system for each fuel or mixing of the two fuels before they reach a single injector. As an initial attempt at using dual fuels, a large diesel truck engine has been fitted with an additional fuel injection pump for methanol so that it could be mixed with diesel fuel before the single injector [7]. Hydrocarbon emissions were generally higher than for pure diesel operation and emission control under transient conditions was difficult. To overcome these problems a completely separate injection system was installed for the methanol. In general, if the diesel fuel was injected first so that it would act as a pilot to ignite the methanol, HC emissions could be kept at levels comparable to pure diesel operation, but transient control of HC was still difficult. Other emissions as CO and NOx were not measured in these tests but smoke emissions were only one fifth those of pure diesel operation.

Other investigators have indicated that a similar dual injection system could be used to provide the same low level of hydrocarbon emissions with the correctly phased injection of 15% to 40% methanol [27]. In addition, CO and NOx emissions were reported to be the same or lower than for pure diesel operation while smoke was reduced.

EMULSIONS AND MIXTURES- Since alcohols are not very miscible in diesel fuel, only alcohol-diesel fuel solutions or emulsions stabilized by surfactants may be used in unmodified diesel fuel injection systems. Unfortunately, no data exists on the emissions from alcohol-diesel fuel emulsions. However, the use of a 20% water-diesel fuel emulsion showed only a minor effect on NOx emissions, less CO and smoke and higher HC emissions [8]. All the observed effects for the emulsion could be achieved by limiting the maximum engine power with pure diesel fuel to that available from the reduced energy emulsified fuel. Whether alcohol-diesel fuels will behave similarly is not known.

While emulsions may be necessary to use alcohol and diesel fuel in the original fuel metering system, one test was conducted where methanol was supplied by a separate fuel metering pump then mixed with diesel fuel before the single injector. Hydrocarbon emissions were difficult to control and rose to very high levels during transient operation [7].

THERMAL EFFICIENCY AND POWER- DIESEL FUEL

The thermal efficiency of a diesel engine is higher than that of a spark ignition engine for several reasons. First, diesel engines operate at much higher compression ratios (CR) and thermal efficiency is strongly influenced by CR as shown by the following relationship for the ideal Otto cycle with constant volume heat addition

$$\eta_T = 1 - \frac{1}{CR^{(\gamma-1)}}$$

where η_T is the thermal efficiency and γ is the ratio of specific heats for air. Secondly, the diesel engine is not throttled so there are less air pumping losses. Finally, diesels can operate at very lean equivalence ratios and thermal efficiency is known to increase with leaner mixtures in a real engine due to increases in γ for the air-fuel mixture. The thermal efficiency of an actual diesel engine varies with the speed and load but it is typically 20% to 50% higher than that of a spark ignition engine.

The power output of a diesel engine increases with the amount of fuel which is injected but is usually limited by the increased smoke emissions as stoichiometric mixture ratios are approached. Because the fuel is delivered inside each cylinder, diesel engined vehicles do not have the driveability problems that gasoline fueled cars experience as a result of their air-fuel mixing and distribution problems.

THERMAL EFFICIENCY AND POWER- ALCOHOLS

Pure alcohols have been used only in the case of ethanol in a high compression ratio engine [5]. A decrease in thermal efficiency compared to diesel fuel operation at normal compression ratios was attributed to the intake air preheating which was necessary to decrease the alcohol's ignition delay. Intake air preheating also reduced the intake air density decreasing the mass of air which could be inducted and causing a loss in maximum power. It was conjectured that the power output could be restored by turbocharging.

In other tests with alcohols containing ignition improvers, it was found that 10% cyclohexanol nitrate (Kerobrisol) allowed ethanol to be used in an unmodified diesel engine [5]. Here the thermal efficiency was the same for the ethanol as for diesel fuel but power output was not reported. In a similar way methanol has been used in a standard diesel engine by adding 16% hexyl nitrate by volume. In this case power output was limited by the lower energy content of the methanol compared to diesel fuel but thermal efficiency was not determined [8]. Other methanol tests with an ignition improver called Cetonox did not report thermal efficiency or power [7].

THERMAL EFFICIENCY AND POWER- ALCOHOL-DIESEL FUEL "BLENDS"

FUMIGATION- Tests with ethanol and methanol in a fumigated diesel show that thermal efficiency was improved by as much as 30% in a direct injection engine at certain combinations of alcohol-diesel fuel ratios and overall equivalence ratio [19]. Power output generally increased as alcohols were added to the intake air unless the alcohol-air mixture was too lean for normal combustion. In some cases the evaporative cooling effect of the alcohols quenched the diesel fuel combustion. Increases in power output with diesel engine fumigation are not surprising since there must be excess air in pure diesel operation to avoid the excessive hydrocarbon and particulate emissions which result from rich mixtures. The alcohols which are inducted can then utilize this excess air for combustion.

Similar increases in power and thermal efficiency have been seen when methanol was fumigated in an open chamber turbocharged diesel. Thermal efficiency increased slightly (5%) at 2/3 and full load with alcohol to diesel mass ratios up to 1.2 but efficiency decreased substantially (25%) at 1/3 load [12]. Ethanol induction showed the same trends as methanol with slight increases in efficiency at 2/3 and full load. With this particular engine, the reduction in smoke with fumigation allowed power output at a given smoke level to be increased by as much as 36% for methanol and 10% with ethanol depending on the alcohol-diesel fuel ratio.

EMULSIONS- If emulsions are used in unmodified diesel engines, the replacement of some diesel fuel with alcohol will result in less volumetric fuel energy and maximum power may fall accordingly. This trend has been demonstrated in tests with a 20% methanol-diesel fuel emulsion where maximum power dropped by 6% to 10% [28]. However, efficiency increased by a few percent.

Other tests with 20% methanol or ethanol emulsions with 5% surfactant showed the same specific energy consumption as diesel fuel at high loads but somewhat increased consumption at 1/2 and 3/4 load while a solution of 20% ethanol and diesel fuel provided the same specific energy consumption and thermal efficiency as diesel fuel for all speeds and loads [13].

DUAL INJECTION- Tests with a large diesel truck engine with separate injection systems for methanol and diesel fuel showed slightly higher efficiencies at low speeds but lower efficiencies at high speeds at full load [7]. Power was increased throughout the speed range at full load.

CONCLUSION

The evidence to date clearly indicates that alcohols can extend diesel fuel supplies without major engine and fuel system modifications. This can be done by fumigation, by injection of diesel-alcohol fuel emulsions, or by dual injection of alcohols and diesel fuels. The ultimate effect on engine life has yet to be assessed but it appears that there need be no loss of power and efficiency or increase in emissions. Further work is needed if alcohols are to be considered a total replacement for traditional diesel fuels. Also the potential of the alcohols for reducing NOx and particulate emissions needs to be fully assessed.

REFERENCES

1. Adelman, H.G., D.G. Andrews and R.S. Devoto, "Exhaust Emissions From a Methanol-Fueled Automobile", SAE Transactions, Volume 81, Paper No. 720693, 1972.
2. Pefley, R.K., et. al., "Characteristics and Research Investigation of Methanol and Methyl Fuels", DOE Report Contract No. EY-76-S-02-1258, University of Santa Clara, Santa Clara, CA August 1977.
3. Adelman, H.G., L.H. Browning and R.K. Pefley, "Predicted Exhaust Emissions From a Methanol and Jet-Fueled Gas Turbine Combustor", AIAA Journal, Volume 14, No. 6, June 1976, pp. 793-798.
4. Pullman, J.B., "Methanol, Ethanol and Jet Fuel Emissions Comparison From a Small Gas Turbine", SAE Paper No. 781013, November 1978.
5. Bandel, W., "Problems in the Application of Ethanol as Fuel for Utility Vehicles", Proceedings of the International Symposium on Alcohol Fuel Technology, Methanol and Ethanol, Wolfsburg, Federal Republic of Germany, November 1977. English Translation published by the U.S. Department of Energy, CONF-T11175, July 1978.
6. Cummings, D.R. and W.M. Scott, "Dual Fueling the Truck Diesel with Methanol", ibid. Paper 2-5.
7. Holmer, E., "Methanol as a Substitute Fuel in the Diesel Engine", ibid. Paper 2-4.
8. Marshall, W.F., "Experiments With Novel Fuels for Diesel Engines", BERCITPR-7718, Bartlesville Energy Center, U.S. DOE, Bartlesville, Oklahoma, February 1978.
9. Mullins, B.P., "Studies on the Spontaneous Ignition of Fuels Injected into a Hot Air Stream", National Gas Turbine Establishment, England, Report Nos. 89, 90, 97, 107, 1952. Also Fuel, 32, 343, 1953.
10. Havemann, H.A., M.R.K. Rao, A. Nataryan, T.L. Narasimhan, "Alcohol in Diesel Engines", Automobile Engineer, June 1954, pp. 256-262.
11. Springer, K.J. and T.M. Baines, "Emissions from Diesel Versions of Production Passenger Cars", SAE Paper No. 770818, SAE Transactions, Volume 86, Section 4, 1977.
12. Barnes, K.D., D.B. Kittleson and T.E. Murphy, "Effect of Alcohols as Supplementary Fuel for Turbocharged Diesel Engines", SAE Paper No. 750469, Automotive Engineering Congress and Exposition, Society of Automotive Engineers, Detroit, Michigan, February 1975.
13. Moses, C.A., D.W. Naegeli, E.C. Owens and J.C. Tyler, "Engine Performance of Alcohol/Diesel Fuel Blends", Southwest Research Institute, to be published.
14. Dryer, F., "Water Addition to Practical Combustion Systems- Concepts and Applications", Sixteenth Symposium (International) on Combustion, The Combustion Institute, Pittsburgh, 1976, pp. 279-296.
15. Alperstein, M., W.B. Swim and P.H. Sweitzer, "Fumigation Kills Smoke- Improves Diesel Performance", SAE Transactions, Volume 66, 1958.
16. Panchapakesan, N.R., Gopalakrishnan, K.V., B.S. Murthy, "Factors that Improve the Performance of an Ethanol-Diesel Oil Dual-Fuel Engine", Proceedings of the International Symposium on Alcohol Fuel Technology, Methanol and Ethanol, West Germany, Nov. 21-23, 1977.
17. Keller, J.L., G.M. Nakaguchi and J.C. Ware, "Methanol Fuel Modification for Highway Vehicle Use", Final Report, Department of Energy Contract, No. EY-76-C-04-3683, Union Oil Company of California, Brea, California, July, 1978.
18. Amano, M., H. Sami, S. Nakagawa and H. Yoshizake, "Approaches to Low Emission Levels for Light-Duty Diesel Vehicles", SAE Paper No. 760211, February, 1976.
19. Klaus, B., P.S. Pederson, "Alternative Diesel Engine Fuels: An Experimental Investigation of Methanol, Ethanol, Methane and Ammonia in a D.I. Diesel Engine with Pilot Injection", SAE Paper No. 770794, September 1977.
20. Heydrich, J., "Exhaust Emissions and Noise from Truck Diesel Engines", Cummins Engine Company, Columbus, Indiana, February, 1977.

21. Lipkea, W.H., J.H. Johnson and C.T. Vuk, "The Physical and Chemical Character of Diesel Particulate Emissions- Measurement Techniques and Fundamental Considerations", SAE Special Publications 430, February 1978.

22. Garbe , R., U.S. Environmental Protection Agency, Ann Arbor, Michigan, Personal Communication

23. Vogh, J.W., "Contribution of Some Carbonyl, Phenol, and Hydrocarbon Components to Diesel Exhaust Odor", U.S. Bureau of Mines, Report of Investigation 7632, 1972.

24. Khatri, N.J., J.H. Johnson and D.G. Teddy, "The Characterization of the Hydrocarbon and Sulfate Fractions of Diesel Particulate Matter", SAE Paper No. 780111, February 1978.

25. "On the Trail of New Fuels- Alternative Fuels for Motor Vehicles", Federal Ministry for Research and Technology, Bonn, Germany, 1974, translated by Addis Translations International, Portola Valley, California,for Lawrence Livermore Laboratory, U.S. DOE, Livermore, California.

26. Murthy, B.S. and L.G. Pless, "Effectiveness of Fuel Cetane Number for Combustion Control in Bi-Fuel Diesel Engine", Journal of the Institution of Engineers (India), Volume 45, No. 7, Pt. ME 4, March, 1965, pp. 155-183.

27. Pischinger, F. and H. Stutzenberger, "Zweistoffbetrub Gasol-Methanol fur Diesel-strassenfahzeuge-Methanolzumischung", Paper presented at the BMFT-Stalusseminar, 1977.

28."Liquid Fuels from Renewable Resources: Feasibility Study", Volume A, Demand Studies prepared for Government of Canada by Intergroup Consulting Economists, Winnepeg, Manitoba, March, 1978.

Driving Cycle Economy, Emissions and Photochemical Reactivity Using Alcohol Fuels and Gasoline*

Richard Bechtold
U.S. Dept. of Energy

J. Barrett Pullman
Univ. of Santa Clara

THE TECHNICAL FEASIBILITY of using alcohols to replace hydrocarbons as fuels for current internal combustion engines has been established [1,2,3,4,5,6,]*. Moreover, of the several options for using alcohols, the use of neat alcohol is very attractive for reasons of improved energy economy with reduced oxides of nitrogen emissions.

There have been many studies of the dependence of gasoline exhaust hydrocarbon composition upon fuel composition [7,8,9,10]. The importance of engine stoichiometry and emission control systems have also been studied and recognized as the more dominant variables influencing both hydrocarbon mass emissions and composition [11, 12,13,14].

However, relatively few investigators have reported the detailed composition of alcohol fuels' exhaust hydrocarbons and oxygenates [1,6, 15,16,17]. The quantity and composition of these fuel and combustion derived emissions will determine their role as oxidant precursors in urban atmospheres.

The above studies have identified the mass emissions of exhaust aldehydes as one class of oxygenates which are increased with alcohol fuels relative to gasoline. However, the mass emissions of photochemically reactive olefins and aromatics have generally been observed to diminish with the use of alcohol fuels.

This paper describes some of the efforts supported by the Department of Energy to assess both the energy economies and the relative air pollution impact from the end use of neat alcohol

---

* Numbers in brackets designate references at end of paper.

*Paper 800260 presented at the Congress and Exposition, Detroit, Michigan, February 1980.

ABSTRACT

An oxidation catalyst equipped vehicle and several three-way-catalyst (TWC) equipped vehicles were modified to operate on the Federal Test Procedure using gasoline or alcohol fuels. Unburned (hydro)carbon emissions were generally lowest when methanol fuel was used. Oxides of nitrogen (NO_x) were reduced an average of more than 50% by using alcohol fuels in contrast to gasoline. Photochemical reactivity comparisons of unburned fuel emissions were made by calculation and also with a 100 cu. ft. smog chamber. Synthetic reproductions (surrogates) of stoichiometric methanol exhaust were less photochemically reactive than gasoline exhaust surrogates for the 8.5:1 compression ratio engine conditions. This effect was observed even though methanol exhaust surrogates were tested at higher hydrocarbon-to-NO_x ratios (20:1 vs. 13.8:1) than were the gasoline exhaust surrogates. The exhaust from the stoichiometric TWC-equipped vehicles was extremely low in calculated and experimental reactivities for both methanol and gasoline fuels. This was due to their very low mass emissions and low exhaust hydrocarbon-to-NO_x ratios.

SAE/PT-80/19/$02.50

fuels under a variety of engine/vehicle conditions.

EXPERIMENTAL

TEST CARS- The major data was obtained using a 1976 Dodge (4500 lb. inertia weight) with 8-cylinder engine and 3-speed automatic transmission. Engine displacement was 318 cubic inches (5.2 liter). Both an 8.5:1 compression ratio (CR) and a 13:1 CR configuration were tested on the vehicle at Bartlesville. An experimental fuel induction system was used in an effort to provide operation at nearly constant equivalence ratio for all fuels during 1975-FTP driving cycle operation.

Spark timing was restricted to one schedule optimized for gasoline operation (at 8.5:1 CR) for all fuels and equivalence ratios tested. An oxidation catalyst of monolithic construction was used for those test conditions which indicate catalyst. Secondary air injection was used only during stoichiometric operation with the 13:1 CR. Exhaust gas recirculation (EGR) was used where indicated. Stoichiometric (1.0) and lean (0.8) equivalence ratios were matched for all fuels tested.

Additional data were obtained using four 1978 Pintos with three-way-catalyst (TWC) emission control systems and closed loop feedback carburetor control.

Three of the Pintos were methanol-fueled and were modified and tested at the University of Santa Clara. Carburetor fuel passageways and jets were enlarged to provide the increased volumetric flowrate of methanol needed to maintain stoichiometric operation. A modified camshaft with reduced overlap was also installed on the methanol-fueled Pintos. The unmodified fourth Pinto was gasoline-fueled and was tested at Bartlesville.

THE FUELS- Two gasolines were used in the Dodge vehicle depending on the compression ratio. The first gasoline was a low-octane, lead-free, summer-grade gasoline with hydrogen/carbon (H/C) ratio of 2.04 and a specific gravity of 0.718. This fuel was also used in the single Pinto tested at Bartlesville. The high-octane gasoline used for the 13:1 CR tests was lead-free with a H/C ratio of 1.898 and a specific gravity of 0.740.

Commercially available methanol, "methyl fuel", 190-proof ethanol and an ethanol solvent were tested. The methanol contained less than 0.05 wt% water. The methyl fuel consisted of 75% methanol, 5% ethanol, 7.5% η-propanol and 12.5% isobutanol by volume. The 190-proof ethanol contained no denaturants and was similar to ethanol fuel used in Brazil. The ethanol solvent consisted of 90.1% by volume 200-proof ethanol, 4.5% methanol, 4.5% ethyl acetate and 0.9% methyl isobutyl ketone.

The relatively dry methanol was used in the Dodge at Bartlesville and in the three Pintos at Santa Clara. The three other alcohol test fuels were used in the Dodge at Bartlesville only.

EMISSION TEST PROCEDURE- Exhaust emissions were measured in accordance with the 1975 Federal Test Procedure (FTP), however, no evaporative emissions measurements were obtained. A brief preconditioning procedure was followed on days when fuels or equivalence ratios were changed. Steady-state dynamometer operation during modifications to the vehicle also served as the preconditioning treatment.

Exhaust hydrocarbons were measured in accordance with the FTP-CVS procedures using a flame ionization detector (FID). The FID measured the amount of ionizable carbon present in the exhaust sample and was calibrated by referencing its response to a known concentration of propane. However, in the FTP calculations it is assumed that all of the ionizable carbon comes from unburned fuel molecules with a predetermined hydrogen to carbon ratio.

On that basis, using the H/C ratio of the fuel, the response of the FID was used to calculate the mass amount of unburned fuel present in the exhaust gas.

The FTP-prescribed method of calculation presents some problems when comparing exhaust hydrocarbon mass emissions from substantially different fuels. The gasoline used in this study had a H/C ratio of 2.04, which resulted in calculated exhaust hydrocarbons consisting of 85% carbon and 15% hydrogen (mass basis). Methanol, however, is only 37.5% carbon— the rest being hydrogen and oxygen— while ethanol is 52% carbon. To eliminate this arbitrary mass calculation difference, only the FID measured mass of ionizable carbon in the exhaust from each fuel will be reported in the following sections.

An additional problem with the exhaust hydrocarbon measurements results directly from using the standard propane calibration technique. The FID does not respond equally to a carbon atom in an alcohol and to a carbon atom in the alkane class of hydrocarbons.

The response of the FID (Beckman Model 402) to methanol has been determined to be only 73% as great as the FID response to propane on a parts per million carbon (PPMC) basis [18]. Similarly, the response of the FID to ethanol was found to be 83% of its PPMC response to propane. Thus, the FID response could be corrected for the amount of alcohol in the exhaust only if the exhaust hydrocarbon to alcohol ratio were reliably known. This adjustment procedure appears trustworthy in the present case of methanol exhaust where over 97% of the non-aldehyde mass of total (hydro)carbon* emissions are methanol (see Table 2). However, the denatured ethanol exhaust was found (by gas

*(hydro)carbon = unburned fuel emissions, including hydrocarbons, alcohols and aldehydes, mass calculated as carbon.

chromatography) to contain variable amounts of light hydrocarbons as well as unburned ethanol (see Table 2).

Ethanol mass emissions ranged from 43% to 84% of the total (hydro)carbon FTP emissions. Therefore, no single correction factor could be relied upon throughout the ethanol tests.

Through GLC analysis of each phase of the cycle, the ratios of hydrocarbons to ethanol were determined. The FID response of each phase of the driving cycle was corrected according to the percentage of hydrocarbons and ethanol with the assumption that hydrocarbons have 100% response in the FID. Accurate determination of ethanol exhaust gas (hydro)carbon content depends on analysis of the hydrocarbon species present and the FID response to ethanol.

SAMPLE COLLECTION PROCEDURE AND ANALYSIS FOR INDIVIDUAL HYDROCARBONS, ALCOHOLS AND TOTAL ALDEHYDES- One sample collection procedure was used for the 1976 experimental vehicle. A constant-volume sampler (CVS) supplied dilute exhaust gases to three bags in the usual manner. However, one additional cold-start sample was collected during the first 163 seconds of the cold transient phase. Each of these four bags was analyzed separately via GLC capillary and packed column techniques [19]. Total aliphatic aldehydes were measured and reported as formaldehyde during each phase using the MBTH wet chemistry technique [20]. The masses of over 60 hydrocarbons were calculated individually for each phase of the FTP. The three phases were combined with the prescribed FTP weighting factors (.43, 1.0, .57), and a resultant FTP-weighted mass was determined for each of over 60 detectable hydrocarbons, alcohols and total aldehydes.

The 1978 TWC Pinto exhaust was collected and analyzed in the same manner except that only the customary three phases were sampled.

The three 1978 TWC methanol-fueled Pintos were tested in accordance with the 1975 FTP with the exception that no evaporative emissions procedures were used. The MBTH technique was used to measure aldehydes. The total FID response was assumed to be methanol.

PHOTOCHEMICAL SMOG CHAMBER- A chamber of aluminum and pyrex construction was used for experimental reactivity comparisons between synthetic reproductions (surrogates) of methanol and gasoline exhaust. The chamber formerly was used at the Bureau of Mines in Bartlesville, Oklahoma. The chamber has a volume of 100 cubic feet and a 1.25 ft^{-1} surface to volume ratio.

Irradiation was provided by 84 blacklights (F30T8B1) and six sunlamps (FS20). The initial light intensity was measured by the method of Wu and Niki[21] to correspond to a K_1 value of .26 min^{-1}. One blacklight was replaced every 12 hours so that the maximum use of any lamp was limited to about 1000 hours.

Light intensity in the near ultraviolet was monitored by three photobodies with peak responsitivity at a wavelength of 335 nanometers.

Initial humidity was controlled at .008 pounds of water per pound of dry air. Steady state operating temperature was controlled at 88°F ± 3°F.

RESULTS

Vehicle FTP emissions results are presented in the following sections for total unburned (hydro)carbon, carbon monoxide (CO), oxides of nitrogen (NO_x) and unregulated aldehyde emissions. Energy economy results are also reported for the Dodge for all fuels, equivalence ratios, compression ratios and EGR conditions.

LOW COMPRESSION RATIO RESULTS- The results of the 8.5:1 CR tests with EGR show that use of the oxidation catalyst caused the expected large decreases in unburned (hydro)carbon emissions at the fuel-lean condition (Fig. 1). The differences in unburned (hydro)carbon between the fuels appear insignificant and suggest no apparent difficulty for the oxidation catalyst when using alcohol fuel. Some reduction in unburned (hydro)carbon was observed at the stoichiometric condition limited by the fact that no supplemental air was injected into the exhaust gas.

The unburned (hydro)carbon results from the "best" condition (fuel-lean with catalyst) were higher than were permitted for 1976-model vehicles sold in the United States. This was the consequence of a fuel induction/emission system tailored to provide a direct comparison between different fuels rather than optimum emission control. This will apply to the CO and NO_x emissions as well. The focus of this study was on evaluating these fuels under conditions of equal stoichiometry and fuel preparation. The reported tests will probably be inferior in terms of emissions and efficiency to a system optimized for a given fuel.

Carbon monoxide emissions were mainly a function of equivalence ratio (Fig. 2). Little difference in CO output between the fuels was seen at the fuel-lean condition.

The test conditions using the oxidation catalyst show reduced CO output at the fuel lean condition. The CO emissions at stoichiometric with oxidation catalyst were about the same as without catalyst due to the lack of excess oxygen available at that condition (no air pump).

Measurements of NO_x (Fig. 3) show that the alcohols all had a distinct advantage over the gasoline at the fuel-lean and stoichiometric conditions. This difference is inherent to the alcohols due to reduced peak flame temperatures and less favorable concentration of the species necessary for NO_x formation [22] in the reaction zone. Addition of the oxidation catalyst had little effect on NO_x output.

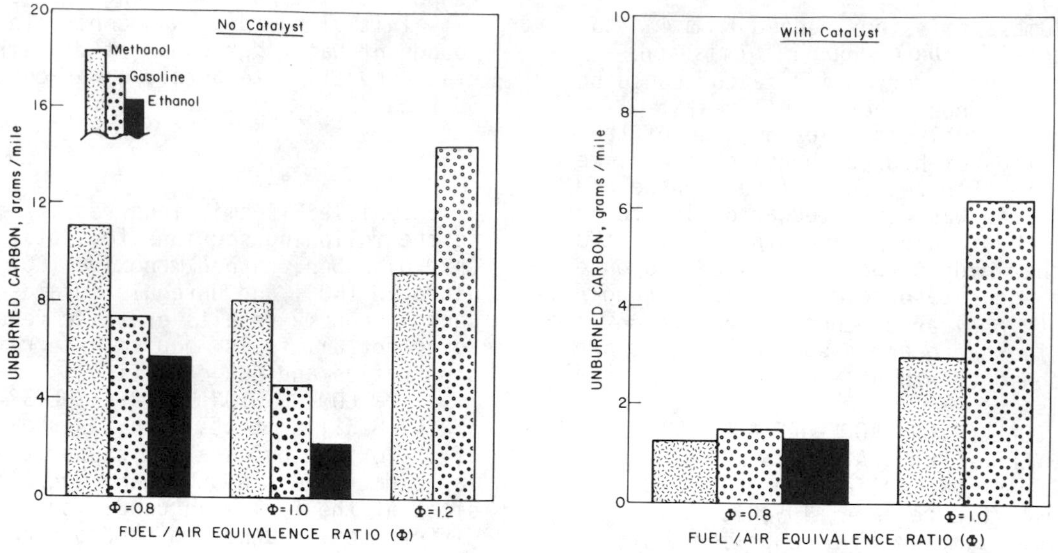

Fig. 1 - FTP unburned (hydro)carbon emissions as a function of fuel, equivalence ratio and catalyst use

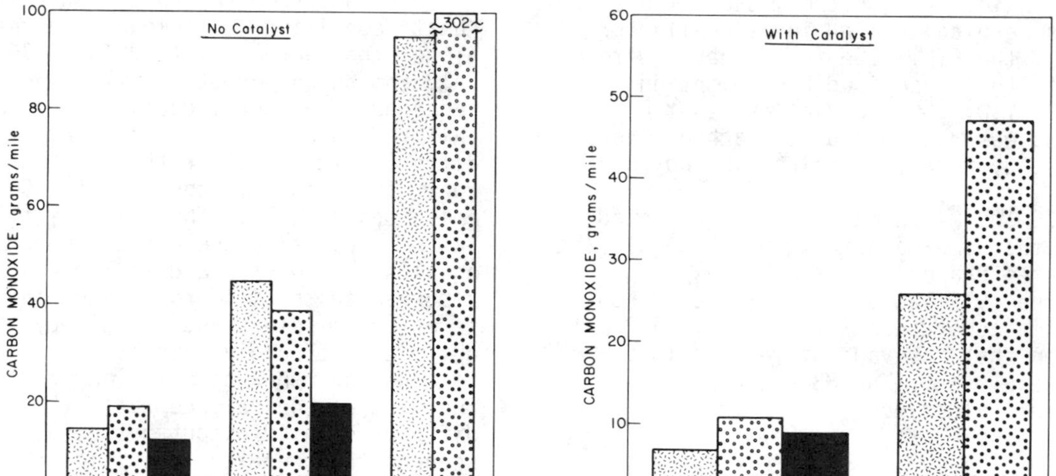

Fig. 2 - FTP carbon monoxide emissions as a function of fuel, equivalence ratio and catalyst use

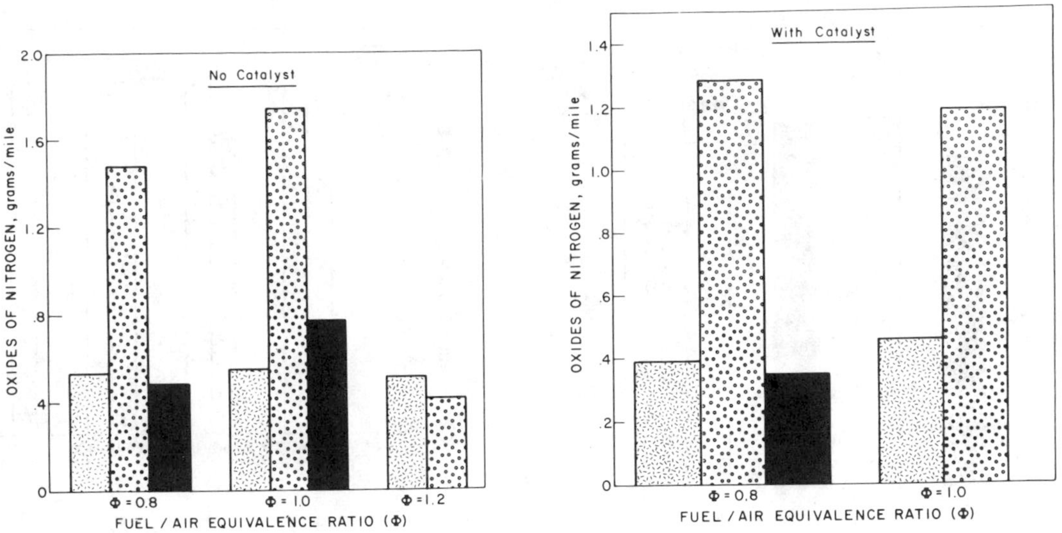

Fig. 3 - FTP emissions of oxides of nitrogen as a function of fuel, equivalence ratio and catalyst use

Exhaust aldehyde measurements show that approximately twice the amount of aldehyde was generated from the alcohols as compared to the gasoline (Fig. 4). The trend of increasing aldehydes from stoichiometric to lean equivalence ratios is in agreement with most published work. The addition of an oxidation catalyst greatly reduced the total mass of aldehydes emitted. A larger relative decrease of aldehydes occurred at the fuel-lean condition than at the stoichiometric condition. This is probably due to the oxygen available at the fuel-lean condition enabling the catalyst to function. With an oxidation catalyst, the concentration of aldehydes in the exhaust was greatly reduced during the stabilized and hot transient phases. Without oxidation catalyst, the concentrations of aldehydes were nearly the same during all phases of the driving cycle.

Comparisons of energy economies (Fig. 5) shows that methanol has a good tolerance for operation at various equivalence ratios; its energy economy changed little between the conditions tested. Operation with methanol at the fuel-lean condition resulted in the best energy economy obtained.

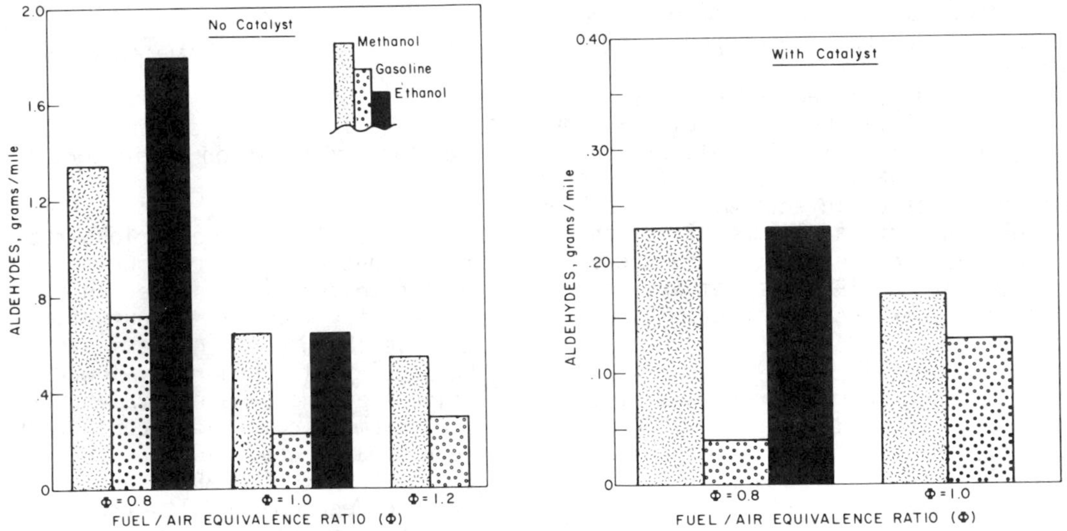

Fig. 4 - FTP aldehyde emissions as a function of fuel, equivalence ratio and catalyst use

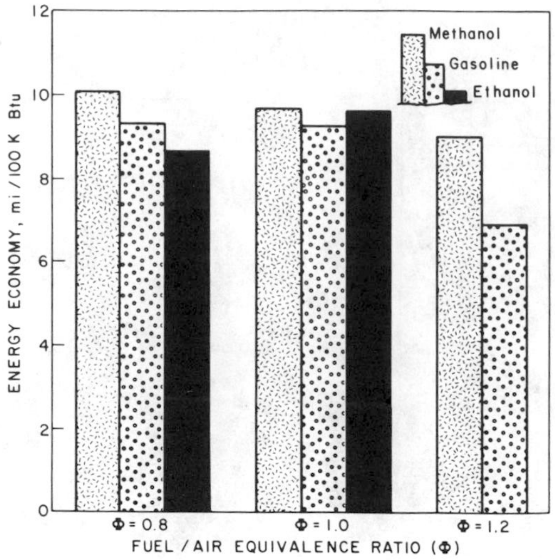

Fig. 5 - FTP energy economies as a function of fuel and equivalence ratio but with constant spark timing schedule

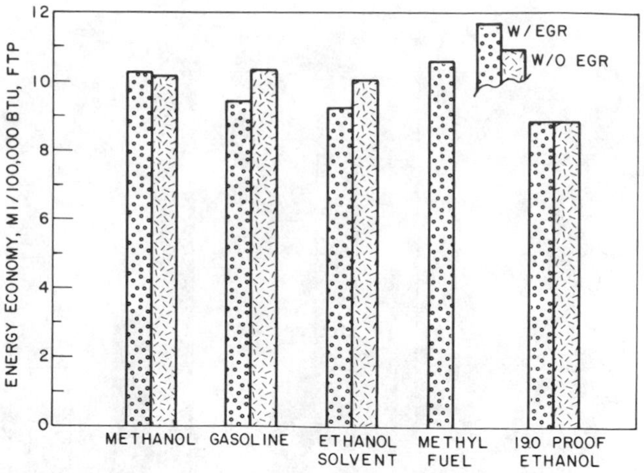

Fig. 6 - FTP energy economy comparison among fuels for the lean, high compression ratio condition

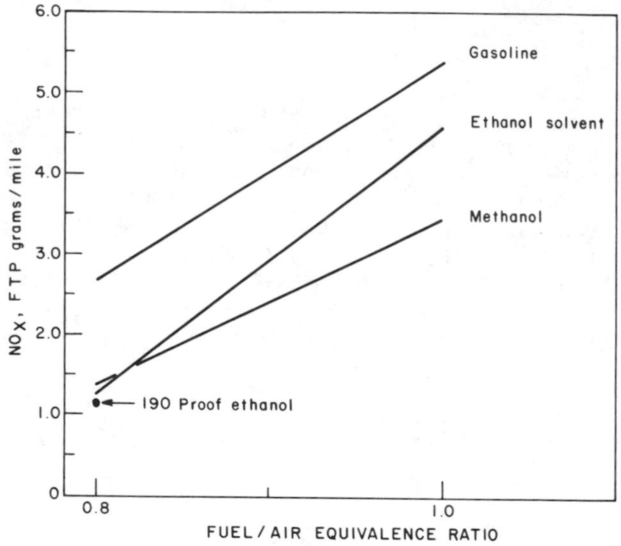

Fig. 7 - FTP NOx emission comparison among fuels for high compression ratio condition without EGR

INCREASED COMPRESSION RATIO RESULTS- The engine was converted to 13:1 CR through the use of pistons having increased dome height. A wide variety of fuels were tested--methanol, methyl fuel, gasoline, ethanol solvent, and 190-proof ethanol.

The focus of the test results with the vehicle at 13:1 was on NO_x and energy economy. Carbon monoxide emissions were not much different from the results obtained at 8.5:1 CR. Hydrocarbon mass emissions and composition data will be presented and discussed in a later section. Spark timing was unchanged from the 8.5:1 CR tests and was not optimized. Slight knock occurred under some full-throttle accelerations when using gasoline; no knock occurred using any of the alcohols.

The best driving cycle energy economy was obtained using the methyl fuel (Fig. 6). Without EGR, the energy economies of methanol, ethanol, and gasoline were virtually the same. The use of EGR caused some reduction in economy for gasoline and ethanol solvent, but had little effect on methanol or the 190-proof ethanol. The decrease in energy economy of the 190-proof ethanol in comparison to the other fuels is probably due to the presence of water.

Oxides of nitrogen were increased at 13:1 CR as compared to 8.5:1 CR. Without EGR, NO_x increased greatly as the mixture strength was increased from lean to stoichiometric (Fig. 7). The rate of increase was highest for the ethanol solvent but nearly the same for gasoline and methanol. The 190-proof ethanol was tested only at the lean condition and had NO_x emissions slightly less than methanol or ethanol solvent.

A ranking of the various fuels with respect to NO_x output at the two CR's and equivalence ratios tested is presented in Fig. 8. The use of EGR greatly reduced NO_x output. Gasoline proved to have the highest NO_x emissions as expected. Methanol yields generally lower NO_x than ethanol solvent at the same equivalence ratio, but stoichiometric methanol had higher NO_x than the lean ethanol condition.

TWC PINTO COMPARISONS- The FTP emissions and fuel economy results from one gasoline fueled and three methanol fueled Pintos are presented in Table 1.

The average mass of unburned fuel (hydro) carbon emissions (including both methanol and

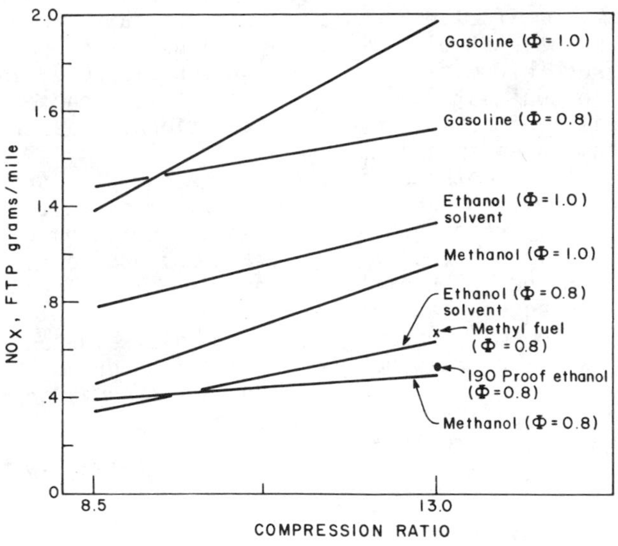

Fig. 8 - FTP NO_x emission comparison among fuels as a function of compression ratio and equivalence ratio with EGR

Table 1: Summary of Three-Way-Catalyst Pinto FTP Emissions and Mileage for One Gasoline and Three Methanol-Fueled Vehicles

FUEL	HC*	CO	NO_x	ALDE-HYDES	URBAN MPG.	URBAN mi/10^6 BTU	HWY MPG.	HWY mi/10^6/BTU
METHANOL	.152	2.16	.623	.0184	10.8	191.	13.0	230.
Number of tests	(28)	(28)	(28)	(21)	(28)	(28)	(12)	(12)
GASOLINE	.381	4.89	.68	.0037	19.9	169.	27.2	231.
Number of tests	(3)	(3)	(3)	(3)	(3)	(3)	(3)	(3)
CERTIFI-CATION	.355	9.	1.5	N/A	21.	178.	29.	246.

*(Hydro)carbon mass emissions were calculated using a molecular weight of 12. throughout rather than 32. or 13.85 for methanol or gasoline unburned fuels respectively.

formaldehyde) was 56% less for the methanol fueled Pintos than for the gasoline fueled Pinto tested. Previous engine dynamometer studies [4] also reported decreased unburned (hydro)carbon emissions from methanol fuel in contrast to gasoline. Carbon monoxide emissions were also reduced over 50%. Oxides of nitrogen emissions were reduced only 8% with the use of methanol in the Pintos which also had modified (reduced overlap) camshafts. The previously reported 50% NO_x reductions from methanol in the Dodge tests would be expected also for the TWC Pintos with OEM camshaft. It should be noted that in another recent study [23] of Pinto emissions the average NOx emission rate was 1.2 grams per mile with Indolene fuel.

Urban and highway fuel economies indicate a possible advantage for methanol fuel (with the camshaft change) on the urban test. However, a slight decrease in average fuel economy was observed for methanol during highway test conditions. The high variability and low number of fuel economy tests renders these trends statistically uncertain.

COMPOSITION AND REACTIVITY ANALYSES- The mass of exhaust and evaporative hydrocarbon emissions from current production vehicles is subject to tightening Federal regulations. The composition of those emissions is widely recognized to influence their contributions as precursors of photochemical oxidants. Indeed, many hydrocarbon reactivity classification schemes have been proposed on that basis [24,25, 26,27].

The detailed hydrocarbon and oxygenated hydrocarbon measurements described earlier have been reduced to seven classes of (hydro)carbons including the alcohols and aldehydes. These detailed data are presented in the following sections corresponding to the various engine and vehicle conditions tested. Six of the seven classes were chosen to correspond to the traditional classifications of organic chemistry.

A computer program was developed to perform calculations on the detailed (hydro)carbon exhaust emission data. Table 2 presents the range of NO_2 reactivities for each of the seven exhaust (hydro)carbon classes measured in this study.

Mole-weighted, NO_2 reactivities were summed for all observed (hydro)carbon species using the following equations:

$$\text{total reactivity} = \Sigma n_i R_i \qquad (1)$$

$$\text{total reactivity per gram} = \frac{\Sigma(n_i R_i)}{\Sigma m_i} \qquad (2)$$

$$\text{total reactivity per mole carbon} = \frac{\Sigma(n_i R_i)}{\Sigma(n_i C_i)} \qquad (3)$$

Where: n_i = FTP weighted moles of ith species
R_i = rate of formation of NO_2 photochemical reactivity normalized to ethylene [28]
m_i = mass of the ith species
C_i = number of carbon atoms per molecule of the ith species.

Table 2: Organic Hydrocarbon Classes and Their Range of NO_2 Reactivities (Normalized to Ethylene)

ORGANIC HYDROCARBON CLASS	LOWEST REACTIVITY SPECIES OBSERVED	ETHYLENE EQUIV. NO_2 REACTIVITY	HIGHEST REACTIVITY SPECIES OBSERVED	ETHYLENE EQUIV. NO_2 REACTIVITY
ALKANES	METHANE	0		
NON-METHANE ALKANES	ETHANE	0.15	2,2-DIMETHYLPENTANE	5.5
ALKENES	ETHYLENE	1.0	2,3-DIMETHYL-2-BUTENE	34.7
AROMATICS	BENZENE	0.19	1,3,5-TRIMETHYLBENZENE	3.0
ALCOHOLS	METHANOL	0.12	ETHANOL	0.38
ALKYNES	ACETYLENE	0.0	METHYLACETYLENE	0.0
ALDEHYDES			FORMALDEHYDE	5.0

Associated with each set of detailed class (hydro)carbon exhaust emissions data are presented sets of calculated exhaust surrogates. The surrogates are five compounds used in photochemical smog chamber tests to synthetically represent the exhaust (hydro)carbon reactivity of the many exhaust compounds within the five most reactive classes. Detailed (hydro)carbon class surrogates are reported in Appendix Tables A1, A2, A3 and A4 for each major test condition. Within each of five classes the total mole-weighted reactivities of the many observed exhaust species were replaced by the reactivity equivalent quantity of the class surrogates.

$$\text{Moles of Class Surrogate} = \frac{\Sigma(n_i R_i)\ \text{class J}}{\text{R class J surrogate}} \qquad (4)$$

Where: class J = one of the five surrogate classes

This surrogate replacement method also resulted in a close approximation to observed exhaust NMPPMC/NO_x ratios. The experimental and urban reactivities of hydrocarbon-NO_x mixtures are known to be strongly influenced by their non-methane PPMC/NO_x ratios. Therefore, the achievement of close agreement among the observed exhaust calculated surrogates and smog chamber non-methane PPMC/NO_x ratios was an internal objective of this study.

EMISSION ANALYSIS AT 8.5:1 CR WITHOUT OXIDATION CATALYST- The mass fraction of FTP exhaust (hydro)carbon within each class is considered a useful characterization for reactivity assessments. Table 3 presents detailed non-catalyst emission data from the Dodge for the seven observed classes on a percent carbon or carbon mass fraction basis. The total mass of (hydro)carbon (in grams) for the 7.5 mile tests is also reported in Figure 9 for all the test conditions.

It is evident that methanol exhaust (hydro) carbons were predominantly unburned methanol with minor amounts of formaldehyde. However, it must be recognized that in the non-catalyst condition the mass of emissions was substantial and that the hydrocarbon to NO_x ratio of 67:1 was extremely high.

The gasoline-fueled tests resulted in a characteristic broad distribution of exhaust hydrocarbons among all seven classes reported. The mass of unburned (hydro)carbon was reduced relative to methanol and the NMPPMC/NO_x ratio was more characteristic of urban atmospheres.

The ethanol solvent exhaust (hydro)carbons were dominated by unburned ethanol. In addition, alkene and aldehyde fractions were also significant. The total mass of (hydro)carbon emissions was comparatively low.

The comparison between stoichiometric and lean operation indicated a slight increase in (hydro)carbon mass emissions during non-catalyst lean operation for all three fuels. The aldehyde carbon percent was increased by lean operation for methanol and gasoline. Lean operation also caused an increase in the mass of aldehyde emissions for all three fuels relative to stoichiometric test conditions.

Associated with the exhaust (hydro)carbon composition data in Table 3 are calculated class surrogates in Appendix Table A1 for use in smog chamber simulations of vehicle exhaust. The calculated total reactivities from Table A1 are also presented in Figure 10 to facilitate comparisons among all test conditions. Figure 10 indicates that for non-catalyst (NC) operation both alcohols resulted in higher calculated total reactivities than the gasoline. This was the result of an increased mass of total (hydro)carbon emissions during lean operation and the result of a much larger aldehyde carbon percent during stoichiometric operation.

TABLE 3: Exhaust (Hydro)carbon Composition for 8.5:1 CR Operation with EGR but Without Oxidation Catalyst

		EXHAUST (HYDRO)CARBON MASS FRACTION (%)								
FUEL	FUEL AIR EQUIVA-LENCE	TOTAL ALKANE	NON-METHANE ALKANE	ALKENE	AROMATIC	ALCOHOL	ALKYNE	ALDEHYDE	MASS OF TOTAL FTP (HYDRO)CARBON (GRAMS)	RATIO OF TOTAL NMPPMC/NO_x
Methanol	1.0	1.0	0.05	0.1	0.02	95.0	0.2	3.7	67.4	68.2
Gasoline	1.0	41.5	33.60	33.2	9.70	0.5	13.2	2.1	34.9	9.4
Ethanol Solvent	1.0	9.5	1.80	16.1	0.40	58.5	4.8	10.7	17.1	10.5
Methanol	0.8	0.3	.03	.03	.02	94.9	0.1	4.6	76.6	73.6
Gasoline	0.8	51.3	48.70	29.50	10.10	0.9	4.5	3.6	55.8	18.8
Ethanol Solvent	0.8	3.5	0.70	9.70	.09	76.7	1.2	8.8	57.1	59.1

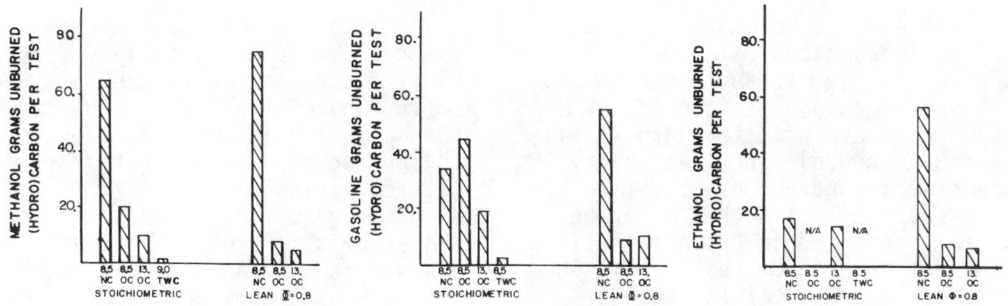

Fig. 9 - FTP exhaust (hydro)carbon mass emissions among alcohol and gasoline fuels for various engine and emission control conditions (grams per test)

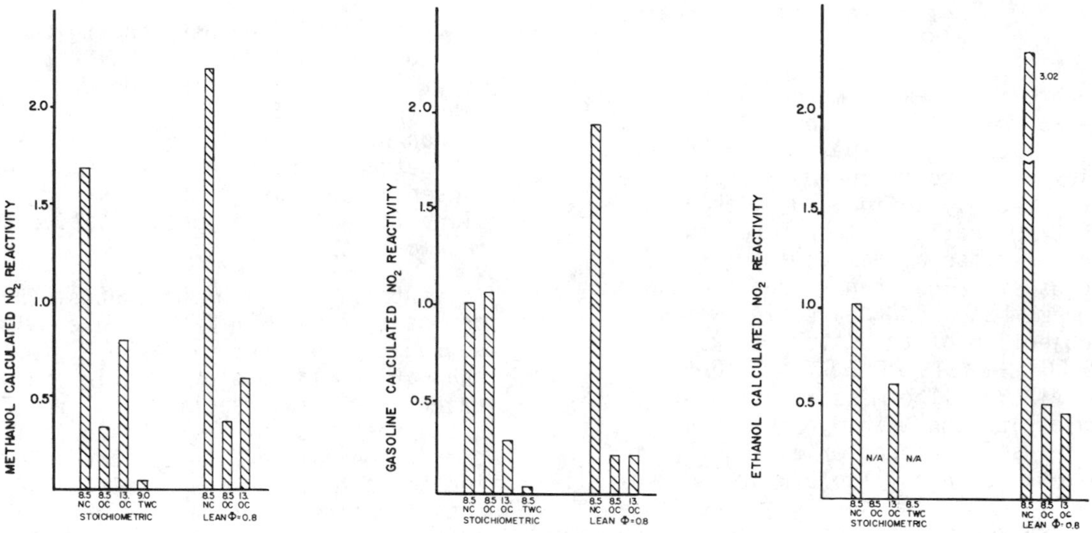

Fig. 10 - FTP exhaust (hydro)carbon calculated NO_2 reactivities among alcohol and gasoline fuels for various engine and emission control conditions (mole-ethylene equivalent units per test)

The calculated NO_2 reactivity per mole of carbon (presented in Table A1) was slightly less for methanol in comparison with gasoline at both equivalence ratios. Ethanol's reactivity was found to be higher than that of gasoline. This result means that if both fuels resulted in equal (hydro)carbon mass emissions then the (hydro)-carbon reactivity of methanol exhaust would be less than for gasoline exhaust based solely on their relative compositions.

The calculated reactivity per gram of unburned methanol was less than half the reactivity per gram of the gasoline. This was the result of their nearly equal reactivity per mole of carbon combined with the fact that the mass of unburned methanol is about 2.3 times greater per carbon atom than that of gasoline exhaust.

During lean non-catalyst operation the total calculated NO_2 reactivities were increased for all fuels relative to stoichiometric operation. This was caused by an increased mass of (hydro)carbon emissions from all fuels tested, plus the fact that the aldehyde carbon percents were also increased.

During lean operation the calculated NO_2 reactivity per gram and per mole carbon were also increased for methanol and gasoline, though not for ethanol fuel exhaust. Although lean combustion offers potential advantages in energy economy, as shown in Figure 5, some additional form of emission control appears to be necessary for alcohol fuels as well as for gasoline.

EMISSION ANALYSIS AT 8.5:1 CR WITH OXIDATION CATALYST- Detailed composition data in the form of (hydro)carbon class carbon percents are reported in Table 4 for Dodge tests using three fuels at two equivalence ratios. The total FTP mass of unburned (hydro)carbon emissions is presented for comparisons in Figure 9.

The mass of unburned (hydro)carbon was substantially reduced for all fuels and equivalence ratios except the stoichiometric gasoline condition. Carbon monoxide emissions were shown in Figure 2 to be about 80% greater for the gasoline fuel stoichiometric condition, therefore, a slightly richer equivalence ratio was probably used with the gasoline. The composition of stoichiometric methanol exhaust was predominantly unburned fuel with a smaller carbon percent of formaldehyde at stoichiometric conditions relative to the non-catalyst condition. However, during lean catalyst-equipped operation the carbon percent of formaldehyde was almost doubled relative to the non-catalyst condition. Overall, though, the mass of aldehyde emissions was reduced over 80% by the addition of an oxidation catalyst at both the lean and stoichiometric conditions.

The aldehyde and alkene carbon percents were reduced by the catalyst for gasoline exhaust at both equivalence ratios. The carbon percent of the alkane class was increased relative to the non-catalyst condition at both equivalence ratios.

The lean ethanol solvent exhaust composition decreased in its carbon percent alkenes but was increased in the total alkane class.

The comparison between stoichiometric and lean catalyst-equipped operation indicates a substantial reduction in carbon mass emissions when sufficient excess air was supplied to the catalyst. The composition of the lean exhaust indicates methanol's expected fourfold carbon percent increase in aldehydes. However, on a mass basis the aldehydes were less than doubled for the lean condition relative to stoichiometric.

The calculated smog chamber exhaust surrogate carbon percents are presented in Appendix Table A2. Total calculated reactivity comparisons are presented in Figure 10. The addition of an oxidation catalyst generally reduced the calculated total reactivities relative to the non-catalyst condition. A very substantial total reactivity reduction was evident for the lean condition among all fuels.

The calculated reactivity per gram and per mole carbon were reduced by the use of a catalyst at the stoichiometric condition only. Under lean conditions methanol increased in reactivity per gram and per mole carbon while gasoline decreased substantially. These latter effects were largely due to the increased carbon percent of aldehydes for methanol at lean conditions but no change for the gasoline composition.

EMISSION ANALYSIS AT 13:1 CR WITH OXIDATION CATALYST- The detailed (hydro)carbon composition data are presented in Table 5. The aldehyde carbon percents were generally increased for all fuels at the high CR condition. Stoichiometric methanol exhaust at high CR increased in aldehyde carbon percent by a factor of eight. An air pump was used for that condition and partial oxidation of unburned methanol in the exhaust system was the most probable cause for the increased carbon percent of aldehyde. The mass of aldehyde emissions was correspondingly increased by a factor of five relative to low CR, stoichiometric operation without an air pump. Gasoline exhaust aromatics were increased in carbon percent at 13:1 CR, however, this simply reflected the increased aromatic content of the high octane gasoline. Complementary decreases in gasoline's alkane and alkene carbon percents and mass emissions were also observed for the high CR condition.

The mass emissions of exhaust (hydro)carbon were decreased for the alcohol fuels tested at all high CR conditions. This was also true for all fuels tested at the stoichiometric condition for which an air pump was used. During lean operation, however, gasoline produced a slight increase in mass emissions of (hydro)carbon with the 13:1 CR. Figure 9 summarizes these effects for comparison.

A five fuel comparison was conducted at the lean stoichiometry condition. The mass of (hydro)carbon emissions from methanol and ethanol

TABLE 4: Exhaust (Hydro)carbon Composition for 8.5:1 CR operation with EGR and Oxidation Catalyst

FUEL	FUEL AIR EQUIVALENCE	EXHAUST (HYDRO)CARBON MASS FRACTION (%)							MASS OF TOTAL FTP (HYDRO)CARBON (GRAMS)	RATIO OF TOTAL NMPPMC / NO_x
		TOTAL ALKANE	NON-METHANE ALKANE	ALKENE	AROMATIC	ALCOHOL	ALKYNE	ALDEHYDE		
Methanol	1.0	2.6	.20	.10	.05	95.4	.09	1.7	20.2	22.4
Gasoline	1.0	52.0	42.5	28.0	11.6	0.7	6.8	0.9	45.5	17.4
Methanol	0.8	2.3	.06	.04	.02	89.8	.01	7.8	8.7	11.1
Gasoline	0.8	60.4	47.0	21.3	9.5	0.8	7.1	1.1	10.9	3.8
Ethanol Solvent	0.8	15.7	1.8	5.1	.05	68.6	2.6	8.0	10.4	13.7

Table 5: Exhaust (Hydro)carbon Composition for 13:1 CR Operation with EGR and Oxidation Catalyst (Supplemental air injection was used during stoichiometric operation only)

FUEL	FUEL AIR EQUIV.	EXHAUST (HYDRO)CARBON MASS FRACTION %							MASS OF TOTAL (HYDRO)CARBON (GRAMS)	RATIO OF TOTAL NMPPMC / NOx
		TOTAL ALKANE	NON-METHANE ALKANE	ALKENE	AROMATIC	ALCOHOL	ALKYNE	ALDEHYDE		
Methanol	1.0	2.1	0.1	0.1	0.1	82.7	0.2	14.9	11.3	5.9
Gasoline	1.0	54.6	35.0	13.1	19.7	-	11.7	0.9	19.1	4.0
Ethanol Solvent	1.0	22.2	9.6	6.5	0.7	60.8	3.3	6.4	14.7	6.4
Methanol	0.8	1.7	0.9	0.2	0.6	75.7	0.06	21.7	5.9	6.1
Methyl Fuel	0.8	7.6	0.95	5.7	0.3	69.3	0.97	16.1	6.9	5.1
Gasoline	0.8	60.0	54.0	15.9	21.9	-	1.0	1.2	11.7	3.7
Ethanol Solvent	0.8	29.6	12.9	10.5	2.5	43.6	1.0	12.9	7.0	4.8
190-Proof Ethanol	0.8	12.6	2.0	12.6	0.4	64.0	0.6	9.8	14.6	13.1

solvent fuels was decreased by the increase in compression ratio. Unburned fuel mass emissions from methyl fuel were 17% greater than those from methanol. The use of pure 190-proof ethanol increased the mass emissions of unburned (hydro)-carbon by 109% relative to the anhydrous ethanol solvent fuel.

The calculated vehicle exhaust surrogates for smog chamber tests are presented in Table A3. The calculated total reactivity data are presented in Figure 10 for convenient comparison. Under stoichiometric conditions methanol exhaust increased in calculated total reactivity and gasoline exhaust decreased in calculated reactivity as a result of the increased compression ratio and use of an air pump. Similarly, during lean operation, gasoline and ethanol solvent total reactivities were decreased at high CR, however, lean methanol exhaust reactivity was increased.

Comparisons among the five fuels at high CR and lean conditions indicate that methyl fuel resulted in a lower total reactivity than methanol due to reduced aldehyde emissions. The 190-proof ethanol resulted in higher calculated reactivity due to the doubled mass emissions of (hydro)carbon. The gasoline exhaust yielded a lower calculated reactivity than either alcohol at the lean high compression ratio condition.

EMISSION ANALYSIS OF TWC PINTOS AT 8.5:1 CR- The FTP (hydro)carbon emission results from three methanol-fueled Pintos and one gasoline-fueled Pinto are presented in Table 6. The total mass of (hydro)carbon from the 7.5 mile test average was substantially reduced for these vehicles in contrast with the best results from the oxidation catalyst-equipped experimental Dodge.

The methanol exhaust contained half the emissions of unburned carbon in comparison with the gasoline-fueled tests.

The class carbon percents for methanol fuel exhaust were 10.5% aldehyde, with the remainder assumed to be unburned methanol. This is a conservative assumption, since some of the exhaust hydrocarbons are certain to be methane and very few alkenes or aromatics have been observed in the methanol exhaust data reported earlier.

The composition of the gasoline exhaust, which was measured in detail at the Bartlesville Energy Technology Center, shows that over 70% of the carbon is in the reactive alkane, alkene or aromatic classes. Exhaust (hydro)carbon class surrogates are presented in Appendix Table A4.

Figure 10 presents the calculated total reactivities for the TWC Pintos. The methanol Exhaust total reactivity is calculated to be twice as great as that of gasoline exhaust. This

Table 6: Exhaust (Hydro)carbon Composition for 8.5:1 CR Operation with EGR on Three-Way-Catalyst Equipped Pintos

FUEL	FUEL AIR EQUIV.	EXHAUST (HYDRO)CARBON MASS FRACTION (%)							MASS OF TOTAL FTP (HYDRO)CARBON (GRAMS)	RATIO OF TOTAL NMPPMC / NOx
		TOTAL ALKANE	NON-METHANE ALKANE	ALKENE	AROMATIC	ALCOHOL	ALKYNE	ALDEHYDE		
Methanol	1.0	-	-	-	-	89.2	-	10.8	1.28	1.12
Gasoline	1.0	61.5	38.6	11.2	20.8	0.2	6.0	0.3	2.89	1.59

is the result of formaldehyde's high NO_2 reactivity and the present linear summation technique used for assessing total reactivity.

In contrast with the 8.5:1 CR oxidation catalyst vehicle results, the total reactivity calculated for the TWC Pintos has been reduced over 80% for both fuels. This was due to the very low mass emissions of unburned (hydro)carbon for both fuels with the TWC emission control system.

The NO_2 reactivity per gram and reactivity per mole carbon of exhaust were calculated to be greater for methanol than for gasoline due primarily to the presence of formaldehyde.

OVERVIEW OF CALCULATED REACTIVITY COMPARISONS- Cumulative mole-weighted NO_2 reactivities are presented in Figure 10 for the exhaust from the several fuel/engine/emission control conditions tested. The linear summation technique used to calculate the exhaust's total reactivity is dependent upon both the quantity of exhaust hydrocarbons and their individual NO_2 reactivities. Several exceptions to this mass emissions/NO_2 reactivity correspondence are noted, particularly for exhausts with relatively large aldehyde carbon percents. The aldehyde carbon percents were generally highest for the alcohol fuels during lean engine operating conditions and for the high compression ratio conditions tested (as shown in Tables 3, 4, and 5). The linear summation of mole-weighted NO_2 reactivity is one technique which has been used in this and previous reports in the literature to characterize the comparative smog-producing potential among exhaust and evaporative hydrocarbon samples. The NO_2 reactivity of a compound manifests itself early in the photochemical smog formation process. Hence, the NO_2 reactivity measure does not necessarily provide a good indication of the peak ozone yield or dosage over time. Similarly the many other attributes of smoggy urban atmospheres such as eye irritation and aerosol formation were not necessarily predictable from NO_2 reactivity summations.

An example should be made regarding the NO_2 reactivity of formaldehyde. Formaldehyde has a relatively high NO_2 reactivity in comparison with a reference standard compound such as ethylene.

$$R_{ethylene} = 1.0; R_{formaldehyde} = 5.0$$

The presence of formaldehyde in an urban surrogate smog chamber experiment has the effect of increasing the initial rate of formation of NO_2 from the primary emission species nitric oxide (NO). However, depending upon the quantity and composition of the non-aldehyde (hydro)-carbons present, there may not be a sufficient conversion of NO into NO_2 to result in a build-up of ozone. Similarly the NO_2 reactivity of formaldehyde is a very imperfect measure of its contribution to the formation of photochemical smog or ozone. Therefore, we have conducted a series of photochemical smog chamber experiments in which vehicle exhaust (hydro)-carbon class surrogates were utilized to match both the calculated NO_2 reactivities within each of five hydrocarbon classes and to match as nearly as possible the non-methane PPMC to NO_x ratios ovserved within each hydrocarbon class of the exhuast. By this method, comparisons between alcohol and gasoline exhaust surrogate mixtures could be made which, in contrast to the NO_2 reactivity summation technique, would more thoroughly reflect the complex and non-linear photochemical reactions occurring in the original exhaust samples.

PHOTOCHEMICAL SMOG CHAMBER EXPERIMENTAL COMPARISONS- Two separate types of smog chamber comparisons were made among the alcohol fuel exhaust and gasoline exhaust surrogate mixtures. In the first type the exhaust surrogates were compared at equal initial NO_x concentrations in the smog chamber. The NMPPMC/NO_x ratios and (hydro)carbon class surrogates were reproduced in accordance with the calculated data in Appendix Tables A1, A2, A3, and A4. The initial NO_x concentration used was 0.40 PPM in the chamber, all of which was nitric oxide at the start of irradiation.

The second type of comparison was made at simulated equal dilution ratios in the atmosphere. This type of experiment recognizes the low NO_x emissions from the combustion of alcohol fuels. The calculated surrogate class carbon percents and NMPPMC/NO_x ratios were preserved, but the initial NO_x concentration in the chamber was varied in proportion to the FTP NO_x emissions observed from the vehicle.

The first type of "equal NO_x" experiments will be presented initially. A summary of smog chamber experimental ozone maxima obtained by irradiation of exhaust (hydro)carbon surrogates are presented in Figure 11. These summary results cover a broad range of FTP vehicle engine operating conditions and include several types of emission control technologies applied to methanol gasoline and ethanol solvent fuel exhausts. The irradiation times required to yield these ozone maxima are presented along with other smog chamber experimental data in Appendix Table A5. The maximum ozone yields, at a constant initial chamber NO_x concentration, were generally comparable between the gasoline and methanol exhaust surrogates for each of the fuel/engine/emission control conditions reported in Figure 11. The largest effect upon ozone yield was not caused by changing fuels, but rather by lowering the (hydro) carbon emissions via changes in compression ratio or sophistication of the emission control technology.

The exhaust emissions from vehicles certified to meet current 0.41 grams per mile hydrocarbon and 2.0 grams per mile NO_x standards result in single vehicle hydrocarbon to NO_x ratios of about 0.7 PPMC/NO_x. Such "NO_x rich"

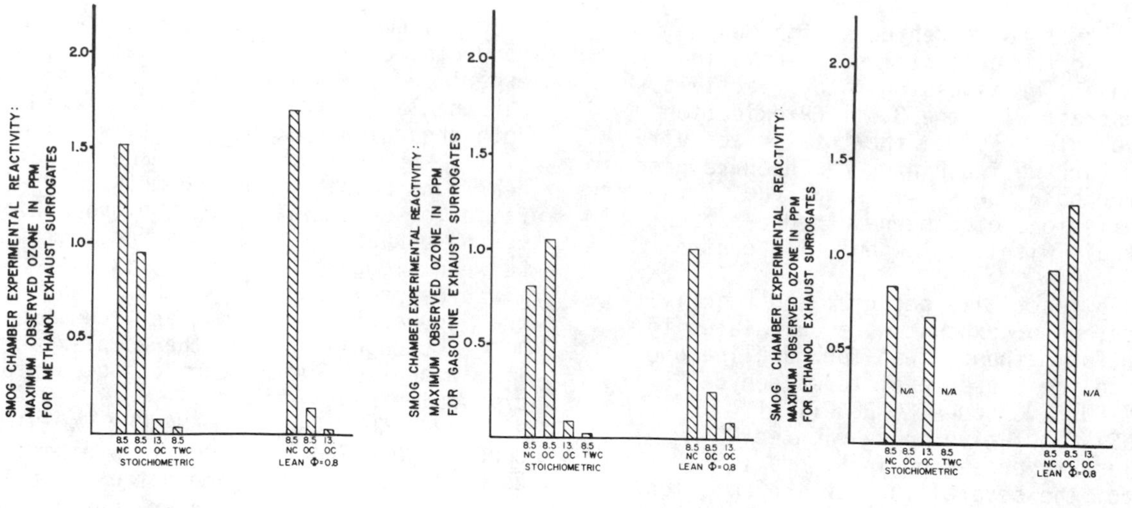

Fig. 11 - Comparative ozone maxima from photochemical smog chamber experiments among alcohol and gasoline fuel exhaust surrogates for various engine and emission control conditions (PPM)

emissions are generally not able to produce substantial ozone without interacting with hydrocarbon emissions from other sources. This was one of the major controlling variables observed for the ozone yields during the smog chamber experiments.

It should be noted that NO_x emissions were changed over a more limited range than were (hydro)carbon emissions by changing fuels and/or catalyst technologies. Hence, there was a direct correspondence between the mass of (hydro)carbon emissions in Figure 9 and the ozone maxima reported in Figure 11 which was primarily due to the exhaust PPM-carbon to NO_x ratios.

The (hydro)carbon class composition of exhaust from the three fuels produced a second order effect relative to the primary influence of the PPM-carbon to NO_x ratios. It is important to note that the composition of stoichiometric methanol exhaust with a 23.5:1 PPM-carbon to NO_x ratio was less photochemically reactive by all criteria than the composition of stoichiometric gasoline exhaust at a 13.8:1 PPM-carbon to NO_x ratio. The evidence reported in Table 7 supports the expected low reactivity of methanol exhaust (even at a higher PPM-carbon to NO_x ratio) in comparison with gasoline exhaust surrogates.

The conversion of nitric oxide into nitrogen dioxide was another reactivity criterion for which comparative evidence is summarized in Figure 12. The initial NO_x concentration was 0.40 PPM nitric oxide. The relative completeness of its conversion into nitrogen dioxide at the time of the NO_2 peak was a good indicator of the overall mixture reactivity which included combined effects of composition and the PPM-carbon to NO_x ratio. A general pattern was observed; if less than half the initial nitric oxide was converted into nitrogen dioxide at the time of the NO_2 peak, then very little ozone would be produced within the first 15 hours of continuous irradiation. Such a pattern is evident in comparing Figure 12 with Figure 11.

During stoichiometric operation methanol exhaust surrogates produced consistently lower NO_2 maxima in comparison with gasoline surrogates. Lean stoichiometry conditions reversed the above trend with less peak NO_2 being produced by gasoline exhaust surrogates. Lean ethanol exhaust surrogate experiments also produced more NO_2 than the gasoline surrogates.

Table 7: Photochemical Smog Chamber Comparisons of Stoichiometric Methanol and Gasoline Exhaust (Hydro)carbon Surrogates at 8.5:1 CR With Oxidation Catalyst

	CHAMBER INITIAL CONDITIONS, PPM MOLAR							CHAMBER RESULTS; T IN HOURS, [ ] IN PPM								
FUEL	BUTANE	PROPENE	TOLUENE	METHANOL	FORMALDEHYDE	NO_x	$\frac{PPMC}{NO_x}$	$T_{\frac{NO_2}{2}}$	T_{NO_2}	T_{HCHO}	T_{O_3}	$[NO_2]_m$	$[HCHO]_m$	$[O_3]_m$	$\frac{[NO_2]_m}{[NO_x]_o}$	$\frac{[HCHO]_m}{[HCHO]_o}$
METHANOL	-	-	-	9.26	.154	.40	23.5	.61	2.45	0.1	15.8	.26	.36	.97	.65	1.22
Gasoline	.548	.706	.158	.056	.061	.40	13.8	.43	1.05	2.3	4.2	.325	.58	1.05	.81	9.5

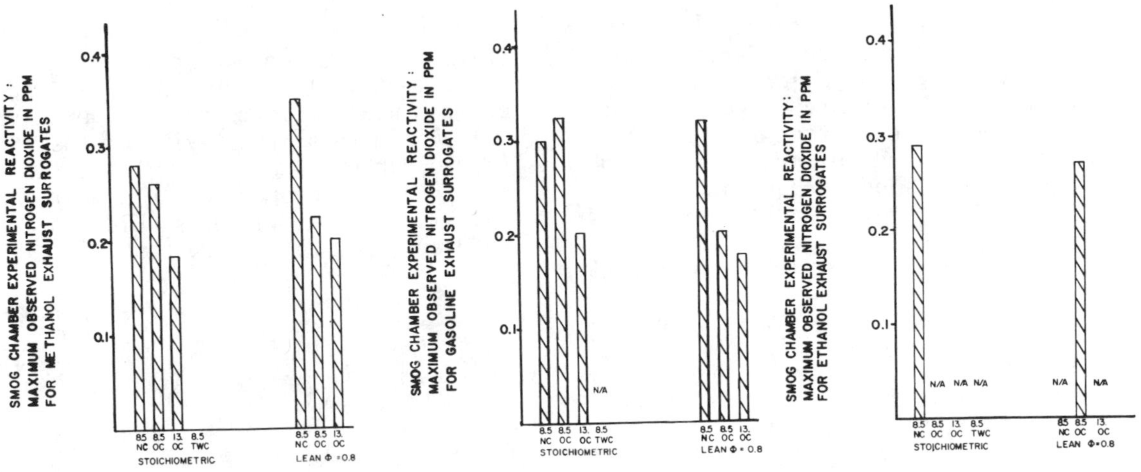

Fig. 12 - Comparative nitrogen dioxide maxima from photochemical smog chamber experiments among alcohol and gasoline fuel exhaust surrogates for various engine and emission control conditions (PPM)

The tailpipe emissions of aldehydes presented in Figure 4 were not the only automotive contributions to urban formaldehyde air pollution. During the smog chamber experiments with gasoline exhaust surrogates, it was evident that photochemical product formaldehyde (resulting from alkene oxidation) added substantially to the initial concentrations. Table A5 reports that gasoline surrogates produced increased formaldehyde ranging from 100% to 850% above starting amounts. Table 7 indicates that for one stoichiometric with catalyst comparision, methanol exhaust formaldehyde was initially 150% higher than gasoline surrogate formaldehyde. However, during the smog chamber irradiation the gasoline's photochemical product formaldehyde increased 850% compared with only a 22% formaldehyde increase from methanol exhaust. The maximum formaldehyde concentration during the gasoline surrogate experiment reached 0.58 PPM compared to only 0.36 PPM maximum from the methanol exhaust surrogates. The above vehicle and exhaust dilution conditions using gasoline produced a larger concentration and a greater mass of formaldehyde than resulted from the use of methanol fuel. This occurred despite the fact that the gasoline surrogate (hydro)carbon emissions had been diluted almost three times more than the methanol surrogates (4225 vs. 1571) in order to compare at equal NO_x concentrations. Some overlap in maximum formaldehyde yield between fuels was observed when corrections were made to equalize exhaust surrogate dilution ratios for other comparisons.

The photochemical reactivity comparisions of ethanol solvent exhaust surrogates were conducted in a slightly different manner from the methanol or gasoline experiments. Chui, Anderson and Baker recently reported results [6] of detailed aldehyde emissions from an ethanol fueled vehicle operating without an oxidation catalyst. About 80% of the exhaust aldehydes were observed to be acetaldehyde and the remainder were formaldehyde. Therefore, the MBTH aldehyde calculations were adjusted for that molar composition and both acetaldehyde and formaldehyde surrogates were used for the smog chamber experiments.

The vehicle tests using ethanol solvent fuel under stoichiometric, non-catalyst conditions resulted in an exhaust (hydro)carbon to NO_x ratio comparable to that of gasoline but much lower than that of methanol exhaust (see Table 3). The composition of the (hydro)carbon surrogates were over 70 carbon percent ethanol and 13 carbon percent aldehyde. The experimental reactivity results indicate (in Table A5) that the ethanol exhaust had a shorter "time to the NO_2 peak" than gasoline but a nearly equal "time to the ozone peak." No equal dilution comparisons were made of photochemical product yields since the gasoline surrogates were diluted by a factor of over 6000 while the ethanol surrogates were diluted about 2700 times. However, it appears very likely that photochemical product yields of nitrogen dioxide, formaldehyde and ozone would be less from ethanol solvent fuel emissions relative to gasoline when tested at equal dilution ratios.

One other stoichiometric ethanol solvent vehicle test was conducted at the 13:1 CR condition including the use of an oxidation catalyst. All three test fuels resulted in exhaust hydrocarbon to NO_x ratios below 6.5:1 (see Table 5). These "NO_x-rich" exhaust and surrogate mixtures were observed during the smog chamber tests to convert about half the nitric oxide into nitrogen dioxide at the time of the NO_2 peaks. Moderate to low ozone production was observed during the subsequent ten hours of irradiation.

The lean ethanol solvent tests were conducted for all three vehilce conditions. The non-catalyst condition resulted in an extremely high non-methane part-per-million carbon to NO_x ratio of 59:1. This high ratio of reactants exhibited a high photochemical reactivity during smog chamber tests. The detailed results are presented in Appendix Table A5.

The lean ethanol vehicle tests with an oxidation catalyst resulted in a 13.7:1 NMPPMC/NO_x ratio shown in Table 4. The corresponding class surrogates were irradiated and found to be moderately reactive within the smog chamber. The net effect of the surrogate composition and NMPPMC/NO_x ratio resulted in sufficient nitrogen dioxide production to lead to a substantial build-up of ozone.

The high CR ethanol solvent vehicle tests resulted in a 4.8:1 NMPPMC/NO_x ratio for FTP emissions. The smog chamber irradiation of (hydro)carbon class surrogates resulted in an incomplete conversion of nitric oxide into nitrogen dioxide for this "NO_x-rich" condition. No appreciable build-up of ozone was observed as is reported in Table A5.

The series of "equal NO_x" experiments in the smog chamber provided one means for comparing the reactivities of the (hydro)carbon compositions and NMPPMC/NO_x ratios observed in vehicle tests. The second available technique involved smog chamber exhaust surrogate comparisons at equal dilution ratios relative to the NO_x emissions. By the latter method the effects of the low NO_x emissions characteristic of alcohol fuel combustion can be directly evaluated in the smog chamber.

PHOTOCHEMICAL REACTIVITY AND PRODUCT YIELDS AT EQUAL EXHAUST DILUTION- Equal dilution ratios in the chamber were chosen for the second type of smog chamber experiments conducted. Stoichiometric gasoline and methanol exhaust surrogates were tested in the smog chamber at both high and low dilution ratios. The exhaust dilution ratio was determined by the following simple equation.

$$\text{Dilution Ratio} = \frac{(\text{Moles FTP NOx})\ (10^6)}{(\text{Moles Chamber Air})(\text{PPM NOx Initially in Chamber})}$$

The NMPPMC/NO_x ratios were closely preserved within each hydrocarbon surrogate class.

Smog chamber experiments were also conducted for lean gasoline and methanol exhausts using the same procedures. Mixtures of gasoline and methanol exhaust surrogates were also tested for any evidence of an adverse interaction. Table A6 presents the detailed results of these"equal dilution" ratio experiments for all the above conditions.

The "time to the NO_2 peak" is one reactivity criterion observed during all smog chamber experiments. In Figure 13 is plotted the experimental "time to the NO_2 peak" versus the calculated total reactivity of the exhaust (hydro)carbon components. These two variables are inversely proportional in that large calculated reactivities should and do result in reduced times to the NO_2 peak.

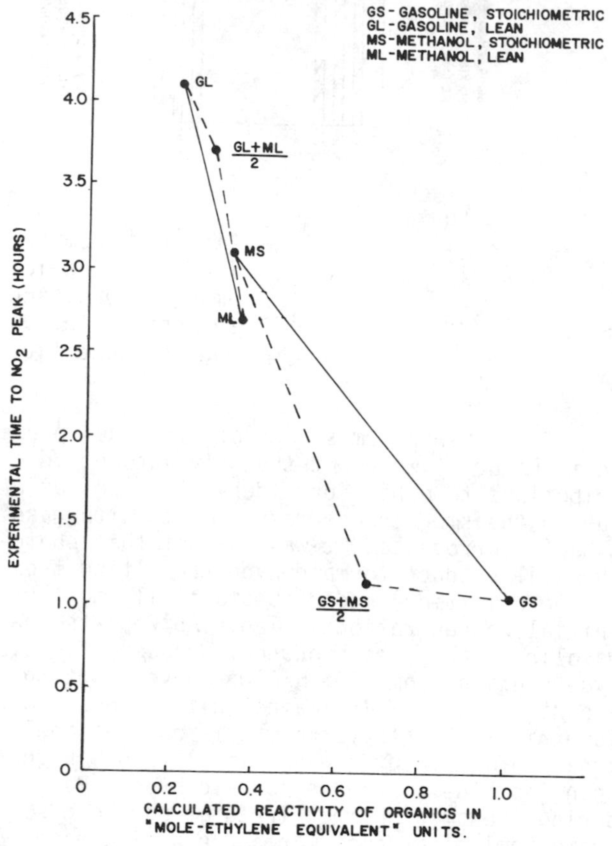

Fig. 13 - Methanol and gasoline surrogate "time to the NO_2 peak" reactivity comparisons for FTP exhaust under high chamber dilution conditions

The highest calculated reactivity among these fuels and conditions was for the stoichiometric gasoline condition. The smog chamber confirms this by yielding the shortest time to the NO_2 peak. Stoichiometric methanol exhaust was calculated to have a moderately low NO_2 reactivity value. The smog chamber experiment resulted in a correspondingly longer time to the NO_2 peak. The mixture of one half gasoline and one half methanol exhaust surrogates resulted in an intermediate calculated NO_2 reactivity and an experimentally intermediate "time to NO_2 peak".

The results were inverted for the lean stoichiometry condition. Here the gasoline exhaust was calculated to have the lowest NO_2 reactivity, and it did require the longest times to achieve the NO_2 peak. The lean methanol exhaust had a larger NO_2 reactivity by the linear summation technique, and it also had an experimentally higher reactivity than the lean gasoline surrogates. The lean mixture of one-half gasoline and one-half methanol surrogates resulted in intermediate calculated and experimental results as above for the stoichiometric mixtures.

It appeared somewhat fortuitous that the calculated and experimental results achieved such a consistent correspondence. No consideration was given in the NO_2 reactivity linear summation technique to the hydrocarbon to NO_x ratio of the exhaust or of the surrogates. Since the non-methane PPM carbon to NO_x ratio is known to strongly influence experimental NO_2 reactivities, it is noteworthy that the PPM-carbon to NO_x ratio did not dominate the experimental results.

The interpretation is that stoichiometric methanol exhaust, which had a consistently higher NMPPMC/NO_x ratio, was so much less reactive due to its composition that the high NMPPMC/NO_x ratios were not sufficient to dominate the photochemistry and achieve shortened times to the NO_2 peak.

Further evidence of the low compositional reactivity of stoichiometric methanol exhaust is presented in Figure 14. The "time to the ozone peak" for methanol exhaust surrogates at a 20:1 PPM-carbon to NO_x ratio was much longer than the time required for the gasoline surrogates' peak at an even lower hydrocarbon to NO_x ratio. The

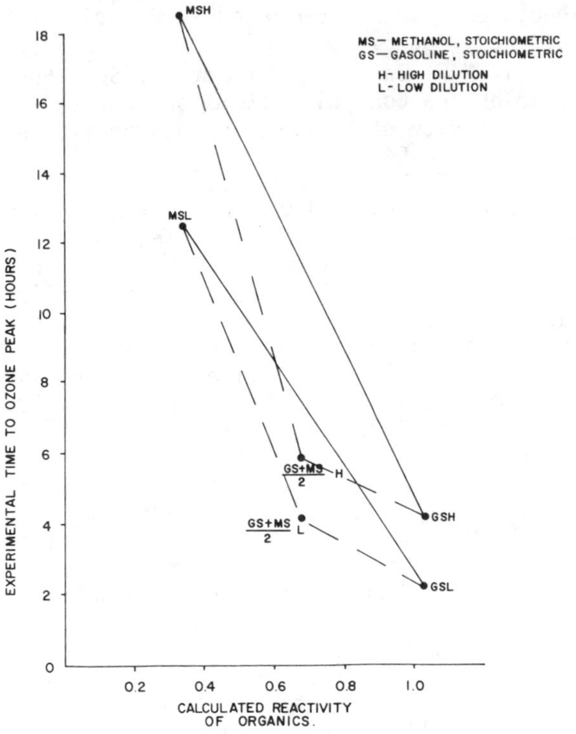

Fig. 14 - Methanol and gasoline surrogate "time to the ozone peak" reactivity comparisons for FTP exhaust under catalyst-equipped conditions

mixtures of one-half gasoline surrogates and one-half methanol surrogates resulted in a less reactive composite mixture relative to gasoline.

The stoichiometric vehicle exhaust experiments, conducted at equal dilution ratios, resulted in substantial advantages for methanol surrogates in terms of their lower photochemical product yields relative to gasoline. Maximum NO_2 concentrations from the methanol experiments averaged only 30% of the yields from the gasoline surrogate experiments. Methanol's formaldehyde yield averaged only 26% of gasoline's photochemical product formaldehyde. Methanol's ozone production was only 53% of the stoichiometric gasoline ozone yields.

The lean vehicle exhaust simulations were dominated by the very low PPM-carbon to NO_x ratios observed with the use of gasoline. The NO_2 maxima remained lower for the methanol exhaust surrogates, however, the formaldehyde and ozone yields were lower with the gasoline exhaust simulations.

CONCLUSIONS

1. FTP driving cycle emissions of carbon monoxide were comparable among the alcohol fuels and gasoline for a variety of catalyst-equipped test conditions.

2. FTP NO_x emissions were reduced by over 50% with the use of methanol or ethanol in non-catalyst or oxidation catalyst-equipped tests. However, the use of methanol in three TWC-equipped vehicles with modified (reduced overlap) camshafts resulted in only an 8% NO_x reduction relative to gasoline in an unmodified vehicle.

3. FTP emissions of total aldehydes were higher for the alcohol fuels tested, particularly during lean operation. They were also higher during stoichiometric high compression ratio tests using airinjection. They ranged from 2% to 22% of the total mass of unburned (hydro)-carbon emissions. The gasoline exhaust aldehyde carbon fraction ranged from 0.3% to 3.6% of the total (hydro)carbon emissions.

4. The addition of an oxidation catalyst substantially reduced all (hydro)carbon emissions, especially at the lean $\Phi = 0.8$ equivalence ratio condition.

5. Calculated NO_2 photochemical reactivity summations were generally found to correspond to the mass of (hydro)carbon emissions. In effect, the (hydro)carbon reactivity summation was lowest for the TWC Pintos and highest for the non-catalyst vehicle condition.

6. Photochemical smog chamber experiments were conducted with reactivity equivalent (hydro)carbon class surrogates. The maximum ozone yields in the chamber experiments were found to closely correspond to the mass emissions of (hydro)carbon for all fuels tested. It should be noted that gasoline exhaust surrogates were diluted over twice as much as alcohol

exhausts for these comparisons at equal chamber NO_x.

7. Mixtures of gasoline exhaust surrogates and methanol exhaust surrogates were experimentally found to be less reactive than the gasoline exhaust for the stoichiometric engine conditions. However, during lean operation the methanol/gasoline mixtures of exhaust were photochemically more reactive than the gasoline surrogates alone.

8. The photochemical reactivity of stoichiometric methanol exhaust was found to be lower than that of gasoline exhaust, even with methanol exhaust at high NMPPMC/NO_x ratios. During lean operation the gasoline exhaust, with a very low NMPPMC/NO_x ratio, was less reactive than the methanol exhaust.

9. The exhaust from the TWC-equipped Pintos was observed to have extremely low NMPPMC/NO_x ratios for both methanol and gasoline fuels. These resulted in only partial conversions of nitric oxide into nitrogen dioxide and therefore no build-up of ozone was observed during smog chamber tests.

10. Poorly controlled (hydro)carbon emissions from methanol or ethanol solvent fueled vehicles can be as photochemically reactive as the emissions from comparable gasoline fueled vehicles. However, for the range of exhaust compositions observed, the NMPPMC/NO_x ratio of the alcohol fuel exhaust must be higher in order to produce an equally reactive mixture.

11. The combination of lower NO_x emissions and less photochemically reactive (hydro)carbon emissions from methanol fueled vehicles suggests that a beneficial impact upon urban atmospheres should result from its use.

ACKNOWLEDGEMENT

Appreciation is expressed to Mr. Lou Browning for his valued assistance in preparing the data reduction program. Mr. Chris Edwards and Mr. Steve Espinola were instrumental in conducting smog chamber experiments. Their key efforts are greatly appreciated.

REFERENCES

1. H.G. Adelman, D.G. Andrews, and R.S. Devoto, "Exhaust Emissions from a Methanol-Fueled Automobile", SAE Paper No. 720693, presented at the SAE National West Coast Meeting, San Francisco, CA., Aug. 21-24, 1972.

2. R.D. Fleming and T.W. Chamberlain, "Methanol as Automotive Fuel Part 1- Straight Methanol", SAE Paper No. 750121, presented at the Automotive Engineering Congress and Exposition, Detroit, MI Feb. 24-28, 1975.

3. H. Menrad, W. Lee and W. Bernhardt, "Development of Pure Methanol Fuel Car", SAE Paper No. 770790, presented at the Passenger Car Meeting, Detroit Plaxa, Detroit, MI., Sept. 26-30, 1977.

4. R.K. Pefley, L.H. Browning, M.L. Hornberger, W.E. Likos, M.C. McCormack and J.B. Pullman, "Characterization and Research Investigation of Methanol and Methyl Fuels", Final Report EPA Grant No. R803548-01 and Progress Report ERDA Contract No. EY-76-S-02-1258. Reprinted by U.S. Department of Energy COO/W1258-01.

5. N.D. Brinkman, "Vehicle Evaluation of Neat Methanol- Compromises Among Exhaust Emissions, Fuel Economy and Driveability", Fourth International Symposium on Automotive Propulsion Systems, Washington, D.C. Vol. II, April 17-22, 1977.

6. G. Chui, R. Anderson and R. Baker, "Brazilian Vehicle Calibration for Ethanol Fuels", Alcohol Fuels Technology Third International Symposium, Asilomar, CA., May 28-31, 1979.

7. R.K. Stone and B.H. Eccleston, "Vehicle Emissions vs. Fuel Composition- API- Bureau of Mines- Part II", API Preprint No. 41-69, American Petroleum Institute, New York, NY, May 13, 1969.

8. K.T. Dishart, "Exhaust Hydrocarbon Composition, its Relation to Gasoline Composition", 35th Midyear Meeting American Petroleum Institute, Houston, TX, May 14, 1970.

9. B. Dimitriades et. al., "The Association of Automotive Fuel Composition with Exhaust Reactivity", Bartlesville Energy Technology Center, Bartlesville, Oklahoma, Report of Investigation #7756, 1973.

10. H. Hosaka, T. Onodera and E.G. Wigg, "The Effect of Fuel Hydrocarbon Composition on Exhaust Emissions from Japanese Vehicles", SAE Paper No. 780625.

11. M.W. Jackson, "Effects of Some Engine Variables and Control Systems on Composition and Reactivity of Exhaust Hydrocarbons", SAE Paper No. 660404, 1966.

12. W.E. Morris and K.T. Dishart, "The Influence of Vehicle Emission Control Systems on the Relationships Between Gasoline and Vehicle Exhaust Hydrocarbon Composition", Effect of Automotive Emission Requirements on Gasoline Characteristics, ASTM STP487, American Society for Testing and Materials", pp. 63-93, 1971.

13. E.E. Wigg, "Reactive Exhaust Emissions from Current and Future Emission Control Systems", SAE Paper No. 730196.

14. M.W. Jackson, "Effect of Catalytic Emission Control on Exhaust Hydrocarbon Composition and Reactivity:, SAE Paper No.780624

15. M. Matsuno et. al, "Alcohol Engine Emissions- Emphasis on Unregulated Compounds", Alcohol Fuels Technology Third International Symposium, Asilomar, CA May 28-31, 1979.

16. R. Bechtold and B. Pullman, "Driving Cycle Comparisons of Energy Economies and Emissions from an Alcohol and Gasoline Fueled Vehicle", Alcohol Fuels Technology Third International Symposium, Asilomar, CA, May 28-31, 1979.

17. D. Hilden and F. Parks, "A Single-Cylinder Engine Study of Methanol Fuel- Emphasis

on Organic Emissions", SAE Paper No. 760378, presented at the Automotive Engineering Congress and Exposition, Detroit, MI., Feb. 23-27, 1976.

18. C.J. Raible and F.W. Cox, "Chromtographic Methods of Analysis for Methanol and Ethanol in Automotive Exhaust", SAE Paper No. 790690, presented at the SAE Passenger Car Meeting, Dearborn, MI, June 1979.

19. B. Dimitriades, C.J. Raible, and C.A. Wilson, "Interpretation of Gas Chromtographic Spectra in Routine Analysis of Exhaust Hydrocarbons", U.S. Dept. of Energy, BERC/RI-7700/1972.

20. Coordinating Research Council, "Oxygenates in Automotive Exhaust Gas: Part 1. Techniques for Determining Aldehydes by the MBTH Method", Report No. 415, June 1968.

21. C. Wu and H. Niki, "Methods for Measuring NO_2 Photodissociation Rate", Environmental Science and Technology, Vol. 9, No. 1, Jan. 1975.

22. C. LaPointe and W. Schultz, "Comparison of Emission Indexes Within a Turbine Combustor Operated on Diesel Fuel or Methanol", SAE Paper No. 730669 presented at the National Power Plant Meeting, Chicago, IL., June 18-22, 1973.

23. N. Sheth and T. Rice, "Quantification and Reduction of Sources of Variability in Vehicle Emissions and Fuel Economy Measurements", SAE Paper 790232, 1979.

24. K. Darnall, A. Lloyd, A. Winer and J. Pitts, "Reactivity Scale for Atmospheric Hydrocarbons Based on Reaction with Hydroxyl Radical", Environmental Science and Technology, Vol. 10, No. 7, July 1976.

25. B. Dimitriades, "The Concept of Reactivity and its Possible Application in Control", Proceedings of the Solvent Reactivity Conference, RTP, NC, Nov. 1974. EPA-650/3-74-010.

26. F. Farley, "Photochemical Reactivity Classification of Hydrocarbons and Other Organic Compounds", Proceedings of the Solvent Reactivity Conference, RTP, NC, Nov. 1974. EPA-650/3-74-010.

27. B. Dimitriades and S. Joshi, "Application of Reactivity Criteria in Oxidant-Related Emission Control in the USA", Research Triangle Park, N.C., Nov. 1976.

28. W.A. Glasson and C.S. Tuesday, "Inhibition of the Atmospheric Photooxidation of Hydrocarbons by Nitric Oxide", GMR-475, General Motors Corp., March 1965.

APPENDIX

TABLE A1: Calculated Exhaust Surrogate Data; Surrogate Carbon Mass Fractions by Class and Mole-Weighted Reactivities for the Data in Table 3

FUEL	FUEL AIR EQUIVALENCE	CALCULATED SURROGATE CARBON MASS FRACTION (%)					RATIO OF TOTAL NMPPMC / NO_x	$\Sigma n_i R_i$ CALC. TOTAL REACTIVITY	$\frac{\Sigma n_i R_i}{\Sigma m_i}$ REACTIV. PER GRAM EXHAUST	$\frac{\Sigma n_i R_i}{\Sigma n_i C_i}$ REACTIV. PER MOLE EXH. CARBON
		BUTANE	PROPYLENE	TOLUENE	METHANOL ETHANOL	FORMALDEHYDE				
Methanol	1.0	.04	.10	.03	96.1	3.7	68.1	1.68	.0094	.30
Gasoline	1.0	35.00	44.40	16.80	0.7	3.1	10.2	1.00	.0237	.34
Ethanol Solvent	1.0	1.60	14.20	.40	70.8	12.9	9.4	1.04	.0344	.73
Methanol	0.8	.02	.02	.02	95.3	4.6	73.5	2.20	.0108	.35
Gasoline	0.8	42.70	34.30	17.10	1.2	4.7	14.9	1.93	.0279	.42
Ethanol Solvent	0.8	0.50	7.60	.09	82.3	9.4	56.6	3.02	.0283	.63

TABLE A2: Calculated Exhaust Surrogate Data; Surrogate Carbon Mass Fractions by Class and Mole-Weighted Reactivities for the Data in Table 4

FUEL	FUEL AIR EQUIVALENCE	CALCULATED SURROGATE CARBON MASS FRACTION (%)					RATIO OF TOTAL NMPPMC / NO_x	$\Sigma n_i R_i$ CALC. TOTAL REACTIVITY	$\frac{\Sigma n_i R_i}{\Sigma m_i}$ REACTIV. PER GRAM EXHAUST	$\frac{\Sigma n_i R_i}{\Sigma n_i C_i}$ REACTIV. PER MOLE EXH. CARBON
		BUTANE	PROPYLENE	TOLUENE	METHANOL ETHANOL	FORMALDEHYDE				
Methanol	1.0	.15	.08	.06	97.9	1.8	22.3	.34	.0064	.20
Gasoline	1.0	39.6	37.6	20.5	1.0	1.3	13.8	1.06	.0193	.28
Methanol	0.8	.05	.03	.02	91.9	8.0	11.1	.36	.0159	.50
Gasoline	0.8	48.1	31.7	17.3	1.2	1.6	2.8	.22	.0165	.24
Ethanol Solvent	0.8	1.5	4.5	.06	84.1	9.8	13.0	.49	.0256	.56

Table A3: Calculated Exhaust Surrogate Data; Surrogate Carbon Mass Fractions By Class and Mole-Weighted Reactivities for the Data in Table 5

FUEL	FUEL AIR EQUIV.	Calculated Surrogate Carbon Mass Fraction (%)					RATIO OF TOTAL NMPPMC/NOx	$\Sigma n_i R_i$ CALC. TOTAL REACTIVITY	$\frac{\Sigma n_i R_i}{\Sigma m_i}$ REACTIV. PER GRAM EXHAUST	$\frac{\Sigma n_i R_i}{\Sigma n_i C_i}$ REACTIV. PER MOLE EXHAUST CARBON
		BUTANE	PROPYLENE	TOLUENE	METHANOL ETHANOL	FORM-ALDEHYDE				
Methanol	1.0	0.1	0.07	0.1	84.5	15.2	5.9	.79	.0269	.84
Gasoline	1.0	39.7	16.3	42.5	-	1.4	3.1	.27	.0119	.17
Ethanol Solvent	1.0	9.3	5.9	0.9	75.8	8.0	5.9	.59	.0230	.48
Methanol	0.8	0.7	0.2	0.7	76.5	21.9	6.1	.58	.0378	1.18
Methyl Fuel	0.8	0.8	4.6	0.3	76.5	17.8	4.9	.52	.0314	.92
Gasoline	0.8	47.4	15.3	35.9	-	1.4	3.3	.22	.0157	.23
Ethanol Solvent	0.8	13.2	9.9	3.2	56.8	16.8	4.4	.46	.0389	.79
190-Proof Ethanol	0.8	1.7	10.8	0.5	75.6	11.5	12.4	.82	.0313	.68

Table A4: Calculated Exhaust Surrogate Data; Surrogate Carbon Mass Fractions By Class and Mole-Weighted Reactivities for the Data in Table 6.

FUEL	FUEL AIR EQUIV.	Calculated Surrogate Carbon Mass Fraction (%)					RATIO OF TOTAL NMPPMC/NOx	$\Sigma n_i R_i$ CALC. TOTAL REACTIVITY	$\frac{\Sigma n_i R_i}{\Sigma m_i}$ REACTIV. PER GRAM EXHAUST	$\frac{\Sigma n_i R_i}{\Sigma n_i C_i}$ REACTIV. PER MOLE EXHAUST CARBON
		BUTANE	PROPYLENE	TOLUENE	METHANOL	FORM-ALDEHYDE				
Methanol	1.0	-	-	-	89.5	10.5	1.12	.069	.0199	.633
Gasoline	1.0	42.3	14.9	42.0	0.3	0.4	1.42	.035	.0104	.152

TABLE A5: Alcohol and Gasoline Surrogate Smog Chamber Experiments at Variable Exhaust Dilution Ratios But Equal Initial Chamber NO_x Concentrations

VEHICLE TEST CONDITIONS				CALCULATED SURROGATES, PPM MOLAR									CHAMBER INITIAL CONDITIONS, PPM MOLAR								CHAMBER RESULTS; T IN HOURS, [ ] IN PPM									
FUEL	CATALYST	CR	Φ	BUTANE	PROPENE	TOLUENE	METHANOL	FORMALDEHYDE	NO_x	SURRO. DILU.	CALC. REACT.	PPMC/NO_x	BUTANE	PROPENE	TOLUENE	ALCOHOL	FORMALDEHYDE	ACETALDEHYDE	NO_x	PPMC/NO_x	$T_{NO_2/2}$	T_{NO_2}	T_{HCHO}	T_{O_3}	$[NO_2]_m$	$[HCHO]_m$	$[O_3]_m$	$[NO_2]_m/[NO_x]_o$	$[HCHO]_m/[HCHO]_o$	
METHANOL	NO	8.5	1.0	.003	.009	.001	26.2	1.01	.40	1746	1.68	68.0	-	-	-	26.5	1.0		.38	73.		.65		3.6	.27		1.53	.71		
METHANOL	OXY	8.5	1.0	.003	.002	.001	8.75	.158	.40	1571	.34	22.3	-	-	-	9.26	.154		.40	23.5	.61	2.45	.0	15.8	.26	.36	.97	.65	1.00	
METHANOL	OXY	13.0	1.0	.001	.001	.000	2.00	.360	.40	3317	.79	5.9	-	-	-	2.04	.37		.407	5.9	.48	2.13	.12	>14.	.189	.43	.075	.46	1.17	
METHANOL	TWC	8.5	1.0	.000	.000	.000	.402	.047	.40	2085	.069	1.30																		
METHANOL	NO	8.5	0.8	.002	.002	.001	28.0	1.37	.40	1853	2.20	73.	-	-	-	29.2	1.64		.39	79.		.42		3.0	.34		1.70	.87		
METHANOL	OXY	8.5	0.8	.001	.000	.000	4.09	.356	.40	1364	.36	11.1				3.84	.34		.39	10.7	.36	2.1	.15	>10.	.225	.36	.13	.58	1.06	
METHANOL	OXY	13.0	0.8	.005	.001	.002	1.87	.536	.40	1711	.58	6.1	-	-	-	1.90	.47		.41	5.8	.33	2.0	.13	>13.	.206	.65	.03	.50	1.38	
GASOLINE	NO	8.5	1.0	.238	.404	.065	.020	.084	.40	6076	1.00	6.8	.24	.40	.073	.03	.099		.406	7.0	.58	1.78	2.9	10.0	.30	.31	.81	.74	3.1	
GASOLINE	OXY	8.5	1.0	.547	.692	.162	.055	.069	.40	4225	1.06	13.8	.548	.706	.158	.056	.061		.40	13.8	.43	1.05	2.3	4.2	.325	.58	1.05	.81	9.51	
GASOLINE	OXY	13.0	1.0	.123	.068	.075	.000	.018	.40	6844	.27	3.1	.158	.082	.110		.021		.406	4.1	1.0	4.6	1.2	>10.	.203	.054	.071	.50	2.6	
GASOLINE	TWC	8.5	1.0	.060	.028	.034	.002	.003	.40	2386	.035	1.42																		
GASOLINE	NO	8.5	0.8	.637	.682	.146	.069	.278	.40	5168	1.93	14.9	.64	.69	.14	.079	.28		.447	13.3	.31	.86	N/A	5.0	.36	N/A	1.01	.81	N/A	
GASOLINE	OXY	8.5	0.8	.136	.120	.028	.014	.018	.40	4470	.22	2.8	.148	.126	.033		.031		.431	2.86	1.03	4.1	2.5	>15.0	.216	.12	.25	.50	3.87	
GASOLINE	OXY	13.0	0.8	.155	.066	.067	.000	.019	.40	5343	.22	3.3	.157	.069	.074		.022		.40	3.45	2.0	4.6	2.	>12.0	.177	.044	.07	.44	2.	
ETHANOL	NO	8.5	1.0	.015	.179	.002	1.34	.487	.40	2673	1.04	6.1	.018	.187	-	1.43	.06	.48	.407	11.1	1.0	1.25	.78	9.8	.29	.27	.83	.72	4.5	
ETHANOL	OXY	8.5	1.0	N/A	N/A	N/A	N/A	N/A	N/A	N/A	N/A	N/A																		
ETHANOL	OXY	13.0	1.0	.055	.046	.003	.886	.188	.40	3597	.59	5.8	.055	.055	-	.97	.07	.37	.416	7.5	N/A	2.0	1.5	16.4	N/A	.15	.66	N/A	2.2	
ETHANOL	TWC		1.0	N/A	N/A	N/A	N/A	N/A	N/A	N/A	N/A	N/A																		
ETHANOL	NO	8.5	0.8	.030	.574	.003	9.33	2.14	.40	1676	3.02	33.3	.03	.57	-	9.3	.5	2.1	.40	56.0	N/A	<.65	N/A	8.	N/A	N/A	.96	N/A	N/A	
ETHANOL	OXY	8.5	0.8	.020	.079	.000	2.19	.510	.40	1171	.49	13.0	.046	.093	-	2.11	.46		.40	12.8	.34	1.31	.3	8.	.27	.55	1.28	.68	1.2	T=100°F
ETHANOL	OXY	13.0	0.8	.058	.058	.008	.497	.293	.40	2182	.46	4.4																		

TABLE A6: Methanol and Gasoline Surrogate Smog Chamber Experiments at Two Constant Exhaust Dilution Ratios, Including Gasoline and Methanol Mixtures

VEHICLE TEST CONDITIONS				CALCULATED SURROGATES, PPM MOLAR									CHAMBER INITIAL CONDITIONS, PPM MOLAR							CHAMBER RESULTS; T IN HOURS, [ ] IN PPM								
FUEL	CATALYST	CR	Φ	BUTANE	PROPENE	TOLUENE	METHANOL	FORMALDEHYDE	NO_x	SURRO. DILU.	CALC. REACT.	PPMC/NO_x	BUTANE	PROPENE	TOLUENE	METHANOL	FORMALDEHYDE	NO_x	PPMC/NO_x	$T_{NO_2/2}$	T_{NO_2}	T_{HCHO}	T_{O_3}	$[NO_2]_m$	$[HCHO]_m$	$[O_3]_m$	$[NO_2]_m/[NO_x]_o$	$[HCHO]_m/[HCHO]_o$
GASOLINE	YES	8.5	1.0	.561	.710	.166	.056	.071	.41	4121	1.06	13.8	.548	.706	.158	.056	.061	.40	13.8	.43	1.05	2.3	4.2	.325	.58	1.05	.81	9.51
METHANOL	YES	8.5	1.0	.001	.001	.000	3.34	.060	.153	4121	.34	22.3				3.27	.068	.167	20.0	.61	3.1	1.9	>18.5	.098	.13	.55	.59	1.91
(G+M)/2	YES	8.5	1.0	.281	.356	.083	1.70	.065	.281	4121	.70	16.1	.274	.351	.088	1.73	.065	.272	16.8	.38	1.14	2.75	5.9	.22	.44	.99	.81	6.77
GASOLINE	YES	8.5	1.0	1.44	1.83	.43	.14	.18	1.056	1600	1.06	13.8	1.385	1.82	.415	.145	.150	1.043	13.6	.51	.99	N/A	2.25	.87	1.2	1.59	.83	8.0
METHANOL	YES	8.5	1.0	.002	.002	.000	8.60	.15	.394	1600	.34	22.3				9.26	.154	.396	23.8	.75	2.7	2.5	12.5	.24	.34	.85	.61	2.21
(G+M)/2	YES	8.5	1.0	.723	.917	.214	4.38	.167	.724	1600	.70	16.1	.719	.868	.203	5.43	.147	.723	17.3	.52	1.16	2.4	4.15	.58	.89	1.37	.80	6.05
GASOLINE	YES	8.5	0.8	.148	.130	.030	.015	.020	.434	4121	.22	2.82	.148	.126	.033		.031	.431	2.86	1.03	4.1	2.5	>15.	.216	.12	.25	.50	3.87
METHANOL	YES	8.5	0.8	-	-	-	1.35	.118	.132	4121	.36	11.1	-	-	-	1.42	.117	.131	11.7	.50	2.7	0.0	>19.	.07	.12	.45	.53	1.03
(G+M)/2	YES	8.5	0.8	.074	.065	.015	.684	.069	.283	4121	.29	4.77	.094	.080	.019	.825	.066	.350	4.69	.86	3.5	2.1	12.4	.19	.12	.40	.54	1.82
GASOLINE	YES	8.5	0.8	.381	.335	.077	.039	.052	1.12	1600	.22	2.82	.394	.340	.079	.043	.048	1.11	2.92	1.00	3.7	2.6	>10.	.45	.22	.06	.41	4.58
METHANOL	YES	8.5	0.8	-	-	-	3.48	.304	.34	1600	.36	11.1				4.75	.36	.40	12.8	.61	2.45	0.0	15.8	.26	.36	.97	.65	1.00
(G+M)/2	YES	8.5	0.8	.191	.167	.039	1.76	.178	.729	1600	.29	4.77	.180	.167	.027	1.74	.151	.727	4.54	1.00	2.75	1.8	>12.	.31	.22	.21	.43	1.46

Evaporative and Exhaust Emissions from Cars Fueled with Gasoline Containing Ethanol or Methyl tert-Butyl Ether*

Robert L. Furey
Fuels and Lubricants Dept.
General Motors Research Labs.
Warren, MI

Jack B. King
Advance Product Engrg.
General Motors Engrg. Staff
Warren, MI

THE ADDITION OF some alcohols and ethers to gasoline may extend petroleum supplies and improve unleaded gasoline octane quality. As a result of recent U.S. Environmental Protection Agency decisions, ethanol, tert-butyl alcohol, and methyl tert-butyl ether (MTBE) may now be added to gasolines in the United States at maximum concentrations of 10, 7, and 7 percent by volume, respectively. Thus far, these blending components have not been used extensively because their availability has been limited, but their use is expected to increase.

Some technical assessments of the use of alcohols as automotive fuels have been made (1-3)*, but the major barrier to their commercialization has been their high cost relative to petroleum fuels. Recently, however, Federal and State tax incentives have greatly increased the interest in marketing "gasohol" which is a blend of 90 percent unleaded gasoline and 10 percent agriculturally-derived anhydrous ethanol.

The use of MTBE as a gasoline blending component (4-6) provides a means of using methanol (MTBE is produced from methanol and isobutylene) without some of the technical disadvantages of adding the methanol directly to gasoline.

As part of our evaluation of these alternative automotive fuels (7-10), evaporative and exhaust emissions from current production and prototype cars, using fuels containing ethanol or MTBE, were measured. Fuel economy over the EPA city driving schedule was also determined, and vehicle driveability was briefly evaluated. The major objective of the program was to investigate the effect of fuel composition on evaporative emissions. There-

* Numbers in parentheses designate References at end of paper.

*Paper 800261 presented at the Congress and Exposition, Detroit, Michigan, February 1980.

ABSTRACT

Vehicle tests showed that evaporative emissions were increased significantly by adding 10 percent ethanol to gasoline, but were increased less with 15 percent MTBE in gasoline. The quantity of ethanol or MTBE in evaporative emissions was investigated in laboratory tests. Exhaust HC, CO, and NO_x emissions from a car without closed-loop fuel control were significantly lower with the ethanol and MTBE fuel blends than with gasoline. For cars equipped with closed-loop carburetors, the absolute differences in exhaust emissions among the fuels were small. Fuel economy and driveability were worse with ethanol and MTBE fuel blends than with gasoline.

SAE/PT-80/19/$02.50

fore, laboratory studies of the quantity and composition of the vapor generated from the fuels were also conducted.

EXPERIMENTAL PROGRAM

TEST CARS - The three cars described in Table 1 were used for the exhaust emissions and fuel economy tests. Only the first two cars were used for the evaporative emissions and driveability tests. Although two of the cars were equipped with prototype emissions control systems, variations of these systems are now in production.

TEST FUELS - The four test fuels and some of their properties are listed in Table 2. Unleaded Indolene, the base fuel, is a full-boiling-range gasoline which is widely used for emissions testing. The second fuel was a blend of 10 volume-percent absolute ethanol in Indolene. Since this fuel had a higher vapor pressure than the gasoline alone, another test fuel containing 10 percent ethanol in Indolene was prepared. The Reid vapor pressure (RVP) of the gasoline was reduced by purging with nitrogen, so that a blend of 10 percent ethanol in this low-volatility fuel had the same RVP as the base gasoline. The fourth test fuel was a blend of 15 volume-percent MTBE in Indolene. This concentration of MTBE was used because at the time the emissions tests were run, the EPA was considering a request for a waiver from the Clean Air Act Amendments to permit the use of

Table 1 - Test Cars

- 1978 Production Car

  5.7-L V8 engine
  4-barrel carburetor
  1814 kg (4000 lb) inertia weight
  Oxidizing catalytic converter
  Production evaporative emissions control system
  79.5-L (21-gallon) fuel tank

- Prototype Closed-Loop Dual-Bed Catalyst Car

  4.3-L V8 engine
  2-barrel electromechanical closed-loop carburetor
  1588 kg (3500 lb) inertia weight
  Dual-bed catalytic converter
  Prototype evaporative emissions control system (includes activated carbon component in the air cleaner)
  66.2-L (17.5-gallon) fuel tank

- Prototype Closed-Loop Three-Way Catalyst Car

  4.3-L V8 engine
  2-barrel electromechanical closed-loop carburetor
  1814 kg (4000 lb) inertia weight
  Three-way catalytic converter
  (Evaporative emissions and driveability tests were not run with this car.)

Table 2 - Test Fuels

	Indolene (gasoline)	10 vol. % ethanol in Indolene (gasohol)	10 vol. % ethanol in Indolene with adj. volatility (adjusted gasohol)	15 vol. % MTBE in Indolene (MTBE fuel)
Reid vapor pressure, kPa	62.1	66.9	62.1	62.7
ASTM distillation, °C at % evaporated				
IBP	38.5	38.5	40.5	34
10%	61.5	55	58.5	56
20%	79.5	64	65.5	67.5
30%	96.5	69.5	70	79
50%	114	107.5	107	102
90%	142	158	158	156.5
End	212.5	208.5	200.5	207
Sp. Gr. at 15.6°C	0.747	0.752	0.744	0.744
Atomic H/C ratio	1.81	1.89	1.95	1.88
Atomic O/C ratio	0	0.033	0.034	0.024
Oxygen content, mass %	0	3.7	3.7	2.7
Stoichiometric A/F	14.52	13.94	14.01	14.11

15 percent MTBE. That waiver request was denied, but a later request for 7 percent MTBE was approved. The four test fuels will be referred to as gasoline, gasohol, adjusted gasohol, and MTBE fuel, respectively.

TEST PROCEDURES - Vehicle evaporative emissions were measured using the Sealed Housing for Evaporative Determination (SHED) procedure (11), which was modified as described in Appendix A. Exhaust emissions and fuel economy were also measured according to the official Federal test procedures (11-12). The results were corrected, where necessary, to account for differences in fuel properties. Details are presented in Appendix A. Each car was tested at least twice on each fuel. (However, the closed-loop, three-way catalyst car was not tested with adjusted gasohol.) The fuels were used in the following order: gasoline, gasohol, adjusted gasohol, MTBE fuel, and gasoline.

Driveability of the 1978 production car and the closed-loop, dual-bed catalyst car was evaluated at an ambient temperature of 5°C using the Coordinating Research Council's cold-start driveaway procedure (13). Warmed-up driveability at an ambient temperature of 15°C was subjectively evaluated under cruise conditions and at light-load accelerations from idle. Single tests were run with each fuel.

To obtain a better understanding of the effects of fuel composition on evaporative emissions than was possible with the vehicle tests, the diurnal and hot soak parts of the evaporative emissions test were simulated in the laboratory. Details of the test procedure are provided in Appendix B.

VEHICLE EVAPORATIVE EMISSIONS

DIURNAL, HOT SOAK, AND TOTAL EVAPORATIVE EMISSIONS - The evaporative emissions data from each test are presented in Tables C-1 and C-2 of Appendix C, and the average changes in emissions with the oxygen-containing fuels compared to gasoline are summarized in Figure 1. When a change in emissions was statistically significant at the 90 or 95 percent confidence level, that fact is indicated in the figure. (In this report, the words "statistically significant" are used to mean that the effect was statistically significant at a confidence level of 90 percent or higher.) Results are presented separately for the diurnal and hot soak parts of the test, in addition to the total emissions.

Figure 1 shows that, for both the production car and the prototype closed-loop dual-bed catalyst car, the diurnal emissions were higher with the three oxygen-containing fuels than with the gasoline. However, only the increases with gasohol and 15 percent MTBE fuel were statistically significant. A significant increase was not expected with the adjusted gasohol, because it had the same vapor pressure as the gasoline.

Thus, in the diurnal part of the emissions test, the two cars responded similarly to the various fuels. This is understandable, because diurnal emissions result from heating the fuel in the tank, and the design of the fuel tanks and vapor collection systems of the two cars was similar.

In the hot soak part of the test, the emissions are due primarily to vaporization of fuel in the carburetor. Therefore, the two cars might be expected to respond differently to fuel changes, because of the different carburetor designs, and because the prototype car had a ring of activated carbon in the air cleaner. The purpose of this carbon component was to adsorb vapor from the carburetor vents during the hot soak test.

Figure 1 shows that the hot soak emissions from the production car were higher with the three oxygen-containing fuels than with the gasoline. The increases with the gasohol and adjusted gasohol were statistically significant, but the increase with the MTBE fuel was not. The hot soak emissions from the prototype closed-loop car were higher with the adjusted gasohol than with the gasoline, but the emissions were lower with the gasohol and the MTBE fuel. Only the increase with the adjusted gasohol was statistically significant. Surprisingly, the hot soak emissions from both cars were higher with the adjusted

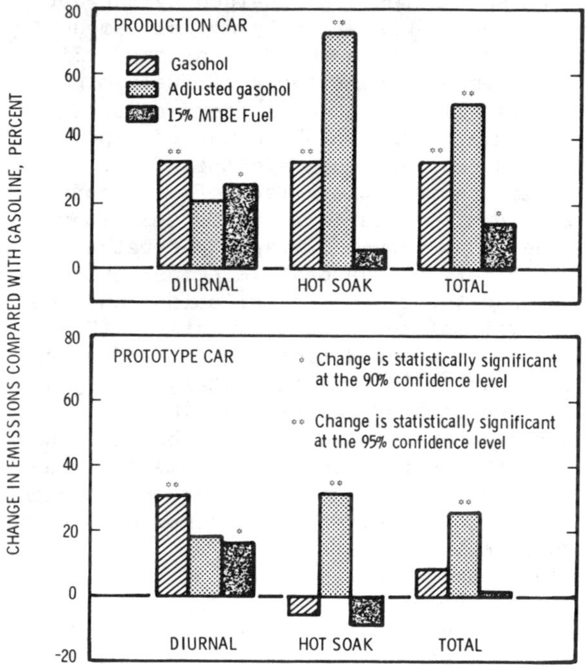

Fig. 1 - Effect of ethanol and MTBE on evaporative emissions

gasohol than with the gasohol. This unexpected result will be discussed in more detail later.

The total evaporative emissions are the sum of the diurnal and hot soak emissions. The total emissions from the production car were higher with the three oxygen-containing fuels than with gasoline, and these increases were statistically significant. The total emissions from the prototype car were higher with the gasohol and adjusted gasohol than with the gasoline, but essentially unchanged with the MTBE fuel. Only the increase with the adjusted gasohol was statistically significant.

The total evaporative emissions from both cars, with all four fuels, were well below the present Federal maximum limit of 6 grams per test. However, the limit is reduced to 2 grams per test in California in 1980, and nationwide in 1981. Unfortunately, the closed-loop carburetor, being a prototype design, allowed vapor leakage around shafts and linkages which would be sealed in a production version. Thus the evaporative emissions control performance of the prototype system was not sufficient to meet a 2-gram standard.

TOTAL VAPOR GENERATED - The total amount of evaporative emissions is certainly of primary interest, and is the only consideration in determining whether a car meets the emissions standards set by the government. However, our objective was to determine how various fuel blends affect emissions. Most of the vapor which is generated during an evaporative emissions test is retained by the activated carbon canister; only a small amount escapes and is measured as emissions. Since vapors from different types of fuels might affect emissions control systems differently, a knowledge of the total amount of vapor generated can help separate the effect of fuel vaporization from the effect of emissions control system deterioration on emissions.

The total amount of vapor generated during each part of the evaporative emissions test is defined as the sum of the emissions and the weight gain of the carbon canister. (The canister was removed from the car and weighed to the nearest 0.01 g immediately before and after each part of the test.) This definition is satisfactory for the production car. However, with the prototype car, some vapor could be adsorbed on the activated carbon component in the air cleaner, and that component was not weighed in these tests.

The canister weight gain and the total amount of vapor generated during each part of each test are listed in Tables C-1 and C-2. The average changes in vapor generated by the oxygen-containing fuels compared to gasoline are shown in Figure 2. In botn cars, the gasohol produced significantly more vapor than gasoline during the diurnal part of the test. This result was expected because of the higher vapor pressure of gasohol. The adjusted gasohol produced about the same amount of diurnal vapor as the gasoline, because its vapor pressure was the same. (The small changes shown in Figure 2 were not statistically significant.) The 15 percent MTBE fuel produced more diurnal vapor than gasoline in both cars, and this result is in agreement with the slightly higher vapor pressure of the MTBE fuel.

Even with ethanol or MTBE in the gasoline, vapor pressure appears to be a good indicator of the amount of vapor generated during the diurnal part of the test, as has been reported for gasolines (14). However, the vapor pressures of the fuels in this study covered a very narrow range; therefore, these results should not be used to derive a mathematical correlation between vapor pressure and the amount of vapor generated.

Figure 3 shows that the blends of 10 percent ethanol and 15 percent MTBE used in this study have vapor pressures very close to the maximum observed for blends ranging from 0 to 100 percent of the alcohol or ether in the gasoline. Therefore, other concentrations of ethanol or MTBE in Indolene would not be expected to produce significantly more diurnal vapor than that found in this study.

In the hot soak part of the emissions test, the vapor arises primarily from vaporization of fuel in the hot carburetor, and the amount of hot soak vapor is therefore related to the distillation characteristics of the fuel. The ASTM distillation curves for three of the four test fuels are shown in Figure 4. (The curves for the gasohol and adjusted gasohol are very similar, and therefore the adjusted gasohol curve is not shown.) It is obvious from Figure 4 that at any given distillation temperature, especially in the range of about 50° to 120°C, more gasohol or MTBE fuel is distilled than gasoline. Figure 2 shows that all of the fuels containing ethanol or MTBE produced significantly more vapor during hot soak than the gasoline.

HIGH HOT SOAK EMISSIONS WITH ADJUSTED GASOHOL - An important observation from Figure 2 is that the adjusted gasohol did not produce more vapor than the gasohol, even though the emissions were higher with the adjusted gasohol than with the gasohol, as discussed earlier. This indicates that the evaporative emissions control systems on the cars retained less of the vapor generated by the adjusted gasohol. The problem does not appear to be related to the capacity of the carbon canister. Less vapor was generated during the hot soak than during the diurnal part of the test, so the capacity of the canister should be more than adequate to handle the hot soak vapors.

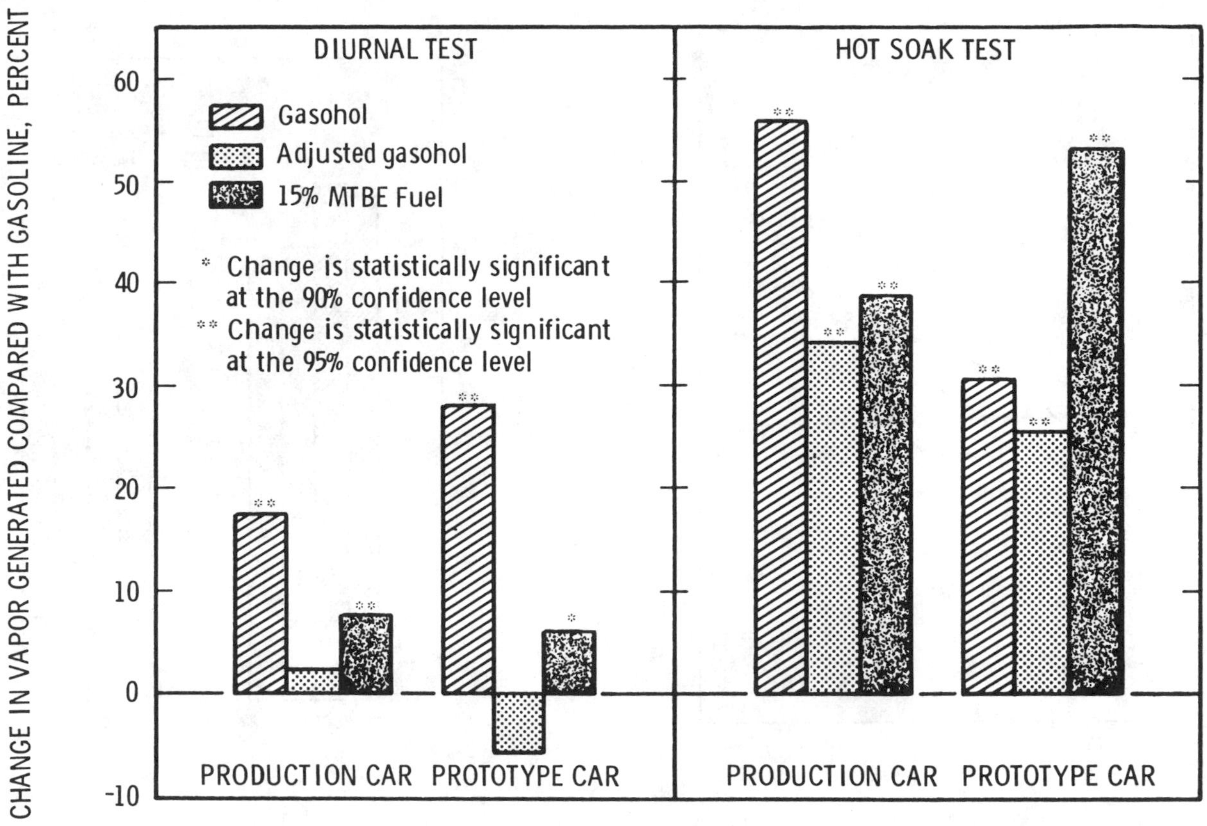

Fig. 2 - Effect of ethanol and MTBE on total vapor generated during evaporative emissions tests

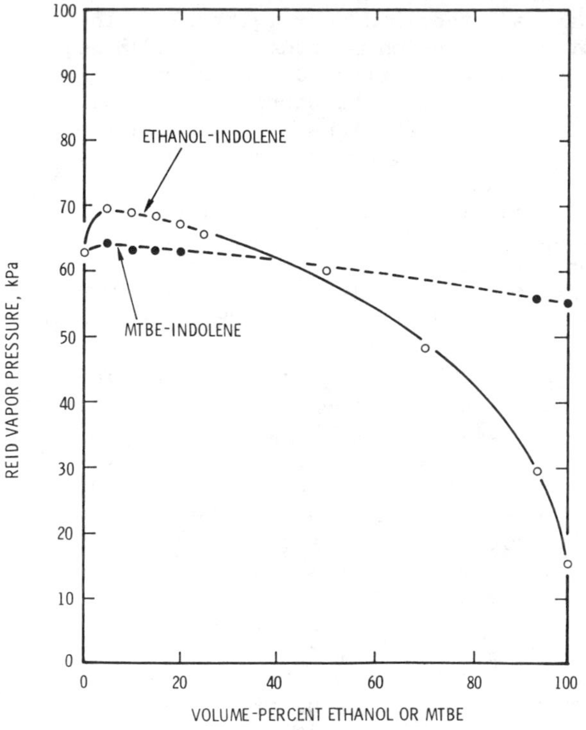

Fig. 3 - Vapor pressures of ethanol-Indolene blends and MTBE-Indolene blends

The higher hot soak emissions with the adjusted gasohol compared to the gasohol may have been due to progressive deterioration of the rubber (nitrile) hose which vents fuel vapor from the carburetor fuel bowl to the canister during a hot soak. The deterioration of the hose may have allowed increased permeation of fuel vapor through the hose. [Permeability of fuel hoses has been reported to be greater with alcohol-gasoline blends than with gasoline alone (15).] Figure 5 shows the hot soak emissions and quantity of vapor generated for four successive tests, the first two with gasohol and the last two with adjusted gasohol. With each successive test (from left to right in Figure 5), hot soak emissions increased even though the total amount of vapor produced did not always increase. It is unlikely that the adjusted gasohol was any more detrimental to the bowl vent hose than the gasohol. Rather, each successive test with an alcohol-containing fuel apparently increased the permeability of the hose, thus increasing evaporative emissions. Condensed fuel was found in the hose after each test with an alcohol-containing fuel, and this liquid probably hastened the deterioration of the hose. Fuel hose deterioration and swell with alcohol-gasoline blends have been reported

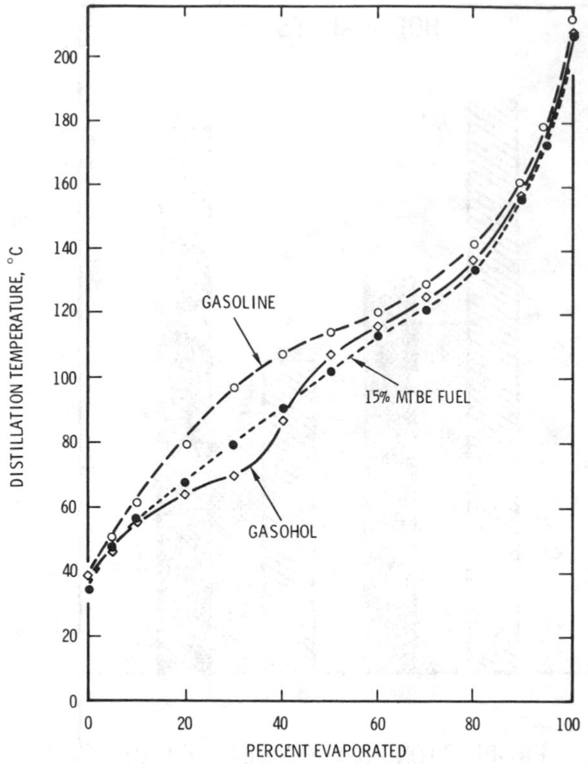

Fig. 4 - ASTM distillation curves of three test fuels

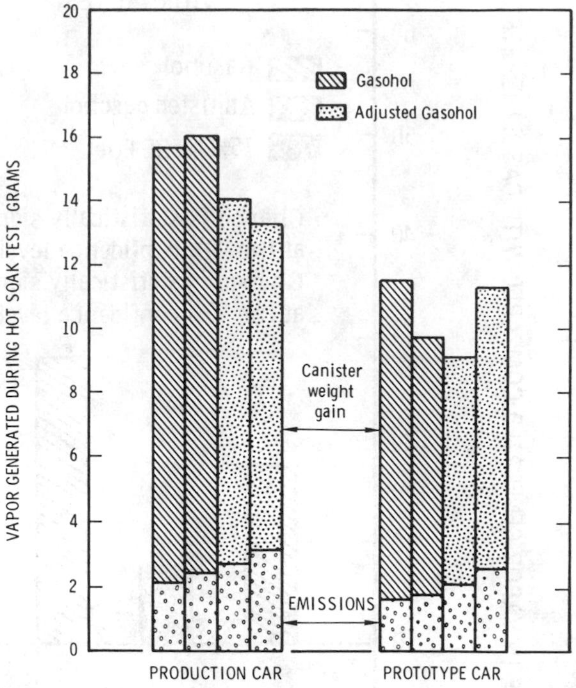

Fig. 5 - Progressive increase in hot soak emissions with fuels containing ethanol

(15-16), and the same type of deterioration probably occurred in the bowl vent hose. Although the bowl vent hose appears to be the most likely source of increased hot soak emissions, permeation of fuel through the fuel hoses cannot be ruled out.

These tests have shown that the evaporative emissions were higher with gasoline containing 10 percent ethanol than with gasoline alone. This result is in agreement with other reports of increased evaporative emissions with 10 percent ethanol fuels (17). The increase in emissions with 15 percent MTBE in gasoline was not as large as with the ethanol fuels.

LABORATORY STUDIES OF EVAPORATIVE EMISSIONS

The vehicle evaporative emissions tests showed how the emissions were affected by the addition of 10 percent ethanol or 15 percent MTBE to gasoline. To obtain a better understanding of these effects, laboratory tests were conducted. These tests were specifically aimed at determining the amount of ethanol or MTBE in the vapor, and whether activated carbon adsorbs ethanol and MTBE more or less readily than hydrocarbons. The first question is important because ethanol has been reported to be moderately reactive in the photochemical formation of smog (18), while MTBE is expected to have a rather low reactivity (9). The relative adsorption of ethanol or MTBE and hydrocarbons on activated carbon is of interest in situations where the canister capacity is insufficient to contain all of the vapor which is generated, and vapor break-through occurs.

The maximum concentration of MTBE permitted in commercial gasolines is now 7 percent. Therefore, the laboratory test program included a blend of 7 percent MTBE in Indolene, in addition to the four fuels used in the vehicle tests.

COMPOSITION OF VAPOR FROM DIURNAL TESTS - The diurnal part of the evaporative emissions test was simulated in the laboratory with 400 mL of fuel in a one-liter round-bottom flask, as described in Appendix B. To determine how well the laboratory test simulated the vehicle test, the mass of fuel vapor generated from each fuel in the laboratory test was determined by adsorption of the vapor on activated carbon. From these laboratory data, using a simple proportion of fuel volumes, an estimate was made of the amount of vapor which would be generated in each of the two test cars. The estimates were then compared with the actual amount of vapor measured during the SHED tests discussed previously.

The comparison is shown in Figure 6, where the diagonal line is the line of perfect correlation. Most of the data points are reasonably close to this line, indicating that the laboratory test adequately simulated the vehicle SHED test.

One of the major objectives of the laboratory work was to determine the amount of ethanol or MTBE in the diurnal vapor. The vapor composition cannot be readily calculated from the liquid composition and the vapor pressures of the pure components, because ethanol forms non-ideal solutions with hydrocarbons, i.e. the solutions do not follow Raoult's law (19). Solutions of MTBE in hydrocarbons are also expected to deviate from ideal behavior because of the difference in polarity of the components (20). Therefore, the vapor compositions of these systems must be determined experimentally.

The vapor generated by the test fuels during the laboratory diurnal tests was collected in Teflon ® * bags containing hydrocarbon-free air, and analyzed by gas chromatography. The results presented in Table 3 are the averages of at least two tests with each fuel. The gasohol produced vapor composed of 7.6 mass-percent ethanol and 92.4 percent hydrocarbons. The adjusted gasohol had 8.6 mass-percent ethanol in the vapor. The higher ethanol content of the vapor from the adjusted gasohol was expected because some of the most volatile hydrocarbons had been removed from the base gasoline to reduce its vapor pressure. With both of the gasohols, the vapor contained a smaller mass fraction of ethanol than the liquid. However, with the MTBE-containing fuels, the vapor contained a larger mass fraction of MTBE than the liquid.

In the absence of evaporative emissions control systems, the ethanol or MTBE content of the diurnal vapor emissions from a vehicle would correspond to that shown in Table 3. However, most of the vapor which is generated in a vehicle's fuel system during the diurnal part of the SHED test is adsorbed and stored on activated carbon. Therefore, it is important to determine how the composition of the vapor might be affected by exposure to activated carbon.

COMPOSITION OF DIURNAL VAPOR FOLLOWING EXPOSURE TO ACTIVATED CARBON - Under some extreme conditions, the capacity of the activated carbon in the canister may not be sufficient to retain all of the vapor, and some of the vapor passes through the canister and is emitted into the atmosphere. If the activated carbon preferentially adsorbs hydrocarbons, the emissions will be richer in ethanol or MTBE than indicated by the data in Table 3. On the other hand, if the polar compounds are preferentially adsorbed, the emissions will contain less ethanol or MTBE. Information available in the literature (21-24)

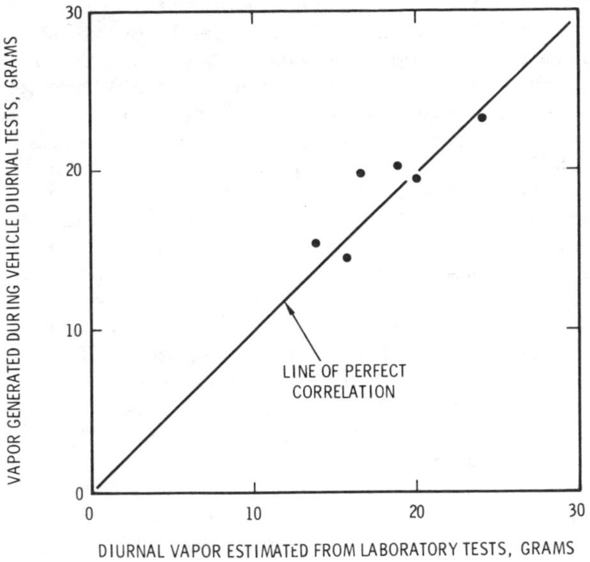

Fig. 6 - Correlation of laboratory and vehicle diurnal test results

Table 3 - Ethanol or MTBE Content of Diurnal Vapor from Laboratory Tests

Fuel	Fuel Additive	Concentration of Additive in Liquid vol. %	mass%	Concentration of Additive in Vapor mass %
Gasohol	Ethanol	10.0	10.7	7.6
Adjusted gasohol	Ethanol	10.0	10.7	8.6
7% MTBE Fuel	MTBE	7.0	7.0	9.0
15% MTBE Fuel	MTBE	15.0	15.1	18.2

suggests that ethanol is adsorbed in preference to some (but not all) hydrocarbons. However, it is important that tests be conducted under conditions which exist in the system of interest.

In our tests to determine how vapor composition is affected under break-through conditions, the vapor generated during the laboratory diurnal simulation was adsorbed on a small quantity of activated carbon. However, the amount of carbon was intentionally insufficient to retain all of the vapor. The vapor which escaped from the carbon was collected in a bag containing air, and was then analyzed. These tests were conducted only with the gasohol and with the 7 percent MTBE fuel. The ethanol and MTBE content of the break-through vapor is shown in Table 4, along with that of the inlet vapor, from Table 3.

The data in Table 4 show that a much smaller proportion of ethanol or MTBE was present in the break-through vapor than in the inlet vapor. Thus, the activated carbon preferentially adsorbed these polar compounds. This result leads to the conclusion that the

* Teflon is a registered trademark of E. I. DuPont de Nemours and Co.

composition of the diurnal evaporative emissions in the SHED test depends on whether the emissions occur because of a leak in the fuel system or because of vapor break-through at the canister. If a leak occurs, the diurnal emissions with gasohol would contain about 7.6 percent ethanol, and the emissions with the 7 percent MTBE fuel would contain about 9 percent MTBE, as reported in Table 3. However, if the emissions are primarily due to canister break-through, they will contain much less ethanol or MTBE, as shown in Table 4.

As shown in Table 4, the break-through vapor from the gasohol contained more of the oxygenated species than the break-through vapor from the MTBE fuel. This is because the gasohol generated more vapor than the MTBE fuel, although the same quantity of activated carbon was used in both tests. Thus, these results do not necessarily mean that the carbon had a greater capacity for MTBE than for ethanol.

RETENTION OF ETHANOL AND MTBE ON ACTIVATED CARBON DURING PURGING - Another aspect of the interaction of the fuel vapor with activated carbon is the desorption process. Even if the carbon retains all of the vapor, and break-through does not occur, some components of the vapor may subsequently be desorbed more easily than others when the canister is purged during engine operation. As a hypothetical example, the ethanol might be so strongly adsorbed that it would not be desorbed during the normal purge cycles. Then, the canister would gradually become saturated with ethanol, so that eventually it would not adsorb a significant amount of hydrocarbons, and evaporative emissions would increase.

In the tests designed to study the retention of ethanol and MTBE on the carbon during purging, an excess of activated carbon was used, so that all of the diurnal vapor was adsorbed. Then, the carbon was purged with hydrocarbon-free air. (See Appendix B for details.) The purged vapor-air mixture was collected in a bag for analysis. The two fuels used for these tests were the gasohol and the 7 percent MTBE fuel.

The concentration of ethanol or MTBE in the purged vapor from any given charge-purge cycle was proportional to the ratio of the mass of vapor purged to the mass of vapor adsorbed during that cycle. The correlation is shown in Figure 7. When the mass of vapor purged in a given cycle was equal to the mass of vapor adsorbed in that cycle, the purged vapor contained the same concentration of ethanol or MTBE as the adsorbed vapor (7.6 percent for ethanol, and 9.0 percent for MTBE). However, if less vapor was purged than adsorbed, the concentration of ethanol or MTBE in the purged vapor was lower. Thus, the activated carbon retained the ethanol or MTBE more tenaciously than the hydrocarbons, and the

Table 4 - Ethanol or MTBE Content of Inlet and Break-through Vapor During Laboratory Simulation of Diurnal Test

Fuel	Fuel Additive	Concentration of Additive in Liquid vol. %	Concentration of Additive in Liquid mass%	Concentration of Additive in Vapor, mass% Inlet	Concentration of Additive in Vapor, mass% Break-through
Gasohol	Ethanol	10.0	10.7	7.6	1.3
7% MTBE Fuel	MTBE	7.0	7.0	9.0	0.2

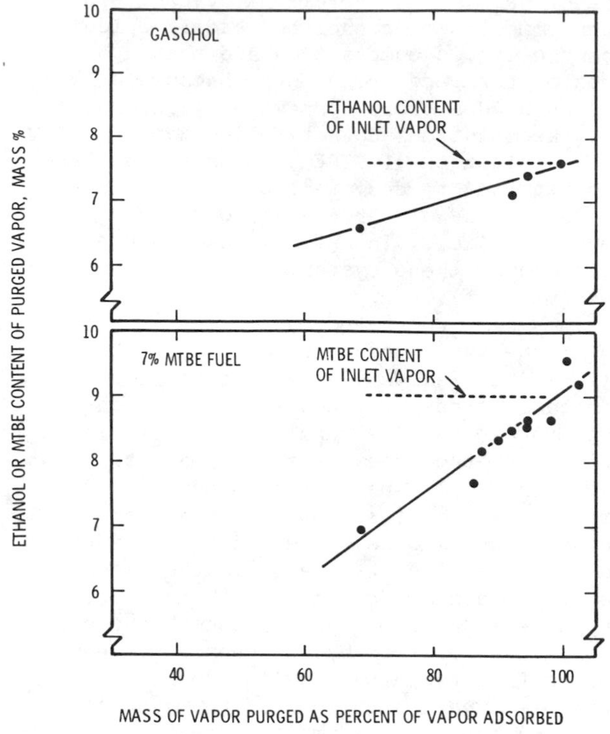

Fig. 7 - Effect of the quantity of vapor purged from activated carbon on the ethanol or MTBE content of the purged vapor

residual vapor* on the activated carbon was enriched in ethanol or MTBE compared with the inlet vapor. It then follows that if <u>more</u> vapor is purged than adsorbed in a given cycle, i.e. some of the residual vapor is also removed, the purged vapor should be richer in ethanol or MTBE than the inlet vapor. In a vehicle, this might occur with prolonged highway driving, with a hot engine compartment and long purge time. This was investigated in two of the laboratory tests with the MTBE fuel by heating the activated carbon during purging.

* Not all of the vapor adsorbed on activated carbon during the first few charge-purge cycles is purged. Some vapor, which is hereafter called the residual vapor, remains on the carbon. Eventually, an equilibrium is reached, where the amount of vapor purged in a given cycle is about equal to the amount adsorbed during that same cycle. This phenomenon is illustrated in Figure 8.

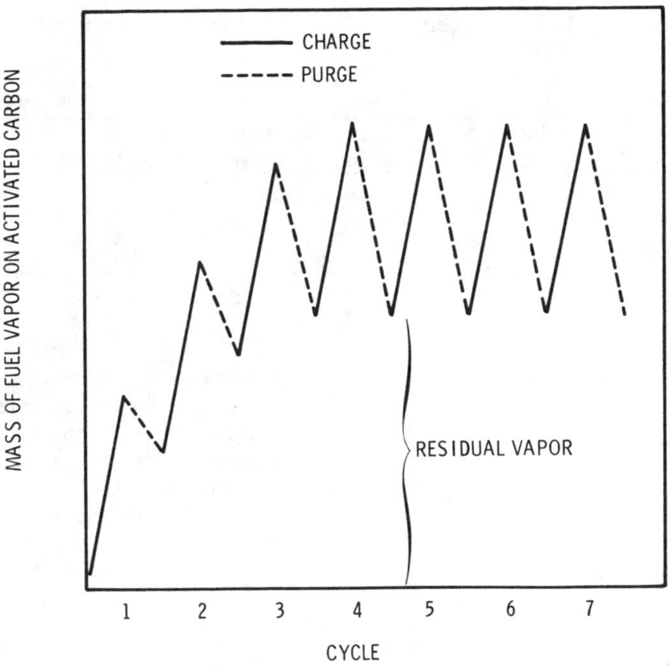

Fig. 8 - Illustrative plot of adsorption and desorption of fuel vapor on activated carbon

Indeed, the purged vapor contained more MTBE than the inlet vapor, as shown in Figure 7.

These studies support the earlier conclusion that activated carbon adsorbs ethanol and MTBE more strongly than some hydrocarbons. However, ethanol and MTBE can be purged, and therefore it does not appear that they will accumulate on the activated carbon to such an extent that the working capacity of the carbon will be severely degraded.

COMPOSITION OF VAPOR FROM HOT SOAK TESTS - The hot soak part of the evaporative emissions test was also simulated in the laboratory by distilling the fuel, as described in Appendix B. The objective of the work was to determine the proportion of ethanol or MTBE in the vapor generated during a hot soak. The tests were run with gasohol and with the 7 percent MTBE fuel. Since carburetor fuel bowl temperatures do not increase at the same rate for all cars, four time-temperature distillation profiles were used with each fuel in the laboratory tests. The mass-percent and volume-percent of fuel distilled during each test were recorded, and the composition of the distilled fuel was determined by gas chromatography. The results are presented in Table 5.

In all of the laboratory hot soak tests, the distillate contained a substantially higher proportion of ethanol or MTBE than the original fuel. The fraction of ethanol or MTBE in the distillate increased with temperaure, as shown in Figure 9.

These laboratory tests have shown that hot soak evaporative emissions from a car, with fuels containing 10 percent ethanol or 7 percent MTBE, will contain a substantially higher proportion of ethanol or MTBE than is present in the fuel. The concentration of ethanol or MTBE in the hot soak emissions will depend upon the maximum fuel temperature in the carburetor. However, if the vapors pass through the activated carbon, the concentration of ethanol or MTBE in the emissions will be reduced.

COMPARISON OF LABORATORY AND SHED RESULTS - To compare the results of the laboratory tests with results from a vehicle SHED test, a complete evaporative emissions test was conducted with a 1978 production car equipped with a 5.7-L V8 engine, 4-barrel carburetor, and a 92.7-L (24.5-gallon) fuel tank. Gasohol was used as the test fuel. Bag samples of the atmosphere in the SHED were taken at the beginning and end of the diurnal and hot soak parts of the test. The contents of the bags were analyzed by gas chromatography to determine the concentration of ethanol in the evaporative emissions.

The concentration of ethanol in the diurnal emissions was only 1.9 percent, which is much lower than the 7.6 percent ethanol in the total diurnal vapor (see Tables 3 and 4). Therefore, the diurnal emissions must have been due primarily to canister break-through. Pressure checks of the fuel tank and lines, both before and after the evaporative emissions test, confirmed the absence of leaks.

The concentration of ethanol in the hot soak emissions was 22 percent, which suggests

Table 5 - Ethanol or MTBE Content of Distillates from Laboratory Hot Soak Tests

	Gasohol			7% MTBE Fuel		
Max. Liquid Fuel Temp., °C	Volume % Distilled	Mass % Distilled	Mass % Ethanol in Distillate	Volume % Distilled	Mass % Distilled	Mass % MTBE in Distillate
68.0	22.2	18.8	15.7	12.2	9.1	12.1
72.0	28.3	25.0	19.7	15.0	11.4	13.4
76.0	34.2	30.8	22.9	17.8	14.6	13.8
83.5	40.5	36.5	25.5	22.8	19.4	15.3

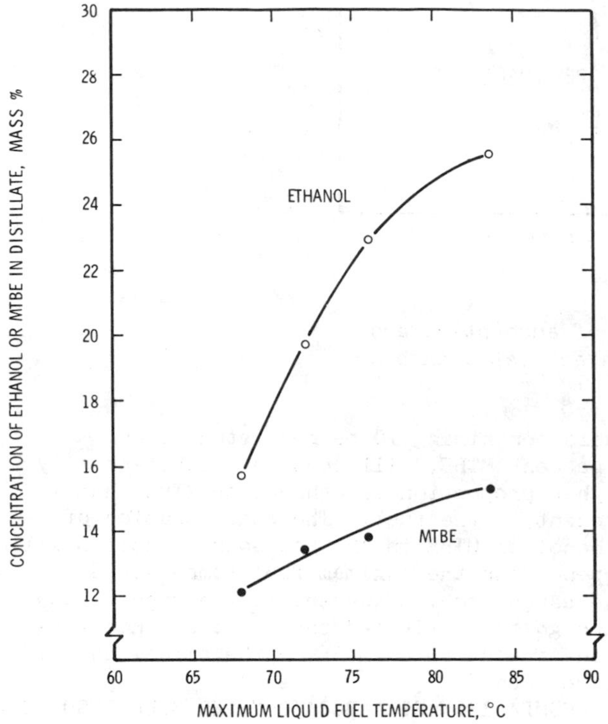

Fig. 9 - Effect of fuel temperature on the composition of the distillates from laboratory hot soak tests with gasohol and 7 percent MTBE fuel

that those emissions were due primarily to leaks, rather than to canister break-through. Leaks may have occurred at several locations in the carburetor, such as the throttle blade shaft, accelerator pump, air horn, and gaskets.

The sources of emissions found in this test, viz. canister break-through for the diurnal emissions and leaks for the hot soak emissions, would probably not be the same for all cars. Thus, the ethanol concentrations reported here may not necessarily be typical of emissions from all cars.

VEHICLE EXHAUST EMISSIONS

Although the primary objective of this test program was to evaluate the effects of ethanol and MTBE on evaporative emissions, the complete vehicle emissions test procedure includes an exhaust emissions test between the diurnal and hot soak parts of the evaporative emissions test. Therefore, exhaust emissions data were obtained. The engine-out (before-catalyst) emissions and tailpipe emissions for the three test cars are listed in Tables C-3, C-4, and C-5 of Appendix C.

The engine-out CO and NO_x emissions and the tailpipe HC, CO and NO_x emissions from the production car without closed-loop fuel control were significantly lower with the gasohol, adjusted gasohol, and 15 percent MTBE fuels than with gasoline. This is due to the leaning effect of ethanol and MTBE, which have lower stoichiometric air-fuel ratios than gasoline.

The absolute changes in exhaust emissions from one fuel to another in the closed-loop dual-bed catalyst car and the closed-loop three-way catalyst car were quite small, even though some of those changes were statistically significant.

FUEL ECONOMY

EPA city fuel economy was calculated from the exhaust emissions results, using the carbon balance method. Details are given in Appendix A. Data were obtained with the same test cars and test fuels used for the exhaust emissions tests. The results are tabulated in Table C-6 of Appendix C, and the effects of the oxygen-containing fuels on fuel economy are summarized in Figure 10. With the addition of 10 percent ethanol or 15 percent MTBE to gasoline, the fuel economy decreased 0.9 to 3.7 percent. All of the decreases except the 0.9 percent reduction were statistically significant. In no case was fuel economy improved with the ethanol or MTBE fuels. This result was expected because the volumetric energy contents of the gasohols and the 15 percent MTBE fuel are 3.4 and 2.8 percent, respectively, below that of the gasoline. These decreases in heating value are within the range of fuel economy changes.

DRIVEABILITY

The oxygen-containing fuels adversely affected the cold-start driveability of the production car at an ambient temperature of 5°C. The driveability demerits were higher with those fuels than with gasoline, as shown in Figure 11. The 15 percent MTBE fuel produced the worst cold-start driveability rating in the production car. This result was unexpected, and was due largely to a stall during a wide-open-throttle acceleration. (In the Coordinating Research Council test procedure, stalls are assigned a relatively large number of demerits.) Only one test was run with each fuel, so test repeatability is not known.

Warmed-up driveability of the production car at an ambient temperature of 15°C was noticeably worse with the oxygen-containing fuels than with the gasoline. Objectionable surge was observed during light-load accelerations from idle. The poorer driveability with the oxygen-containing fuels was most likely due to the leaning effect.

Cold-start driveability tests with the prototype closed-loop dual-bed catalyst car showed slight impairment in driveability with the oxygen-containing fuels. Results are not reported because the car was not tailored to have commercial driveability with gasoline at low temperatures. It is expected, however, that production closed-loop cars will also have poorer cold-start driveability with oxygen-containing fuels than with gasoline because they operate in an open-loop mode for a short time after a cold start. During warmed-up operation, the closed-loop system may compensate for the changes in fuel composition with oxygen-containing fuels.

SUMMARY

Vehicle evaporative and exhaust emissions, fuel economy, and driveability were determined using gasoline, and blends of 10 percent ethanol or 15 percent MTBE in gasoline, as fuels. Laboratory studies of evaporative emissions were also conducted. The major findings are as follows:

1. Evaporative emissions were as much as 51 percent higher with 10 percent ethanol fuel blends than with gasoline. The increase in emissions was due primarily to higher fuel volatility, but an increase in hose permeability caused by the ethanol may also have been partly responsible.

2. Evaporative emissions were as much as 15 percent higher with the 15 percent MTBE fuel than with gasoline.

3. The activated carbon used to control evaporative emissions adsorbed ethanol and MTBE more readily than light hydrocarbons. Therefore, emissions which resulted from overloading the canister contained a

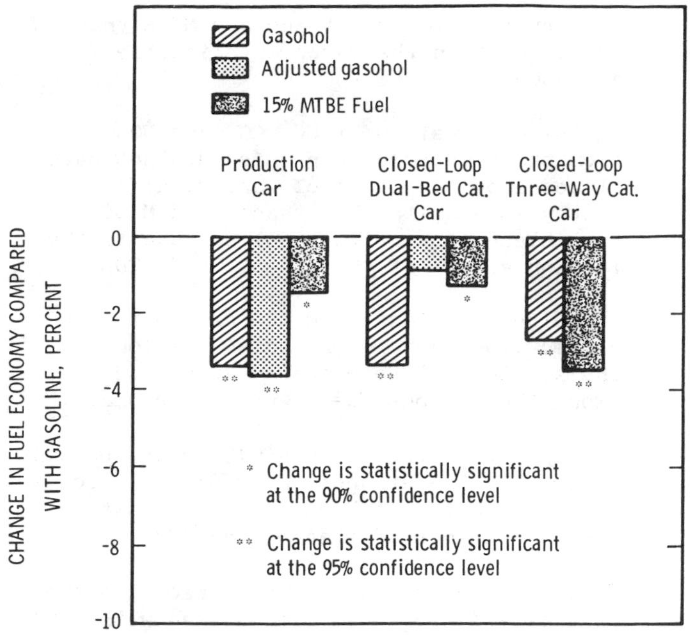

Fig. 10 - Effect of ethanol and MTBE on EPA city fuel economy

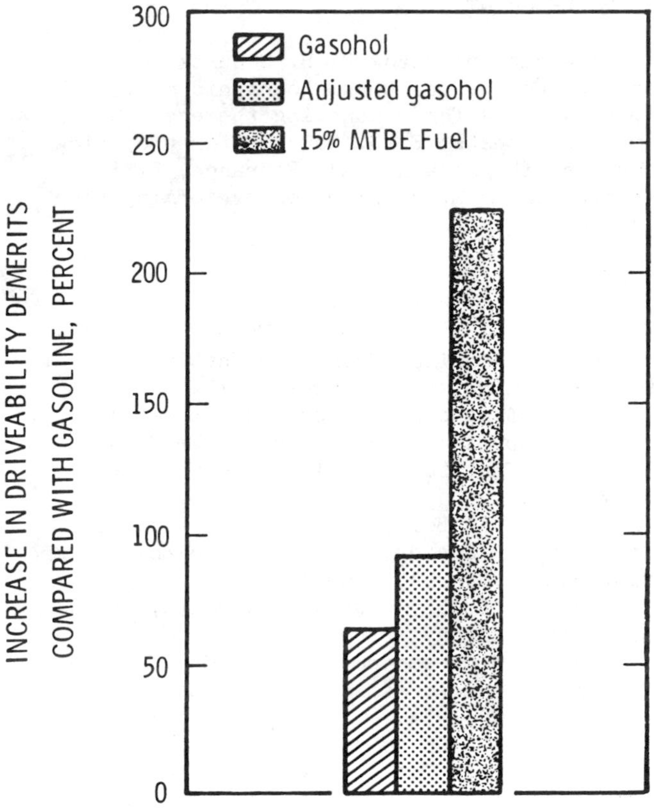

Fig. 11 - Effect of ethanol and MTBE on cold-start driveability of the production car

smaller fraction of ethanol or MTBE than was present in the vapor entering the canister.

4. Tailpipe emissions of HC, CO, and NO_x from the production car which did not have closed-loop fuel control were significantly lower with the ethanol and MTBE fuel blends than with gasoline, due to the leaning effect of the ethanol and MTBE.

5. Tailpipe emissions from the cars which had closed-loop carburetors were very low, and the absolute differences in emissions among the various fuels were quite small.

6. EPA city fuel economy was 0.9 to 3.7 percent lower with the ethanol and MTBE fuel blends than with gasoline because of the lower energy contents of ethanol and MTBE.

7. Cold-start driveability was worse with the ethanol and MTBE fuel blends than with gasoline, probably due to the leaning effect of ethanol and MTBE. The ethanol and MTBE fuels also adversely affected the warmed-up driveability of the car without closed-loop fuel control, but not the car equipped with a closed-loop carburetor.

ACKNOWLEDGMENT

The authors thank P. H. Schultz for performing the gas chromatographic analyses; P. E. Normile for conducting the evaporative emissions tests; E. G. Malzahn for preparing the test fuels; and W. A. Florance, E. S. Halstead, and M. J. Klim for performing the fuel analyses.

REFERENCES

1. "Use of Alcohol in Motor Gasoline - A Review," American Petroleum Institute, Publication No. 4082, August 1971.

2. "Alcohols - A Technical Assessment of their Applications as Fuels," American Petroleum Institute, Publication No. 4261, July 1976.

3. J. A. Bolt, "A Survey of Alcohol as a Motor Fuel," Society of Automotive Engineers Publication SP-254, June 1964.

4. R. W. Reynolds, J. S. Smith, and I. Steinmetz, "Methyl Ethers as Motor Fuel Components," presented at the 168th National Meeting of the American Chemical Society, September 10, 1974.

5. J. D. Chase and H. J. Woods, "Processes for High Octane Oxygenated Gasoline Components," Am. Chem. Soc., Division of Petroleum Chemistry Preprints, 23, 1072 (1978).

6. S. C. Stinson, "New Plants, Processes Set for Octane Booster," Chemical and Engineering News, June 25, 1979, p. 35.

7. R. L. Furey and M. W. Jackson, "Exhaust and Evaporative Emissions from a Brazilian Chevrolet Fueled with Ethanol-Gasoline Blends," Proceedings of the Twelfth Intersociety Energy Conversion Engineering Conference, Vol. 1, p. 54, August 28, 1977.

8. N. D. Brinkman, N. E. Gallopoulos, and M. W. Jackson, "Exhaust Emissions, Fuel Economy, and Driveability of Vehicles Fueled with Alcohol-Gasoline Blends," SAE Preprint 750120, February 1975.

9. R. T. Johnson and B. Y. Taniguchi, "Methyl Tertiary-Butyl Ether - Evaluation as a High Octane Blending Component for Unleaded Gasoline," Am. Chem. Soc., Division of Petroleum Chemistry Preprints, 23, 1083 (1978).

10. M. W. Jackson, "Exhaust Hydrocarbon and Nitrogen Oxide Concentrations with an Ethyl Alcohol-Gasoline Fuel," SAE Publication SP-254, June 1964.

11. Federal Register, Vol. 42, No. 124, June 28, 1977.

12. Federal Register, Vol. 39, No. 200, October 15, 1974.

13. "Driveability Performance of 1977 Passenger Cars at Intermediate Ambient Temperatures - Paso Robles," Coordinating Research Council Report No. 499, September 1978.

14. D. T. Wade, "Factors Influencing Vehicle Evaporative Emissions," SAE Preprint 670126, February 1967.

15. J. R. Dunn and J-P. Sandrap, "Improved Sour Fuel and Fuel Permeability Resistance in NBR Vulcanizates," presented at Scandanavian Rubber Conference, Copenhagen, Denmark, April 2-3, 1979.

16. "Machine Design," Vol. 51, No. 10, May 10, 1979, p. 23.

17. "Characterization Report - Analysis of Gasohol Fleet Data to Characterize the Impact of Gasohol on Tailpipe and Evaporative Emissions," U.S. Environmental Protection Agency, December 1978.

18. A. J. Haagen-Smit and M. M. Fox, "Ozone Formation in Photochemical Oxidation of Organic Substances," Ind. Eng. Chem., 48, 1484 (1956).

19. G. M. Barrow, "Physical Chemistry," McGraw-Hill, New York, 1961, p. 472.

20. S. Glasstone, "Textbook of Physical Chemistry," 2nd ed., D. Van Nostrand Co., New York, 1946, p. 711.

21. F. G. Tryhorn and W. F. Wyatt, "Adsorption. I. Adsorption by Coconut Charcoal from Alcohol-Benzene and Acetone-Benzene Mixtures," Trans. Faraday Soc., 21, 399 (1925).

22. F. G. Tryhorn and W. F. Wyatt, "Adsorption. II. The Adsorption by a Coconut Charcoal of Saturated Vapors of Some Pure Liquids," Trans. Faraday Soc., 22, 134 (1926).

23. C. L. Mantell, "Adsorption," 2nd ed., McGraw-Hill, New York, 1951.

24. J. W. McBain, H. P. Lucas, and P. F. Chapman, "Sorption of Organic Vapors by Highly Evacuated, Activated Sugar Charcoal," J. Am. Chem. Soc., 52, 2668 (1930).

25. EPA-MSAPC Advisory Circular No. 50A, December 16, 1976.

26. S. H. Mick and J. B. Clark, Jr., "Weighing Automotive Exhaust Emissions," SAE Preprint 690523, May 1969.

27. EPA-MSAPC Advisory Circular No. 49A, January 19, 1976.

Appendix A

VEHICLE EMISSIONS AND FUEL ECONOMY TEST PROCEDURES

The official EPA procedures for determining evaporative and exhaust emissions and fuel economy are published in the Federal Register (11, 12). The following discussion provides a brief summary of the procedures, with particular emphasis on how our procedures differed from the official procedures.

EVAPORATIVE EMISSIONS - Prior to the first test with each fuel, the car was driven about 8 km to expose all fuel system parts to the test fuel. Before each test, the activated carbon canister was removed from the car and purged with a vacuum pump for several hours. About 8 L of fresh test fuel was heated slowly in a metal container, and the fuel vapor was collected in the canister. Charging was continued until 2 g of fuel vapor had broken through the activated carbon. This 2-g breakthrough was determined by monitoring the weight of a small container of carbon which was connected by flexible tubing to the bottom of the canister. The canister was then installed on the car. This canister conditioning procedure is not part of the official EPA emissions test procedure, but was used to simulate the condition of a canister when a car is parked for as much as five days before testing (25). It represents a worst-case situation. For the remainder of the preconditioning and testing, our procedure was the same as the official procedure, except that in our tests the canister was removed from the car and weighed before and after each part of the evaporative emissions test.

Correction Factor for Oxygen-Containing Fuels - The mass of evaporative emissions in the SHED was calculated from vapor concentration using the following equation (11):

$$M = \frac{KVCP \times 10^{-4}}{T}$$

where M = mass of emissions in grams
K = 1.2(12 + H/C)
H/C = hydrogen-carbon ratio of emissions
V = net SHED volume in m^3
C = vapor concentration in ppm carbon
P = barometric pressure in kPa
T = absolute temperature in the SHED in °K

The term (12 + H/C) is the average molecular weight per carbon atom, which is calculated from the molecular species of the vapor. The Federal Register (11) uses H/C = 2.33 for the diurnal emissions and H/C = 2.20 for hot soak emissions when using gasoline as the fuel. However, the average molecular species of the vapor is somewhat different when the vapor contains ethanol or MTBE, and a correction should be applied to the emissions results. The correction factor is the ratio of the molecular weight of the vapor with ethanol or MTBE to the molecular weight of the vapor with gasoline alone.

To calculate the correction factor for the 10 percent ethanol fuels, it was assumed that the vapor, if condensed, would also contain 10 volume-percent ethanol. If the density of the condensed hydrocarbon vapor at 20°C is 0.61 g/mL and the density of ethanol is 0.789 g/mL, the vapor contains 4.3 mole-percent ethanol and 95.7 mole-percent hydrocarbons. For the diurnal emissions, the H/C of the hydrocarbon component of the vapor is 2.33. Therefore, the average molecular species of the entire vapor can be represented as $CH_{2.39}O_{0.04}$. The average molecular weight per carbon atom of the vapor containing ethanol is therefore 15.03. Since the average molecular weight of the vapor from gasoline alone ($CH_{2.33}$) is 14.33, the correction factor is the ratio of molecular weights:

$$15.03/14.33 = 1.05$$

All of the evaporative emissions data obtained with the 10 percent ethanol fuels were multiplied by this factor.

In a similar manner, a correction factor of 1.034 was obtained for use with the fuel containing 15 percent MTBE.

EXHAUST EMISSIONS - The exhaust emissions were measured according the 1975 Federal test procedure (11), with a few exceptions. In addition to collecting diluted exhaust gas in bags and measuring the concentration of emissions, the emissions of HC, CO, CO_2, and NO_x were also measured continuously both before the catalytic converter and in the diluted exhaust from the tailpipe. The before-catalyst and after-catalyst analyses were done simultaneously, using two separate banks of analyzers. The continuous measurement technique yields much more detailed information than bag sampling, but the final results are equivalent (26).

Hydrocarbon mass emissions were calculated in the usual manner as the product: (exhaust volume) x (hydrocarbon concentration) x (hydrocarbon density). Exhaust hydrocarbon density was calculated from the ideal gas law and equaled 1.177 times the molecular weight of the exhaust hydrocarbon CH_xO_y species. In agreement with the assumption on which the official EPA calculation procedure is based, it was assumed that the exhaust hydrocarbon CH_xO_y species was the same as the fuel CH_xO_y species. Although the official procedure specifies a hydrocarbon species of $CH_{1.85}$ and a density of 16.33 g/ft^3 for Indolene, the composition of the Indolene used in our tests yielded a $CH_{1.81}$ species with a density of 16.29 g/ft^3. An exhaust species of $CH_{1.89}O_{0.03}$ and a density of 17.00 g/ft^3 were calculated for the fuels containing 10 percent ethanol, while the 15 percent MTBE fuel had an exhaust species of $CH_{1.88}O_{0.02}$ with a density of 16.82 g/ft^3. Rather than changing the computer program to reflect these changes, the hydrocarbon mass emissions were calculated from the values specified in the Federal Register, and the results were multiplied by the following correction factors:

Indolene	0.997
Indolene + 10% ethanol	1.041
Indolene + 15% MTBE	1.030

No correction was made for the low response of the FID hydrocarbon analyzer to ethanol.

FUEL ECONOMY - EPA city fuel economy was calculated by the carbon balance method from the emissions of HC, CO, and CO_2 measured during the exhaust emissions test (12, 27). The following equation was used:

$$MPG = \frac{F \times D}{F \times E_{HC} + 0.4288 \times E_{CO} + 0.2729 \times E_{CO_2}}$$

where MPG = fuel economy in miles per gallon
D = fuel density in grams per gallon
E = exhaust emissions in grams per mile
F = carbon mass fraction of fuel and exhaust hydrocarbon species

The values of F for the various fuels are as follows:

Indolene	0.8681
Indolene + 10% ethanol	0.8317
Indolene + 15% MTBE	0.8405

Appendix B

LABORATORY EVAPORATIVE EMISSIONS TEST PROCEDURES

DIURNAL TEST - A one-liter round-bottom flask containing 400 mL of fuel was used to simulate a fuel tank filled to 40 percent of capacity, as specified for the evaporative emissions test. The flask of fuel was cooled to about 15°C in a water bath and then removed from the bath. When the fuel temperature rose to 15.6°C, the flask was connected to the vapor collection device, and the heating was started.

The flask of fuel was heated in a heating mantle controlled by a manually operated variable transformer. The fuel was heated from 15.6° to 29°C at a uniform rate of 1.1°C every 5 minutes, as specified for the emissions test. The fuel vapor was collected either in a bag or on activated carbon. For bag collection, a 25 x 25-cm Teflon® bag was previously charged with about 2.5 L of hydrocarbon-free, dry air. Although the capacity of the bag was 4.7 L, it was filled to only about half its capacity to avoid pressurizing the bag, which would affect the vaporization of the fuel. The fuel vapor was carried from the flask to the bag by a syringe needle which passed through a Teflon® septum in the bag.

When activated carbon was used to collect the fuel vapor, the carbon was first dried in an oven at 120°C for 2 hours, then cooled in a desiccator. The activated carbon is the same type presently used in most General Motors evaporative emissions control canisters. Thirteen mL (3.6 g) of activated carbon was placed in a 14-mm i.d. glass tube and held in place with glass wool. This quantity of carbon had more than enough capacity to adsorb all of the vapor generated from any of the fuels in the laboratory diurnal test. For the vapor break-through tests, 1.2 g (about 4 mL) of activated carbon was used. The container of activated carbon was connected to the flask of fuel by a short U-shaped piece of glass tubing.

After each charging cycle, the activated carbon was purged with hydrocarbon-free, dry air at the rate of 10 bed volumes per minute for 23 minutes. (One bed volume is the volume of charcoal in the container, either 13 mL or 4 mL for these tests.) The purge air flow rate was monitored by a calibrated rotameter. The direction of purge flow was opposite that of charging. When the purge air-vapor mixture was to be collected in a bag for analysis, the tube of activated carbon was connected to the bag by a syringe needle.

HOT SOAK TEST - The vaporization of fuel from a carburetor during a hot soak was simulated in the laboratory by heating a 100-mL round-bottom flask containing 90 mL of fuel. The flask was completely immersed (except for the neck) in a beaker of silicone oil which was heated and stirred by a hot plate magnetic stirrer. The distilled fuel was collected in a trap cooled in a dry ice-acetone bath. A calcium chloride drying tube was used on the exit arm of the trap to prevent moisture from being drawn into the trap. Before the test began, the tubing and trap were purged for several minutes with dry air. By taking these precautions to eliminate water from the system, the ethanol-hydrocarbon distillate collected in the trap was a homogeneous solution, even at dry ice temperature (-78°C).

The carburetor fuel temperature as a function of time during the one-hour vehicle hot soak test is not specified by regulation, because the temperature in each car is controlled by the engine and vehicle design. For the laboratory tests to properly simulate the vehicle tests, a realistic time-temperature profile must be used. Therefore, carburetor fuel temperatures were measured in two cars during a one-hour hot soak following a 1975 FTP exhaust emissions test. Car A was equipped with a 5.7-L V8 engine and 4-barrel carburetor, while Car B had a 4.3-L V8 engine and 2-barrel carburetor. (These were not the same cars described in Table 1.) Car A was tested twice -- once at an ambient temperature of 21°C and once during a SHED test where the SHED temperature was 31°C. Car B was tested only at 21°C.

The data in Table B-1 show that the maximum carburetor fuel temperatures were 83.5°, 76°, and 72°C, and these maximum temperatures occurred 45 to 50 minutes after the start of the hot soak. With some engines, however, the fuel temperature might not reach even 72°C, because of more rapid heat dissipation from the intake manifold and better insulation of the carburetor from the manifold. Therefore, another time-temperature profile was assumed, with a maximum temperature of 68°C, and it is also shown in Table B-1. Those four time-temperature profiles were used in the laboratory simulation of the hot soak test, and they are identified in the body of the report by the maximum temperatures.

Table B-1 - Carburetor Fuel Temperatures During a 1-Hour Hot Soak

Hot Soak Time, min.	Fuel Temperature, °C: Car A*	Car A**	Car B**	Other (assumed)
0	47	41	46.5	40
5	57	50.5	50.5	46
10	64.5	58	56.5	51
15	69.5	64	61.5	55
20	73.5	68	65	59
25	76.5	71	68	62
30	79.5	73.5	70	65
35	81.5	75	71	67
40	82	75.5	71.5	67.5
45	83	76	72	68
50	83.5	76	71.5	67.5
55	83	75.5	71	67
60	82	74.5	70.5	66

* 31°C ambient temperature
** 21°C ambient temperature

GAS CHROMATOGRAPHIC ANALYSIS - The analyses to determine the fraction of ethanol or MTBE in the fuel vapor were done with a Perkin Elmer Model 3920 dual-column (compensating) gas chromatograph, which was connected to a Hewlett-Packard Model 3380A integrator. The columns were 3.05 m x 3.2 mm stainless steel, packed with 15 percent 1,2,3-tris(2-cyanoethoxy)propane on 100/120 Chromasorb PAW. The column temperature and programming rate for each type of analysis were selected to adequately separate the ethanol or MTBE from the hydrocarbons. The gas chromatograph was calibrated with liquid mixtures of ethanol or MTBE and toluene, diluted with m-xylene.

Appendix C

EMISSIONS AND FUEL ECONOMY DATA

Table C-1 - Evaporative Emissions from a 1978 Production Car

	SHED Evaporative Emissions, g			Canister Weight Gain, g		Total Vapor Generated, g	
	Diurnal	Hot Soak	Total	Diurnal	Hot Soak	Diurnal	Hot Soak
Gasoline	1.00	1.59	2.59	18.78	8.65	19.78	10.24
	1.16	1.64	2.80	18.67	8.30	19.83	9.94
	1.21	1.60	2.81	18.97	8.21	20.18	9.81
	1.61	1.92	3.53	19.06	7.73	20.67	9.65
	1.26	1.80	3.06	17.86	9.50	19.12	11.30
Average	1.25	1.71	2.96	18.67	8.48	19.92	10.19
Gasohol	1.51	2.14	3.65	20.74	13.48	22.25	15.62
	1.83	2.44	4.27	22.60	13.61	24.43	16.05
Average	1.67	2.29	3.96	21.67	13.55	23.34	15.84
% Change*	+33.6	+33.9	+33.8	+16.6	+59.8	+17.2	+55.4
Adjusted gasohol	1.38	2.74	4.12	18.80	11.33	20.18	14.07
	1.65	3.18	4.83	18.94	10.10	20.59	13.28
Average	1.52	2.96	4.48	18.87	10.72	20.39	13.68
% Change*	+21.6	+73.1	+51.4	+1.6	+26.4	+2.4	+34.2
15% MTBE Fuel	1.69	1.90	3.59	20.90	11.60	22.59	13.50
	1.47	1.76	3.23	19.03	13.97	20.50	15.73
Average	1.58	1.83	3.41	19.97	12.79	21.55	14.62
% Change*	+26.4	+7.0	+15.2	+7.5	+50.8	+8.2	+43.5

* Change relative to gasoline.

Table C-2 - Evaporative Emissions from a Prototype Car

	SHED Evaporative Emissions, g			Canister Weight Gain, g		Total vapor Generated, g	
	Diurnal	Hot Soak	Total	Diurnal	Hot Soak	Diurnal	Hot Soak
Gasoline	1.15	2.22	3.37	13.86	6.14	15.01	8.36
	1.06	1.92	2.98	14.83	5.13	15.89	7.05
	1.38	1.57	2.95	13.20	6.98	14.58	8.55
	1.18	1.55	2.73	14.73	7.26	15.91	8.81
Average	1.19	1.82	3.01	14.16	6.38	15.35	8.19
Gasohol	1.45	1.64	3.09	18.11	9.90	19.56	11.54
	1.67	1.80	3.47	18.10	8.00	19.77	9.80
Average	1.56	1.72	3.28	18.11	8.95	19.67	10.67
% Change*	+31.1	-5.5	+9.0	+27.9	+40.3	+28.1	+30.3
Adjusted gasohol	1.15	2.16	3.31	14.00	7.00	15.15	9.16
	1.67	2.64	4.31	12.13	8.73	13.80	11.37
Average	1.41	2.40	3.81	13.07	7.87	14.48	10.27
% Change*	+18.5	+31.9	+26.6	-7.7	+23.4	-5.7	+25.4
15% MTBE Fuel	1.33	1.61	2.94	15.32	10.48	16.65	12.09
	1.47	1.70	3.17	14.52	11.27	15.99	12.97
Average	1.40	1.66	3.06	14.92	10.88	16.32	12.53
% Change*	+17.6	-8.8	+1.7	+5.4	+70.5	+6.3	+53.0

* Change relative to gasoline.

Table C-3 - Exhaust Emissions from a 1978 Production Car

	Tailpipe Emissions, g/mi			Engine-Out Emissions, g/mi		
	HC	CO	NO_x	HC	CO	NO_x
Gasoline	.59	8.61	2.08	2.87	10.57	1.83
	.76	10.68	1.68	2.88	12.04	1.75
	.82	8.51	1.64	2.81	9.99	1.52
	.54	7.18	1.85	2.53	9.30	1.54
	.59	7.99	1.80	2.64	10.15	1.55
	.57	8.14	1.70	2.65	10.32	1.53
Average	.65	8.52	1.79	2.73	10.40	1.62
Gasohol	.53	6.33	1.39	3.05	8.21	1.21
	.52	5.92	1.39	2.98	7.85	1.19
Average	.53	6.13	1.39	3.01	8.03	1.20
Adjusted gasohol	.52	6.22	1.30	3.02	7.76	1.13
	.51	5.59	1.31	2.96	7.39	1.13
	.50	5.38	1.35	2.97	7.72	1.16
Average	.51	5.73	1.32	2.98	7.62	1.14
15% MTBE Fuel	.54	6.73	1.34	3.17	8.16	1.24
	.50	5.88	1.46	3.04	7.85	1.25
Average	.52	6.31	1.40	3.11	8.01	1.25

Table C-4 - Exhaust Emissions from a Prototype Closed-Loop Dual-Bed Catalyst Car

	Tailpipe Emissions, g/mi			Engine-Out Emissions, g/mi		
	HC	CO	NO_x	HC	CO	NO_x
Gasoline	.20	1.22	.62	2.24	14.70	1.28
	.18	.71	.53	2.23	14.13	1.24
	.22	1.35	.63	2.16	14.37	1.32
	.19	.93	.51	2.25	14.41	1.17
Average	.20	1.05	.57	2.22	14.40	1.25
Gasohol	.20	.80	.54	2.28	13.49	1.13
	.20	1.02	.52	2.31	14.13	1.11
Average	.20	.91	.53	2.30	13.81	1.12
Adjusted gasohol	.21	1.14	.55	2.08	12.36	1.23
	.24	1.65	.58	2.00	13.10	1.26
Average	.23	1.40	.57	2.04	12.73	1.25
15% MTBE Fuel	.16	.86	.64	2.13	13.21	1.31
	.16	1.09	.62	2.05	13.07	1.29
Average	.16	.98	.63	2.09	13.14	1.30

Table C-5 - Exhaust Emissions from a Prototype Closed-Loop Three-Way Catalyst Car

	Tailpipe Emissions, g/mi			Engine-Out Emissions, g/mi		
	HC	CO	NO_x	HC	CO	NO_x
Gasoline	.23	2.63	.83	2.62	16.70	1.86
	.27	2.64	.78	2.70	15.97	1.82
	.28	2.69	.84	2.68	16.08	1.94
Average	.26	2.65	.82	2.67	16.25	1.87
Gasohol	.31	2.84	.86	2.70	16.19	1.75
	.32	3.14	.91	2.72	15.97	1.75
Average	.32	2.99	.89	2.71	16.08	1.75
15% MTBE Fuel	.29	2.88	.84	2.69	15.62	1.71
	.27	2.86	.87	2.91	15.53	1.77
Average	.28	2.87	.86	2.80	15.58	1.74

Table C-6 - EPA City Fuel Economy for Three Cars

	Fuel Economy, miles/gallon		
	Production	Dual Cat.	3-Way Cat.
Gasoline	14.46	16.09	15.48
	14.40	15.99	-
	14.54	-	-
	14.65	16.30	15.50
	14.81	16.02	15.89
	14.72	-	-
Average	14.60	16.10	15.62
Gasohol	14.11	15.65	15.26
	14.09	15.47	15.13
Average	14.10	15.56	15.20
% Change*	-3.4	-3.4	-2.7
Adjusted gasohol	14.11	15.89	-
	14.00	16.00	-
	14.08	-	-
Average	14.06	15.95	-
% Change*	-3.7	-0.9	-
15% MTBE Fuel	14.52	15.97	15.02
	14.23	15.81	15.14
Average	14.38	15.89	15.08
% Change*	-1.5	-1.3	-3.5

* Change relative to gasoline.

Surface Ignition Initiated Combustion of Alcohol in Diesel Engines - A New Approach*

B. Nagalingam, B. L. Sridhar,
N. R. Panchapakesan, K. V. Gopalakrishnan
and B. S. Murthy
Internal Combustion Engines Lab.
Indian Inst. of Technology
(Madras/India)

ABSTRACT

The self-ignition temperature of alcohols is so high that abnormally high compression ratios would be required to use them in conventional diesel engines. This paper presents a novel approach of force igniting methanol or ethanol alone in a diesel engine at normal compression ratios. The well established proneness of methanol to pre-ignite in SI engine is made use of in the present method by employing a heated and insulated surface to initiate ignition. A conventional single cylinder diesel engine was modified to work on this principle. The engine operates satisfactorily at the rated speed (1500 RPM) on methanol and ethanol with thermal efficiencies comparable to the normal diesel engine of the same configuration. The operational experience further shows that it is possible to design a self-sustaining hot surface to initiate ignition. The engine also exhibits multi-fuel capability. A new direction for the use of methanol in diesel engines can follow from this technique.

THE EVER-INCREASING COST as well as rapid depletion of petroleum fuels have led to an intensive search for alternative fuels to power Internal Combustion Engines. Methanol and Ethanol offer the most promising substitutes for petroleum as engine fuels at present. These alcohols can be readily made from a number of non-petroleum sources. Methanol can be produced from coal, a relatively abundant fossil fuel. Technology for methanol production in industrial quantities from coal is already available. Ethanol can be produced by fermentation of carbohydrates which occur naturally and abundantly in some plants like sugar cane and from starchy materials like corn and potatoes. Hence, these fuels can be produced from highly reliable and long lasting raw material sources.

The important properties of methanol and ethanol are given in Table 1 (1)+ with the properties of gasoline also being given for comparison. Both these alcohols have high knock resistance which makes them excellent spark ignited (SI) engine fuels. Their latent heat of vaporisation is high leading to

+ Numbers in parentheses designate references at the end of the paper.

*Paper 800262 presented at the Congress and Exposition, Detroit, Michigan, February 1980.

SAE/PT-80/19/$02.50

Table 1 - Important Properties of Alcohols and Gasoline

Property	Methyl Alcohol	Ethyl Alcohol	Gasoline+	Units
Density	0.795	0.790	0.74-0.75	g/cc
Latent heat of vaporisation	265	216	70-100	kcals/kg
Lower calorific value	4700	6400	10500	kcals/kg
Mixture heating value	734	711	714	kcals/kg-air
Stoichiometric air requirement	6.4	9.0	14.9	kg-air/kg-fuel
Ignition limits (air-fuel ratio)	2.15-12.8	3.5-17.0	6.0-22.0	kg-air/kg-fuel
Self-ignition temperature	478	420	300-450	°C
Research Octane Number	>110	>100	92-98	--
Motor Octane Number	92	89	84-88	--
Cetane Number	3	8	8-14	--

+ The values would vary to some extent depending on the composition of a particular sample of gasoline.

excellent volumetric efficiency but also to starting problems. The calorific value of alcohols is substantially lower than that of gasoline, which would lead to a larger fuel tank volume and also increased jet sizes in carburettor. Their low air requirements for complete combustion, however, compensates this quality and leads to practically the same mixture enthalpy content. Hence, alcohol fueled SI engines can produce the same or even slightly higher power than on gasoline.

However, due to their low cetane rating and high self-ignition temperature, alcohols make poor compression ignition (CI) engine fuels without the use of auxiliary fuels. Considering the diesel's advantages as regards thermal efficiency and part load fuel economy, it would be highly desirable to have a CI engine working on alcohol alone. This task is of particular importance to the developing countries since the CI engine plays an important role in their economy. In this paper, the development and testing of an engine using methanol or ethanol alone at diesel engine compression ratios, employing assisted ignition through a heated and insulated surface initially supplied with electrical energy for heating is described.

DIFFICULTIES IN USING ALCOHOLS IN DIESEL ENGINES

Relatively little work has been done on the use of alcohols in CI engines. The self-ignition temperature of alcohols is so high that abnormally high compression ratios would be required for their use in conventional CI engines. Nevertheless, the basic attractiveness of the diesel engine has led to persistent efforts to utilise alcohols in diesel engines. One solution tried has been the mixing of ignition accelerators to alcohols (2,3). Economic production of ignition accelerators in sufficient quantities is the requirement for any large scale introduction of this method. Spark ignition diesels demonstrated by MAN and Deutz is the second possible method of using alcohols in a CI engine. However, it will be difficult to achieve satisfactory reliability for a commercial engine of this type because of the requirement of higher ignition voltage for the high compression ratio of the diesel engine. There would be frequent need of service like spark plug replacement etc. (3).

The final alternative for the use of alcohols in CI engine is the well-known dual fuel approach employing different methods. In one method, alcohol-air mixture is inducted through a carburettor, supplying the major portion of the heat release and is ignited by a pilot spray of diesel oil injection (4,6). Another method is to inject methanol into the combustion chamber after the diesel fuel injection in order to avoid the methanol cooling of the charge in the cylinder to a degree which would jeopardise the ignition of the diesel fuel (3,5). However, this design calls for two complete and separate fuel systems with tank, feed pump, injection pump and injectors. The regular injection system is used for the methanol and must be modified for this purpose. In an earlier work (6) the present authors have investigated some of the factors which affect the performance of an alcohol-diesel oil dual fuel engine, in which alcohol forms the principal fuel through carburetion. The effectiveness of injection timing, compression ratio and ignition accelerating additives on the performance of the engine have been assessed on the basis of thermal efficiency, degree of use of alcohol, combustion characteristics and exhaust emissions. The results show that it is possible to derive as much as 70-80 per cent of total energy requirement of the engine from alcohol for most of the load range. Turbo-charged diesel engines have also been tried out with this concept at the University of Minnesota (7). Air compression supplies the heat of vaporisation necessary for good fuel distribution. The resulting charge cooling benefits power output and engine life.

All the dual fuel systems described above have the basic disadvantage of requiring two different types of fuels and associated components. Hence in the present work effort was concentrated on developing a CI engine working on alcohols alone. Since alcohols have a high tendency to pre-ignite in SI engines, it was decided to make use of this property in this engine by using a hot surface to initiate ignition.

THE SURFACE IGNITION PHENOMENON

Before and during the second world war, it was found that piston failure in many aero-engines was due to pre-ignition. Ricardo (8) as early as in 1942, started research into the effect of fuel type on the tendency to pre-ignite. A special 'Pre-igniter' whose temperature could be controlled was fitted in the cylinder on the side opposite to the spark plug in a E6 variable compression engine. He found that the xylenes are the least prone to pre-ignite, even better than iso-octane, whereas methanol is very prone to pre-ignition. Bowditch and Yu (9) also carried out investigation on hot spot ignition resistance of fuels in order to understand deposit mechanism. They used a chromel wire of 0.04 inch diameter to simulate the hot spot in an engine. They found that for surface ignition to occur, the fuel-air mixture must be heated above its ignition temperature and at the same time its composition must be within the limits of flammability. While studying the deposit ignition resistance rating of fuels, they found that methanol has a very low resistance to deposit ignition. With a high mixture temperature (by increasing compression ratio), the surface temperature required to heat the mixture to its ignition point became lower.

Another very interesting work relevant to the present paper is the computer model developed by Browning et al (10) to study the surface ignition phenomena in methanol fueled SI engines. Their work showed that methanol dissociates easily through formaldehyde to carbon monoxide and hydrogen. The hydrogen then breaks down into various radical species which trigger pre-ignition and early combustion. From experimental engine studies, they confirmed that methanol pre-ignites at a lower surface temperature than gasoline.Since the heat of dissociation of methanol to carbon monoxide and hydrogen is only 13 per cent of methanol's lower calorific value (11), it is easily seen why methanol has a comparatively low surface ignition temperature. Browning et al also studied in detail the effects of the temperature of the hot surface, cylinder pressure, equivalence ratio, presence of diluents such as recirculated exhaust gas and water addition on the surface ignition process of methanol. They found that the temperature of the hot surface and the pressure of the gas mixture were the primary variables in pre-ignition. Pressure also had a strong effect on minimum ignition delay. As the pressure was increased the minimum surface temperature for ignition decreases.

The same phenomenon was observed by Livengood (1) as shown in Fig.1. The tests were carried out on a motored CFR engine with the cylinder wall temperature being varied. The surface temperature required to initiate pre-ignition is seen to decrease markedly with increased compression temperature. Furthermore, methanol is found to have a greater tendency than iso-octane to pre-ignite. Thus there is wide agreement that methanol has high knock resistance but also very poor resistance to pre-ignition. This easy tendency of methanol to pre-ignite is undoubtedly a disadvantage in SI engines resulting in increased heat flow to the piston and

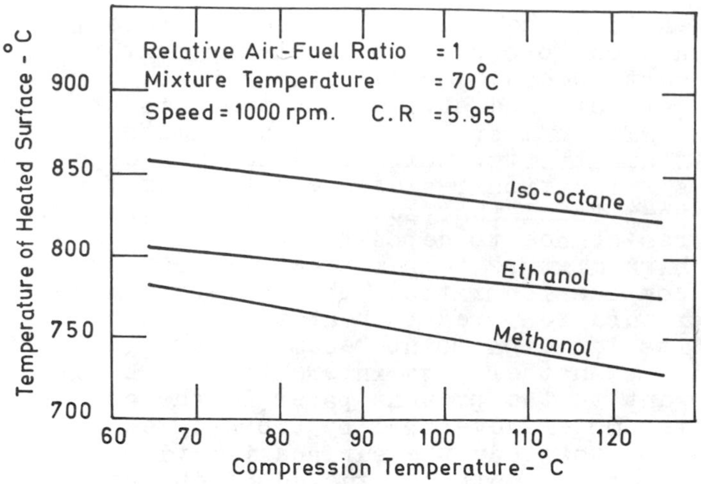

Fig. 1 - Surface temperature for occurrence of pre-ignition as a function of compression temperature (1)

cylinder walls. But it was reasoned that this same disadvantage of methanol can be turned into an advantage in CI engines by initiating the ignition of the sprayed methanol on a heated and insulated surface in the combustion chamber. The design and operation of the engine developed on this line is described below.

CONCEPT OF SURFACE IGNITION ENGINE

The basic concept of the system is as follows. A slab of insulator material, wound with a few strands of heating wire is fixed in the combustion chamber with the wire running on the face exposed to the gases. The fuel injector is located such that a part of the spray impinges on this surface. Ignition is thus initiated. The combustion chamber is made relatively narrow so that combustion spreads quickly to the rest of the space. Since a part of the fuel burns on the insulator surface and since the heat losses from the plate are low, the surface, after some minutes of operation reaches a temperature sufficient to initiate ignition without the aid of external electrical supply.

Before trying out this concept on an engine, bench tests were carried out to study its feasibility. Fuel was injected at the desired frequency by a motor-driven diesel injection pump. Asbestos cement surfaces wound with a few strands of 'Kanthal' (Cr = 20 %, Co = 1.5 - 3 %, Al = 5-6 %) heating element wire of 0.5 mm thickness were placed at various distances and angles from the injector. The frequency of injection was varied from 200-600 times/min. Methanol was used as the fuel.

It was found, by qualitative judgement, that the best results were obtained when the heated surface was flat, with the coil temperature about 1000°C and with the fuel sprayed from a pintle nozzle of cone angle less than 15° (in the open). It was also found that there has to be an optimum interaction between the periphery of the expanding spray and the heated surface. Both excessive interference and too little interference between the surface and the spray led to failure of ignition. The combustion was also seen to be more complete when the burning mixture was confined by a tube immediately after the heated surface.

Based on the findings of the bench tests a diesel engine was converted to work on this process. A conventional, open chamber, single cylinder diesel engine developing 5 H.P. at 1500 RPM was used for this investigation. Details regarding the engine and ambient conditions are given in the Appendix. The original injector was removed and in its place a specially designed plug was fitted. Fig.2 shows the construction of this plug. The plug had a square inner section.

One of the inner walls of the plug was used to mount a rectangular piece of cement-asbestos, which carried a few strands of 'kanthal' heating element wire on the face exposed to the gases. Asbestos was chosen since the right quality of ceramics were not available. The asbestos pieces last about 6-10 hrs. of engine operation. However, for ultimate use, ceramics will have to be employed. A pintle nozzle of spray angle 12° (in the open) was used. The power consumption of the coil was 48 W at 6 V. This was sufficient for all the engine conditions tested.

EXPERIMENTS

The engine test set up is shown in Fig.3 schematically. A photographic close up of the surface ignition chamber is given in Fig.4. The power output was measured by a swinging field dynamometer with Ward-Leonard control. Fuel consumption and exhaust temperature were measured with standard instrumentation. The gas pressure in the cylinder was measured by a Kistler (701 A) piezo-electric pressure transducer.

The pressure-crank angle diagram was displayed on a Tektronix 501 A, dual beam oscilloscope. Photographs of the pressure-crank angle diagrams were taken with C27 Tektronix Polaroid Camera. Peak pressure, delay period and rate of pressure rise were measured from the photographed traces.

The main fuel used for the investigations was methanol. However, the engine, due to the powerful aid to ignition provided by the heated surface, was also able to burn ethanol and gasoline, thus exhibiting a true multi-fuel capability. The engine has a normal compression ratio of 16.5. When the plug was fitted the compression ratio fell to 8.84 on account of the extra space in the plug. When the part-spherical cavity present in the piston was filled up the compression ratio increased to 14.65. Tests were carried out with both 8.84 and 14.65 compression ratios to study their effect on the engine performance. Variable load tests were conducted at the constant speeds of 1000, 1500 and 2000 rpm. The injection timing for all the tests was kept constant at 31° before TDC, static setting.

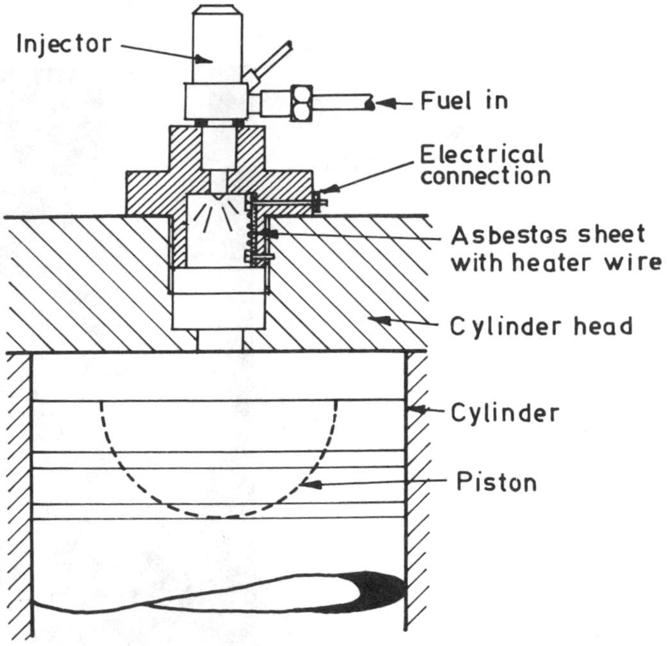

Fig. 2 - The surface ignition plug mounted on engine

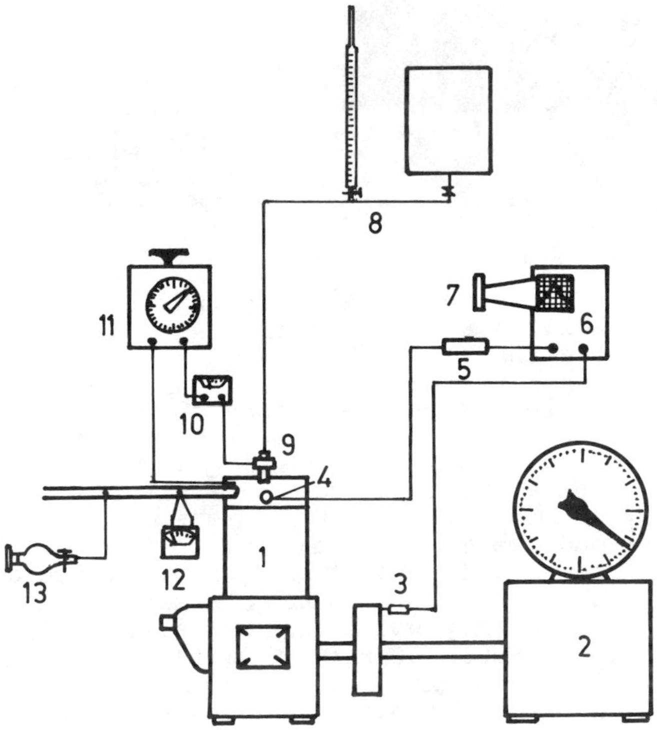

1 Engine
2 Electrical dynamometer with Ward-Leonard control
3 Crank-angle signal pick-up
4 Pressure pick-up
5 Charge amplifier
6 Oscilloscope
7 Oscilloscope camera
8 Fuel flow line with tank and burette
9 Specially designed plug with injector and heated surface
10 Ammeter
11 Auto-transformer
12 Exhaust temperature thermocouple and meter
13 Trichter flask for exhaust sample collection for Aldehyde measurement by MBTH method

Fig. 3 - Schematic diagram of test set-up

Fig. 4 - Close up view of surface ignition chamber

RESULTS AND DISCUSSIONS

ENGINE PERFORMANCE WITH COMPRESSION RATIO OF 8.84 - The brake thermal efficiency of the engine when operating on methanol is shown in Fig.5. The gradual rise and fall of the efficiency with increasing output is noticeable. As the calorific value of methanol is roughly half that of diesel oil the expected output of the engine would be 2.5 H.P. at 1500 RPM, since the fuel injection pump capacity is designed for diesel oil. It can be seen that engine more than meets this expectation. The thermal efficiency is seen to generally decrease with increasing speed. Fig.6 shows the variation of delay period, peak pressure and rate of pressure rise with output. Higher speeds produce longer delay periods, lower peak pressures and lower rates of pressure rise, clearly indicating slowing down of combustion with increasing speed. Two of the contributing reasons could be the shape of the combustion chamber formed in this particular case and the narrow connecting passage between the chamber and the cylinder causing throttling losses. With a more suitable combustion chamber shape, optimised spray pattern and the introduction of controlled turbulence in the chamber it is expected that the combustion rate and thermal efficiency would improve substantially.

ENGINE PERFORMANCE WITH THE COMPRESSION RATIO OF 14.64 - For the next series of tests, the piston cavity was filled up, raising the compression ratio to 14.65. The thermal efficiency is shown in Fig.7. Rather surprisingly, there is no substantial improvement in the thermal efficiency in spite of the large increase in compression ratio. The reasons for this are not clear at

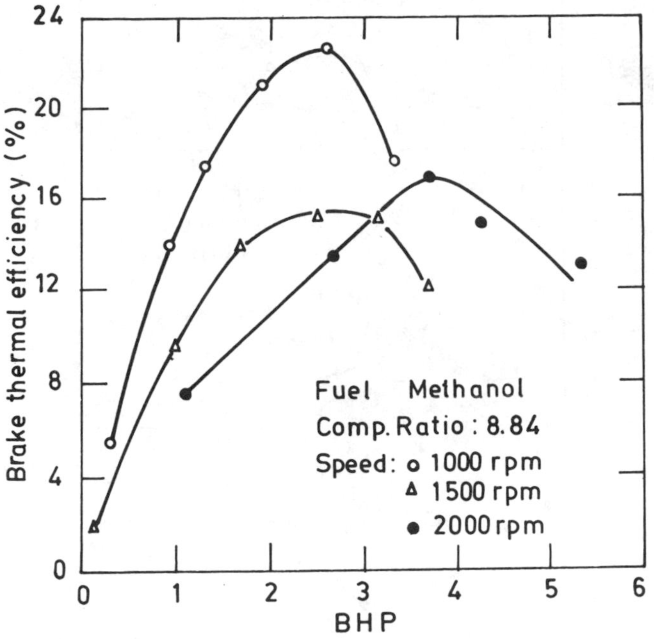

Fig. 5 - Variation of thermal efficiency

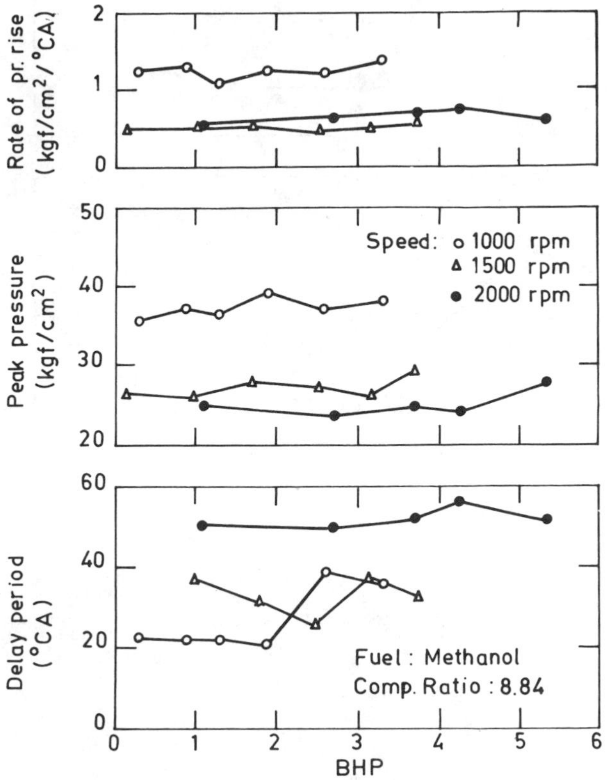

Fig. 6 - Variation of delay period, peak pressure and rate of pressure rise

the moment. Otherwise, the general trend is the same as with the lower compression ratio. At the speed of 2000 RPM it proved impossible to operate the engine with low fuel flow rates. The reason for this also has to be studied.

Fig.8 shows the variation of delay period, peak pressure and rate of pressure rise with the output. As is to be expected, the delay period decreased and peak pressure increased with the increased compression ratio. There is no significant change in the rate of pressure rise.

COMPARISON OF PERFORMAÑCE ON DIFFERENT FUELS - Preliminary runs showed that this surface ignition engine was capable of operating with ethanol and gasoline also. Hence, some comparison tests were made.

The thermal efficiency of the engine with the compression ratio of 8.84 is found to be essentially the same with methanol and ethanol (Fig.9). The delay period (Fig.10) is found to be higher with ethanol. This could be due to the greater proneness of methanol to pre-ignite from a hot surface. At this low compression ratio gasoline could not be ignited.

Fig.11 shows the variation of thermal efficiency between gasoline, ethanol and methanol at the compression ratio of 14.65. Gasoline is found to

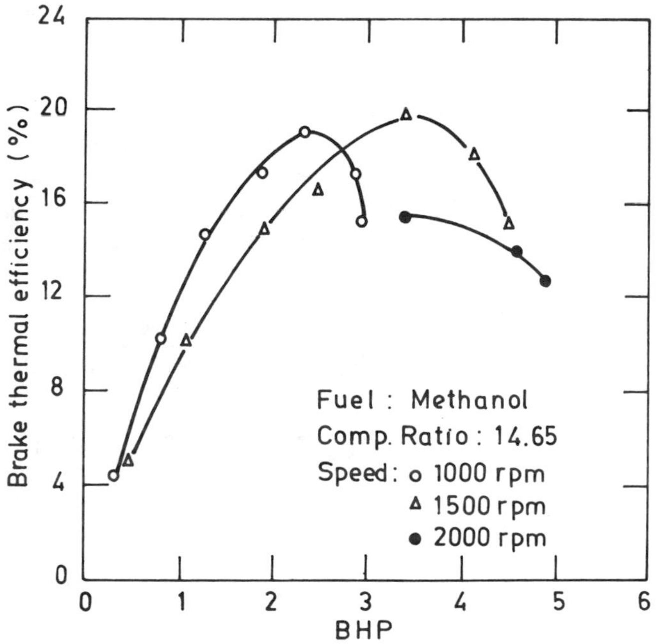

Fig. 7 - Variation of thermal efficiency

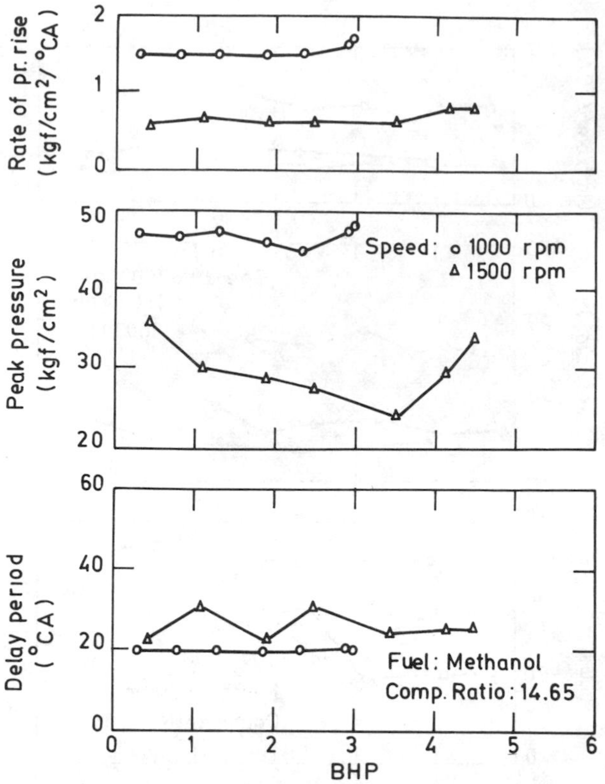

Fig. 8 - Variation of delay period, peak pressure and rate of pressure rise

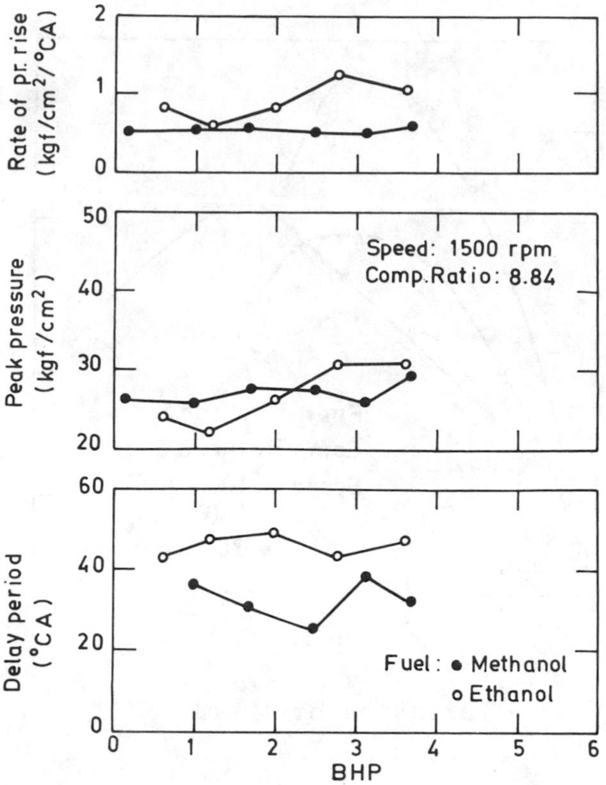

Fig. 10 - Variation of delay period, peak pressure and rate of pressure rise

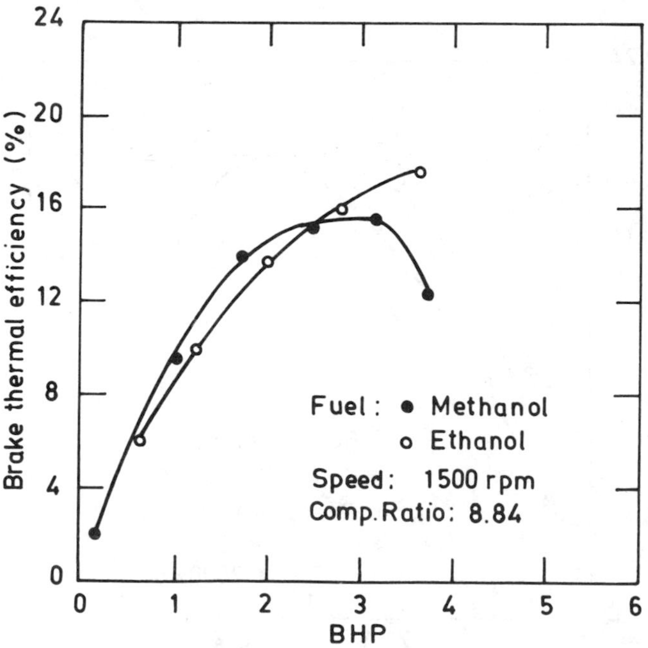

Fig. 9 - Variation of thermal efficiency

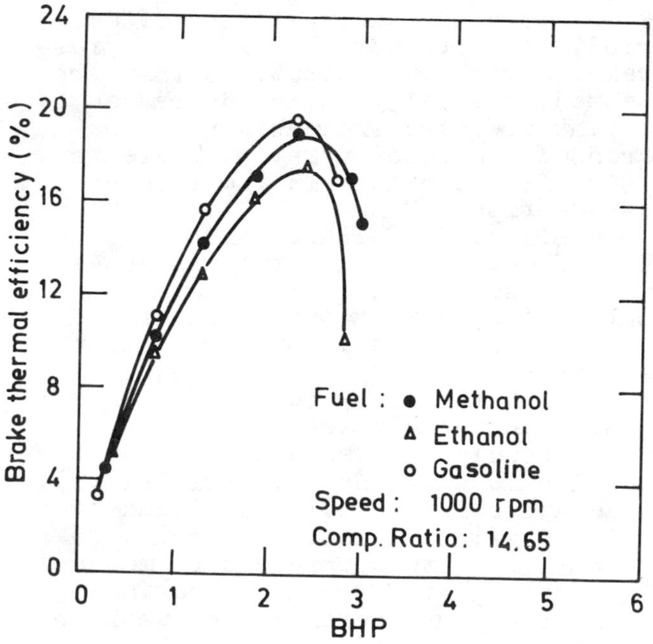

Fig. 11 - Variation of thermal efficiency

have a slightly superior efficiency. The delay period is found to be highest for gasoline (Fig.12). The ethanol used was commercial rectified spirit containing 5 % water.

PRESSURE-CRANK ANGLE DIAGRAMS - Fig.13 shows some typical pressure crank angle diagrams obtained. A,B and C are for methanol, ethanol and gasoline operation at 1000 rpm and output of 2.35 HP. Methanol and ethanol yield practically the same value of peak pressure, with gasoline producing a lower peak pressure. It can also be seen that the delay period progressively increases from methanol to ethanol to gasoline. The lower peak pressure with gasoline is clearly due to the long delay period, with bulk of the combustion occurring after TDC.

Fig.13 D shows the pressure variation with methanol at 2000 rpm. There is a steep increase in the delay period (in crank angles) and a drastic fall in the peak pressure compared to 1000 rpm, clearly indicating slow-developing combustion. This could possibly be remedied by advancing the injection timing.

COMPARISON WITH PERFORMANCE OF THE ENGINE AS A CONVENTIONAL DIESEL ENGINE - In order to judge the performance of the surface ignition engine from a different angle, the engine was operated on commercial diesel fuel at the compression ratio of 14.65 without the surface being heated electrically. Figs.14 and 15 present the data obtained in these tests.

Fig.14 shows that the thermal efficiency is essentially the same with diesel oil and methanol as fuels. The maximum thermal efficiency of the engine is found to be about 23 %, relatively lower than the conventional diesel engines of this size. There are specifically two reasons for this. The compression ratio is only 14.65 against 16.5, the compression ratio of the unmodified diesel engine. The compression ratios for the present investigations were deliberately chosen to be low, in order to demonstrate the feasibility of using methanol in a surface ignition engine even with relatively low compression ratios. But the engine can also operate with higher compression ratios, leading to improved thermal efficiency. The second reason for the low thermal efficiency is the throttling losses in the passage connecting

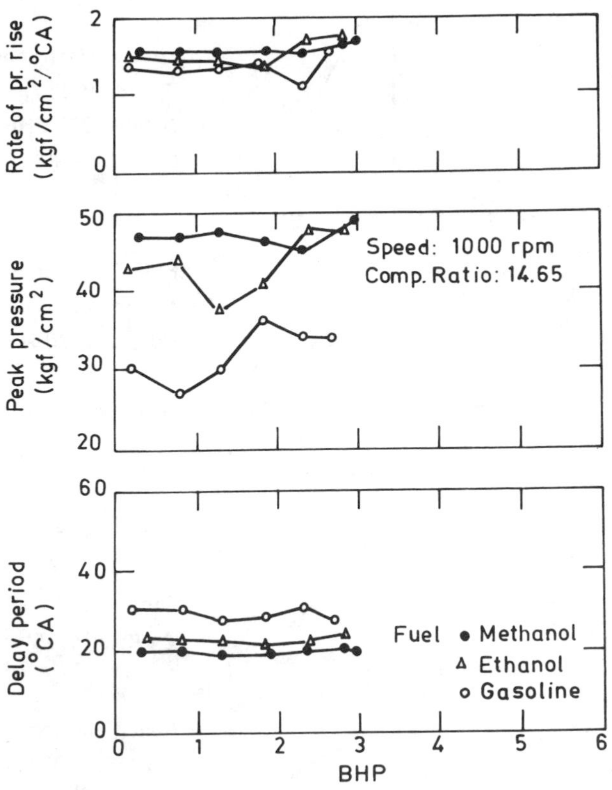

Fig. 12 - Variation of delay period, peak pressure and rate of pressure rise

the surface ignition chamber with the main cylinder. This passage is designed to accommodate the lower portion of the injector and it could not be enlarged since the valve seats are very close by. With an engine specifically designed for this process this loss would be greatly reduced, improving the thermal efficiency.

Comparison of Figs.8 and 15 shows that the peak pressure and rate of pressure rise are markedly higher for diesel operation though the differences in delay period are less significant. The reason could probably be a higher rate of burning for diesel oil, once combustion is initiated.

The above results show that the performance of the surface ignition engine on methanol is quite comparable in thermal efficiency to a corresponding conventional diesel engine.

EXHAUST EMISSIONS OF ALDEHYDES - In SI engines methanol, as is well established, produces higher exhaust emissions of aldehydes than gasoline. Hence, the aldehyde emissions in particular were checked for the surface ignition engine. The MBTH method (3 methyl-2-benzo-thiozolone Hydrazone Hy-

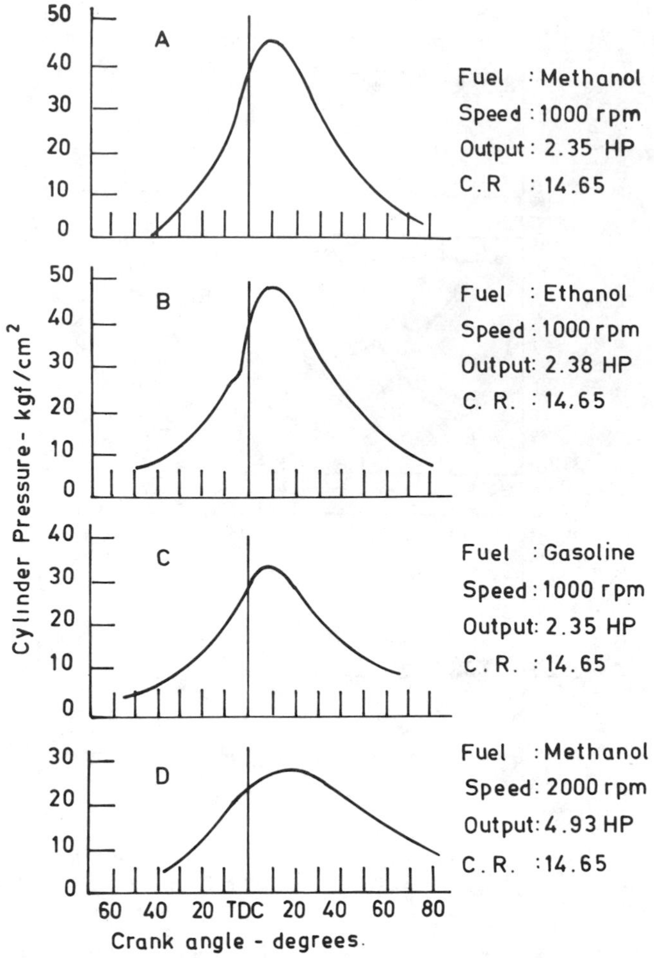

Fig. 13 - Typical pressure-crank angle diagrams under various operating conditions

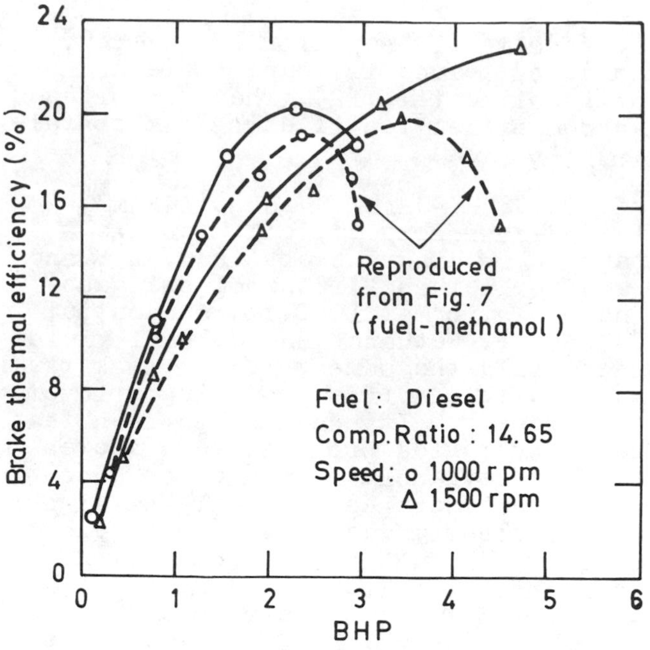

Fig. 14 - Variation of thermal efficiency without surface heating

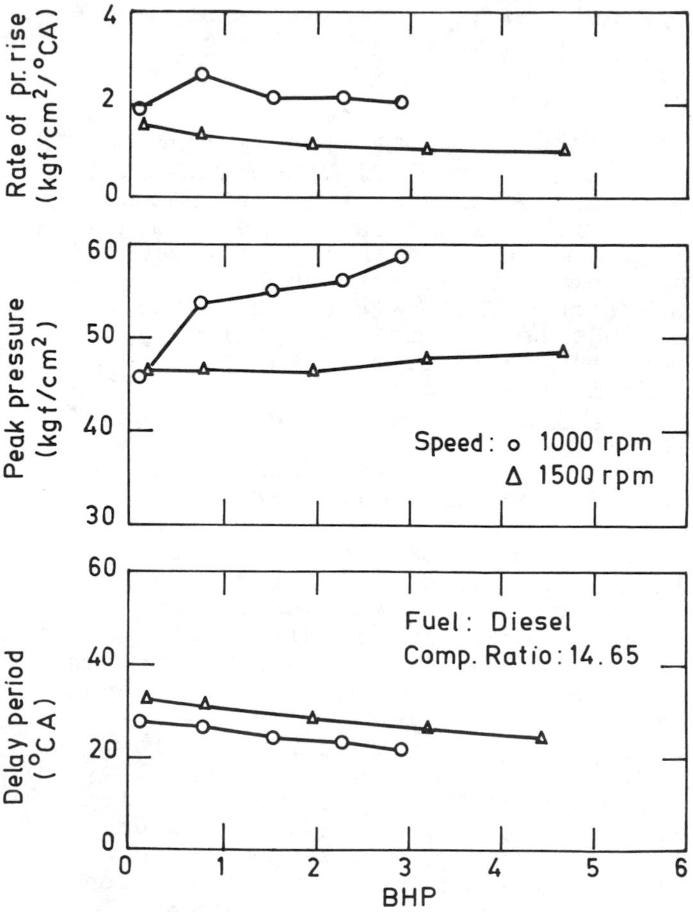

Fig. 15 - Variation of delay period peak pressure and rate of pressure rise without surface heating

drochloride) was used for this purpose. From the analysis of the results, it was observed that the aldehyde emissions were very low, in all cases well below 50 ppm.

CONCLUSION

The present investigation has clearly established the feasibility of operating an engine with a heated and insulated surface used for initiating ignition of the injected fuel. This type of engine lends itself easily to the use of wide variety of fuels, including methanol, ethanol and gasoline.

Further observations from this work are given below:

A) The engine operates smoothly on methanol with a performance comparable to diesel operation.

B) The engine operates more smoothly at lower speeds than at higher speeds.

C) At the same speed the engine operates best on methanol followed by ethanol.

D) Though the tests were carried out with the coil externally heated for ensuring uniform results, it was found that after a few minutes of operation the engine could operate well with the external power supply switched off, particularly at the higher speeds and higher fuel flow rates. There was, however, a slight increase in the delay period in this case.

E) The compression ratio can be as high as in the diesel engine.

F) It is possible, by suitable choice of materials and combustion chamber configuration, to make the ignition surface self-sustaining after an initial period of energy supply.

G) The thermal efficiency can also be improved substantially by further development work.

In conclusion, the surface ignition engine opens up a new direction for diesel engines to work on alcohols alone as fuels.

REFERENCES

1. 'Neuen Kraftstoffen auf der Spur, Alternative Kraftstoffe für kraftfahrzeuge.' Bundesministerium für Forschung und Technologie-Bonn 1974, p.151.

2. Wolfgang Bandel, 'Problems posed by the use of Ethanol as Fuel for Commercial Vehicles.' Second International Symposium on Alcohol Fuel Technology, Wolfsburg, Nov.21-23, 1977, Vol.I.

3. E.Holmer, 'Methanol as a Substitute Fuel in the Diesel Engine.' Second International Symposium on Alcohol Fuel Technology, Wolfsburg, Nov.21-23, 1977, Vol.II.

4. W.M.Scott and D.R.Cummings,' 'Dual Fuelling the Truck Diesel with Methanol.' Second International Symposium on Alcohol Fuel Technology, Wolfsburg, Nov.21-23, 1977, Vol.II.

5. F.Pischinger and C.Havemith, 'A New Way of Direct Injection of Methanol in a Diesel Engine.' Third International Symposium on Alcohol Fuel Technology, Asilomar, U.S.A., May 28-31, 1979.

6. N.R.Panchapakesan, K.V.Gopalakrishnan and B.S.Murthy, 'Factors that improve the performance of an Ethanol-Diesel Oil Dual-Fuel Engine.' Second International Symposium on Alcohol Fuel Technology, Wolfsburg, Nov.21-23, 1977, Vol.II.

7. K.D.Barnes, D.B.Kittelsan and T.E.Murphy, 'Effect of Alcohols as Supplemental Fuel for Turbocharged Diesel Engines.' SAE Paper 750469.

8. H.R.Ricardo, 'The High Speed Internal Combustion Engine.' Blackie & Son Ltd., 1953.

9. F.W.Bowditch and T.C.Yu, 'A Consideration of the Deposit Ignition Mechanism.' SAE Transactions 1960.

10. L.H.Browning and R.K.Pefley, 'Thermokinetic Modeling of Methanol Combustion Phenomena with Application to Spark Ignition Engines.' Third International Symposium on Alcohol Fuels Technology, Asilomar, U.S.A., May 28-31, 1979, Vol.III.

11. D.L.Hagen, 'Methanol as a Fuel.' A Review with Bibliography.' SAE Paper No.770792.

APPENDIX

SPECIFICATIONS OF THE TEST ENGINE

Name of the engine	..	KIRLOSKAR AVI
General details	..	Four-stroke, single cylinder, compression ignition, vertical, water cooled.
Bore	..	80 mm
Stroke	..	110 mm
Swept volume	..	553 cc
Compression ratio-normal	..	16.5:1
Rated power	..	5 BHP at 1500 RPM
Injection timing	..	27° bTDC by static setting
Combustion chamber	..	Open combustion chamber with hemispherical cavity in piston
Average Ambient Pressure During Tests	..	755 mm of mercury
Average Ambient Temperature during Tests	..	$30^{o}C$

BIBLIOGRAPHY

The literature listed in the bibliography references additional material on alcohols as motor fuels. Copies of SAE papers are available in original or photocopy format from SAE. For ordering information contact the Publications Division, SAE, 400 Commonwealth Drive, Warrendale, PA 15096.

G. C. Lawrason and P. F. Finigan, Ethyl Alcohol and Gasoline as a Modern Motor Fuel, SAE Special Publication 254 (SP-254), 1964.

An investigation is being conducted to explore the use of ethanol-gasoline blends in unmodified present-day automotive engines. Results of the program thus far show that, when 25% ethanol was used in lieu of tetraethyl lead in a catalytically cracked base gasoline, the percentage of unburned hydrocarbons in the exhaust effluent was significantly reduced. Combustion chamber deposit weights were sharply reduced with the ethanol-gasoline blends under the endurance conditions selected. Performance of the 25% ethanol blend closely approximated that of the same base gasoline blended with tel, but specific fuel consumption was generally increased at part throttle.

M. W. Jackson, Exhaust Hydrocarbon and Nitrogen Oxide Concentrations with an Ethyl Alcohol-Gasoline Fuel, Special Publication 254 (SP-254), 1964.

The exhaust hydrocarbon and nitrogen oxide concentrations of a single-cylinder engine, operating on a 25% (wt.) ethyl alcohol - 75% gasoline fuel, are compared to those operating on gasoline. For comparisons at the same air-fuel ratio but lower than 15.3, the addition of ethyl alcohol to gasoline reduces the exhaust hydrocarbon concentrations and increases the nitrogen oxide concentrations. At the same air-fuel ratio but higher than 15.3, the addition of ethyl alcohol reduces both the hydrocarbon and nitrogen oxide concentrations. However, tests with automobiles, operating at the same air-fuel ratio with both fuels, indicate that the addition of ethyl alcohol causes an increase in "surge" and, in some cases, results in a power loss. To overcome these performance problems, the ethyl alcohol-gasoline fuel should be operated at about the same percent theoretical air as gasoline. For comparisons at the same per cent theoretical air, the addition of ethyl alcohol to gasoline has little effect on the exhaust hydrocarbon and nitrogen oxide concentrations. On this basis, addition of ethyl alcohol to gasoline appears to have no practical advantage from the standpoint of reducing automobile exhaust hydrocarbon and nitrogen oxide concentrations.

J. A. Bolt, Air Pollution and Future Automotive Powerplants, SAE Paper 680191, 1968.

The automotive gasoline engine has been under heavy attack as a source of air pollution, and is now the subject of a very large program of research and development to reduce its undesirable vehicle emissions. The quantity of emissions that can reasonably be tolerated in different areas of the U.S. is presently unknown because of lack of information concerning air movements and air quality standards for man and plants. It is important that this information be made available as quickly as possible because the cost of emission controls of all types will rise rapidly. With rapidly rising costs for air pollution control from all sources, cost-value analyses are urgently needed for economy.

Major reductions of the undesirable exhaust emissions of present powerplant systems have been made during the last few years and will continue to be accomplished, under the impetus of air pollution requirements and regulations. Improved versions of automotive piston engines will continue for many years to be the main source of power for highway transportation.

There is a place for battery-powered vehicles in small size and short-range applications, especially for areas with serious air pollution. They will, however, require batteries with much greater energy and power density, and greatly reduced cost, before they will achieve significant use. The increased use of central station atomic energy in the future will give added advantages to battery-electric vehicles.

Any successful program for reduction of exhaust emissions from vehicles must include an effective engine inspection and maintenance program. From the standpoint of trained personnel and necessary facilities and instrumentation, this constitutes a major problem. Industry, government, the technical societies, and individuals all have key roles to play in the rapidly growing effort to reduce vehicular emissions.

R. K. Pefley, M. A. Saad, M. A. Sweeney, and J. D. Kilgroe, Performance and Emission Characteristics Using Blends of Methanol and Dissociated Methanol as an Automotive Fuel, SAE Paper 719008, 1971.

Tests were concluded to evaluate engine performance and exhaust emissions using blends of methanol and dissociated methanol as a spark ignition engine fuel. The effects of air-fuel ratio, compression ratio, and spark advance were investigated using a constant-speed full throttle CFR engine. Comparative tests with gasoline indicated that the use of methanol reduced hydrocarbon and carbon monoxide emissions without major changes in engine power and efficiency. Nitrogen oxide levels were similar to those obtained with gasoline. Aldehyde emissions could be controlled by changing the air-fuel ratio and spark advance settings. Preliminary design consideration for a methanol reformer using exhaust energy is discussed.

H. G. Adelman, D. G. Andrews, and R. S. Devoto Exhaust Emissions From a Methanol-Fueled Automobile, SAE Paper 720693, 1972.

An American Motors Gremlin has been converted to low-pollution operation on methanol through the use of an exhaust-heated intake manifold, a rejetted carburetor with heat exchanger for heating of fuel-air charge, a catalytic muffler, and an exhaust-port air injector. Tests carried out at EPA laboratories demonstrated that this car surpasses the 1975-1976 federal standards for unburned HC, CO, and NO_X. The low levels of HC and CO are due to lean operation and the use of an oxidizing catalyst. The low NO_X emissions are due partially to retarded spark-timing and lean operation, and, as indicated in a chemical kinetic model of NO formation, to properties of methanol that are favorable to low NO levels. Results of gas chromatograph and chemical analyses of the exhaust for organics, aldehydes, and ammonia are also discussed.

D. P. Gregory and R. B. Rosenberg, Synthetic Fuels for Transportation and National Energy Needs, SAE Paper 730520, 1973.

The United States petroleum supplies cannot keep up with the demands made upon them by the use of automobiles. Increased importation of oil is not a satisfactory long-term solution. Supplies of coal, nuclear, and solar energy, however, are abundant. We suggest that "clean" fuels could be synthesized from these resources by using these abundant materials. This paper examines the possibilities of making methanol, ethanol, hydrogen, and ammonia for use as vehicle fuels. In the short term, methanol and methanol-gasoline blends appear attractive. In the long term, hydrogen is ideal if its handling problems can be solved.

R. T. Johnson, Energy and Synthetic Fuels for Transportation: A Summary, SAE Paper 740599, 1974.

In light of the probable shortage of liquid petroleum fuels in the United States for the foreseeable future, much attention is being focused on possible fuels as substitutes for conventional gasolines and distillate-type fuels. This paper surveys the presently available synthetic or substitute fuels being considered for transportation uses. The survey includes a brief review of current energy use patterns in the transportation sector and a projected future for the automobile and other transportation modes. Specialized fuel needs for the transportation sector are described including attributes of the ideal fuel.

Discussion of possible synthetic fuel candidates is broken down into categories—hydrocarbons, hydrogen, inorganic hydrogen compounds, and electrochemical systems—and general properties. The properties of an idealized synthetic fuel are formulated and correlated with the properties of the ideal transportation fuel, and an extensive comparison of the selected synthetic fuel candidates is made. A significant portion of this comparative information is presented in graphical and tabular form with appropriate discussion of the data and its presentation.

The paper concludes with a discussion of the application potentials of the various synthetic fuel candidates to various modes. Interim and long-term fuel possibilities are proposed for further consideration.

R. G. Jackson, The Role of Methanol as a Clean Fuel, SAE Paper 740642, 1974.

The use of coal as a prime energy source is examined. The author focuses on the process of coal gasification to produce a synthesis gas for subsequent conversion to either gaseous or liquid products, and particularly on the methanol process. The production of methanol is described, as are ways of using it commercially.

R. W. Hurn and T. W. Chamberlain, Fuels and Emissions--Update and Outlook, 1974, SAE Paper 740694, 1974.

Profound change has been observed in automotive emissions since the late 1960's. This change is marked both in the amount and in the chemical makeup of the controlled emissions. In general, the trend toward decreased levels of the several categories of emissions is well known, but changes in the chemical character of the hydrocarbon component are less well recognized. The nature of these changes and implications for air pollution effects are discussed.

As the absolute levels of emissions are reduced, small incremental changes become relatively more important. Such incremental changes are brought about by change in ambient temperature, and therefore the sensitivity of emission control systems to changes in ambient temperature takes on added significance. Data are presented to show that emissions control may deteriorate seriously at some temperatures at which vehicles normally operate.

Will emissions considerations force change in fuel usage, and if so, how are emissions affected? The question is examined both from the viewpoint of a switch from gasoline to other conventional fuels and from the viewpoint that gasoline composition may change. Moreover, with supplies of motor fuel from petroleum sources in long-term jeopardy, there is growing interest in the character of emissions to be expected in using alternative fuels. The more viable options for alternative fuels are identified, and emissions characteristics of these fuels are described.

H. Davitian, Energy Carriers in Space Conditioning and Automotive Applications: A Comparison of Hydrogen, Methane, Methanol and Electricity, SAE Paper 749037, 1974.

Petroleum and natural gas serve not only as energy resources but also as convenient energy carriers. Replacement of these resources by large, stationary power plants based on energy from fission, fusion, solar, geothermal, etc., sources creates a greatly expanded need for alternative energy carriers. Hydrogen, methane, methanol, and electricity are compared as potential future energy carriers supplying energy for two applications -- space conditioning and automobiles. Several methods of employing each energy carrier in each application are investigated. The comparison is based primarily on the net efficiency of energy use and the operating costs (for energy) for each method of utilization which are computed for technology anticipated to be available in the year 2000. Electricity is found to provide the most efficient overall utilization of energy in virtually all cases considered. Cost considerations tend to favor synthetic fuels when electricity distribution is underground. Cost and efficiency factors do not appear to favor hydrogen sufficiently to encourage conversion of existing fuel distribution systems to hydrogen unless coal is unavailable for synthetic fuel production and a low cost hydrogen production process is developed.

T. B. Reed, R. M. Lerner, E. D. Hinkley, and R. E. Fahey, Improved Performance of Internal Combustion Engines Using 5-30% Methanol in Gasoline, SAE Paper 749104, 1974.

A number of unmodified cars have been tested over a fixed course using mixtures of methanol and gasoline. It was found that mixtures between 5 and 15% increased the fuel economy and performance, and lowered the CO emissions and exhaust temperatures. In addition, knock was eliminated on one engine and "Diesel operation" ceased with 5% or greater mixtures. The improved performance of methanol mixtures is attributed to chemical leaning plus the dissociation of methanol near 200°C which can absorb energy during the compression stroke of the engine and release up to 40% hydrogen for a 10% mixture.

R. M. Tillman, O. L. Spilman, and J. M. Beach, Potential for Methanol as an Automotive Fuel, SAE Paper 750118, 1975.

Multiple cyl engine tests demonstrated that lean methanol operation gave improved tail pipe emissions and better Btu efficiency than gasoline. The reductions in NO_X emissions were particularly significant. Some loss in maximum power output was experienced, but the loss was less than that for gasoline operation at equivalent NO_X emissions. This program was directed toward making the minimum number of engine modifications to simplify retrofitting of existing vehicles to utilize methanol.

R. D. Fleming and T. W. Chamberlain, Methanol as Automotive Fuel Part 1 – Straight Methanol, SAE Paper 750121, 1975.

A study of methanol as an automotive fuel was conducted using a single-cylinder research engine, a 4-cylinder 122-CID (2,000 cc) engine, and an 8-cylinder 350-CID engine. Results showed that when using methanol as fuel, the single-cylinder engine could operate leaner than the multicylinder engines. This difference is attributable to air-fuel mixture maldistribution associated with the multicylinder engines.

Steady-state fuel economy and emissions data are presented and discussed. Results indicate that fuel economy (on an energy input basis) using methanol fuel, is about 5% improved as compared to gasoline fuel economy, and with substantially lower nitrogen oxides emissions for methanol.

T. Powell, Racing Experiences with Methanol and Ethanol-Based Motor-Fuel Blends, SAE Paper 750124, 1975.

Methanol and ethanol have interesting possibilities as motor fuels and have been used as such, mostly in blends, since the early days of the automobile. However, specialized experience with alcohols as high-performance racing fuels more or less parallels World War I era results with substitute motor fuels. These indicate several types of practical operational problems with water solubility, plastics solvent action, metal corrosion and galvanic effects, low air-fuel ratios and low calorific content, and high latent heat. Simply switching to alcohol-gasoline blends in conventional automotive fuel systems and engines is not as straightforward a matter as some short-term laboratory tests might tend to indicate.

Some of the modern-day racing techniques for handling alcohol fuels include: anodizing and plating of nonferrous alloy fuel system and engine castings; using solvent-resistant plastics and corrosion-resistant metals; draining and flushing out the fuel system and engine with hydrocarbon-fuel oil mixes after running; using alcohol-soluble synthetic lube oils; sealing and storage of fuel containers and tanks so as to reduce atmospheric moisture absorption; using higher-energy ignition systems to better fire the "wet" alcohol fuels at high C.R.'s; and nearly tripling and doubling the fuel system capacity.

K. D. Barnes, D. B. Kittelson, and T. E. Murphy, Effect of Alcohols as Supplemental Fuel for Turbocharged Diesel Engines, SAE Paper 750469, 1975.

Alcohols are examined as supplemental carbureted fuels for highspeed turbocharged diesels as typified by the White Motor/Waukes ha F310 DBLT (6 cylinder, 310 cu. in.). Most of the work was with methanol; ethanol and isopropanol were compared at a few points. Fumigation (dual-fueling) with alcohol significantly reduced smoke and intake manifold temperature. These effects were largest at high load. Efficiency and HC emissions were essentially unchanged. Cylinder pressures and rise rates were examined for possible adverse effects on engine structure. The range of speed and load favorable to alcohol dual-fueling are such that, should alcohols become economically competitive as fuels, a practical dual-fuel system could be applied to existing diesel engines.

J. R. Allsup, Methanol/Gasoline Blends as Automotive Fuel, SAE Paper 750763, 1975.

An experimental program was conducted by the Energy Research & Development Administration's Bartlesville (Okla.) Energy Research Center to determine the effect on exhaust emissions and fuel economy when methanol is added to conventional motor fuel, that is, gasoline. Ten 1974 and 1975 model vehicles were used in the study. Ambient temperature was varied from 20° to 100°F to determine temperature effects while using methanol/gasoline blends. Emissions were generally modified as a consequence of the fact that the addition of methanol to gasoline alters both the fuel vapor pressure and the stoichiometry of the air-fuel mixture. Fuel economy was generally decreased by methanol addition.

Moderate mileage accumulation using 10% methanol fuel showed no deterioration either in emissions control or of fuel-related engine components. Driveability differences between methanol 10% blends and gasoline were detectable but were judged not to be objectionable.

C. H. Gonnerman, J. S. Moore, and P. W. McCallum, Fueling Automotive Internal Combustion Engines with Methanol – Historical Development and Current State of the Art, SAE Paper 759127, 1975.

The use of methanol as an internal combustion engine fuel is rich in proven potential but has many major obstacles to its widespread use. The work described in this paper was based on two premises: The first was that to obtain a grasp of the issues, it was necessary to place them in an historical perspective. Were most of the current obstacles present in the distant past? Were promising solutions unearthed in the past but not developed to their full potential due to technology limitations? The second objective was to document potential innovations from a previously isolated sector – the automotive specialty equipment and aftermarket sector.

It was found that, indeed, many of the current obstacles were present in 1913 and that unfortunately, no concepts of any tangible potential were uncovered. On the other hand, the aftermarket and specialty equipment sector was found to be one rich in hardware of considerable potential.

J. C. Gillis, J. B. Pangborn, and K. C. Vyas, The Technical and Economic Feasibility of Some Alternative Fuels for Automotive Transportation, SAE Paper 759128, 1975.

The use of domestic energy resources to make a synthetic alternative or supplemental automotive fuel that would lessen the need for imported petroleum is under active investigation. This paper discusses alternative fuels that are the most promising on technical and economic grounds. Assessments are made for three time frames: 1975–1985, 1985–2000, and beyond 2000. The maximum potential quantity or rate of energy supply is estimated for both renewable and nonrenewable raw energy sources. To estimate the synthetic-fuel market, an energy supply-demand model is presented. The model assumes an all-out effort to reduce fuel imports, but shows a large deficit in the energy supply of the automotive sector. Coal, oil shale, and nuclear fuels appear to be the most promising domestic resources to support a synthetic automotive fuels industry. Nine promising fuels derived from the available resources are listed, and their synthesis processes are briefly discussed. The compatibility of these fuels with existing fuel-distribution systems is considered. Changes in automotive fuel storage and engine design required to accommodate these fuels are identified, and potential fuel costs and fuel synthesis process efficiencies are estimated. On the basis of these considerations, the most worthy alternative (or supplemental) fuels for automotive transportation are liquid hydrocarbons made from coal or oil shale. Methanol made from coal is another possibly useful fuel in the near and middle time frames. In the far term, beyond the turn of the century, hydrogen may become important.

R. T. Johnson and R. K. Riley, Single Cylinder Spark Ignition Engine Study of the Octane, Emissions, and Fuel Economy Characteristics of Methanol-Gasoline Blends, SAE Paper 760377, 1976.

A two phase test program was carried out on a single cylinder, fuel research engine (CFR) to determine the octane, emissions, and fuel economy characteristics of methanol-gasoline blends. The first phase of the work was an evaluation of the octane characteristics of methanol blended with unleaded gasoline. Blends ranging from 2% to 100% by volume methanol for four different base gasolines were knock rated. A simplified mathematical model of the results was developed to aid in comparing the effects of methanol on the octane ratings of the various base fuels. The results indicate that methanol can substantially increase both the Research and Motor Ratings of a relatively low octane unleaded gasoline. However, as the octane rating of the base gasoline is increased, the octane increase produced by the addition of methanol is reduced. For base fuels with Motor Ratings over 85, the addition of methanol has little effect on the Motor Octane rating. Comparisons and data for base fuels having Research Octane Numbers ranging from 81 RON to 98 RON are given.

The second phase of the test program was an evaluation of the emissions and fuel economy of a 10% by volume blend of methanol with two base fuels in the single cylinder engine. Clear Indolene and a 95 RON commercial gasoline were used for these tests. The results for the Indolene blend are presented since both fuels behaved in a similar manner. Where differences did exist appropriate comments are included. For the blends examined, the specific emissions (gm/ihp-hr) were changed very little by the addition of methanol when the engine was operating at equivalent spark, speed, and stoichiometry. Although the addition of methanol increased the ISFC, the thermal efficiency of the engine was not significantly changed.

General conclusions were that a blend of 10% methanol and gasoline demonstrated no significant change in the emissions or energy efficiency over the gasoline fueled engine when operated at equivalent conditions. Under some operating conditions, the addition of methanol could increase the octane rating of the base fuel. This increase could conceivably be enough to reduce knock problems in some vehicles.

A. Koenig, W. Lee, and W. Bernhardt, Technical and Economical Aspects of Methanol as an Automotive Fuel, SAE Paper 760545, 1976.

The results of basic test series conducted on a single-cylinder engine are used to establish the reasons that advocate the introduction of methanol or methanol-gasoline blends as alternative fuel for motor-vehicle operation: lower exhaust emission concentrations, improved energy utilization, higher engine output, and improved knock-resistance. Fuel and energy consumption data are submitted for four vehicle concepts operated on methanol fuels, to establish data on the economic aspects of methanol-operated vehicles. It is apparent that the utilization of the anti-knock effects of methanol can lead to competitive gasoline-methanol blend vehicle operation at the present cost of gasoline and methanol. Vehicle operation of pure methanol would offer economic advantages over gasoline operation if gasoline is derived from coal.

R. T. Johnson, R. K. Riley, and M. D. Dalen, Performance of Methanol-Gasoline Blends in a Stratified Charge Engine Vehicle, SAE Paper 760546, 1976.

A series of driveability and chassis dynamometer tests were performed using various blends of methanol and gasoline in a stratified charge engine vehicle. The vehicle used was a 1975 Honda Civic CVCC. This vehicle is powered by a prechamber type stratified charge spark ignition engine.

The basic intent of this effort was to characterize how methanol-gasoline blends behave in a prechamber stratified charge engine vehicle. Since the stratified charge engine was designed to operate at overall lean air fuel ratios, the leaning effect of methanol blended with gasoline might not produce the general degradation of vehicle performance, emissions, and fuel economy reported for late model vehicles.

The test program was separated into two phases. The first phase was to determine the effect of methanol-gasoline blends on the driveability of the stratified charge engine vehicle. Blends containing 10% to 40% by volume of methanol in gasoline were tested. A weighted demerit system was used to evaluate driveability. Based on this system, vehicle driveability improved slightly with a 10% by volume blend of methanol in gasoline. A 15% by volume blend of methanol yielded vehicle driveability roughly comparable to the base gasoline alone. Further increases in the methanol content produced increased degradation of the driveability. The vehicle would operate on the 40% by volume blend but it was essentially undriveable. Cold weather tests using winter grade gasoline as a base fuel demonstrated that 10% methanol had approximately the same effect on driveability for winter grade fuels as for summer grade fuels.

Emissions and fuel economy of the vehicle operating on base fuels and base fuels blended with methanol were evaluated using a chassis dynamometer and the federal urban driving schedule. The addition of 10% methanol to the base fuels produced only very minor changes in the emissions and fuel economy of the vehicle. Not all indicated changes were statistically significant. Generally, HC emissions increased, CO emissions decreased slightly, NO_x emissions decreased, and volumetric fuel economy decreased slightly.

General conclusions are that the CVCC vehicle tested suffered only slight degradation in driveability using a 10% methanol-gasoline fuel and vehicle emissions and fuel economy are not significantly changed by use of the 10% methanol blend.

L. S. Caretto, Other Engines, Other Fuels: An Overview, SAE Paper 760608, 1976.

The current state of development of several alternative engines for light-duty ground transportation is reviewed. The development of such engines will affect the introduction of various types of fuels, and the availability of fuels may place restrictions on the type of engine used. The interdependence of fuel and engine types is considered. Although the introduction of new engines will change the fuel-use pattern, continued use of present spark-ignition engines will require a continued supply of high-octane fuels.

W. V. Bush, Cost of Fuels for Fuel Cell Automobiles, SAE Paper 770380, 1977.

The cost of operation projected for a fuel cell automobile could be as much as 1–2¢/mi less than for a comparable internal combustion engine car or an all-battery car, assuming energy raw material cost equilibrated at 10$/bbl crude oil equivalent. Fuel cost is a small but significant element in total operating cost. Therefore, the projected cost advantage of the fuel cell automobile increases with increasing cost of fossil fuel, because of the higher energy efficiency projected for the fuel cell automobile.

A Weir, Jr., Alternate Fuels for Power Generation, SAE Paper 770672, 1977.

The potential for electric utility utilization in an environmentally acceptable manner of various fossil fuels is discussed. Fuels and fuel processes considered include Alaskan oil with either desulfurization of the fuel, or stack gas treatment of the combustion products, shale oil, and coal and coal derived synthetic fuels. The latter category includes solvent refined coal, low and medium Btu gas, methanol from coal, as well as "true" liquid fuel from coal and the direct utilization of coal either with stack gas scrubbing and a conventional boiler, or in a fluidized bed boiler.

J. A. Gething and S. S. Lestz, Knocking and Performance Characteristics of Low Octane Primary Reference Fuels Blended with Methanol, SAE Paper 780079, 1978.

Single cylinder SI engine tests were conducted with low octane primary reference fuel (PRF) blends and methanol. The investigated parameters were equivalence ratio, and PRF octane number ranging from 100 to 53 blended with weight percentages of methanol to 75 percent. In order to evaluate the performance of these different fuels, knock intensity, efficiency, and emissions were considered.

The results indicate that even a 53 octane number PRF blended with methanol displays some performance advantages over pure isoctane. This is best detected from the observed efficiency increase for all of the PRF blends except the 91 octane PRF blend. The improved efficiency is also higher than that of pure methanol.

E. E. Ecklund, Methanol and Other Alternative Fuels for Off-Highway Mobile Engines, SAE Paper 780459, 1978.

Considerable insight into alternatives to petroleum fuels for earthmoving and other off-highway vehicles can be gained from investigations related to highway vehicles. At the same time, off-highway equipment has the potential of greater flexibility in choice of fuels because of simpler fuel logistics. Uncertainties in both petroleum resources and alternative supplies are reviewed, and implications for the remainder of this century are highlighted. However, technology to make alcohols is well established and a multiplicity of resources is available. Thus, the technical benefits and problems of using alcohols as a petroleum extender or substitute are detailed.

G. Publow and L. Grinberg, Performance of Late Model Cars with Gasoline-Methanol Fuel, SAE Paper 780948, 1978.

A test program has been completed to determine how the incorporation of 15% methanol in gasoline affects the driveability, exhaust emissions and fuel consumption of three 1978 model passenger cars. The study has shown that use of methanol containing fuel deteriorates the driveability of the cars to what is considered to be an unacceptable performance by the present standards. The exhaust emissions are not significantly affected, except for a moderate reduction in the carbon monoxide levels. The volumetric fuel consumption, however, is increased appreciably.

Relatively simple adjustments to the carburetor and ignition timing can restore the driveability to an acceptable level for operation with or without methanol. These adjustments, however, fail to restore the fuel consumption to its original level and may result in the escalation of hydrocarbon and carbon monoxide emissions. The increase in emissions is even more pronounced when the modified car is reverted back to operation with gasoline.

E. E. Ecklund and R. D. Fleming, The Petroleum Mentality–Bane or Boon, SAE Paper 790425, 1979.

Over the last century, petroleum has become the major source of energy for transportation and the raw material for petro-chemical products. As a result, there are tens of thousands of experts who support the technology and contribute to the economy of this country. For people who were born, brought up, educated, and matured in a climate of using and depending on petroleum, it is most logical to think in terms of the resources, processes and products that are used to produce petroleum. If the resulting experience, knowledge and expertise is put to use, this "petroleum mentality" can be a tremendous asset in our search for and the development of new energy sources. However, only a small fraction of this pool of know-how is now being applied to new energy. Obstacles result from assuming too much and from being too comfortable and/or myoptic about the status, the potential, and the needs regarding petroleum substitutes. Open mindedness, proper attention, and rational consideration can and should change this to a positive approach where the know-how can be far more effective.

Alternate Fuels Committee of the Engine Manufacturers Association, Alternate Portable Fuels for Internal Combustion Engines, SAE Paper 790426, 1979.

This report summarizes a committee investigation and analysis of the supply/demand situation for fuels; short and long term changes in fuels as they impact on portable internal combustion engines; potential areas of research on alternate fuels; and identification of priorities for alternate portable fuel research and development.

G. Purohit and J. Houseman, Methanol Decomposition Bottoming Cycle for IC Engines, SAE Paper 790427, 1979.

This paper presents the concept of methanol decomposition using engine exhaust heat, and examines its potential for use in the operation of passenger cars, diesel trucks, and diesel-electric locomotives.

Energy economy improvements of 10–20% are calculated over the representative driving cycles without a net loss in power. Some reductions in exhaust emissions are also projected.

R. Bertodo, Energy Economics of Alternate Fuels, SAE Paper 790430, 1979.

The energy crisis of the mid-1970's released a frantic search for alternative fuels. The present paper reviews the studies undertaken by the Author's Company and outlines experience with broad specification fuels, vegetable oils and alcohols. Tests were undertaken mainly with the diesel engine and its derivatives in mind.

It is concluded that, in the medium term, the most effective engine/fuel combination is an injected stratified charge engine burning "wide-cut" fuel oils. Such oil could be obtained by a modification to present natural crude refining practices, or from shale and tar sand distillation, or by coal gasification and hydrogenation, or from oil bearing vegetation. Unfortunately, the energy scene is currently confused by the conflict between short term economic gain and long term conservation needs. As a result attention is being focussed on gasolene-like alternatives, notably methyl and ethyl alcohol.

As a consequence it is thought that carburetted stratified charge engines, burning alcohol-based or alcohol extended fuels, are likely to become dominant in the mobile prime mover field. It is to be hoped that progressive depletion of natural crudes will promote the gradual introduction of a more efficient combination based on an injected stratified charge engine. In any case, eventual shortages of natural fuels will have far reaching implications on the choice of materials for both engine and vehicle manufacture. Government legislation and taxation policies will also be affected.

M. S. Harrenstien, K. T. Rhee, and R. R. Adt, Jr., Determination of Individual Aldehyde Concentrations in the Exhaust of a Spark Ignited Engine Fueled by Alcohol/Gasoline Blends.

Individual aldehyde (and acetone) emissions were measured from the exhaust gas of a premixed multicylinder spark ignition engine fueled with Indolene and blends of Indolene and either methanol or ethanol. The engine was operated at constant speed (2000 RPM) and MBT spark advance with fuel-air equivalence ratios (Φ) of 0.96, 0.90 and 0.82. During operation at $\Phi = 0.82$, the engine experienced lean-limit misfiring.

The DNPH method with a gas chromatographic finish was employed to obtain exhaust gas concentrations of aldehydes and acetone. Also, the methods used in the past for measuring engine exhaust aldehyde and acetone data were compared to each other and briefly discussed.

Use of the alcohol blends increased the total aldehyde emission level. Formaldehyde was the largest component, exhibiting a continual increase with increasing alcohol blend level. Acetaldehyde, while exhibiting a small decrease with increasing methanol percentage, showed a marked increase when the ethanol blend was used.

Acetone was the only specie measured with the DNPH method that showed a marked increase when the engine was operated in the lean misfire region ($\Phi = 0.82$).

The C_3 aldehydes were successfully separated and quantified. It was found that acrolein emissions did not increase when the blends were used; in fact, a slight decrease was observed. Propionaldehyde exhibited a decrease when the methanol blends were used.

Samples taken in the exhaust manifold, upstream of an oxidizing catalyst, and downstream of the catalyst, indicated that aldehydes and acetone are partially destroyed in the exhaust system and virtually completely destroyed by the catalyst.

C. P. Chiu and L. H. Hong, The Effects of Methanol Injection on Emission and Performance in a Carbureted SI Engine, SAE Paper 790954, 1979.

A single cylinder carburetor SI engine with a modified L type cylinder head and equipped with a fuel injector was used to inject methanol into the combustion chamber.

Four series of runs at an engine speed of 2500 rpm were made. One series with different values of λ* was performed with gasoline, and the other three series methanol was injected into the gasoline-air mixture. Each series of methanol injected run was supplied with the same energy input per cycle to correspond to that of the gasoline. The results of the experiments indicated that the performance (BHP) of the engine was proportional to the energy input of the mixture up to about 50% methanol injected.

With the equivalence air-fuel ratio above $\lambda = 1.15$ (lean mixture), it was found the NO_X concentration decreased more than 30% while the BSFC remained constant

E. N. Cart, Jr., and J. Percival, Alternate Fuels and Alternate Prime Movers in Non-Highway Transportation, SAE Paper 790957, 1979.

At some point in the future, the transportation system will have to start to use non-petroleum fuels. One area of the transportation system that is critical is the non-highway (marine, railroad, pipeline and aircraft) since it accounts for about 80% of the freight-ton miles moved in the U.S. With the long-life times of the engines used in this area, and the time it takes to introduce a new prime mover or new fuel, it is important to consider the use of alternate fuels now. This paper covers the factors considered in evaluating alternate fuels and prime movers for non-highway transportation and presents a future outlook on possible fuels and prime movers.

J. Strait, J. J. Boedicker, and K. C. Johansen, Diesel Oil and Ethanol Mixtures for Diesel–Powered Farm Tractors, SAE Paper 790958, 1979.

Absolute ethanol can be mixed with No. 1 diesel oil without separation if the mixture remains free of water. Mixtures of diesel oil and up to 30 percent by volume of ethanol were used in laboratory and field experiments in diesel tractor engines. When engines were operated on fuels containing ethanol, the maximum power output of the engines was reduced, fuel consumption increased, engine noise increased, and the delay period was extended. These changes in engine operating characteristics increased in magnitude as the proportion of ethanol in the fuel increased.

No significant performance characteristic of the diesel engine measured in this study was improved by the use of a diesel oil-ethanol blend to replace diesel oil to fuel the engine. Based upon this study, there appears to be no valid technical reason to recommend that diesel-powered farm tractors be operated on a mixture of diesel oil and ethanol in preference to diesel oil.

---

*Defined in paper.

R. J. Nichols, Investigation of the Octane Rating and Autoignition Temperature of Methanol-Gasoline Blends, SAE Paper 800258, 1980.

The octane rating and autoignition temperature of methanol-gasoline blends were measured. Large increases in autoignition temperature and ignition delay time were found for small percentages of added methanol. These increases correlated well with the increases in research octane rating throughout the range of blends. The evaporative cooling effect of the methanol was concluded to be the major mechanism suppressing detonation.

M. D. Leshner, C. A. Luengo, and F. Calandra, Brazilian Experience with Self-Adjusting Fuel System for Variable Alcohol-Gasoline Blends, SAE Paper 800265, 1980.

A fuel control system has been developed which allows fuels of various stoichiometries to be used interchangeably, without suffering a fuel consumption penalty, allowing a more efficient use of the combustion energy.

This Adaptive Lean Limit Control system uses a single, digital sensor and an electronic circuit to detect lean limit engine operation, and feeds back information to the fuel system to maintain the best economy mixture, regardless of the fuel blend being used.

This paper will describe the hardware, and include the results of extensive vehicle testing, using 20 percent and 50 percent ethanol-gasoline blends.

Additional References Regarding Use of Alcohols as Motor Fuels

The topic of alcohols as motor fuels has been addressed by many other organizations in addition to the Society of Automotive Engineers. The bibliography included with this special program lists only papers presented at SAE meetings. To broaden the reader's perspectives, listed below are several other publications and sources of information regarding the use of alcohols as motor fuels.

1. B. H. Eccleston and F. W. Cox, *Physical Properties of Gasoline/Methanol Mixtures,* BERC/RI-76/12, Bartlesville Energy Research Center, Bartlesville, Oklahoma, January, 1977.

2. N. D. Brinkman, "Vehicle Evaluation of Neat Methanol—Compromises Among Exhaust Emissions, Fuel Economy and Driveability," *Energy Research,* Vol. 3, 1979, pp. 243-274.

3. Proceedings: First International Symposium, *Methanol as a Fuel,* Stockholm, Sweden, March, 1976.

4. Proceedings: Second International Symposium, *Alcohol Fuel Technology—Methanol and Ethanol,* Wolfsburg, Federal Republic of Germany, November, 1977.

5. Proceedings: Third International Symposium, *Alcohol Fuels Technology,* Asilomar, California, May, 1979.

*6. J. L. Keller, G. M. Nakaguchi, and J. C. Ware, *Methanol Fuel Modification for Highway Vehicle Use,* Prepared for U.S. Dept. of Energy, HCP/W 3683-18, July, 1978.

7. Phone Call to Alternative Fuels Data Bank, Division of Utilization, Bartlesville Energy Technology Center, Dept. of Energy, Bartlesville, Oklahoma.

**Item 6 is an example of one of the many reports on use of alcohols that were prepared under contract to the Department of Energy or its predecessor agencies. The federal government has also published many reports on the generation of alcohols. Additional information on the processes for alcohol production can be found in items 3, 4, and 5, and the references contained in these publications.*

INDEX